T L29 0034

Handbook of Measurement Science

Volume 1 Theoretical fundamentals

Handbook of Measurement Science

Volume 1 Theoretical Fundamentals

Edited by
P. H. Sydenham
School of Electronic Engineering
South Australian Institute of Technology

A Wiley-Interscience Publication

JOHN WILEY & SONS

Chichester · New York · Brisbane · Toronto · Singapore

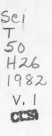
Library of Congress Cataloging in Publication Data

Main entry under title:

Handbook of measurement science.
   A Wiley–Interscience publication.
   Bibliography: p.
   Contents: v.   1.   Theoretical fundamentals
   1. Measuring instruments   I. Sydenham, P. H.
T50.H26   620′.0044   81-14628
ISBN 0 471 10037 4 (v. 1)   AACR2

British Library Cataloguing in Publication Data:

Handbook of measurement science.
   Vol. 1: Theoretical fundamentals
   1. Measuring instruments
   I. Sydenham, P. H.
   620′.0044   QC100.5
ISBN 0 471 10037 4

Set in Monophoto Times by
COMPOSITION HOUSE LTD., SALISBURY

Printed by Page Bros. (Norwich) Limited.

To Ludwik Finkelstein
for example, encouragement and friendship

Contributing Authors

P. ATKINSON   *Department of Cybernetics, University of Reading, UK*

A. G. BOLTON   *School of Electrical Engineering, South Australian Institute of Technology, Australia*

A. VAN DEN BOS   *Department of Applied Physics, Delft University of Technology, The Netherlands*

R. P. W. DUIN   *Department of Applied Physics, Delft University of Technology, The Netherlands*

L. FINKELSTEIN   *Department of Systems Science, The City University, London, UK*

F. A. GERRITSEN   *Department of Applied Physics, Delft University of Technology, The Netherlands*

R. W. GRIMES   *Customer Networks and Equipment Section, Telecom, Australia*

F. C. A. GROEN   *Department of Applied Physics, Delft University of Technology, The Netherlands*

D. HOFMANN   *Department of Measurement Engineering, Friedrich–Schiller University, Jena, GDR*

J. C. JOOSTEN   *Department of Applied Physics, Delft University of Technology, The Netherlands*

W. J. KERWIN   *Department of Electrical Engineering, University of Arizona, Tucson, USA*

M. J. MILLER   *School of Electronic Engineering, South Australian Institute of Technology, Australia*

D. M. MUNROE   *Princeton Applied Research Corporation, New Jersey, USA*

P. H. SYDENHAM   *School of Electronic Engineering, South Australian Institute of Technology, Australia*

P. W. VERBEEK   *Department of Applied Physics, Delft University of Technology, The Netherlands*

C. J. D. M. VERHAGEN   *Department of Applied Physics, Delft University of Technology, The Netherlands*

E.-G. WOSCHNI   *Department of Theory of Information Techniques, Technische Hochschule Karlmarx Stadt, GDR*

E. L. ZUCH   *Datel-Intersil Inc., Mansfield, Massachusetts, USA*

Contents

7. Pattern recognition 277

C. J. D. M. Verhagen, R. P. W. Duin, F. A. Gerritsen, F. C. A. Groen,
J. C. Joosten, and P. W. Verbeek

12. Signal data conversion   489
E. L. Zuch

13. Transmission of data   539
R. W. Grimes

Editor's Preface

Measurement systems have been in use since the times of earliest man. However, only in recent decades has it become clear to many that there exists an underlying collection of fundamentals that is applicable, in part or whole, to all measurement situations regardless of how diverse the applications appear to be.

The 1960's saw the start of many discussions, at conferences and the like, on what constitutes this set of knowledge. By the 1970's the situation was such that there was debate only on the finer points of what to include, not the core material.

Around 1976 I felt the need to resolve the issue, in the spirit of measurement, by producing a printed statement that would act as a standard for future use and comparison. Subsequently I put a proposal to my colleagues on the Higher Education Committee (TC-1) of the International Measurement Confederation IMEKO that the Committee should produce such a statement of the fundamentals involved. It was decided that I should personally produce such a Handbook using committee-men and other persons to help me design the structure and assist with contributions.

In 1977 I circularized a possible structure of some 36 chapters, along with brief abstracts, for comment. (This was realized by sorting my personal collection of texts, reprints, lecture notes, and experience after removal of works concerned with application only.) The fifteen experts who studied the proposal agreed, with a few minor changes, that the format represented the fundamental material of measurement systems. In this way it was resolved that measurement science had reached the degree of stability of content needed for a topic to become teachable and usable in a systematic rigorous manner. A simultaneous endeavour worthy of mention here as it supports the sentiment of this Handbook is the *Instrument Science* series of solicited review articles run in the *Journal of Physics E: Scientific Instruments*. (Now published (1982) as

'Instrument Science and Technology Volume 1' B. E. Jones (ed.) Adam Hilger, Bristol). After incorporation of the suggestions received a general outline document was prepared and sent to invited contributors.

This Handbook provides what the majority of the many hundreds of books on measurement do not—the basic enduring fundamentals of design. It does not seek to provide as its main theme, a summary or a catalogue, of instrument application. D. M. Considine, in the preface for his *Process Instruments and Control* Handbook stated 'the work was a step forward (in 1957) in the maturation of a mature science'. That publication related very much to the practice of instrumentation. It is my belief that this book will be another key document in the maturity of measurement systems design, one laying down fundamental design building principles.

The contents are structured according to the philosophical sequence in which a measurement system is first conceived, then designed, made, installed, and maintained. The material divides conveniently into subjects concerned principally with theoretical principles needed to design measurement systems—this constitutes Volume 1—and fundamental material concerned with more specific design, application, and maintenance of measurement systems— Volume 2.

Volume 1 begins with the theoretical basics of measurement, the basic interface existing between the problem and the designer. Conceptual understanding is then continued in Chapter 2 followed, in Chapter 3, by an introduction to the terminology used.

Theory of signals is given in Chapter 4 providing a point of reference to much of the signal theory involved in realizing measurement system stages. The increasing importance of digital signals is recognized by the inclusion of Chapter 5. Error in measurement is covered in Chapter 6.

In a measurement situation the fundamental need is to be able to map one parameter of the system of interest into a mathematical equivalent. In pattern recognition (Chapter 7) the need is to establish a many-to-one mapping and as such this subject is most important in measurement systems design. Chapter 8, on parameter estimation, is concerned with realization of measurement values in situations where randomness is prevalent. Signals of the analog or digital form invariably need some form of selective spectral processing as they are conveyed through the measurement system chain and Chapters 9 and 10 deal with this.

Despite the existence of many texts on instrumentation almost none of them address the problem of extracting signals in the presence of noise sources at any useful depth. Chapter 11 is, in this way a unique contribution. Chapter 12 introduces the Handbook user to the methodology of converting signals between the analog and digital signal format domains.

A most valuable feature of electronically based measurement systems, compared with the alternatives, is that the sensors can be virtually any distance away from the data output. This feature means greater freedom of system layout than for mechanical and other earlier signalling methods. Thus transmission of data is an important subset of knowledge in measurement systems design. Chapter 13 focuses primarily on digital data networks as these are rapidly becoming the norm. To close the volume the final chapter concentrates on feedback systems design because the use of closed loops can greatly improve the performance of measurement stages.

Volume 2 covers the design of sensing and transformation stages at both the linguistic and mathematical levels. It also discusses human factors of instrument design: measurement of electrical quantities and the three main design regions (electronic, mechanical, and optical).

Several chapters review, from a theoretical rather than a practical viewpoint, how common variables are transduced. These chapters are included partly for completeness and to enable readers to see how material in other chapters is applied. Three chapters deal with the design, manufacture, and management of measurement systems, and a chapter is also devoted to the calibration, evaluation, and accreditation aspects of measurement hardware.

Compiling this Handbook has highlighted the multidisciplinary nature of measurement. One area of difficulty that could not be resolved ideally was that of terminology and symbols used. Any attempt to unify symbols to a common set would have meant that the uninitiated reader of a given chapter would, in many cases, have had to relearn the symbols when referring to the cited literature. It was deemed more satisfactory to retain the accepted symbols for a discipline in the chapter describing it. Wherever possible, alternatives have been brought to the reader's attention.

Despite any amount of personal enthusiasm for such a handbook it must be well recognized that its realization is only due to willing cooperative people. Notably I wish to thank sincerely two people for helping the project get under way: Professor L. Finkelstein of the City University, London, and Professor B. Veltman of the Technical University of Delft. Mr Hank Daneman of HLD Associates, USA also assisted in the early stages.

Next for acknowledgment must be the many contributors who have given their valuable time to summarize their belief of what 'every-young-man-should-know' about their special area of interest. Behind them also are their support staff who deserve a mention. Several librarians, and their staff, at the National Measurement Laboratory, Sydney, at City University, London and at the Levels Campus of the South Australian Institute of Technology, also helped me with the addition of references and with checking bibliographic material. On the day-to-day work front Mr Mike Miller, Principal Lecturer in the School of Electronic Engineering, helped me as a valuable sounding-board and advisor when I felt the need to check material submitted. Secretarial staff of my School, Elaine Milsom, Sue Wilkins, and Rachel Littler, each helped with retyping of text where needed.

Converting a sheaf of original typescript and diagrams into a consistent well produced Handbook is a task equally as great as writing it. My appreciation is extended to the production team of John Wiley and Sons at their Chichester, UK, location.

P. H. SYDENHAM

Adelaide
October, 1981

Handbook of Measurement Science, Vol. 1
Edited by P. H. Sydenham
© 1982 John Wiley & Sons Ltd.

Chapter

1

L. FINKELSTEIN

Theory and Philosophy
of Measurement

Editorial introduction

Measurement is a fundamental procedure in obtaining knowledge and in controlling systems. Numerous texts and published papers are devoted to its practice, a pursuit that dates since earliest man and is followed in all manner of human enterprise. Yet surprisingly we are only just beginning to try to understand the philosophical processes occurring at the measurement interface expressing these in terms of a formalized mathematical model.

Formalization with mathematics enables a topic to be mechanized with man-made machines—such as 19th century Boolean algebra that eventually led to today's digital computer systems. It also enables refinement of procedures. When we have developed a satisfactory formal approach to all measurement, and when that is widely understood, we should be able to be more efficient in designing and using measurement systems and in reworking data collected in the past.

This chapter reviews the advances made in measurement theory that can be readily understood and applied by the practitioner. It also analyzes why measurement is a fundamental method of science and presents a critical review of what constitutes a measurement. The gap between the abstract philosopher's approach and that of the pragmatic instrument designer is gradually being bridged. This aspect of measurement systems is probably the least taught, least understood, and least heeded by physical hardware designers. It is the key to real success if advances in application (especially in hardware capability) are to be given freedom to be used. Philosophical design strategies are needed urgently, strategies that can be applied by non-specializing engineers and scientists.

Of the several possible ways in which a measurement interface can be modelled, the mathematical one appears to be the most basic and useful in the long term. In this case, modelling is done in terms of set theory. To assist, an appendix is presented in which the relevant theory is outlined; Lin and Lin (1974) is also useful.

As will be seen in this chapter just what constitutes a measurement and what kinds of events can be measured are still very much matters for debate. This reflects the embryonic nature of the philosophy of measurement and its, as yet, paucity of exposure in areas of measurement practice. Other subsequent chapters also depict other descriptions of the measurement process. Each has something to contribute to assist design or application of measurement systems.

1.1 INTRODUCTION

Measurement is the process of empirical, objective assignment of numbers to the properties of objects and events of the real world in such a way as to describe them.

Measurement is the most fundamental method of science. Firstly, science aims at an objective empirical description of the universe and thus measurement of what is observed is the goal towards which scientific investigation is directed. Galileo Galiliei expressed this when he made his programmatic statement: 'Count what is countable, measure what is measurable and what is not measurable, make measurable.' Measurement enables the laws and theories of science to be expressed in the precise and concise language of mathematics. Science aims at describing complete domains of knowledge using measured data expressed in mathematical formalism. The mathematical description of knowledge is said to be the hallmark of true science. This view is usually expressed in the often quoted statement of Lord Kelvin: 'I often say that when you can measure what you are speaking about, and express it in numbers you know something about it; but when you cannot measure it, when you cannot express it in numbers your knowledge is of a meagre and unsatisfactory kind: it may be the beginning of knowledge, but you have scarcely, in your thoughts, advanced to the stage of science whatever the matter may be.' This strict formulation of the essentiality of measurement is often disputed by those engaged in such fields as the social and behavioural sciences, where the problems of measurement are difficult and there is much objective empirical observation and qualitative theorizing without measurement being possible. It may be questioned whether the physical sciences which are totally based on measurement and mathematical formalism are suitable paradigms for other domains of knowledge. However the universal importance of measurement cannot be disputed. These, and other fundamental issues are also discussed in Sydenham (1979).

When the property of an object or event is characterized by a number, this number carries information about the property. Modern technology has made immense strides in the development of instrumental means of information acquisition from physical objects and events. The information is encoded in the form of a physical signal and can be processed by a variety of information machines. The information can be output in the form of a number representing a physical property, in other words a measure, or used for decision or control. These powerful modern means of information acquisition and processing, constitute the nerves and brains of an immense variety of modern technical systems from chemical and electricity generating plant to aircraft and space vehicles. Measurement and related processes have thus acquired a vital technological importance. Finkelstein (1977) further expands these aspects.

Measurement is thus universal and all pervasive. As such the process of measurement of physical properties seems intuitively obvious. We learn to measure properties such as length or mass as children. We find no conceptual problem in measuring the length of an object as say 15 mm or interpreting or handling the information. For this reason most textbooks of the physical sciences or technology gave, refer to Sydenham (1979), virtually no attention to the definition or analysis of the concept of measurement. It may well be questioned what place a discussion of the philosophy of measurement has in a handbook such as this, substantially devoted to practical problems of physical measurement. However the foundational concepts of measurement are both of basic importance and interest as well as of some practical significance.

It is only necessary to look at some simple physical measurements to see that the concept of measurement is not trivially obvious. If we state that the volume of one object is two times that of another, what is the interpretation of two times, since division is an operation defined for numbers not objects? What about the statement that the temperature of one object is twice that of another?

Besides those physical properties for which there are well defined measurement scales, there are others for which suitable measures are more problematic. Hardness is an obvious example. There are many properties of materials, for instance, which are of technical importance and for which suitable scales of measurement are difficult to establish. To give a few examples, we have the 'spreadability' of butter, the 'foldability' of paper and the 'strength' of coal, that is the ease with which it can be cut or worked by tools. The theory of measurement may give an approach to the procedures for the setting up of suitable scales of measurement in such cases.

It is the social and behavioural sciences and their managerial application, which give particularly good examples of practical and philosophical problems in the formation of scales of measurement. Starting from psychological properties like taste and smell, the measurement of which may be of great technical importance, we can quote others such as intelligence or alienation of great theoretical and practical interest. The level of conflict in a community or the standard of living are basic concepts in our study of society, the measurement of which presents problems. Finally, even the apparently exact managerial accountancy techniques present us with conceptual difficulties in the measurement of the level of profit.

In addition to the practical significance of the theory of measurement, it has an intrinsic philosophical importance. As the basic means for our understanding of the universe, it is essential that the nature of measurement be understood. The last hundred years have seen a powerful and fruitful development of the understanding of the foundations of mathematics and logic, sciences which in the past were based on vague and intuitive foundations. The theory and philosophy of measurement can be seen as part of that development.

This chapter gives a brief outline of the foundational concepts of measurement and their philosophical background. It will as far as possible attempt to be simple in its presentation of concepts, while giving the essential aspects of formal measurement theory.

1.2   OUTLINE OF HISTORICAL DEVELOPMENT OF MEASUREMENT THEORY

The ancient Greeks were the first to investigate the philosophical foundations of measurement through the practical pursuit of measurement in crafts and trade that arose several millenia earlier in the course of the Urban Revolution in Mesopotamia and Egypt. The school of Pythagoras was concerned with the philosophy of the relation between numbers and the real world and hoped to make arithmetic the fundamental study in physics. Plato's Academy developed an extensive theory of magnitudes, though it was concerned more with the nature of numbers than with an analysis of the nature of measurement. Aristotle studied the concepts of measurement in his *Metaphysics*. (Greek writings are extensively translated into the English language in such works as the Loeb classics.)

The Middle Ages saw much scholarly study of the theory of measurement though not concerned with the application of measurement to scientific observation. With the rise of modern science, Newton in his development of mechanics provided the first comprehensive mathematical theory of a domain of physics. In his *Arithmetica Universalis* he developed a theory of magnitudes based on arithmetic.

The true foundations of the modern theory of measurement were laid by Helmholtz (1887) in a thorough logical analysis of the epistemology of counting and measuring. The work was part of the beginning of studies in the logical foundations of mathematics with which the theory of measurement is closely connected. An important development of the work of Helmholtz was the axiomatization of measurement of additive quantities by Hölder (1901).

The British physicist, N. R. Campbell provided a lucid and thorough analysis (Campbell, 1920) of the fundamental basis of the measurement of physical quantities in his *Physics: the Elements*. The book can still be read with pleasure and profit. Campbell's theory was based on the measurement of physical properties for which an empirical operation of addition could be constructed, what is now known as extensive measurement. The theory became generally accepted under the influence of logical positivism (Cohen and Nagel, 1934; Hempel, 1952).

The theory of measurement from Helmholtz to Campbell and those who developed their work, was concerned with physical measurements, which are based on additive quantities and quantities derived from them. In the social and behavioural sciences, however, there are many properties which cannot be empirically added nor derived from additive quantities. This created great philosophical difficulty. A report of a committee of the British Association for the Advancement of Science which considered quantitative methods rejected measurements not based on additivity, and hence the possibility of psychological measurements, and even questioned the status of the thermodynamic scale of temperature.

These rigid positions of the classical theory of measurement were broken down by work in the social and behavioural sciences. The concern of the social sciences with the concept of utility led through the work of the early writers Bentham and Pareto to the classical work of von Neumann and Morgenstern (1944). Their axiomatic formulation of the theory of utility has been the basis of much work in the theory and practice of measurement in the social sciences. In psychology S. S. Stevens carried out much fundamental work on developing an appropriate analysis of the nature of measurement (Stevens, 1946, 1951, 1959). Other important workers who should be mentioned are Torgerson (1958), who presented an excellent exposition of the fundamentals of measurement and scaling, and Coombs (1964) who developed a theory of data.

The proceedings of a conference in the United States (Churchman and Ratoosh, 1959) presented a review of the classical approaches to measurement as extended to needs of the social and behavioural sciences. Ellis (1966) is a most useful and interesting philosophical analysis of measurement. It is not formal in approach and is principally concerned with physical measurement though it takes into account non-classical theory.

Modern measurement theory may be said to originate from the work Tarski on relational systems and model theory (Tarski, 1954). The theory which may be termed representational theory of measurement considers measurement, loosely speaking as the establishment of a correspondence between a set of manifestations of a property and the relations between them and a set of numbers and the relations between them. Suppes and Zinnes (1965) provide a clear exposition of the theory, in the development of which Suppes has been one of the key workers. Pfanzagl (1968) is the first work devoted to the theory of measurement with a representational approach. Krantz et al. (1971) published a very detailed and thorough account of the representational foundations of measurement, though only part of their work has been published. At the time of preparation the most recent release on this subject is Roberts (1979).

While the theory of measurement finds a place in most modern texts on quantitative psychology, sociology and the like, it receives very little, if any, attention in the literature of physical measurement and instrumentation. Surveys of the theory (Finkelstein, 1973, 1975a) have stimulated some awareness in the foundations of measurement among measurement and instrumentation engineers. They have begun to bridge the large epistemological gap currently existing between theorists and hardware designers' understanding of measurement systems.

1.3 THE NATURE AND PROPERTIES OF MEASUREMENT: AN INFORMAL DISCUSSION

Before embarking on a formal definition and analysis of the measurement process it is first examined informally. The purpose is to highlight the principal aspects of its nature and properties without employing the rigorous but not always familiar concepts and symbolism of the logical foundations of mathematics.

The outset of the discussion will be the informal definition of measurement presented at the beginning of the chapter:

> 'Measurement is the process of empirical, objective, assignment of numbers to properties of objects or events of the real world in such a way as to describe them.

The definition will now be analysed and discussed.

Firstly, measurement is the assignment of numbers to properties of objects and events. It is thus the description of properties of objects or events and not of the objects or events. One measures the length of an object, the temperature of an object and so on. Measurement presupposes the existence of a clear concept of a property as an abstract aspect of a whole class of objects of which individual instances or manifestations are the subject of measurement.

The definition states that the assignment of numbers in measurement is such that the numbers describe the property of the object or event. The meaning can be explained as follows. Consider that a number, or measure, is assigned by measurement to the property of an object, and other numbers are assigned by the same process to other manifestations of the property. Then the numerical relations between the numbers or measures, imply and are implied by empirical relations between the property manifestations. Thus if the numbers, assigned to the manifestations of a particular property in two objects by measurement are equal, this implies that the two property manifestations are empirically indistinguishable. Conversely empirical indistinguishability implies the equality of measures. Again if the numbers assigned by measurement to the manifestations of a particular property, in a series of objects, can be placed in order of increasing magnitude, this implies that there is an empirical relation which would result in the placing of the objects in the same order in respect of the property. Conversely, an empirical order among manifestations of the property, implies the same order among the measures. This correspondence between the numerical relation among measures and the empirical relation of the corresponding property manifestations is all that is basically expressed by the more rigorous formal definition to be given later.

The above clearly indicates that measurement is a process of comparison of a manifestation of a property, with other manifestations of the same property. This is a common part of many informal definitions of measurement. Many definitions however go further, to state that the measure of a property expresses the ratio of the magnitude of the property to a standard magnitude taken as unity. As indicated in the introduction to this chapter, this begs the essential question of what measurement is. The statement is untrue for many scales of measurement and would make the measurement of many properties impossible.

There can be no ratio of two properties, only a ratio of their measures. The empirical relation corresponding to the ratio of two measures must be analysed, if it is to be meaningful. For many scales of measurement, measures do not correspond to a ratio to the unit magnitude. To take an obvious example, the temperature of a body on the Celsius scale is not the ratio of the temperature to a unit degree Celsius. As another example, intelligence of a person, for instance, cannot be measured as a ratio, to the intelligence of a person having unit intelligence.

There is a divergence of views as to whether any descriptive assignment of numbers is adequate for the process to qualify as measurement. At one extreme, broadly in the social and behavioural sciences, there is the view that any empirical, objective assignment of numbers which describes a property manifestation, can be termed measurement. At the other extreme is the view that only numbers which reflect in some way a ratio to a unit magnitude of a property are true measures. This is the classical view and that of most informal definitions of measurement in physics. Many other definitions imply that to be true measurement, the assignment of numbers must imply at least an empirical order among the property manifestations, corresponding to a concept of ordering according to magnitude. The author is an advocate of the first view (Finkelstein, 1975a, b) though without wishing to impose the view on others.

The matter will be considered later in the chapter in which it will be argued that the assignment of symbols other than numbers may essentially be very close to measurement.

The next aspect of the definition of measurement which requires discussion is the fact that measurement is an objective process. By this is meant that the numbers assigned to a property by measurement must, within the limits of error, be independent of the observer. It is not uncommon for properties such as the suitability of candidates for a post to be valued by judges on a scale of 0–10. Numbers resulting from such a valuation cannot be considered measures, unless it is established that the same numbers would, within acceptable limits of error, result from any valuation process of the subject using the same procedure.

The informal definition of measurement proposed stresses the fact that measurement is an empirical process. This means first that it must be the result of observation and not, for example, of a thought experiment. Further, the concept of the property measured must be based on an empirical relation. For example, the use of the Universal Decimal System of library classification, results in the assignment to documents of numbers which describe their content. The assignment is essentially objective, for classification by trained observers would, with a low error rate, result in the same number being assigned to the same document. Let us accept, just for the sake of this argument, that assignment of numbers which describe class membership can be measurement. Even then the above library classification cannot be measurement. The reason is that the classification of knowledge on which it is based is substantially an arbitrary convention.

It is convenient to discuss at this stage the properties of measurement which give it the key role in science.

As indicated in the introduction to the chapter there is first and foremost the objectivity of measurement. A measure is an objective description and hence a proper scientific datum. Conversely, it can be claimed that if we can arrive at a totally objective description of a property manifestation, the most vital step towards measurement has been taken.

The second property of measures is that they are descriptions of great conciseness. A single number tells us what it would take many words to express.

Measurement gives, further, a description which is precise, pinpointing by a single number a particular entity, where a verbal description indicates a range of similar but differing things.

A measure of a property gives us an ability to express facts and conventions about it in the formal language of mathematics. Without the convenient notation of this language, the complex chains of induction and deduction by which we describe and explain the universe would be too cumbersome to express.

It follows from what has been said that description by numbers is not good in itself. The only value of measurement lies in the use to which the information is put. Science is not just the amassing of numerical data, it depends upon the way in which the data are analysed and organized.

Finally, we return to the fact already mentioned that measurement enables the measurand to be expressed in signals which can be handled by machines. This will be analysed later in the chapter.

1.4 THE ELEMENTS OF THE FORMAL THEORY OF MEASUREMENT

It is now necessary to look at the formal representational theory of measurement. While the preceding general discussion is adequate for insight, a full presentation of the modern philosophy of measurement requires recourse to rigorous symbolism.

The brief presentation which follows is a development of accounts given in Pfanzagl (1968) and Krantz et al. (1971).

A representational theory of measurement has four parts:

  (i)   an empirical relational system corresponding to a quality;
 (ii)   a number relational system;
(iii)   a representation condition;
(iv)   a uniqueness condition.

These will now be considered.

(i) *Quality as an empirical relational system*

Consider some quality (for example length, hardness, etc.) and let $q_1, q_2, \ldots,$ $q_i, \ldots$ represent individual manifestations of the quality so that we can define a set of all possible manifestations:

$$Q = \{q_1, q_2, \ldots, q_i, \ldots\} \tag{1.1}$$

$$\Omega = \{w_1, w_2, \ldots, w_i, \ldots\} \tag{1.2}$$

represent the class of all objects manifesting elements of Q.

Consider further that there exists on Q a set $\mathscr{R}$ of empirical relations $R_1,$ $R_2, \ldots, R_i, \ldots, R_n$ and let us denote

$$\mathscr{R} = \{R_1, R_2, \ldots, R_i, \ldots, R_n\} \tag{1.3}$$

Then the quality is represented by an empirical relational system

$$\mathscr{Q} = \langle Q, \mathscr{R} \rangle \tag{1.4}$$

(ii) *Numerical relational system*

Let N represent a class of numbers and let

$$\mathscr{P} = \{P_1, P_2, \ldots, P_i, \ldots, P_n\} \tag{1.5}$$

be a set of relations defined on N so that:

$$\mathscr{N} = \langle N, \mathscr{P} \rangle \tag{1.6}$$

represents a numerical relational system.

Commonly $\mathscr{N}$ is just the real number line.

(iii) *Representation condition*

The representation condition requires that measurement be the establishment of a correspondence between quality manifestations and numbers in such a way that the relations between the referent property manifestations imply and are implied by the relations between their images in the number set.

Formally, measurement is defined as an objective empirical operation

$$M : Q \to N \tag{1.7}$$

such that $\mathscr{Q} = \langle Q, \mathscr{P} \rangle$ is mapped homomorphically into (onto) $\mathscr{N} = \langle N, \mathscr{P} \rangle$ by M and F. F is a one-to-one mapping, with domain $\mathscr{R}$ and range $\mathscr{P}$:

$$F : \mathscr{R} \to \mathscr{P} \tag{1.8}$$

so that we can denote

$$P_i = F(R_i); P_i \in \mathscr{P}; R_i \in \mathscr{R} \tag{1.9}$$

P is an n-ry relation if and only if it is the image under F of an n-ry relation.

By a homomorphic mapping we mean that for all $R_i \in \mathcal{R}$ and all $P_i \in \mathcal{R}$ and $P_i = F(R_i)$,

$$R_i(q_1, \ldots, q_i, \ldots, q_n) \leftrightarrow P_i(M(q_1), \ldots, M(q_i), \ldots, M(q_n)) \qquad (1.10)$$

Measurement is a homomorphism because M is not one-to-one, it maps separate but indistinguishable property manifestations to the same number. Then

$$\mathscr{S} = \langle \mathcal{Q}, \mathcal{N}, M, F \rangle \qquad (1.11)$$

constitutes a scale of measurement for $n_i = M(q_i)$. The image of q_i in N under M is called the measure of q_i on scale $\mathscr{S}$.

(iv) *Uniqueness condition*

The representation condition may be valid for more than one mapping M. One may admit certain transformations from one scale of a property to another without invalidating the representation conditions. The uniqueness condition defines the class of scale transformations to those for which the representation condition is valid.

(v) *Uncertainty*

The above definition has been given in terms of deterministic relations and mappings. However, all experimental observations are accompanied by error. Uncertainty should be introduced into the representational theory of measurement. Leaning and Finkelstein (1979) have presented such a theory based on the concepts of a probabilistic relational system and probabilistic homomorphism.

1.5 QUALITY CONCEPT FORMATION

Measurement presupposes something to be measured. Both in the historical development and logical structure of scientific knowledge, the formulation of a theoretical concept or construct, which defines a quality, precedes the development of measurement procedures and scales.

Thus the concept of 'degree of hotness' as a theoretical construct, interpreting the multitude of phenomena involving warmth, is necessary before one can conceive and construct a thermometer. Hardness must, similarly, first be clearly defined as the resistance of solids to local deformation, before we seek to establish a scale for its measurement. The search for measuring some such conceptual entity as 'managerial efficiency' must fail until the concept is clarified.

The formation of concept in empirical science (Hempel, 1952) is an important and much discussed subject. Here consideration will be confined to the concept of quality.

The basic notion is that of a manifestation of a quality, an abstract, single sensed aspect of an object or event, such as, for example, the smell of a substance. Observation of the real world leads to the identification of empirical relations among these single manifestations. Examples of such relations are similarity, difference and the like. As a result the concept of a quality is formed as an objective rule for the classification of a collection of empirically observable aspects of objects into a single set, together with the family of objective empirical relations on that set. The resulting relational system is a quality and each single member of the set is termed a manifestation of the quality. This was formally set out in Section 1.4 where quality was defined as the set Q, of all manifestations q of the quality, $Q = \{q\}$, together with the set $\mathcal{R}$ of all relations R on Q.

We can thus see that there is a difficulty in the measurement of such qualities as beauty. The existence and meaningful use of the word beauty indicates the usefulness of the concept. However, there is not an objective rule for classifying some aspect of observable objects as manifestations of beauty. Similarly, there are no objective empirical relations such as indistinguishability or precedence, in respect of beauty. The basis for measurement of beauty is thus absent from the outset.

When there exists a clearly defined quality, as a set into which its manifestations can be objectively and empirically classified, together with a set of empirical relations, then we can always find some symbolical relational set by which it can be represented. Thus if a quality is definable as above, we can set up for it a scale of measurement, in its broadest sense. However, this is a relatively late state of development of the concept of quality.

In some cases the concept of a quality arises from invariances in numerical laws arrived at by measurement. An obvious example is Young's modulus. This quality is arrived at from Hooke's law: the observation that for an extensive class of materials, strain is proportional to stress. In general, however, one starts from some direct concept of a quality and then seeks measurement scales.

Usually given a concept of a quality and a scale of measurement based on that quality, the accumulation of data by measurement leads to the clarification and re-evaluation of the quality concept. This in turn leads to improvement in the measurement scale. The process is an ascending spiral. The history of development of thermometry from simple devices and concepts to the modern thermodynamic definition of temperature is an example.

One of the principal problems of scientific method is to ensure that the scale of measurement established for a quality yields measures, which in all contexts describe the entity in a manner which corresponds to the underlying concept of the quality. For example, measures of intelligence must not disagree with our basic qualitative concept of intelligence. It is usual that once a scale of measurement is established for a quality, the concept of the quality is altered to coincide with the scale of measurement. The danger is that the adoption in science of a well defined and restricted meaning for a quality like intelligence, may deprive us of useful insight which the common natural language use of the word gives us.

Finally, let us explain the concept of an empirical quantity. If there is an order relation in the quality relational system, enabling us to order quality manifestations in a way which has formal similarity to the relations equal, greater and less, then the quality is termed a quantity.

1.6 SOME EMPIRICAL RELATIONAL SYSTEMS AND DIRECT SCALES OF MEASUREMENT

1.6.1 Object and Scope of Section

In the present section an attempt will be made to analyse some qualities as empirical relational systems and to explain the logical basis of deriving a scale of measurement for them. Extensive measurement, that is measurement of physical quantities for which we can construct an operation having the formal properties of addition, is the basis of physical measurement and will be considered in detail. Elegance and rigour will be eschewed in favour of as much clarity and simplicity as possible.

1.6.2 Extensive Measurement

The extensive scales of physical measurement are based on establishing for the quality $\mathcal{Q}$ of empirical objects, for which a scale is to be determined, of an empirical ordering with respect to $\mathcal{Q}$ of the class Ω of all objects possessing elements of Q, together with an operation $\circ$ of combining the objects, elements of Ω, which has with respect to $\mathcal{Q}$ the formal properties of addition. Such scales are known as extensive. The theory of the form of construction of such scales of measurement originates from the work of Helmholtz (1887) and Hölder (1901) and is lucidly developed in the work of Campbell (1920). There has, however, been much work concerned with establishing elegant systems of axioms of representation (Pfanzagl, 1968; Krantz *et al.*, 1971).

The above will now be stated more formally: the basis of a scale of measurement of $\mathcal{Q}$ is the definition of the set Q.

Secondly, there must be an operational procedure which establishes on the set of objects Ω possessing $\mathcal{Q}$ an empirical equivalence relation $\sim$ and a transitive empirical relation $\prec$ with respect to $\mathcal{Q}$ such that $\langle Q, \sim, \prec \rangle$ is an order system. Without discussing order in detail we shall use the simple definition of order system given in Appendix 1.A.

Finally, consider objects $w_1, w_2, w_3, w_4 \in \Omega$ exhibiting property manifestations $q_1, q_2, q_3, q_4 \in Q$. For an extensive measurement scale there must be an operation $\circ$ of combining w_1, w_2 with respect to q_1 and q_2 which we shall denote by $q_1 \circ q_2$ with the same formal properties as addition.

For all $q \in Q$

(i)	$q \circ q_2 \in Q$		(1.12)
(ii)	$q_1 \circ q_1 \sim q_2$		(1.13)
(iii)	$q_1 \circ q_2 \sim q_2 \circ q_1$	commutativity	(1.14)
(iv)	$q_1 \circ (q_2 \circ q_3) \sim (q_1 \circ q_2) \circ q_3$	associativity	(1.15)
(v)	if $q_2 \sim q_3$ then $q_1 \circ q_2 \sim q_1 \circ q_3$		(1.16)
	if $q_3 \sim q_2$ then $q_1 \circ q_3 \sim q_1 \circ q_2$		(1.17)
(vi)	if q_1, q_2, q_3,... bear to each other the relation $\sim$ and $q_1 \prec q_1'$, then there is a number n such that $q_1' \prec q_1 \circ q_2 \circ \cdots \circ q_n$	Archimedean postulate	(1.18)

With these definitions the empirical relation system $\langle Q, \sim, \prec, \circ \rangle$ has a structure of the same properties as the numerical relation system $\langle \mathrm{Re}, =, <, + \rangle$.

With an empirical ordering operation and an 'additive combination' thus established one proceeds to the setting up of a scale.

A single object with $s_1 \in Q$ is chosen as standard and assigned the number 1, that is, it is chosen as the unit of the scale. One then constructs or seeks another object with $s_1' \in Q$ such that $s_1' \sim s_1$. One can then construct a standard $s_2 \sim s_1 \circ s_1'$ and assign it the number 2, $s_3 \sim s_2 \circ s_1$ and assign it the number 3, and so on. Fractional standards can be generated by constructing $s_{1/2}, s_{1/2}' \in Q$, $s_{1/2}' \circ s_{1/2} \sim s_1$ and assign $s_{1/2}$ the number $\frac{1}{2}$. Thus we generate

$$S = \{\ldots, s_{1/2}, s_1, s_2, s_3, \ldots\} \tag{1.19}$$

Any $q \in Q$ is then measured by finding the element s_i to which it bears the relation $\sim$, and assigning to it the number corresponding to s_i.

As an example, in the measurement of mass the equipoise balance offers the means of establishing empirical order. If the arm balances, the masses in the two pans are equivalent. The tipping of the balance indicates that one mass is 'heavier' than the other. Thus we can rank a series of weights in order of heaviness. The lumping together of two objects is with respect to mass an operation with the properties of addition.

A variety of problems with respect to the form of measurement discussed above are examined in the literature, such as various forms of necessary and sufficient representation conditions, the possibility of constructing scale based on operations resulting in alternative numerical representation such as for instance, multiplicative scales based in a homomorphism into $\langle \mathrm{Re}, =, <, \rangle$ and conceptual difficulties imposed by the limitations in practice of the size of standard. Readers are referred to Krantz et al. (1971).

1.6.3 Matching Scale

A matching scale is based on the establishment on the set of quality manifestations Q of an empirical indifference relation $\sim$.

Given $\langle Q, \sim \rangle$, a set of differing elements $s_i \in Q$ $(s_i \sim {}'s_j$ if $i \neq j)$ are selected to form a standard set $S = \{s_1, s_2, \ldots, s_k\}$. Numbers (or other symbols) $n_i \in N$ are then assigned to each $s_i \in S$, the same number n_i not being assigned to two differing elements $s_i \in S$ (if $s_i \sim {}'s_j$, $n_j \neq n_j$). The fundamental measurement operation M of the scale consists of an empirical operation in which measurands $q_i \in Q$ are compared with members of the standard S. If $q_n \sim s_i$ it is assigned the number n_i.

An example of this form of scale is a colour code in which the relation 'matches' constitutes the empirical indifference relation.

In the establishment of a colour code we first establish the indistinguishability relation based on a colour match. This indistinguishability relation can be tested for symmetry and transitivity. Reflexivity is implicit. The relation is therefore an equivalence relation. We then select as standards a set of coloured objects, each a distinct colour manifestation. Each is assigned a different number or other symbol such as a word label. Any unknown colour manifestation is compared with the standards and if it matches one of them it is assigned the same number or symbol as the standard.

Matching scales are not generally considered measurement scales, since they are not quantitative in the sense that they do not establish on Q relations formally similar to the relations greater or less. Also it is not generally practical to establish sufficient standard elements to ensure that every $q_i \in Q$ can be matched with a standard and hence assigned a measure.

1.6.4 Ranking Scales

In ranking scales, an empirical order system $\langle Q, \sim, \prec \rangle$ is established on Q. A set of differing standard objects having $s_i \in Q$ is then selected and arranged in an ordered standard series $S = \{s_1, \ldots, s_n\}$ according to $\langle Q, \sim, \prec \rangle$. Numerals are assigned to each s_i say i in such a way that the order of numerals corresponds to the order in S of standards to which they are assigned. Any $q \in Q$ can then be compared with the elements of S in the same way as in nominal measurement. If q bears the relation $\sim$ to any $s_i \in S$ it is assigned the numeral of s_i. If an entity is not equivalent to any $s_i \in S$ one can determine between which two standard elements it lies in the empirical order system.

The best example of a ranking scale of measurement is the Mohs scale of hardness of minerals.

Ten standard minerals are arranged in an ordered sequence so that precedent ones in the sequence can be scratched by succeeding ones and cannot scratch them. The standards are assigned numbers 1 to 10. (The sequence is talc 1, gypsum 2, calcite 3, fluorite 4, apatite 5, orthoclase 6, quartz 7, topaz 8, corundum 9, diamond 10.) A mineral sample of unknown hardness which cannot be scratched by quartz and cannot scratch it, is assigned measure 7.

1.7 INDIRECT MEASUREMENT

The preceding section considered measurement scales formed by direct mapping from a quality relational system to a numerical relational system. Frequently, however, scales of measurement for qualities are constructed indirectly through a relation of the quality to be measured and other qualities, for which measurement scales have been defined. The reason may be, for example, the impossibility of setting up a satisfactory measurement scale directly. Thus for example we cannot set up an extensive measurement scale for viscosity or density, since there is no appropriate combination operation having the properties of addition. Another reason is the wish to set up a consistent set of measurement scales, in which a minimal set of qualities with direct scales is defined, and scales for other qualities are defined by derivation from them.

In its simplest form consider a case when every object that manifests the quality to be measured exhibits a set of other qualities which are measurable. Then to each a manifestation of the measurand quality, there corresponds a set of measures of the associated qualities. These associated or component measures can be arranged in an ordered array. If manifestations of the measurand quality, have identical arrays of component measures, if and only if they are indistinguishable, then the array of component measures characterises the measurand. This will now be expressed formally.

Consider a quality $\mathscr{Q}_0$, for which it is desired to construct a scale of measurement. $\mathscr{Q}_0$ consists of the set Q of all the manifestations of the quality and of $\mathscr{R}_0$ a set of relations among the manifestations. Consider now the class of all objects Ω which exhibit manifestations of the quality $\mathscr{Q}$. Let each element of Ω also exhibit logically independent qualities $\{\mathscr{Q}_1, \mathscr{Q}_2, \ldots, \mathscr{Q}_n\}$.

Let us assume that there exists for each element of the above scale of measurement such as

$$\mathscr{S}_i = \langle \mathscr{Q}_i, \mathscr{N}_i, M_i, F_i \rangle \qquad (1.20)$$

Assume that to each object $w \in \Omega$ there corresponds one and only one $q_0 \in Q_0$ and one and only one element:

$$\mathbf{q} = \langle q_1, \ldots, q_i, \ldots, q_n \rangle \qquad (1.21)$$

($q_i \in Q_i$, etc.) of the product set $\mathsf{X}_{i=1}^n Q_i$. That element corresponds to an ordered array of measures:

$$\mathbf{M}(q_0) = \langle M_1(q_1), \ldots, M_n(q_n) \rangle \qquad (1.22)$$

where $M_i(q_i)$ is the image of $q_i \in Q_i$ in N_i under M_i and so on.

If now for any q_0, $q_0' \in Q_0$, $q_0 \sim q_0'$, $\mathbf{M}(q_0) = \mathbf{M}(q_0')$ where $\sim$ represents empirical indistinguishability, then we can say that $\mathbf{M}(q_0)$ characterizes q_0.

If we can combine the various component measures or in other words map them into a single number so that numbers assigned to the quality manifestations by this process, imply and are implied by empirical relations between the quality manifestations, then this sets up an indirect scale of measurement. This will now be explained formally, referring to the discussions above.

Let there exist a mapping ϕ from the product set $\mathsf{X}_{i=1}^{n} N_i$ into a number set N_0.

$$n_0 = \phi(M_1(q_1), \ldots, M_i(q_i), \ldots, M_n(q_n)) \tag{1.23}$$

is then the image of $M_1(q_1), \ldots, M_i(q_i), \ldots, M_n(q_n)$ in N_0 under ϕ. We can then see that the composition of ϕ, $(M_1, \ldots, M_i, \ldots, M_n)$ and the correspondence between elements of Q_0 and $\mathsf{X}_{i=1}^{n} Q_i$ constitute a mapping from Q_0 to N_0, which we shall denote M_0.

Consider that there is a set of relations $\mathscr{P}_0$ on N_0 giving a relational system $\mathscr{N} = \langle N_0, \mathscr{P}_0 \rangle$ and that there exists a correspondence $F_0 \colon \mathscr{R}_0 \to \mathscr{P}_0$. Then if M_0, F_0 map $\mathscr{Q}_0$ homomorphically into (onto) $\mathscr{P}_0$ then:

$$\mathscr{S}_0 = \langle \mathscr{Q}_0, \mathscr{S}_0, \mathscr{S}_1, \ldots, \mathscr{S}_n, \ldots, M_0, F_0 \rangle \tag{1.24}$$

constitutes an indirect measurement scale for $\mathscr{Q}_0$. Typically $\mathscr{Q}_0 = \langle Q_0, \sim, \prec \rangle$, that is we have an empirical order system on our measurand quality. The mappings M_0, F_0 are such that whenever the objects are ordered according to $\langle \mathscr{N}_0, =, < \rangle$ they are also ordered according to $\langle N_0, =, < \rangle$.

Consider as an example the scale of measurement of density of homogeneous bodies. Each such body possesses mass, say m, and volume v (where m and v are assumed to be measures on already defined scales). It is an empirically established law that objects of the same material, and hence conceptually of the same density, have the same ratio (m/v). When different materials are ordered according to our concept of density they are also ordered according to the respective ratio m/v. Hence a scale of measurement of density is based on the ratio of mass to volume.

A few observations should be made here. The mapping ϕ is not unique in its order preserving properties with respect to density. For instance, $(m/v)^2$ would be an equally valid derived measure of density. The form of ϕ is chosen to result in the greatest simplicity of mathematical relations involving density. The properties of the function ϕ are an idealization of real observations.

In general once an indirect scale of measurement has been defined it is treated as a definition of $\mathscr{Q}_0$.

The description of qualities by multidimensional arrays of measures, where the measures are not combined, is known as multidimensional measurement. An example might be a characterization of a shape by a set of geometrical measures.

Measurement by combining component measures, so that the resultant number characterizes the place of the measurand quality manifestation in an empirical order is termed conjoint measurement. It is based on the conjoint establishment of order on the measurand quality and on the component qualities.

In physical measurement indirect measures of qualities are obtained as multiplicative monomial functions of the measures of component qualities. This is generally known as derived measurement (Krantz *et al.*, 1971), though see a more general statement term of the term derived in Finkelstein (1975a).

1.8 UNIQUENESS: SCALE TYPES AND MEANINGFULNESS

As stated in Section 1.4, the requirement that the fundamental measurement procedure of a scale should map the empirical relational system $\mathscr{Q}$ homomorphically into the numerical relational system $\mathscr{N}$ does not determine the mapping uniquely.

Thus there is an element of arbitrary choice in the setting up of scales of measurement. In the case of scales based on additive combination, for instance, the choice of the unit standard is arbitrary.

In the case of a ranking scale of measurement the actual numbers assigned to the standards are arbitrary, subject only to the requirement that they should be in the required order.

The requirement of homomorphism thus defines a class of scales which may be called equivalent. The class of transformations which transform one member of a class of equivalent scales into another is called the class of admissible transformations. The conditions which admissible transformations must satisfy are known as the *uniqueness conditions*. They specify that a scale is unique up to a specified transformation.

Consider as an example a mapping which maps homomorphically the empirical relational system $\langle \mathscr{Q}, \sim, \prec, \circ \rangle$ into the numerical relational system $\langle \mathrm{Re}, =, <, + \rangle$, that is, it is the fundamental measurement procedure of a scale based on additive combination. If all measures $M(\ )$ be replaced by $\alpha M(\ )$ where α is a real positive number, then the measures $\alpha M(\ )$ preserve the required homomorphism and multiplication by a real positive number is an admissible transformation. In effect there has been a change of unit to $1/\alpha$ on the original scale.

We can classify scales by the classes of transformations admissible for them (Stevens, 1959). Let m be numbers representing measures on a scale and let m' be corresponding numbers on the transformed scale. The generally accepted classification of scales is given in Table 1.1.

The problem of the meaningfulness of statements made about a quality in terms of its measures is important. Such a statement is meaningful if its truth is unchanged by admissible transformations of the scales of measurement, in

Table 1.1   Classification of scales of measurements

Class of admissible transformations	Scale type
$M' = F(M)$ where $F(\quad)$ is any one-to-one substitution	nominal
$M' = F(M)$ where $F(\quad)$ is any monotonic increasing function	ordinal
$M' = \begin{cases} \alpha M + \beta & \alpha > 0 \\ \alpha M & \alpha > 0 \end{cases}$	interval ratio

other words, if it reflects the empirical relational system on which the scale is based and not just the arbitrary conventions of the scale.

We say that a k-ry relation P is meaningful if:

$$P(M(q_1), \ldots, M(q_k)) = P(M'(q_1), \ldots, M'(q_k)) \tag{1.25}$$

where $M \to M'$ is any admissible transformation.

Thus as a very simple example, it is meaningful to speak of the ratio of two masses, since that ratio is invariant with respect to changes of the unit of mass, but it is not meaningful to speak of the ratio of two hardnesses measured on the Mohs scale, since that ratio would be changed by a monotonic increasing transformation of the scale.

Where $P(\quad)$ is an independently empirically established relation among objects, expressed in terms of measures $M(\quad)$ it is still meaningful, even if it is not invariant with respect to admissible transformations of the scale of measurement.

Another view of meaningfulness which can be taken is that only such statements involving measures are meaningful which can be logically traced to the empirical operations on which the measurement is founded.

One aspect of the meaningfulness of interpretations of measurement data is the description by statistics. Table 1.2 presents statistical measures meaningful

Table 1.2   Statistical measures appropriate to measurements made on various classes of scale

Scale type	Measures			
	Location	Dispersion	Association	Significance
nominal	mode	information	transmitted information	chi-squared test
ordinal	median	percentiles	rank-order correlation	sign test
interval	mean	standard, deviation average deviation	product-mean correlation	t-test f-test
ratio	geometric and harmonic mean	% variation		

for measurements on various classes of scale. Statistical measures in a row of the table corresponding to a particular scale are also meaningful for the scale listed below, but are meaningless for those listed above.

1.9 MEASUREMENT AND OTHER FORMS OF SYMBOLIC REPRESENTATION

Measurement is only one form of representation of entities by symbols. It is closely related to other forms of symbolization.

At its simplest level a symbol may be only a name or label which can be used to refer to the object and handle information about it. The number of an item of equipment on an inventory list is such a symbol. The inventory number can be used to handle information about the equipment in a compact and convenient manner without the equipment itself being handled.

The type number or class name of a class of equipment is a symbol which describes the equipment, insofar as it denotes its similarity in certain respects to other equipment bearing the same symbol. Thus the class label 'Concorde' referring to an aircraft describes insofar as it denotes its similarity to other aircraft bearing the label 'Concorde' and its difference from aircraft described by other class names. The statement 'John is flying to New York by Concorde' gives information to the recipient of the message in a compact form, without him or her having to see the aircraft.

Library classification is another example to be considered (see Section 1.2). Taking an imaginary classification system for the contents of books and documents we may have a set of rules by which a combination of letters and numbers is assigned to describe the contents of an item. Two items with the same number say FA 592 have equivalent content in the sense that they deal with the same class of subject matter. The designation FA 592 however may give further information. Depending on the rules of classification and assignment of class symbols, it may be possible to determine the relation of items so designated, to those denoted, for example, by FA 592 or FB ___ and so on.

Properties of objects can be described by the symbols of natural language. For example the words 'red', 'blue', and 'green' are closely similar to labels on a nominal scale of colour. The sequence 'hot–warm–tepid–cool–icy' has some similarities to an ordinal scale of temperature (Pfanzagl, 1968).

The examples given are not in any way exhaustive, nor have they been rigorously analysed. They show that there is a large number of practical cases in which objects of the real world and their attributes and characteristics are represented not by numbers but by conventional symbols. This does not constitute measurement, but shares with measurement its most essential characteristics: namely, the representing symbol designates the entity represented, relations between correspond to relations between represented entities, and symbols can be used to achieve responses corresponding to the entities they represent.

The formal representational theory of measurement based on these definitions given in Section 1.4 can be extended to the more general case of representation by symbol systems.

A symbol will be defined here as an object or event, which has a defined relation to some entity, for the purpose of eliciting a response appropriate to that entity in its absence.

Let Q be some set of entities and let R be some set of relations on Q constituting a relational system:

$$\mathscr{Q} = \langle Q, \mathscr{R} \rangle \tag{1.26}$$

Note that Q may now be a set of objects, events, abstract entities, etc. and $\mathscr{R}$ need not be empirical. Now let Z be a set of objects or events to be used as symbols and let $\mathscr{P}$ be a set of relations defined or existing on Z to constitute a symbol relational system:

$$\mathscr{Z} = \langle Z, \mathscr{P} \rangle \tag{1.27}$$

With F a mapping from $\mathscr{R}$ onto $\mathscr{P}$ as in the case of measurement, we can again define M as a mapping from Q into (onto) Z such that M and F map $\mathscr{Q}$ homomorphically into (onto) $\mathscr{Z}$.

Then:

$$\mathscr{C} = \langle \mathscr{Q}, \mathscr{Z}, M, F \rangle \tag{1.28}$$

Let $z_i = M(q_i)$ be the image of q_i in Z under M. Then z_i is termed the symbol of (or for) q_i under $\mathscr{C}$ and q_i is termed a meaning or referent of z_i under $\mathscr{C}$. $\mathscr{C}$ is termed the symbolism or code.

If we are given z_i and the code $\mathscr{C}$ we have information about any q_i for which z_i is a symbol and about relation between q_i and other members of Q.

All the considerations of the formal theory of measurement can be generalized to the symbolization of any relational systems by any general symbol system.

A simple example will be given to illustrate that application of representational theory to the representation by symbols. Consider a set of aircraft denoted by Q. Consider further that we may divide the aircraft into three types partitioning the set into three subsets. There is thus an equivalence relation $\sim$, which relates two aircraft which are members of Q and are of the same type. Then we may denote this division of aircraft into types as a relational system $\mathscr{Q} = \langle Q, \sim \rangle$. Let us now choose three symbols A, B, C, one corresponding to each of the three aircraft classes, giving us the symbol set $Z = \{A, B, C\}$. There is on the symbol set an identity relation $\equiv$, which relates any two identical symbols of Z. We thus have a symbol relational system. The class labelling is a mapping M which assigns to any aircraft in Q the symbol corresponding to its class name.

F is the correspondence of $\sim$ and $\equiv$.

If we are given the symbol for an aircraft, say A, we know that it is of the same type as any other aircraft with the symbol A. This is the principle on which the type name 'Concorde', discussed previously, is assigned to an aircraft.

Of course, as has already been mentioned, natural language words are used to represent and describe objects and events of the real world as well as their properties. Indeed they are the most generally used method for such symbolic representation or description.

The above representational theory may be used for the analysis of such descriptions. However the comparison of colour description in natural language with colour labels on a colour matching scale serves to illustrate some of the problems. Take the symbols 'muddy brown' and 'A33' which may be given to the same colour using natural language and a colour chart respectively. In addition to the subjectivity and vagueness of the linguistic description as compared with the colour chart label, we note that the expression 'muddy brown' communicates more than a denotation of colour. There is firstly the connotative meaning communicated by association with the properties of mud: dirtiness, stickiness and so on. Further there is an affective component of meaning. Something is communicated by the message originator concerning his or her unfavourable feelings about the colour. (See Leech (1974) for a good introduction to the descriptive properties of natural language.)

To conclude this discussion of representational systems it has been argued that the theory of the foundations of measurement can be extended to embrace forms of representation of entities by symbol systems other than numerical ones. The basic concepts of the foundational theory of measurement, such as representation, uniqueness, meaningfulness and like, can be extended to general symbolic systems (Finkelstein, 1975b).

Non-numerical representations are like measures in that they describe the entity represented and some of its relations in a form which enables information about the entity to be conveniently manipulated.

The most important special feature of measurement is that the assignment of numbers in measurement is objective and represents empirical facts.

1.10   MEASUREMENT, INFORMATION, AND INFORMATION MACHINES

1.10.1   Information

The concept of information will now be explained. The essence of information is symbolic representation. The elements of some set of objects or events, which we term symbols, carry information about the elements of some other set of entities, the set of referents, if a defined relation, preferably one-to-one, exists between the elements of the two sets. More generally there may be relations among the symbols so as to constitute a symbolic relational system and there may be relations among the referents so as to constitute a referent relational system. The symbols carry information about the referents, and the relations they bear to each other, if there is a defined mapping, preferably isomorphic, of the referent relational system into or onto the symbolic relational system.

Information consists of the symbol together with the relation it bears to the referent.

Using the notation introduced before, if we have a referent relational system $\mathcal{Q} = \langle Q, \mathcal{P} \rangle$ a symbolic relational system $\mathcal{Z} = \langle Z, \mathcal{P} \rangle$ and mapping $M: Q \to Z$ and $F: \mathcal{R} \to \mathcal{P}$ giving us a code $\mathcal{C} = \langle \mathcal{Q}, Z, M, F \rangle$, then

$$\mathcal{I} = \langle \mathcal{C}, z_i \rangle \tag{1.29}$$

represents information about a q_i for which z_i is a symbol and about relations between q_i and other members of Q.

Measurement is a special case of representation by symbols in which a measure n_i, together with the scale of measurement $\mathcal{S}$, represents information $\mathcal{I} = \langle \mathcal{S}, n_i \rangle$ about the measurand q_i.

The use of the term 'information' as given here differs from the usage of the term in information theory. There are, however, essential similarities.

In information theory we consider an information transmission channel which transforms elements x_i of a set of inputs X into elements of a set of outputs Y. This transformation is in general many-to-many.

The quantity of information provided by the occurrence of an output y about the occurrence of an input x_k is defined as:

$$I(x_k; y_i) = \log \frac{P(x_k | y_i)}{P(x_k)} \tag{1.30}$$

The base of the logarithm defines the unit of the scale.

If we consider Q as analogous to X, S as analogous to Y, and M as analogous to the X to Y transformation of the communication channel, we see that the underlying concepts of 'information' in this chapter and in information theory are fundamentally similar. In both information is knowledge about and entity provided by an image of the entity under a mapping.

Information theory is concerned with measures of 'quantity of information' for the special problem of transmission through a channel which transforms an input to an output. It measures the quantity of information transmitted through a channel by the change from *a priori* knowledge to *a posteriori* knowledge about the channel input provided by the channel output. This measure conforms to the concept of information as defined in this chapter, though it is designed for a restricted problem class only.

1.10.2 Information Machine

Having defined information we shall consider a class of machines which acquire, process, present and give effect to information. This class includes instruments for measurement, computation, communication, and control.

In the functioning of information machines, the information-carrying symbol is either the instantaneous value of a physical quantity, such as the angle of deflection of a pressure gauge pointer, or some parameter of the variation of a physical quantity with time, such as the instantaneous frequency of a variable-frequency alternating voltage in a frequency modulation system.

The pointer deflection angle together with the gauge calibration law carries information about the input pressure: it is the information-carrying symbol.

The physical quantity, the magnitude or time variation of which carries information, is termed the signal. Thus in the above examples the signals are the pointer deflection and the voltage.

A machine may be defined in general terms as a contrivance which transforms a physical input into a physical output for some definite purpose. Thus, for instance, a lever constitutes a simple mechanical machine transforming input power into mechanical output with the object of achieving mechanical advantage.

An information machine functions by performing a prescribed transformation of an input physical signal into an output physical signal. Thus, the instantaneous magnitude or some parameter of the time variation of the output signal is related through the instrument transformation to input signal magnitude or some time variation parameter. The relevant feature of the output signal, together with the instrument transformation law, carries information about some referent feature of the input signal, as has been shown in the case of the pressure gauge above.

The requirement to maintain prescribed functional relations between input and output distinguishes information machines from other forms of machine, the purpose of which is to generate and transform power, and determines the principles of analysis and design of information machines.

The distinctive features of a class of machines which maintain specified functional relationships between inputs and outputs were recognized in particular by Draper *et al.* (1952) and Kuhlenkamp (1971).

The nature of instruments as information machines determines the approaches to their design, description, and application. This forms the basis of a systematic instrument science (Finkelstein, 1977).

1.11 MEASUREMENT THEORY IN THE PHYSICAL, SOCIAL, AND BEHAVIOURAL SCIENCES

The present section is intended to discuss the state of application of measurement theory in the physical, social, and behavioural sciences. It is not intended to consider specific techniques and methodologies of measurement, a subject obviously too wide in scope.

1.11.1 Measurement Theory in the Physical Sciences

The classical theory of measurement (Helmholtz, 1887; Campbell, 1920) was developed to give an account of measurement in the physical sciences.

In terms of this classical theory, measurement in the physical sciences is based on establishment of direct extensive scales of measurement for a number of physical quantities as described in Section 1.6.1. These quantities are used as the base of a system. Scales for other physical quantities are obtained as derived scales, that is indirect scales in terms of the base quantities, in the form of multiplicative monomial functions of the base quantities.

This account is not totally satisfactory even for classical physics. For example, the SI system of units (NPL, 1973) has seven base quantities: length, mass, time, electric current, thermodynamic temperature, amount of substance, and luminous intensity. The base unit of current for example is defined as a current which if maintained in two parallel conductors of infinite length, negligible circular cross section, and placed a specified distance apart, produces a specified force per unit length. This, then, involves firstly scales of measurement of length and force. Further it involves a theory of electromechanical interaction, that is a law of force between conductors this is essential to enable us to calculate the instrument law of a current balance by which the unit of current must be realized. The actual realization of the definition in terms of infinitely long infinitesimal conductors is not possible. Even for quantities such as length, for which an extensive scale can be established using concatenation of standard lengths, the unit of length is in fact defined in terms of the wavelength of light, which is not a material object but a construct of physical theory.

The analysis could be taken further, but the examples chosen give an understanding of the deficiencies of the classical theory of measurement as an account of the foundations of physical measurement. The situation of physics is that it consists of a number of axiomatized theories such as Euclidean geometry, classical mechanics, thermodynamics, electromagnetism, and so on. The scales of measurement of classical physics are based on the acceptance of these theories as representations of the real world and defining the units on that basis rather than on the individual axiomatization and establishment of scales of particular physical quantities. A formal theory of measurement in the physical sciences based on this view awaits development. Collection of bibliographic material, as a start in this direction, is in progress at least one institution (Van Brakel, 1977).

Special problems arise in quantum and relativistic physics which cannot be handled by the theory of measurement presented above. In quantum physics the interaction between observer and observed system imposes a limit on the certainty of the joint measurement of the attributes of the system such as position and momentum. This imposes a fundamental difficulty for measurement theory. The theory of relativity has a large impact on the theory of measurement in the physical sciences. In terms of the view presented above it attempts to represent reality using a theory different from those used in classical measurement, for example classical mechanics. The rejection of the concept of simultaneity and an upper limit to velocity are examples of such differences. There are thus fundamental developments of application of measurement theory to the physical sciences to be undertaken.

1.11.2 Measurement Theory in the Social and Behavioural Sciences

A complete analysis of the theory of measurement in the social and behavioural sciences would require a detailed discussion of their nature, content and methodology. The subject would be too extensive to tackle in this chapter. However, some of the key problems of the theory and philosophy of measurement in these sciences can be simply summarized.

Firstly one should justify the discussion under one heading of these very different classes of science. The reason is that key problems of application of the theory of measurement in the social and behavioural sciences are essentially the same.

Firstly, as has already been stated, it is the requirements of the social and behavioural sciences which have led to the replacement of the restrictive classical theory of measurement by the more comprehensive modern representational theory.

The social and behavioural sciences are concerned very much with such attributes or qualities as utility, standard of living, alienation, intelligence, and the like. The first problem in attempting to measure them is the difficulty of establishing an adequate objective concept of these qualities based on empirical operations. The conceptual framework is often absent.

When a scale of measurement for a quality such as utility or standard of living is established, there remains a fundamental problem of establishing that the measure and concept correspond. For example, index figures that are often prepared for the purpose of measuring standard of living are disputed by those whom they do not suit, as not measuring what they feel standard of living means.

The empirical operations involved in establishing scales of measurement in the social and behavioural judgement, commonly involve responses by human observers. These are, for example, required to judge whether two stimuli, such as light intensities, pitch of sound, etc., are undistinguishable. As another instance they are used to give an ordinal rating to a number of alternatives. Although the data thus derived may be sufficiently consistent for a population of observers to consider their objective, they are nevertheless subject to considerable statistical scatter.

The scales of measurement, the social and behavioural sciences being frequently based on determination of equivalence and order only, are then only monomial or ordinal.

Conjoint and multidimensional scales of measurement are commonly used. They express some conceptual quality such as the 'alienation' of a work force in terms of measurable quantities, such as worker-days lost through disputes and absenteeism. The difficulty again is the establishment of agreement between the concept of the quality and the measures adopted.

There are no wholly axiomatized theories in the social and behavioural sciences, which correspond to, say, classical mechanics or thermodynamics.

In conclusion it may be stated that in these sciences it is by no means universally agreed that the clear formation of concepts in terms of empirical observation, is possible or desirable. Nor is there agreement that the search for data, through measurement, advances knowledge and understanding. The opponents of quantification would say that human nature and behaviour are too variable to enable the methodologies of the physical sciences to be applicable to them.

1.12 CONCLUSIONS, TRENDS, AND DEVELOPMENTS

The conclusions, trends and developments are best outlined succinctly.

The interest and utility of fundamental measurement theory has been argued at the outset of the chapter. Judgement must now be left to the reader.

There are a number of trends and developments in the field, as well as many unresolved problems. These have been mentioned before, and they may now be usefully listed:

(a) There is a need for a further development of the treatment of uncertainty in fundamental measurement theory.
(b) The exploration of the relation between measurement and other forms of symbolic representation, such as descriptive natural language, is a potentially fruitful field of research.
(c) A theory of measurement based on axiomatized theories, rather than representation conditions for measurement scales of individual quantities, should be developed for the physical sciences.
(d) For the social and behavioural sciences it is right to pursue the Galilei programme: to attempt to measure that which is measurable and to render measurable that which is not. This endeavour will answer doubts about the feasibility and usefulness of measurement in these domains in one way or another.

REFERENCES

van Brakel, J. V. (1977). *Meten in de empirische wetenschappen*, University of Utrecht (Vakgroep Onderzoek van de Grondslagen van de Natuurkunde), Utrecht, Netherlands.

Campbell, N. R. (1920). *Physics: The Elements*, Cambridge University Press, Cambridge. (Reprinted: (1957). *Foundations of Science*, Dover, New York).

Churchman, C. W., and Ratoosh, P. (Eds) (1959). *Basic Concepts of Measurements*, Cambridge University Press, Cambridge.

Cohen, M. R., and Nagel, E. (1934). *An Introduction to Logic and Scientific Method*, Harcourt, Brace, New York.

Coombs, G. M. (1964). *A Theory of Data*, Wiley, New York.

Draper, C. S., McKay, W., and Lees, S. (1952). *Instrument Engineering*, Vols. 1–3, McGraw-Hill, New York.

Ellis, B. (1966). *Basic Concepts of Measurement*, Cambridge University Press, Cambridge.

Finkelstein, L. (1973). *Fundamental Concepts of Measurement*, Acta IMEKO VI, IMEKO, Budapest. pp. 11–18.

Finkelstein, L. (1975a). 'Fundamental concepts of measurement: definitions and scales', *Measurement and Control*, **8**, 105–11.

Finkelstein, L. (1975b), 'Representation by symbols as an extension of the concept of measurement', *Kybernetes*, **4**, 215–23.

Finkelstein, L. (1977), 'Instrument science—introductory article', *J. Phys. E: Sci. Instrum.*, **10**, 566–72.

Helmholtz, H. V. (1887). *Zählen and Messen erkenntnis—theoretisch betrachet*, Philosophische Aufsaetze Eduard Zeller gewidmet, Leipzig (Transl. by C. L. Bryan (1930). *Counting and Measuring*, Van Nostrand, New York.)

Hempel, C. G. (1952). *Fundamentals of Concept Formation in Empirical Science: International Encylopaedia of Unified Science*, Vol. 2, No. 7, Univ. Chicago, Chicago.

Hölder, O. (1901). 'Die Axiome der Quantität und die Lehre vom Mass', *Ver. kom, Sachs. Gessel. Wiss., Math.-phys. Klasse*, **53**, 1–64.

Krantz, D. H., Luce, R. D., Suppes, P., and Tversky, A. (1971). *Foundations of Measurement*, Vol. 1: *Additive and Polynomial Representations*', Academic Press, New York.

Kuhlenkamp, A. (1971). *Konstruktionslehre der Feinwerktechnik*, Carl Hanser, Munich.

Leaning, M. S., and Finkelstein, L. (1979). *A Probabilistic Treatment of Measurement Uncertainty in the Formal Theory of Measurement*, *Acta IMEKO VIII*, to be published.

Leech, G. (1974). *Semantics*, Penguin Books, Harmondsworth.

Lin, S. V. T., and Lin, Y. F. (1974). *Set Theory: An Intuitive Approach*, Houghton-Mifflin, Boston.

von Neumann, J., and Morgenstern, O. (1944). *Theory of Games and Economic Behaviour*, Princeton Univ. Press, Princeton.

NPL (1973). *The International System of Units*, HMSO, London.

Pfanzagl, J. (1968). *Theory of Measurement*, Physica Verlag, Würzburg–Vienna.

Roberts, F. S. (1979). *Measurement Theory*, Addison-Wesley, Reading, Mass.

Stevens, S. S. (1946). 'On the theory of the scales of measurement', *Science*, **103**, 677–80.

Stevens, S. S. (Ed.) (1951). *Handbook of Experimental Psychology*, Wiley, London.

Stevens, S. S. (1959). 'Measurement psychophysics and utility'. In C. W. Churchman, and P. Ratoosh (Eds), *Basic Concepts of Measurement*, Cambridge University Press, Cambridge. Chap. 2.

Suppes, P., and Zinnes, J. L. (1965). 'Basic measurement theory'. In R. D. Luce, R. R. Bush, and E. Galanter, *Handbook of Mathematical Psychology*, Wiley, New York.

Sydenham, P. H. (1979). *Measuring Instruments: Tools of Knowledge and Control*, Peter Peregrinus, London.

Tarski, A. (1954). 'Contributions to theory of models', *Indagationes Mathematicae*, **16**, 572–88.

Torgerson, W. S. (1958). *Theory and Methods of Scaling*, Wiley, New York.

APPENDIX 1.A: BASIC NOTIONS OF SET THEORY

1.A.1   Set, Subset, Family of sets

A set (or class) is a collection of elements. For example: the set of all forces.

Sets are denoted by $A, B, \ldots, Q, \ldots$

Elements of sets are denoted by $a, b, \ldots, q, \ldots$

If A is a set and if a is a member of the set A we denote it by $a \in A$.

A set A may be specified by the listing of its elements $A = \{a_1, a_2, \ldots, a_n\}$; or by specifying a property to be possessed by its elements, $A = \{A \mid P(a)\}$, the set of all a for which $P(a)$ is true.

B is said to be a subset of A, if every element of B is an element of A: we write $B \subset A$.

$$A = B \text{ if and only if } B \subset A \text{ and } A \subseteq B.$$

An empty set contains no elements and is denoted by $\varnothing$.

All sets are considered to be subsets of a larger non-empty set called the universal set U.

A set of sets is called a class of sets and is denoted by $\mathscr{A}, \mathscr{B}, \ldots$

1.A.2   Set Operations

The union of two sets A and B, $A \cup B$ is the set of elements belonging to at least A or B. The union of n sets $A_1, \ldots, A_n$ is denoted by $\bigcup_{i=1}^{n} A_i$.

The intersection of two sets A and B, $A \cap B$ is the set of elements common to both A and B. If $A \cap B = \varnothing$ they are called disjoint.

The difference of two sets A and B, $A - B$, is the set of elements of A which are not elements of B. In particular $U - A$ is called the complement of A denoted by A'.

The Cartesian product of two sets A and B, denoted by $A \times B$ is the set whose elements of all ordered pairs (a, b) for which $a \in A, b \in B$. This can be extended to n sets $A_1 \times A_2 \times \cdots \times A_n$ the set of all elements, n-tuples $(a_1, a_2, \ldots, a_n)$, $a_1 \in A_1$, etc., and denoted by $\mathsf{X}_{i=1}^{n} A_i$.

1.A.3   Relations

Relations and their properties will be defined in terms of sets.

A relation R between elements of sets A and B is a subset of $A \times B$; $R \subset A \times B$.

This is a binary relation. An n-ry relation is $R \subset A_1 \times \cdots \times A_n$. We shall also denote an n-tuple which is an element of R by $R(a_1, \ldots, a_n)$.

Let us denote $R \subset A \times A$ and $a \in A$, $b \in A$. R is called a relation on A.

The set of all a for which $(a, b) \in R$ is termed the domain of R, while the set of all b for which $(a, b) \in R$ is the range of R.

The set of all elements of $A \times B$ which are not elements of R, is denoted by R' and defines the complementary relation of R.

The set of all (b, a) for which $(a, b) \in R$ is the inverse relation of R, denoted by R^{-1}.

A relational system $\langle A, \mathscr{R} \rangle$ is a family of sets consisting of a set and a class of relations $\mathscr{R}$ defined on it. $\mathscr{R}$ is called the structure of the system. If A is a set of extramathematical entities and $\mathscr{R}$ are empirical relations defined on it the system is called an empirical relational system. If A is a set of numbers and $\mathscr{R}$ are relations on them the system is termed a numerical relational system.

1.A.4 Special Types of Relations

R is called symmetric if and only if whenever $(a, b) \in R$, $(b, a) \in R$.

R is called reflexive if and only if for each $a \in A$, $(a, a) \in R$.

R is called transitive if and only if whenever $(a, b) \in R$ and $(b, c) \in R$ then $(a, c) \in R$.

A relation R is called connected in A if for all $a, b \in A$, $a \neq b$, either $(a, b) \in R$ or $(b, a) \in R$.

1.A.5 Equivalence, Order

A relation R which is reflexive, symmetric, and transitive is called an equivalence relation. It will be denoted by $\sim$.

An equivalence relation $\sim$, is a congruence relation for a relational system $\langle A, \mathscr{R} \rangle$ if whenever $a \sim b$, a can be substituted for b in any $R \in \mathscr{R}$. This is called the substitution property of the relation.

Equality, $=$, is an equivalence relation between mathematical entities, which has the substitution property with respect to all relational systems for those entities.

A relational system $\langle A, \sim, < \rangle$ will be called an order system if and only if:

  (i) $a, b \in A$ exactly one of the relations holds $a \sim b$, $a < b$, $b < a$;

  (ii) $\sim$ is an equivalence relation;

  (iii) $<$ is transitive.

An example of an order system is the numerical system $\langle \text{Re}, =, < \rangle$ for real numbers.

1.A.6 Mapping

A mapping of A into B denoted $M: A \rightarrow B$ is a rule, which assigns to each element $Q \in A$ an element of B, called the image of a under M. This element is denoted by $M(a)$.

A is the domain of M and the set of images of the elements of A under M is the range of M.

If the whole of B is the range of M the mapping is called onto M.

If different elements of A have the same image in B the mapping is called many-to-one.

If each different element of A has a different image the mapping is called one to one.

1.A.7 Homomorphism

Given two relational systems $\langle A, \mathcal{R} \rangle$ and $\langle B, \mathcal{P} \rangle$ where A and B are sets and $\mathcal{R}$ and $\mathcal{P}$ are classes of relations on A and B respectively, $\mathcal{R} = \{R_1, \ldots, R_n\}$ and $\mathcal{P} = \{P_1, \ldots, P_n\}$ consider that there is a mapping M from A onto (into) B, $M: A \rightarrow B$. Then M is a homomorphism from $\langle A, \mathcal{R} \rangle$ onto (into) $\langle B, \mathcal{P} \rangle$ if and only if for all $a_1, \ldots, a_k \in A$.

$R_i(a_i, \ldots, a_k) \Leftrightarrow P_i(M(a_1), \ldots, M(a_k))$ for $i = 1, \ldots, n$ ($\Leftrightarrow$ means implies and is implied by).

1.A.8 Operations

Let A denote any set. Then a binary operation $\circ$ on A is a mapping which assigns to each pair of elements $a, b \in A$ an element $a \circ b \in A$.

An operation is *commutative* if:

$$a \circ b = b \circ a$$

An operation is *associative* if:

$$a \circ (b \circ c) = (a \circ b) \circ c.$$

Handbook of Measurement Science, Vol. 1
Edited by P. H. Sydenham
© 1982 John Wiley & Sons Ltd.

Chapter

2   P. H. SYDENHAM

Measurements, Models, and Systems

Editorial introduction

The physically existing world consists of numerous systems each having specific uniquely set parameters. Man's models of these systems aim to be as general as possible so that they can be widely applicable. Thus mathematical and linguistic models require information that restricts them to the specific when they are applied to physically existing problems. This information is obtained by measurement.

This chapter introduces the reader to the relationships existing between measurements, systems, and models of systems. It is intended to provide some understanding of measurement situations in order that an appropriate measurement strategy can be employed or the reasons for difficulties encountered might be explained.

2.1 INTRODUCTION

There has been, in the past, a general tendency for designers and users of measuring equipment to view a measurement within too limited a perspective. Efficient application and design of hardware requires of the user a broad understanding of the role that a piece of measurement hardware plays in the total system in which it is placed.

Current understanding of the measurement process itself, having been summarized in the previous chapter, now enables an introduction to the relationship between measurements, systems, and models of systems.

As measurements help us to produce models of real situations and as understanding of systems is largely conveyed in terms of models of systems, this chapter will concern itself with a general introduction to the systems approach in relation to its relevance to measurement systems design and application.

The concept of the *systems* approach has become 'all things to all men'. In the *hard-sciences* (also called *physical sciences*) of physical science and engineering it has become associated with the use of very exacting (often, however, only appearing to be so) mathematical models. At the other extreme, such as in parts of the social sciences, in economics, in human geography, in environmental studies, and other disciplines of the *soft-sciences* (also called *empirical sciences*), the models so produced may be very vague in definition due to their complexity often even to the point of attempting to allow for the possible existence of as yet unexplained systematic (not metaphysical) factors that produce a system, the whole of which is more than the sum of its parts. (This is known as the *gestalt* concept.)

Debate flows back and forth between both boundaries as each kind of group attempts to make use of the concepts of the others, often without success. This is not the place to attempt to unify systems approaches—many experts have tried. This chapter makes use of the wide range of such concepts because measurement problems span the whole field of human endeavour and some of the necessary understanding of measurement systems is still in the poorly defined state.

Design and application of a measurement system begin with an appraisal of the measurement need extending design outward to finally produce equivalent signals at some convenient location. It will be found in practice that situations in the hard-sciences generally are relatively straightforward to realize conceptually because the parameters to be sensed are clear cut. In contrast the situation for soft-science situations can be such that it is not possible to decide what it is that should be measured, nor how to do it when it is identified.

This chapter provides an appreciation of a wide background of knowledge that is generally helpful to measurement designers, helping them to better appreciate the generalities of situations met.

2.2   SIGNALS

As will be discussed in more detail in Section 2.7.3, measurement information is conveyed in some suitable form via an energy or mass transfer link, the measurement information being conveyed by such a carrier as a changing state or *modulation* of the carrier. The changing medium used in this way is termed the *signal*.

Signals, therefore, can take the form of variations of some parameter of light beams, of fluid pressure, of mechanical movements, and, the most used today, of electrical energy. Figure 2.1 is a useful chart of information carrying techniques.

Where the modulation conveys no useful or desired information it is termed a noise signal or merely *noise*. Noise, when excessive, can mask the true signal providing false information. The energy or mass medium carrying the signal is embodied in the system *instrumentation*.

Once a measurement parameter has been transduced into an equivalent signal the signal can be treated by any one of the numerous signal processing methods

Information about a process

Waveshape	Instantaneous local	Incremental local	Instantaneous regional	Common name	Simple examples
	Individual property	Patterns of properties			
Constant level	amplitude	amplitude frequency	amplitude frequency	analog data / wave shape analysis / signature analysis	mercury-in-glass thermometer / electrocardiograph / spectroscopy / black and white photography / color photography
Sine wave	amplitude frequency phase	frequency	amplitude	AM = amplitude modulation / FM = frequency modulation / PM = phase modulation / FCM = frequency code modulation	AM radio / doppler effect, vibrating-wires / photoelasticity, Kerr cell / temperature-sensitive paints / antenna characteristics
Pulse trains	amplitude frequency position duty cycle	amplitude	amplitude	PAM = pulse amplitude modulation / PFM = pulse frequency modulation / PPM = pulse position modulation / PWM = pulse width modulation or PDM = pulse duration modulation / PCM = pulse code modulation	FM radio or tape recorder / digital data / computer card

Figure 2.1 Common information-carrying methods. (Reprinted from Stein, P. (1969). *J. Metals*, **22**, No. 10, 41. Courtesy of American Institute of Mining, Metallurgical and Petroleum Engineers (AIME))

according to signal transmission concepts. The prime requirement is that the one-to-one mapping made at the transducer is not altered by the transmission system before the signal arrives at its destination. An exception to this rule is when the properties of the transmission system are deliberately used to compensate, in some way, for a known deficiency in the sensing stage or to extract only part of the conveyed information.

2.3 MODELLING

2.3.1 Models Motor Ingenuity

Models of various kinds are used extensively to provide representations of some aspect of the real-world system of interest. They enable us to investigate the real situation without needing to actually produce it or modify an existing situation. Models are developed for many reasons—to enable, for example, investigation of a given system's behaviour, to enable a scenario to be considered, to provide a more convenient medium of discussion of certain features. They provide, as was recently stated, *means to motor ingenuity*.

If a model were developed to represent exactly and completely a given system then clearly it is an exact replica. In practice a model rarely is a complete representation, it only models chosen aspects of the system; this is further expanded in the following sections where various kinds of models are discussed. Other published accounts on this topic are Finkelstein (1977) and Sydenham (1979).

2.3.2 Linguistic Models

The *linguistic* model uses the natural spoken language to express sufficient parameters, and their interactions, of the system of interest such that an adequate level of explanation, of information transfer, takes place. In the empirical sciences and the arts this is the prime form for presenting models of situations and relationships.

Its major shortcoming is the lack of exactitude and speed of communication that can be reached within a given level of effort. It serves as the first level of modelling, providing for reasonably efficient communication between persons, especially in the area of expression of subjective and emotive concepts. It can rarely serve the hard-sciences as well as it has been adopted and used in the soft-sciences. It often plays a complementary role along with the pictorial methods, providing *descriptions* of parameters. It is the *semantic* feature of many words, and groups of words, that makes resolution, of such matters as differences of interpretation of specifications, so often intractable.

2.3.3 Iconic Models

An *icon* (ikon) is an image, figure or pictorial representation of a concept. *Iconography* is the description of a subject by means of drawings or figures.

In engineering and the sciences iconography is used extensively; indeed to such an extent that most communication of ideas uses it. The importance of pictorial representation in the history of technology has been studied (Ferguson, 1968).

Examples of iconography are the use of three-view engineering drawings of objects, circuit diagrams, block schematic diagrams, isometric projection drawings, and graphs of various kinds. Graphs are singled out as a class in Finkelstein (1977) because of the conceptual difference that they represent performance relationships rather than the more direct physical relationships of make-up. Frequency response plots, signal flow diagrams, flow charts, signal graphs to show system nodal interconnections, activity charts, and production resources evaluation charts are each examples of the graphical form of icon.

The iconic level of modelling is usually reached very early in any design or discussion of a measurement system problem. Lines are drawn appropriately, their geometrical relationships and linguistic annotations conveying the aspect of information required.

These considerably motivate the mind and certainly assist such matters as construction and assembly. However, they do not provide the same depth of objective rigor as do the next class of model to be considered—the mathematical model.

2.3.4 Mathematical Models

Mathematics is in itself a modelling discipline for it enables relationships between defined quantities to be expressed in terms of representative mathematical symbols and statements.

Mathematical models can be extremely powerful design aids. In many areas this form of model expresses fundamentals of systems at a conceptual depth well beyond that which can be realized as practical systems. As an example it was by way of a mathematical model of electromagnetic fields that Clerk–Maxwell was able to show that, at the time, unobservable radiation can take place under the correct conditions. Mathematical models are also able to be run faster in time than the physical system they represent. They enable a degree of prediction. A subject can be considered to have reached a level of maturity when it can be modelled mathematically to an adequate level. Often the effort, which can be considerable, to produce the mathematical model is a worthy investment. It was the 19th century philosopher Boole's work that produced Boolean algebra which subsequently enabled the ubiquitous physical digital computer to be developed according to a rigorous design basis. Section 2.5 deals with the generation procedure for these models.

2.3.5 Physical Models

It is often convenient to construct *physical* models of a situation. Examples of these are the scaled-down versions of a proposed process plant or a model of a device for seeking patent approval. These clearly, again, only model selected features of the whole. Given that any system of interest has numerous aspects it can be seen that there could easily be more effort and material construction expended on models than would exist in building the real system itself. Experience is needed to decide which kind of model is the most appropriate and which kind will provide the greatest increase in understanding for a given level of effort input.

It is to be expected that the more sophisticated the model the wider its applicability and usefulness.

Information, signal, and power flow models are distinguished as important to the understanding of measurement systems in Finkelstein (1977).

2.4 SYSTEM STRUCTURE

As stated in Bosman (1978) systems can be *reticulated* or *box-cut*, a process in which the whole is, for purposes of analysis, broken down into more detailed assemblies of *subsystem* blocks, the process being taken as far as is needed, perhaps finally to the basic *elements*. So much has been written about systems, by so many persons, covering such a wide range of applications, as to make the universal unification of terminology perhaps impossible to achieve. With respect to instrument systems, Bosman (1978) and Mesch (1976a, b) appear to be the only guidance statements made on the conjoint subject formed between system theory and measurement systems design. In the main, Bosman discusses it in terms of levels within *extensive* systems whilst Mesch deals with typical *block diagram transfer function* interconnections and the system state. Mathematical modelling of instruments in particular is covered in Finkelstein and Watt (1978) and in Volume 2 of this handbook.

In the summary to Bosman's paper he reticulates an instrument system into *subsystems* and *partial systems*. The former concerns 'separately identifiable conglomerations of equipment (instruments) with connected (partial) functions which are logically geared to the system's main goal'. He defines partial systems as a generic part concerning one aspect. Further partitioning arrangements suggested are to consider *aspects* (examples are performance, operation, physical, economic) and *structures* (functional, organic, activity, information, sociotechnical are suggested).

The approach adopted by Mesch is somewhat complementary to that of Bosman as it provides another aspect of system reticulation. Principally it is based upon the fact that subsystem *blackboxes* can be expressed as mathematical *transfer functions* interconnected to form the whole.

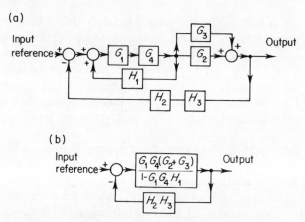

Figure 2.2   Example showing (a) given block diagram and
(b) the reduced single-loop form

Commonly met configurations of transfer functions are the *serial* or *cascade* chain, the *parallel* connection, and the single *feedback* loop. These are dealt with in Chapter 4 (Section 4.3 in particular) and in Section 14.3, where systems are also discussed.

An algebra of block diagrams has been developed (Di Stefano *et al.*, 1967): it is useful for handling more complex situations as it provides means of possible reduction (Figure 2.2) or as a means to isolate a block. Surprisingly, a small number of structural arrangements suffice for much of instrument design, this being because adequate design can be done using *reduced* models wherein only the *dominant* transfer functions need to be retained in the final model.

A transfer function is a mathematical model of an input–output relationship. The physical form of the hardware that provides that relationship is not embodied specifically in a transfer function. It is, therefore, often possible in practice to produce modelled realizations in more than one energy domain. For example the transfer function for a second-order system, can describe, in the mechanical domain, the time–velocity behaviour of a mass and a spring, in the electrical domain the behaviour of current in an inductance–capacitance circuit plus other arrangements such as behaviour in a fluid medium. This universality has become known as the *analogies*. A general understanding of this commonality in physical systems is to be obtained from Shearer *et al.* (1967). As this is discussed at some depth in Section 4.3.2.2, it needs no further expansion here.

2.5   DEVELOPMENT OF MATHEMATICAL MODELS

Mathematical models tend to be those one strives for ultimately because of the in-depth insight they bring to a subject. It is, therefore, of value to describe briefly the steps taken in treatment of a topic where this form of model is sought.

The first step is to consider the subject to be modelled from any and every aspect that appears relevant. Considerable experience and skill assists this vague initial step along. As features, aspects, relationships and the like, are realized they are recorded as some kind of model, generally as symbolic black boxes in which will be at least a linguistic form of description. For example, a temperature sensor would be a black box having temperature labelled as the input and some form of equivalent signal output. The box could be labelled 'temperature sensor'.

A second step (in reality any step of the process that might be advanced as progress is enabled by acquisition of new knowledge) is to sort the general information into systematic classes aiming to identify the various black boxes and their interconnections. Figure 2.3a illustrates this first step which leads to a schematic block diagram in which links are formed and labels given to individual blocks.

This iconograph is then continually refined and broken down until it is possible to see that individual blocks are at such a level of simplicity (Figure 2.3b) that they can be assigned appropriate general mathematical input–output relationships (see Figure 2.3c). Linear equations are preferred but non-linear expressions can also be handled. For example, the second-order, spring–mass system mentioned earlier would be represented as a box in which the transfer function is that of the second-order linear differential equation.

In many cases the arrangement of boxes and the equations selected as the model may only provide the overall transfer function sought with internal operation being quite different at the *nodes* (or *ports*) of the system. It is important to make the distinction between modelling a system at only a total transfer level and as one in which internal operation is also to be modelled.

A model, thus made, is relevant to a very wide range of physical manifestations for the numerical coefficients of the various equations selected are not yet *identified*. To *bound* the general model into an adequately specific one, it is necessary to convert all such parameters into the numerical state. These parameters are, in fact, measurement variables that either have already been evaluated (such as the mass of an electron, the value of gravitational attraction or a conversion coefficient of a sensor) or need to be measured in order to bound the model.

Thus is created a mathematical model of the system of interest. However, to stop here is hazardous for models must be proven, *evaluated* in some way to ensure that they do indeed represent the reality desired.

One testing method, of course, is to build the model in its real-world physical form and compare its performance to excitation inputs with the original system. This is often not practicable because of such factors as: the cost is not warranted, the system cannot yet be built, or the model must run at a different time rate.

The concept of the analogies provides a means of testing for it enables analogous systems to be constructed on analog or digital *simulators*.

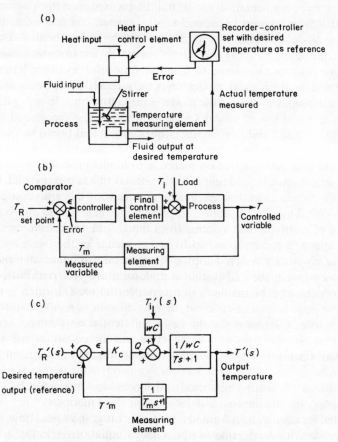

Figure 2.3 Stages in development of mathematical model of an example temperature control system: (a) physical layout in block form with descriptions added; (b) system broken down into recognisable blocks; (c) mathematical equations added with coefficients ready to be identified. (Adapted from Coughanowr and Koppel (1965))

Usually the first prepared models are proven to be inadequate in some respect and the process of refinement continues until one runs out of ideas, effort or, more optimistically, until the model provides an adequate level of simulation.

At this point, provided the limitations of the model are well known, the model can then be used as the basis for further, more efficient, study of the original system. To cite an example, mathematical modelling of an inductive sensor enabled the modeller to investigate frequency of operation, interconnections, and effect of physical geometry on the sensitivity without the need to actually build a single device.

The process of mathematical model building is explained in Coughanowr and Koppel (1965) in relation to general process control systems. In a specifically instrument sense see Finkelstein and Watts (1978) and Volume 2. The aspects of fitting equations and sizing coefficients is a major part of *systems identification* and *parameter estimation*—refer to Chapter 8 for a survey of the latter.

Creating a mathematical model rests very much upon adequate *a priori* knowledge being available for the model is build from this. In many disciplines, especially those of the empirical sciences, although it is recognized that more knowledge is needed and in what area, the practical reality can be that the data needed cannot be obtained.

Many authors have written on methods of modelling *badly defined* systems but in general it must be said that considerable effort has been expended on such problems with less than desired returns. A review of this situation is given in Young (1978). There, the author also presents his generalized approach to the modelling of badly defined systems from the *holistic* viewpoint, one in which one attempts to produce a model from knowledge of the *intact* system. The alternative approach, often adopted, is that according to the *reductionist* philosophy wherein the real system is philosophically broken down into subsystems which are explainable in mathematical terms. The catch is that the *a priori* knowledge is often insufficient, thereby allowing a breakdown to be made as the observer believes it should be rather than what it might actually be. Young presents examples of his suggested systematic procedure applied to several badly defined systems. A guide to development of models in general is also given.

The process of creating mathematical models is often aided by the use of *graph* theory. In this approach the simultaneous equations of the system are represented by iconographs formed from lines that graph signal flow. Figure 2.4 is an example. An introduction to signal flow graphs is given in Di Stefano *et al.*

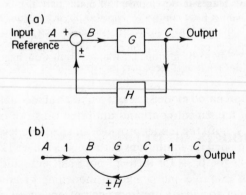

Figure 2.4 Graph theory applied to a feedback loop: (a) block diagram; (b) signal flow graph

(1967) where the various rules of the algebra, definitions, and construction are dealt with in a tutorial manner. Many texts exist on graph theory, of which Mayeda (1972) and Henley and Williams (1973) are written from the engineering viewpoint; the former also contains an extensive bibliography. Graph theory found earlier application in the management and operations control of engineering enterprise. It began to be applied to physical systems modelling in the mid-1960's, a time at which a branch, that of bond graphs, emerged to handle multidisciplinary system elements. Details of this approach are to be found in Karnopp and Rosenberg (1968), Karnopp (1974), and Karnopp and Rosenberg (1975).

2.6 THE PLACE OF MEASUREMENTS IN SYSTEMS

A key purpose of a measurement is to map a physical parameter into (usually) an equivalent number set. In many instances that is all that is needed, but there also exist many situations where the measurement signal is used to continually control a process. It is rarely appropriate to view a measuring instrument as a stand-alone artefact. It is important to know its use in relation to the specific extensive system of which it forms a part. Figure 2.5 depicts one author's model of the entire process (Sydenham, 1979).

A measuring system comprises a *sensing* stage, in which the original parameter to be measured, called the *measurand*, is transduced into an appropriate equivalent signal. The sensor's role is to extract specific information, to act as an information filter, passing information on the state of a particular chosen parameter existing within a possibly infinite set of definable parameters that totally describe the system.

A measuring system, as well as being able to convey internal messages—the signal—without loss of accuracy must also, overall, map variables in a faithful manner. This line of thought is expanded in Sydenham (1979).

Measurements of actual systems—and no other kind exists—provide numerical values for the state variables, thus providing specific bounds to generalized situations. Any mathematical modelling exercise must eventually require measured data when it needs to be applied. In many cases generalized mathematical models need some coefficients to be limited numerically in order to allow reduction of an otherwise intractable general solution.

2.7 MODELS OF THE MEASURING INTERFACE

2.7.1 The Set Theory Model

The previous chapter provides a statement on current considerations aiming to find a basic rigorous mathematical model for what occurs at the interface formed between a sensor and the system to which it is connected.

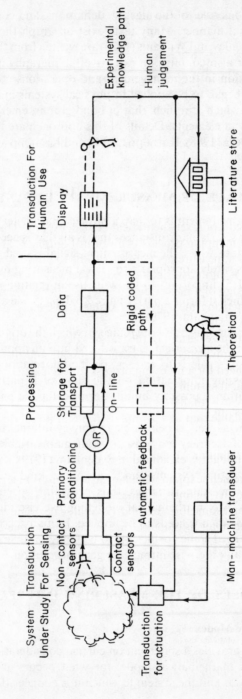

Figure 2.5   Pictorial representation of place of measurement in relation to a generalized system (From Sydenham, P. H. (1979). *Measuring Instruments: Tools of Knowledge and Control*. Reproduced by permission of Peter Peregrinus Ltd.)

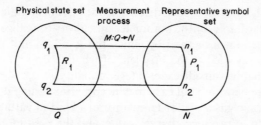

Figure 2.6 Set-theory model of a measurement (From Finkelstein, L. (1975). *Measurement and Control*, **8**, 105–11. Reproduced by permission of Institute of Measurement and Control)

That model is based upon set theoretical considerations. The iconographic representation of the mapping process is given in Figure 2.6, after Finkelstein (1975). This model of a measurement appears to be the most fundamental and being mathematical, it should ultimately pave the way to machine decision-making about the design of a measurement interface. At present, however, it is doubtful if many practising measurement systems designers would find it of great use as an aid to design.

From it we can recognize certain of the features demanded of a measurement system. The need for equivalence between the real scale and the mapped one is evident as is also the need for a mapping process for each stage parameter to be measured in the physical system.

Although they may eventually be seen as no more than learning aids, two other models are available that also stimulate thought about design.

2.7.2 The Popular Definition Model

Most texts on measurement explain what a measurement is in terms of comparing the unknown quantity against a defined standard for that kind of quantity which has embodied in it some form of subdivisional scale. Figure 2.7 shows how this can be represented in terms of measuring the length parameter of an object.

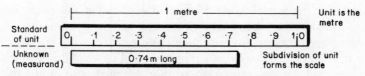

Figure 2.7 Popular definition representation of a measurement (From Sydenham, P. H. (1979). *Measuring Instruments: Tools of Knowledge and Control*. Reproduced by permission of Peter Peregrinus Ltd.)

Like the set-theoretical model, discussed above, it also indicates the need for an agreed standard unit for the parameter and a way to realize it. It also makes apparent the need for a method by which the standard and the unknown are compared. Also revealed is the problem of what to do about any differences between them that lie outside exact integer equality.

Many measurements are not easy to conceptualize in such terms, for example the thermodynamic gas scale for temperature, a kind of variable that does not algebraically add to produce a larger value, and the even more poorly defined concepts such as conflict, love, and pain.

2.7.3   The Information Selection Process Model

The term *information* has two different usages. In common language it relates to a collection of facts, ideas, entities, concepts, and attributes that define a subject or object—for instance an encyclopaedia is 'full of information'; it contains a host of meaningful statements.

In the *information theory* sense used by communication engineers it is concerned with the quantity conveyed in a message passing through a communication channel. In this sense the conceptual idea of *meaning* of that message is not catered for by the theory. An apparently nonsensical message (for example, a coded message) can be transmitted with utmost fidelity.

Thus a measuring system is concerned with both kinds of information. It must map the variable (that is, codify the measurand) and also transmit it according to information theory.

Philosophical thought in this direction is scant, especially by measurement systems designers. One writer (Stein, 1970), suggests that an object possesses *latent* information, that is ready-to-be-tapped information, that the sensor selectively filters out into the measurement channel. Based upon the information concept of Stein it is possible to generate an informational aspect model of the measurement interface. Referring back to Figure 2.5, it can be seen the sensor connects to the system in order to provide an energy or mass transport link. We know from experience that measurement data flow on such carriers but we have little idea of the quantitative relationship between them (Sydenham, 1979).

Nevertheless this model shows clearly that the act of measurement *must* bring about an imbalance of original energy or mass balance conditions.

It is, therefore, necessary to create such a link in order to obtain a measurement and the imbalance must be either negligibly small or well known so that it can be compensated. One measurement method adopted is to adjust the energy flow until it ceases, this is called the null-balance method. The conditions of balance then constitute the measurement at that point.

The information model also shows how stray noise energy can be so disastrously influential on good measurements.

2.8 KINDS OF MEASUREMENT SITUATION

2.8.1 Interaction Between the System and the Measuring Stage

The measuring interface comprises the system of interest and some kind of measurement sensor that codes meaning to what is sensed.

Four classes of this can be identified as an aid to understanding what might take place. The basis of the class into which to place a specific situation is based on the rigidity of the code associated with the meaning allocated.

Class 1. This is the case when the designer has created a hardware sensor and applied it to produce data to which a clearly defined meaning has been stated. An example might be the act of placing an electric thermometer into a water bath, recording the results as indicating changes in bath temperature.

In essence someone has coded the data to mean the temperature unit with some kind of scale. The code is rigidly applied; it is thereafter assumed, until recalibration, that the code remains unchanged. Clearly there always exists a constant danger that malfunction or an additional noise source could occur altering the code. For this reason many measurement systems will have a periodic self-checking feature.

It is important to realize in this class that whatever happens to the measurement link it does not, once set up, materially alter the system, this being the key feature of the classes that follow.

Class 2. When a scientist or engineer wishes to learn about a process a sensor is applied having, at the onset, an ascribed mapping code.

The process is unaltered by the application of a properly designed sensor but here it is the observer who changes as observation continues. In many cases the knowledge obtained by the act of measurement alters the observer's concept of what is being measured. The meaning, which is ascribed by the user in the first place, can unwittingly be changed as measurement proceeds. This situation continually arises in the use of measurements to seek new knowledge.

Class 3. When two or more observers communicate using their natural senses and data processing ability the situation of class 2 extends to both systems possibly changing as observation of each other continues.

This totally interactive state is regularly met in human communications and is just being met between humans and the *semi-intelligent* computing devices. It is to be expected that the occurrence of this kind of problem will increase as time passes and more intelligent machines are produced.

Class 4. The logical extension of class 3 is to the state where two man-created-hardware intelligent systems sense each other and act accordingly. This already arises in the interfacing of computational networks.

2.8.2 Access to System Measurement Nodes

This section deals with another kind of measurement situation that arises, that of access to measurement data points of a system.

Several situations will be met in practice when measurement is desired.

Accessible real parameters. Of the potentially infinite parameters that might need to be measured only certain of them will, at any given time and state-of-the-art, be *accessible*. A part of the practice of the scientific method is very much a pursuit of developing means to make parameters accessible.

Inaccessible real parameters. Although it can be reasoned or, by indirect method, experimentally shown that real parameters exist that would be worth measuring, in some cases it is not possible to do so for practical reasons. As an example measurement of many of the human physiology parameters is not yet possible under the normal operating state conditions because measurement methods would cause permanent damage.

Inaccessible unreal parameters. Models can, as has been said earlier, generate state parameters that have no real physical relationships to the system modelled. Thus any amount of direct measurement effort will not lead to measurements being made. The data sought, however, might be obtained by indirect methods. As a simple example the physical real electrical circuit inductor is often modelled as a pure inductor in series with a pure resistor. It is not possible to measure the voltage at the central node directly. The values can, however, be deduced by properly constructed separate inductance and resistance tests or by use of a dual control alternating current bridge.

In each of the above cases it is tacidly assumed that suitable parameters could be defined. In many situations defining a unique singly variable parameter is the key problem. Often a *many-to-one* mapping is attempted to overcome this. Chapter 7 discusses this in terms of the generalized topic of pattern recognition.

Another difficulty met might be that although a suitable parameter has been identified and made accessible by use of the appropriate sensor, the all-important standard of the unit is not held constant. Examples of this arise in the discipline of *econometrics*, the study of measurement in economics. Analysis methods have been developed there that assume that the standard is changing in a definable statistical manner. In the physical science situation there is a tendency to ignore the likelihood of instability of the standard once it is declared.

REFERENCES

Bosman, D. (1978). 'Systematic design of instrumentation systems', *J. Phys. E.: Sci. Instrum.*, **11**, 97–105.

Coughanowr, D. R. and Koppel, L. B. (1965). *Process Systems Analysis and Control*, McGraw-Hill, New York.

Di Stefano, J. J., Stubberud, A. R. and Williams, I. J., (1967). *Feedback and Control Systems*, Schaum, New York.

Ferguson, E. S. (1968). 'The mind's eye: nonverbal thought in technology', *Science*, **197**, 827–36.

Finkelstein, L. (1975). 'Fundamental concepts of measurement: definition and scales', *Measurement and Control*, **8**, 105–11.

Finkelstein, L. (1977). 'Instrument science: introductory article'. *J. Phys. E: Sci. Instrum.*, **10**, 566–72.

Finkelstein, L. and Watt, R. D. (1978). 'Mathematical models of instruments—fundamental principles', *J. Phys. E: Sci. Instrum.*, **11**, 841–55.

Henley, E. J. and Williams, R. A. (1973). *Graph Theory in Modern Engineering*, Academic Press, New York.

Karnopp, D. and Rosenberg, R. C. (1968). 'Analysis and simulation of multiport systems: the bond graph approach to physical system dynamics'. MIT Press, Cambridge, Mass.

Karnopp, D. (1974). *Symposium on Basic Questions of Design Theory: Synthetic Dynamics: Bond Graphs in Design*, North Holland, Amsterdam.

Karnopp, D. and Rosenberg, R. C. (1975). *System Dynamics: a Unified Approach*, Wiley, New York.

Mayeda, W. (1972). *Graph Theory*, Wiley, London.

Mesch, F. (1976a). 'The contribution of systems theory and control engineering to measurement science', *Survey Lecture* SL5, *IMEKO VII Congress, London, May*.

Mesch, F. (1976b). 'Systemtheorie in der Meßtechnik-eine Übersicht', *Technisches Messen atm*, **4**, 105–12.

Shearer, J. L., Murphy, A. J. and Richardson, H. H. (1967). *Introduction to System Dynamics*, Addison-Wesley, Reading, Mass.

Stein, P. K. (1970). 'The role of latent information in information processing in measuring systems', *Shock and Vibration Bull.*, **41**, 81–107.

Sydenham, P. H. (1979). *Measuring Instruments: Tools of Knowledge and Control*, Peter Peregrinus, London.

Young, P. (1978). 'General theory for modelling badly defined systems', in G. C. Vansteenkiste (Ed.) *Modeling, Identification and Control of Environmental Systems: IFIP Conf.*, North Holland, Amsterdam.

Handbook of Measurement Science, Vol. 1
Edited by P. H. Sydenham
© 1982 John Wiley & Sons Ltd.

Chapter

3
P. H. SYDENHAM

Standardization of Measurement Fundamentals and Practices

Editorial introduction

The preceding chapters began this handbook by laying down an understanding of the measurement process and its relevance to practical systems of both physical and empirical science nature. Measurement of the physical parameters has, in general, been accomplished more easily than in the empirical sciences and for this reason has matured at a faster rate in its procedural and terminological areas.

This chapter is concerned with establishing a proper understanding of standard procedures, practices and concepts in the areas of nomenclature, physical standards, and standards of specification. It rounds out with a section describing the relevant institutional activities that exist to generally aid the practice of measurement. Many of the concepts, although not yet clearly applied to measurement practice in the empirical sciences, are nevertheless largely relevant to those more complex measurement situations.

The material presented in this chapter is of key importance to good measurement practice. It is an area reasonably well developed in standardizing institutions but one that is yet to be taken up at an adequate level by the majority of measurement scientists and technologists. Use of common linguistic terms and set procedures would do much to improve interchange of knowledge.

3.1 INTRODUCTION

The basic concepts explained in the previous chapter are applied in many different disciplines, in many different ways. Putting them to general practical use preferably requires the establishment of agreed procedures which are controlled by the use of agreed primary physical standards of the various defined units. This, in turn, creates a need for standardized nomenclature to describe the concepts concerned.

As well as standards for the physical units there exists another kind of standard that relates to practices and specifications. These standards also possess agreed nomenclature.

Although measurement science has not been seen in the past as a clear-cut discipline existing in its own right there has nevertheless, emerged over the last century a considerable amount of standardization of its terminology and of the physical and specification types of standard. This chapter provides an introduction to these aspects and to the organizations, procedures, and systems of administration used to help ensure that these standards progress toward becoming common methodologies. As will be shown, continual progress is necessary. This introduction will help make it clear as to how to proceed with such matters and how to make allowances for the variations that exist and will, no doubt, continue to exist.

The subject matter that needs comment divides appropriately into groups on nomenclature (Section 3.2), classification of measurement science knowledge (Section 3.3), physical standards representing the agreed units (Section 3.4), standard specifications (Section 3.5), how standards are put to effective use (Section 3.6), and finally (Section 3.7) the institutions that are operating in this area. Material given in other chapters, where more specific mention occurs, is also relevant.

3.2 NOMENCLATURE OF MEASUREMENT

3.2.1 Standard Nomenclature of Measurement Science

The terminology of a discipline develops as the discipline matures. Eventually a stage is reached where the practitioners decide it is time that the terms that they use to describe, in a linguistic manner, the concepts that they use frequently should be standardized to overcome the confusion that has prevailed. If the discipline is followed by only a small number of dominant practitioners it is reasonably easy to bring about uniformity of nomenclature. Measurement, however, is practised in all branches of science and industry and has been developing with little coordination as a described and recorded methodology for over two centuries. As might, therefore, be expected, the terminology of measurement in use is far from fitting a sole common standard set of terms.

The problem is complicated by the universality of the use of the basic concepts in so many different disciplines, in many languages, and because there are so many terms involved. No single and absolutely encompassing terminology has yet evolved. However, the past few decades have seen the emergence of several dominant sets of nomenclature that cover the general terms and definitions involved in measurement. It is not possible, nor sensible (because of the multiplicity of standards existing), in the space permitted here to describe all terms defined. Furthermore, they are in a state of constant flux. This section will introduce the main terms met and used, describing where the various issued standards may be located. It is important to use the standard terminology defined for the task in hand.

Firstly, measurement science is also often referred to as *metrology* this being the field of knowledge concerned with measurement. This statement immediately raises another problem that exists; the use of numerous synonyms to describe the same, approximately equal, concepts.

Standard terminology has been issued at all levels ranging from international down to within a single commercial organization or within a particular author's text.

The Organisation Internationale de Métrologie Légale (OIML) issued a document entitled 'Vocabulary of legal metrology, fundamental terms' in 1969. The official definitions are those expressed there in the French language. The British Standards Institution issued a dual language version in 1971 as PD 6461. The English language listings are not official but must suffice for a very large part of the world. This document gives an extensive list of terms that arise in the course of describing the methodology of metrology. Table 3.1 lists several relevant documents that give general metrological terminology. To a large extent many of the individual countries' standard terminologies are based upon the OIML document mentioned above but there will be small differences.

Table 3.1   Selected standards documents relating to nomenclature of general metrology. (See Tables 3.4 and 3.5 for explanation of the code letters; many are endorsed as other national standards)

Document reference	Title
(OIML)PD 6461 : 1969	Vocabulary of legal metrology—fundamental terms (dual language version from British Standards Institution)
AS 1514. Part 1 : 1980	Glossary of terms used in metrology—Part 1. General terms and definitions
BS 5233 : 1975	Glossary of terms used in metrology
BS 2643 : 1955	Glossary of terms relating to performance of measuring instruments (endorsed as AS Z23)
IEC 50(00) : 1975	General index of the international electrotechnical vocabulary
IEC 50(05) : 1956	Fundamental definitions
IEC 50(07) : 1957	Electronics
IEC 50(12) : 1955	Transductors
IEC 50(20) : 1958	Scientific and industrial measuring instruments
ISO R31 series	On base quantities, mechanics, heat, electricity, light, acoustics, physical and inorganic chemistry, atomic and nuclear, nuclear and ionizing radiations
AS 1384 : 1973	Transducers for electrical measurements
AS 1633 : 1974	Glossary of acoustic terms
AS 1057 : 1971	Glossary of terms used in quality control
AS 1929 : 1977	Glossary of terms used in non-destructive testing

Furthermore the OIML document does not adequately cater for modern auto-metrology practice. (At the time of editing AS 1514 Part 1 is the most recent published revision of the terminology of metrology practice: revision of the international level of standard is currently underway.)

Numerous additional standard specifications exist that state the terms to be used in the various specialist areas of metrology. In all, a considerable number of defined terms is available and it is the duty of the person or group involved to be familiar with those terms that pertain to areas of direct interest to them. However, many people who measure do not make use of these nomenclatures, because they do not work within any definite field, and therefore possibly lack guidance about which standards to use, and because they are not required to use them. As an example, many journals that report measurement technique do not specify a nomenclature to adopt beyond specifying the use of SI metric units.

It is not reasonable to give an encyclopaedic glossary of the terms involved and the reader is directed to consult the relevant standard. There are, however, several terms that predominate in common usage and that appear throughout this handbook. They are given here in the form stated in the BSI English translation of the OIML terminology (PD 6461).

Terms can be grouped into those concerning the characteristic features of the measurement process and of the instrument used, the kinds of errors arising, and the methods of measurement. These are now discussed in turn.

3.2.2   Nomenclature of a Measurement and of Measurement Performance of an Instrument

Three distinctly different concepts arise when conducting a measurement or describing the performance of a measuring instrument. They are often used incorrectly as synonymous terms.

The first relates to the resolving ability of a measurement process. This is commonly called the resolution but the PD 6461 document does not, in fact, include this word, *discrimination* being given as that word to be used to describe 'the quality which characterises the ability of the measuring instrument to react to small changes of the quantity measured'. The quantity to be measured is usually called the *measurand* in English speaking countries, but this term also does not appear in the PD 6461 document.

Ability to discriminate can be enhanced by increasing the gain of the sensing stage. For example, a magnifying glass can be used to sight the position of the index mark against a ruled scale with greater resolution than by the unaided eye; an electronic amplifier can be used to enlarge an electrical metrological signal. Too often this figure is quoted alone, users wrongly assuming that it related to repeatability and accuracy. Having adequate discrimination is a necessary but not sufficient condition for a satisfactory measuring instrument.

Discrimination should be set so that the values obtained for the steady-state measurement value vary a little, thereby indicating that the apparatus is responding. Too great a discrimination level will provide excessive fluctuation of the signal requiring excess capacity data display or collection. A well designed instrument has its discrimination tailored to suit the task. It is usually relatively easy to obtain the discrimination level needed, gain no longer being a problem in instrument design.

A very common error made is to quote the discrimination value in a way that suggests the instrument possesses that degree of *repeatability* and even *accuracy*, the next two terms to be defined here. The three terms, discrimination, repeatability, and accuracy, are quite different descriptions. They are not synonymous.

The 'quality which characterizes the ability of a measuring instrument to give the same value of the quantity measured, not taking into consideration the systematic errors associated with variations of the indications' is defined in the OIML document as the *repeatability*. The term *precision* is very widely used to describe this attribute but, again, the OIML standard does not mention the term.

It is important to distinguish between the closeness of values that define the repeatability of the individual, or the group of values, when measured in the short term with the same apparatus (as defined here) and the same parameter determined by a long-term set of measurements or by different persons with different apparatus. This latter case is called the *reproducibility*.

A considerable part of published accounts on methods of handling data to decide such parameters as the repeatability are based implicitly upon the premise that the same measurement is made several times providing a set of data that can be operated upon in some statistical way. It is, however, often the case that only one measurement can be made. Methods exist that will enable the observer to estimate the repeatability, or reproducibility in this instance.

A third situation needing clarification also arises when the instrument, the observer's performance, and the external perturbing parameters (called the *influence quantities*) are each constant, variation in the values being caused by changes of the measurand itself. This is the repeatability (or reproducibility) of the measurement rather than of the instrument which was discussed above. In practice the two sources of variation occur simultaneously making analysis or reduction of influence effects complex.

Instruments with adequate discrimination and repeatability can often be entirely useful instruments. However, there are circumstances where the third attribute, *accuracy*, becomes necessary. *Accuracy* of an instrument is 'the quality which characterizes the ability of a measuring instrument to give indications approximating to the true value of the quantity measured'. It is an expression of the truthfulness available: lack of accuracy arises from both the instrument and the imperfect standard of the unit. A bent pointer on an indicating voltmeter will give the same level of discrimination and repeatability as one not having a bent pointer but the reading observed may be in error from the true defined value for the volt. It is well to stress here that no instrument yet made is absolutely accurate at indicating the value of a measurand. It is technologically impossible to obtain the absolute case. An instrument that is defined as providing the primary standard of a unit can be such but the influencing quantities and the intercomparison process that must be used to make a determination of a measurand with that apparatus will always introduce error. A philosophy should be adopted whereby instruments are regarded as 'not good enough until proven otherwise'. Measurement is a science of understanding errors as much as measurements.

The accuracy of an instrument depends upon knowledge of how indicated values relate to the agreed standard. The process which ties accuracies from the used instrument to the primary standard is called *traceability* and is covered below in Section 3.6.2. Accuracy is finally assigned to an instrument by agreement: it does not automatically arise from good design alone. The term accuracy is often confused with another, *linearity*, which expresses how values lie on a linear, proportional scale. The scale might be very linear but biased in slope or offset from the assigned value and, therefore, not be accurate.

Accuracy becomes important when many different user groups make use of the same declared unit for expressing and applying their results. They will not be working in a consistent manner unless each has his equipment set to read the same value of output as other users' equipment for the same input as their equipment. Consider, for example, the manufacture of parts for a motor car that come from different countries and must fit together. This implies that measuring instruments must generally be routinely compared against a common standard of the unit, a process called *calibration*.

In order of ease of attainment discrimination is usually the most easily procured feature in an instrument, repeatability comes next with accuracy, very much harder and more costly to obtain, in a third place.

Other terms that describe the characteristics of instruments include transfer function, sensitivity, amplification, gain, hysteresis, magnification factor, damping, response time, drift, and stability: many of these are not defined in the PD 6461 document but can be found described in other documents given in Table 3.1.

3.2.3 Nomenclature Describing Errors

The next group of terms in common usage is that group concerned with description of the errors arising in the measurement and in the instrument. Measurements are never perfect; errors occur as deviations from the perfect case. They arise from numerous sources but can be placed into one of two dominant general groups: *systematic* or *random* errors.

Errors result in the measurement determination having associated with it a certain level of uncertainty about the value obtained. The range within which the true value lies is termed the *uncertainty*. In many cases the limits of range must be estimated; statistical probabilities can be stated where the case allows it, which is not always so.

Past practice has tended to quote a measurement value followed by the numerical value of the error band, calling this latter statement the error of the measurement. Correctly it is the uncertainty of the measurement: the true error may actually be less but the observer can only be certain that it lies within a given range.

Systematic errors are those that can be predicted, from past knowledge of the operation involved, on an individual measurement basis. For example the voltmeter mentioned earlier may have a bent pointer but calibration can determine the amount of bending allowing a correction to be made to all values subsequently taken with it. The cause of a systematic error may be known but may not have been eliminated for reasons of expediency or impossibility. Such errors can also come from an unknown source but may have been established, as a predictable function of certain known related variables in an empirical manner. An example of the latter case is the use of curve fitting to an experimental determination of the phenomenon which allows a correction to be calculated for future readings. This curve is termed the *calibration chart*.

Random errors are those which cannot be predicted on an individual basis but for which a statistical method can yield information about the mean value of a set of data using the theoretical laws of probability. An example is the level of white electrical noise at any required instant of time. Being truly random in nature it is only feasible to predict the mean of a run of signals or set of values, or to state a probability of the next instant of signal having a certain given value. The nature of random signals is often assumed to be Gaussian but this is not always the case. If it is not so then there is a strong chance that most statistical methods used are not entirely applicable (Feinstein, 1971). The handling and manipulation of errors is the subject of Chapter 6.

The two groups of error, systematic and random, are formed from many subclasses of error. The various terminologies used to distinguish errors include error of measurement, systematic error, random error, parasitic error, error of method, observer error, parallax error, interpolation error, error of indication, repeatability error, rounding error, discrimination error, hysteresis error, response error, datum error, zero error, intrinsic error, influence error, temperature error, supplementary error, total error and so on (refer to AS 1514 Part 1, 1980).

3.2.4 Nomenclature Describing Measurement Methodology

Although all measurements are made by comparing the measurand against a defined standard in some way, there exist many ways to achieve this end. The methodology of making measurements also has a defined terminology that enables efficient communication of concepts by the use of generally accepted terms.

A *direct* method is that method 'by which the value of a quantity to be measured is obtained directly, without the necessity for supplementary calculations based upon a functional relation between the quantity to be measured and other quantities actually measured'. Some confusion may arise between this definition and the direct comparison method in which the desired quantity is obtained at its full value by comparison with a quantity of the same unit. For example, reading the length of a piece of metal stock by reference, using direct comparison, against a graduated rule. The required measurement is in the same units as that of the scale used.

An *indirect* method of measurement is that in which the parameter sought is gained by use of intermediate stages of different units which are linked in some positive manner. As an example the method of measuring distance using the transit time of a pulse of radiation is indirect because the distance is calculated from the relationship linking the speed of light with the time of flight which is the actual measurand observed. When the measurement made is based upon the *base quantities* (those agreed physical quantities not related to any other in the SI system of units) used to define the quantity the measurement is said to be using a *fundamental* method.

The *comparison* method of measurement is that 'method of measurement based upon the comparison of the value of a quantity to be measured with a known value of the same quantity, or with a known value of another quantity which is a function of the quantity to be measured'. Into this class of method is placed the *direct comparison* method already mentioned. Another subset of this class is the *substitution* method in which the measurand is replaced by a suitable quantity which is adjusted in value to bring the indicator back to the value indicated initially by the measurand. The *transposition* method is also a direct comparison method, one in which 'the value of the quantity to be measured is initially balanced by a first known value A of the same quantity: next the value of the quantity measured is put in the place of this known value and is balanced again by another known value B. If the balance indicating device reads the same in both cases the value of the quantity measured is $\sqrt{AB}$'.

In the *differential* method comparison yields a slight difference between the standard and the measurand that is used to apportion a small additive value to the standard value. Yet another comparison method relies on *coincidence* occurring between the standard and the measurand such as is used in setting the rate of a clock by observing the coincidence of the clock cycles with a standard source of cycles. The *null* method is somewhat similar to, but not identical with, the differential method. In using the null procedure the measurand and the standard are adjusted until the difference between them is zero. In this case the null indicator is not calibrated for the purposes of giving a small value to be added to the standard; it is there only for seeking the null condition.

Two more comparison methods are the *complementary* method and the *resonance* method. In the former the value of the quantity to be measured is combined with a known value of the same quantity so adjusted that the sum of these values is equal to a predetermined comparison value. The other method is a comparison procedure in which a known relationship between the compared values of the same quantity is established by means of the attainment of a condition of resonance.

Terminology used in describing aspects of the physical standards of units is covered in Section 3.4. Many terms used in the presentation of description of the static and dynamic regimes of an instrument or other system are introduced in chapters of Volume 2 pertaining to these aspects.

Many works on measurement include glossaries ranging from the short to the long. These often include the basic metrology terms in addition to other terms in popular use. These, in general, are not tied to any particular agreed standard of terminology and their use is, therefore, probably less effective than their compilers intended. They are, however, of value in obtaining understanding but they should preferably not be used as sources of published terms, the official standards being safer to rely upon. It is important to prevent proliferation of non-agreed terminology but this is easier said than done. Texts with glossaries include Beckwith and Buck (1969), Bell and Howell (1974), Foxboro-Yoxall (1972), Herceg (1972), NS Corp (1977), O'Higgins (1966), Stata (1969), and Sydenham (1974). Clason (1977) provides dictionary style definitions in eight languages. Dietrich (1973) defines the use of the uncertainty concept of quoting error; many other works are in print on the statistical manipulation of data for which many references can be obtained from the bibliography published by the Higher Education Committee of the International Measurement Confederation (IMEKO, 1980).

3.3 CLASSIFICATION OF MEASUREMENT SCIENCE KNOWLEDGE AND PRACTICE

Although a chapter of Volume 2 deals with the topic in more depth it is necessary at this stage to mention briefly the problem that must be faced in retrieving any aspect of the knowledge pertaining to measurement science.

Information about the terminology used is related, more often than not, to the subject to which it applies. The knowledge of measurement science is scattered over the many areas of its application; only a little is to be found in a few specialist groups devoted to its fundamentals. It is, therefore, usually necessary to search for terminology in the various similar applications as well as the class in which general terms are covered. For example there exist many standards about nomenclature in such fields as, say, acoustic noise, temperature measurement, and flow metering. The nomenclature of these overlaps considerably in concepts but often uses different words for the same concepts. Thus there is often a large and confusing array of terms issued in several standards on the same topic.

If generally agreed classifications had been decided before the current confusion about the material of measurement science had developed—in the 1900's—then, in all probability, it too would now be reasonably well compartmentalized into clear-cut divisions such as are found for the literature about the discipline of physics, for instance. The internal taxonomy of knowledge and its terminology for measurement science are only just beginning to become organized, this handbook being one of the first attempts to order the fundamental issues involved rather than stating their practical outcome at the current state of the art.

Information is stored in one or more of several generic groups. Consider the thermoelectric junction. It might have its information classified under the *principle used*, for example, under thermoelectricity. It could also be placed with other *similar applications* of the instrument or method, such as the thermocouple's use for temperature control. Alternatively it may be found placed in a group representing *uses of the same technological device*, like a group on the uses of the thermocouple in measurements of many kinds. Yet another group where information may be available could be under a classification on the basis of the *energy regime* used; as an example the thermocouple would be found described in a work on electricity or on its subclassifications. Other possible classification locations that would need searching to establish the relevant terminology and other information would be under *systems concepts*, because of the general mathematics concerned, *industrial practice*, because of the use of thermocouples in process plant, *instrument design*, and so on. The number of possible places where one would certainly find definitive information is great.

It should now be clear, from the above example of the thermocouple, that standardization of terminology and literature storage for measurement science topics are not likely to succumb to any simplistic, singular, standards. The problem there is far greater than has been faced in bringing about a coherent system of primary and derived standards for the physical quantities.

The discussion of the various aspects of storage and retrieval of relevant information, given in Volume 2, will assist the organization of an efficient literature search of library stock and computer data bases.

3.4 UNITS OF PHYSICAL QUANTITIES AND THEIR DEFINING STANDARDS APPARATUS

The fundamental concept of what a measurement is has been developed in Chapter 1. The common concept, that most close to the actual working face of the application of measurements, is that a measurand is compared against the defined standard, or something representing that standard, the difference from the actual magnitude of the standard being expressed in subdivisions of the basic unit used according to some form of scaling.

The *standard* is the physical object or characteristic of a physical apparatus that represents the conceptual unit chosen to represent a particular measurable attribute. For example, a particular piece of metal uniquely represents the unit of mass; here the unit is called, by convention, the kilogram. It is represented, physically at the primary level, by a sole piece of material maintained under defined and controlled conditions. This is the physical standard of the *unit* of measurement.

Each unit and its standard are derived by man. The general philosophy adopted for the creation of the primary standards is that they be based upon some physical principle that is known to be as invariant as can possibly be found.

In many cases the associated unit was defined before the principles now maintaining the standard were developed. For this reason many primary units appear to use very awkward values. For example, the metre is currently defined as 1,650,763.73 wavelengths of the radiation of krypton-86 gas established under defined, closely-controlled, conditions. This rather inconvenient number arises because the metre was formerly defined (in the same manner as the kilogram) by a metal bar of a given length, which, in turn, was based upon an unrealistic distance equal to one ten-millionth of the distance of a quadrant of the earth's circumference.

It is more expedient to retain any new standard's value close to the magnitude of the unit adopted than it is to change the whole, traceable, chain of units whenever a new and better physical principle is discovered. At present the krypton length standard appears to be about to be replaced by the use of radiation from an iodine, or methane-stabilized helium–neon, continuous-wave, laser. Even this, however, may be overstepped by adopting the concept of defining length in terms of a chosen and declared value of the speed of light. This will then relate length to the physical standard of time instead of to its own unique primary standard principle.

Most of the primary standards are now based upon natural physical principles but their units are still declared as man chooses. The values are refined as better knowledge is gained. In the past this has led to a bewildering range of units.

The development of physical standards and their units has been a long and tortuous process (see Bendick, 1947; Kisch, 1965; Klein, 1975; Sydenham 1979). Since earliest times rulers have realized the need for legal metrology but were generally unable to do very much about it beyond establishing the rules and standards: they could not adequately police adoption across large areas of land. National units of length and volume were implemented but it was not until comparatively recently that a coherent system of measurement units and their primary physical standards was introduced. This came in the form of the SI metric system, which is almost that sole legal system in use throughout the world. Bailey (1977) is a good introductory account about the various kinds of standard. Sanders (1972) and Verman (1973) are two of the few published texts on the subject of standardization. Other works of relevance are cited in the following section on standards of the specification form.

The development and maintenance of the primary physical standards is a specialist area of measurement science. It is one where the prime task is to develop, maintain, and apply apparatus that will provide definitions of the required base and derived units at the highest possible accuracy and reproducibility. Bell and Clarke (1975), on producing a standard for density (a seemingly easy variable), provide insight into the theoretical and practical levels of sophistication required. Factors such as cost, time to set up and make an observation, size and portability of the apparatus, ability to be mass produced for commercial sale, and other factors important to the industrial user, are of lesser significance in this case than achieving the best possible metrological performance. In each extreme the basic philosophy is much the same: it is the emphasis that is different. It takes many years to develop new standard apparatus and obtain agreement for its general use throughout the cooperating countries. Figure 3.1 shows apparatus maintaining the standard unit of time, the second. To maintain this standard is a measurement science task combined with fundamental physical research. Page and Vigoureux (1975) provide insight in their account of the first century of operation of the BIPM, Paris, the organiza-

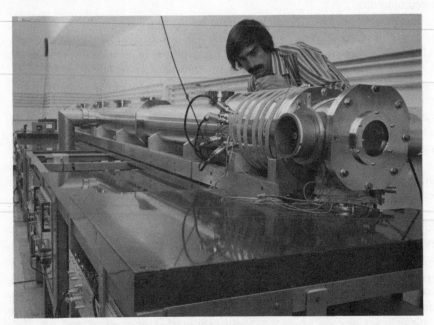

Figure 3.1   Physical apparatus used to maintain the primary frequency standard for SI unit, the second, in the United States of America (National Bureau of Standards, USA)

tion responsible for administration of the world's legal units and standards for the SI system. The *Bureau International des Poids et Mesures* (International Bureau of Weights and Measures—BIPM) has been responsible for metric systems since the 1875 *Convention du Mètre* was officially signed.

Illustrated descriptions of the apparatus used to set up the primary standards of the international network are presented in Page and Vigoureux (1975); wall charts are available from the National Bureau of Standards, Washington, USA (NBS, 1976), and from the National Physical Laboratory, England (HMSO, 1972). Cochrane (1966), in his official history of the National Bureau of Standards, also includes detail. Table 3.2 lists some national laboratories. The capabilities, which cannot be totally extensive, of the various national measurement laboratories are usually described in booklets published periodically, NML (1975) and NBS (1977), being examples. Such booklets often include lists of people and services that the various laboratories can provide. The laboratories usually offer more than standards maintenance, their specialist apparatus and staff being available for solving the more specialized and unusual measurement problems that the general industry cannot meet. The practice of standards and calibration is the subject of a recent series of articles by various authors, published in *Measurement and Control* (1978). Much of the work of the national physical standards group appears in the journal *Metrologia*.

Nomenclatures for the physical standards have not yet achieved a single uniform terminology. Some guidance and historical background are to be found in McNish (1958), and Cochrane (1966), but the definitive sources are, of course, the various international and national glossaries issued on the subject of general terminology of metrology already mentioned in Section 3.2.1.

Several classes of standard exist and confusion is commonplace about the terminology that should be applied. The following definitions are based upon the English language entries of the PD 6461 document referred to earlier.

An *international standard* of measurement is that 'recognized by an international agreement to serve internationally as the basis for fixing the value of all other standards of the given quantity'. This may need to be practically compared with the sole global standard, such as in the case of the kilogram, or it may be capable of development at a national level, as is the case with the standard of length where many countries operate their own krypton interferometers.

The *national standard* of a measurement is that 'recognised by an official national decision as the basis for fixing the value, in a country, of all other standards of the given quantity'. Theoretically, and on occasion practically, this may not be the standard having the highest metrological performance in that country (which is termed the *primary standard*) but it usually is. It is quite possible for better apparatus to be developed than the defined standard but time must elapse before it can be legally instituted to replace that then existing as the national standard. Research establishments will often be able to claim better performance in terms of discrimination, repeatability, and reproducibility for

Table 3.2　Names and location of selected national bodies responsible for legal national physical standards. (Courtesy of National Measurement Laboratory, Sydney)

Country	Name and location
Argentina	The Director General Instituto Argentino de Racionalización de Materiales (IRAM) (Argentine Standards Institute) Chile Road 1192 Buenos Aires, Argentina
Australia	The Director National Measurement Laboratory, CSIRO PO Box 218, Bradfield Road Lindfield, NSW 2070, Australia
Austria	The Director Bundesamt für Eich-und Vermessunswesen 16, Arltgasse 35 1163 Wien (Vienna), Austria
Belgium	Ingénieur en Chef Directeur du Service Belge de la Métrologie 24/26 rue J.A.De Mot B-1040 Bruxelles, Belgium
Brazil	The Director General Instituto Nacional de Pesos e Medidas Rodovia Washington Luiz Quilometro 23-XPrem Municipio de Duque de Caxias Estado Rio de Janiero, Brazil
Bulgaria	The Vice President Comité d'État de Normalisation PO Box 11 1000 Sofia, Bulgaria
Cameroon	The Chief Service Central des Poids et Mesures Ministère de l'Economie et du Plan Boíte-Postal 493 Douala, Cameroon
Canada	The Director Division of Physics National Research Council Ottawa K1A OR6, Canada
Chile	The Chief Division of Metrology Instituto Nacional de Normalización Casilla 995 Correo 1 Santiago, Chile

Table 3.2 (*continued*)

Country	Name and location
Czechoslavakia	The Vice President Urad pro Normalizaci a Mereni Vàclavské nàmesti c. 19 11347 Praha (Prague) 1, Nové Město Czechoslavakia
Denmark	The Director Justervaesenet (Bureau of Weights and Measures) Amager Boulevard 115 DK-2300 Copenhagen, Denmark
Cuba	The Director Centre de Recherches Métrologiques Comité Estatal de Normalizacion 5 ta 306 e/CyD Vedado Habana, 4 Cuba
Egypt	The Director National Institute for Standards National Research Centre al-Tahrir Street, Dokki Cairo, Egypt
Cyprus	Senior Officer Research and Industrial Development Ministry of Commerce and Energy Nicosia, Cyprus
Finland	The Director-General Valtion teknillinen tutkimuskeskus (Technical Research Centre of Finland) Vuorimiehentie 5 02150 Espoo 15, Finland
France	President Comité de direction Bureau National de Métrologie 8-10, rue Crillon 75194- Paris Cedex 04, France
German Democratic Republic	The Vice President Amt für Standardisierung, Messwesen und Warenprüfung Hauptabteilung Gesetzliche Metrologie Wallstrasse 16 1026 Berlin, German Democratic Republic
Federal Republic of Germany	The President Physikalisch-Technische Bundesanstalt Bundesallee 100 33 Braunschweig, Federal Republic of Germany

(*continued*)

Table 3.2   (*continued*)

Country	Name and location
Hungary	President Országos Mérésügyi Hivatal Németvölgyi-ùt 37/39 Budapest XII, Hungary
India	The Director National Physical Laboratory Hillside Road New Delhi 12, India
Indonesia	The Chief Service of Metrology Departemen Perdagangan Direktorat Metrologi-Standardisasi and Normalisasi Djalan Pasteur 27 Bandung, Indonesia
Iran	Director General Institute of Standards and Industrial Research Ministry of Industries and Mines PO Box 2937 Teheran, Iran
Ireland	The Chief Department of Physics Institute for Industrial Research and Standards Ballymun Road Dublin 9, Ireland
Italy	The Chief Ufficio Centrale Metrico Via Antonio Bosio 15 00161 Rome, Italy
Japan	The Director National Research Laboratory of Metrology 10-4, 1-Chome, Kaga, Itabachi-ku Tokyo, Japan
Republic of Korea	The Director National Industrial Standards Research Institute 199 Dongsoongdong Chongno-ku Seoul, Republic of Korea
Mexico	The Director General Consejo Nacional de Ciencia y Tecnologia Insurgentes Sur 1677 Mexico 20 DF, Mexico

Table 3.2 (*continued*)

Country	Name and location
Netherlands	The Director Institute of Theoretical Physics Universiteit van Amsterdam Spui 21 Amsterdam, The Netherlands
Norway	The Director Justerdirektoratet (National Bureau of Weights and Measures) Postbox 6832 St Olavs Plass Oslo 1, Norway
Pakistan	The Director Pakistan Standards Institution 39 Garden Road, Saddar Karachi 3, Pakistan
Poland	The President Polski Komitet Normalizacji i Miar ul. Elektoralna 2 00-139 Warszawa, Poland
Portugal	Director of Quality Secretariat de Estado da Industria Pesada Direkao Geral dos Servicos Industrias Industrial Rua Jose Estevao, 83-A Lisboa 1, Portugal
Romania	The Director Institutul National de Metrologie Sos. Vitan-Birzesti No. 11 Bucharest 5, Romania
Spain	The Secretary Comision Nacional de Metrología y Metrotecnica 3 calle del General Ibañez Ibero Madrid 3, Spain
South Africa	The Director National Physical Research Laboratory Council for Scientific and Industrial Research PO Box 395 Pretoria 0001, South Africa
Sweden	The Director General Statens Provningsanstalt PO Box 857 S-501 15 Boras Stockholm, Sweden

(*continued*)

Table 3.2 (*continued*)

Country	Name and location
Switzerland	The Director Office Fédéral de Métrologie Lindenweg 50 3084 Wabern/Be, Switzerland
Thailand	The Director Division des Poids et Mesures Ministeres des Affaires Economiques Bangkok, Thailand
Turkey	The Director Service des Mesures et des Etalons et Ministere Commerce Ticaret Bakanligi Olculer ve Ayarlar Mudur Vekili Bakanliklar, Ankara, Turkey
USSR	The Chief Gosstandart Leninsky Prospect 9 Moscow 117049, USSR
United Kingdom	The Director National Physical Laboratory Teddington, Middlesex TW11 0LW UK
United States of America	The Director National Bureau of Standards Washington DC 20234, USA
Uruguay	The President Instituto Uruguayo de Mormas Técnicas Avda. Agraciada 1464 Montevideo, Uruguay
Venezuela	Métrologiste en Chef Servico Nacional de Metrologia Legal Ministerio de Fomento Av. Javier Ustariz Edif. Parque Residencial Urb. San Bernardino Caracas, Venezuela
Yugoslavia	The Director Savezni zavod za Mere i Dragocene Metale Mike Alasa 14 11000 Beograd, Yugoslavia

some of their equipment and they may need to adopt it as their own internal standard but that does not make it the legal standard. Carefully controlled scientific and legal processes must be used to monitor any newly proposed apparatus to ensure that a change is certainly for the better in the long term. For example, proposals for the adoption of an absorption-stabilized laser as the new length international standard have been under observation for many years. The term prototype, which was once used to denote an international standard, is now deprecated, except in a historical context.

A *secondary standard* is that arrived at by some form of comparison with the primary standard or reference standard (see below). The term is used to describe either a subsidiary or a hierarchical place in the traceable chain. Thus its use can be somewhat confusing for the term does not inherently imply anything about its quality compared with the primary standard. It is not necessarily inferior. The term substandard is not included in the PD 6461 document and, therefore, should presumably be avoided in use.

Obviously only one location can have the primary standard so practical requirements dictate the use of a local sole standard to which all others in the locality are made traceable. This secondary standard is called the *reference standard*. The reference standard must also be carefully maintained in controlled conditions of ambient and use so as to preserve its calibration. From the reference standard it is necessary to form other standards that can be used at the more hazardous (for the instrument) working face. These are called *working standards*, sometimes also referred to as *field standards*.

Natural processes are often used to standardize measurements, the units being defined as a man-made arbitrary decision. The use of a radiation wavelength has already been mentioned. This form of standard is given the name *reference value standard*. Another example is the use of certain chemico-physical points to give the international practical temperature scale (IPTS). These enable a standard to be established without the need to compare it with a unique defined apparatus. As an example the National Bureau of Standards has declared that one brand of marketed laser interferometer suffices as a reference standard for length.

Breaking down a measurement situation, wherein numerous interconnected attributes are needed to define just one quality, into the appropriate quantities existing in the defined SI system is often prohibitive. Examples are to define a standard for a particular grade of iron, for the various chemicals, and for standard pollution levels. An alternative available is to use *reference materials* (not defined in the PD 6461 document). These are materials, or substances, that are officially recognized as a standard of certain attributes. They are characterized by having a high degree of stability of one or more of their chemical, physical or other metrological properties. They are called *standard reference materials* (SRM's) in the United States (where they largely originated), and *certified reference materials* in some other countries.

Reference materials may be consumed in the measurement process or they may be reusable after periodic calibration. They are also often used to conduct a *quality audit* of a group of cooperating standards laboratories by requesting each to measure the same SRM and declare their results for overall scrutiny. Audits, also called *round robins*, lead to better standard specifications as they give realistic answers to the real capability of that industry. As an example, a set of variously sized objects was circulated for stated dimensional quantities to be determined. The results were studied and a new revised standard for tolerancing was eventually issued.

Within a firm making many parts of the same size and form it is common practice to hold samples of the product that are the firm's local reference. It is more expedient to use these than to laboriously measure each individual parameter, one at a time. The samples act as standards for comparison.

Other forms of reference material are those pure substances used to define such units as density, the IPTS triple points, and the krypton used in the length interferometer. The purity and the properties of the substance are used to maintain the value of a defined quantity.

The introduction of what can be properly called universal systems of physical units and their standards began in earnest in the 18th century with the proposal of the French to use a metric numerical basis and a given set of primary units. The metric SI system evolved from this after many years of partial adoption, change, consideration, and (occasionally) inaction.

Units may arise inherently following the adoption of a freely available natural standard, such as the use of seeds from plants or the length of the human foot. The unit is given a name that describes the standard; in these cases it was the carat and the foot respectively. Once units have been established in this way it remains for them to be refined, adopting better standards (which ultimately are not freely available) as they are devised. In the early days of laser development a commonly used standard of power output was the number of Gillette razor blades that could be vaporized through. This was a very convenient form of standard, the units being as so many *gillettes*. However, to allow standards to arise in this uncontrolled way leads to confusion and lack of accuracy with its inherent lack of interchangeability of results: who can be certain that razor blades of all manufacturers would have the same fusing properties! History has well shown that units and standards need careful and tight control if the community is to get the best rate of progress from them. The need for new units and standards arises continuously as fundamental research discovers new effects. When this happens those concerned should act responsibly giving careful thought to the units and standards they adopt. These should be traceable within the SI system and be adopted after agreement by workers in similar fields. Unfortunately progress must first be made using interim standards which will provide some control whilst clear realization of the units involved is allowed to emerge.

The magnitude of the unit adopted is an important consideration for many reasons. Too small a unit can result in large numbers or much subdivision, the opposite is clearly also inefficient. For this reason the scale of units has tended to be matched to the task. The *ångström* is a convenient magnitude for describing atomic dimensions, the *light-year* is suited to galactic space distances. However, the practice of adopting units for the sake of convenience alone leads to multiplicity of units, the result being that the need for conversion becomes commonplace with its inherent potential for error. The SI system comes close to this ideal using decades of thousands and unity units. The kilogram is, however, an important incorrect departure from the fundamental philosophy for it is not of unit value.

The now obsolete British Imperial, foot–pound–second, system resulted from centuries of chaotic development in which expediency of the local needs seems to have been the main criterion adopted. Early metric systems used a centimetre–gram–second (CGS) basis; the MKSA system used the metre–kilogram–second–ampere basis. The Americans adopted the Imperial system in use at the time of their colonization and today still retain certain units as they were in those times—this is the reason why the US gallon is smaller than the now obsolete British gallon.

The system of units used affects trade probably more than any other sector of the developed country. It was the *Conferénce Générale des Poids et Measures* (CGPM) that, in 1954, agreed to a single uniform and very comprehensive system based upon the various forms of the metric system then in use. After 1954, the BIPM, mentioned earlier, yet again became the centre for revision of the metric system in use. In 1954 it was generally agreed to change from the several systems in use to the 'Practical system of units' which added thermodynamic, temperature, and luminous intensity units to the list of base units. In 1960 it became known as the *Système International d'Unites* or just SI. In 1971 the chemical unit of substance, the mole, was added to the already agreed base units to give the present seven base units.

Although the SI system provides a framework for total uniformity between all nation's systems of units, it is not adhered to in every aspect. Each country has usually retained a few of its formerly used units, the nautical 'knot' being one example. For this reason it is necessary to consult the national standard statement, not the prime international SI document. Table 3.3 lists some of the major standards documents issued on use of the SI metric system.

From the base units of the SI system it is possible to derive most other units that are ever needed by the dimensionally appropriate multiplication and division of the base units. The metric standards also contain statements about the preferred routes to obtain derived units. Figure 3.2 gives a chart showing the base units and derived units having special names which are obtained by appropriate combination. This arrangement allows calibration of a derived unit in a way that is traceable to the primary standards apparatus and that will, if users

Table 3.3   Selected SI metric units standards documents. (See Tables 3.4 and 3.5 for explanation of the code letters)

Document reference	Title
ISO 1000: 1973	SI units and recommendations for the use of their multiples and of certain other units
BS 3763: 1973	The International System of Units (SI)
BS 5555	SI units and recommendation for the use of their multiples and certain other units (endorsed ISO 1000: 1973)
BS PD 5686	The use of SI units
NBS SP-330: 1977	The international system of units (SI) (translation approved by BIPM of the publication entitled *Le Système International d'Unités*)
ANSI Z210.1: 1976 and ANSI/ASIM E380: 1976 and ANSI/IEEE 268: 1976	American national standard metric practice (conversion)
ASTM E380: 1970	Standard metric practice guide (a guide to the use of SI—the International System of Units)
DIN 1301 Teil 2	Units: sub-multiples and multiples for general use
AS 1000: 1974	The international system of units (SI) and its application
See also ANSI	Metric package
ANSI	A bibliography of metric standards—SP11b American National Standards Institute, New York
BIPM	*Le Système International d'Unités*, International Bureau of Weights and Measures (BIPM) and the authorized
NPL NBS	English-language versions from the National Physical Laboratory, UK and National Bureau of Standards, USA

abide by the directives issued, be calibrated by the same path in all places. Knowledge of the theory and practice of dimensional analysis is needed for this operation. The variable must first be identified in its dimensional form which then leads to paths to follow to combine units. For example, to obtain calibration for acceleration transducers correctly requires a system of calibrations that derives velocity from the metre and the second, combined with the second again to obtain acceleration. Texts on dimensional analysis are by Esnault-Pelterie (1950), Massey (1971), and Pankhurst (1964).

Due to the rapid upsurge in adoption of the metric system in recent times many texts and standards about the metric system have appeared; examples include Bradshaw (1975), Chiswell and Grigg (1971), Feirer (1977), Metrication Board, UK (1971), and O'Neill (1976), National standards organizations issue defining documents for use of SI (see Table 3.3).

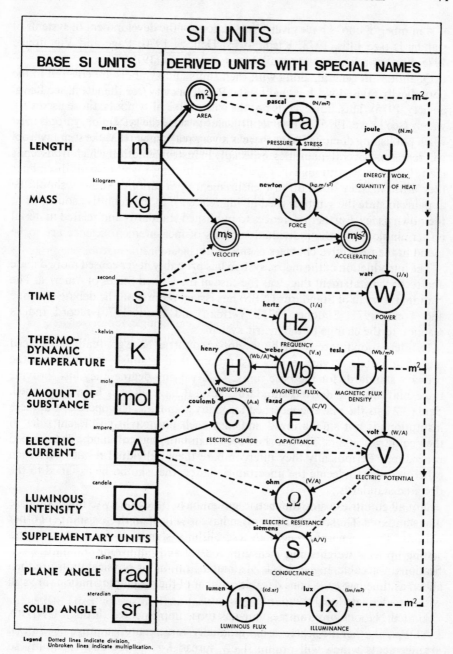

Figure 3.2 Hierarchy of SI base and derived physical measurement units having special names (Courtesy of Metric Conversion Board, Australia (1972), after chart in *Metric Handbook*: SAA MHI-1972, Standards Assoc. of Australia, Sydney)

A number of books have given explanations of the development of systems of standards (see Ellis, 1973; Klein, 1975; Dellow, 1970; Page and Vigoureux, 1975. Further references are available in Sydenham (1979).

Quantities in general, along with their units and symbols are covered in the periodically revised work issued by the Royal Society (for the latest, see Royal Society, 1975). This, however, is not the sole source of symbols; those used vary widely (see Lowe, 1975). It is generally safer to state the system of symbols used when publishing documents for there is a very real chance that the same symbol will represent several quantities, especially in instrumentation which transcends many boundaries. Although they are not likely to come together in the disciplines they arose in, they often do in measurement applications. Authors should be at pains to state the symbols used giving reference to the definitive source or to their own usage. Further references to published standards and related material on terminology for the individual subjects of measurement science are to be found in the respective chapters of this work.

The introduction of the metric system has probably never caused more debate and recorded statement than was the case in the United States of America. The National Bureau of Standards (NBS) has issued reports on the debate (Simone and Treat, 1971; Simone, 1971). Verman and Kaul (1970) record India's reaction to the change to the metric system.

By international agreement the developed countries have gradually introduced into their public service sectors institutions that establish and maintain the primary standards for the base units. To a varying degree, they also provide traceability for the derived standards. Several have already been mentioned. Table 3.2 lists the places concerned for many countries. The operation of these institutions is not standardized and they each play slightly different roles in their countries' affairs. They do, however, maintain physical standards to agreed procedures. Periodically they intercompare their individual results with each other in order to decide the uncertainty figures that should be assigned to the declared standards.

Not all countries using the metric system have the capacity to maintain base unit standards. Those that do not often have to rely upon the assistance of other nations. They generally begin by establishing standards for trade reasons, setting up legal weights and measures control and publishing standard specifications; traceable, higher levels of standardization for physical units are then added as time passes. A considerable number of the developing nations are still without adequate local standards apparatus.

United Nations programmes, such as those implemented through UNIDO, UNDP, and other aid schemes, are gradually implementing a range of bilateral arrangements which will progressively spread knowledge and capability to maintain traceable calibrations chains throughout the world. Additions during the 1970's include the Korean Standards Research Institute (KSRI) with its network and the Brazilian system of weights and measures.

Eventually it is to be presumed that all countries will be party to the *international agreement of the metre* under control of the work of the BIPM, and that most will possess their own base unit facilities where they have a large enough population to warrant a complete traceable system. However, there is a long way to go in this regard (the cubit is still used in the highlands of an island in the Asian region), but already considerable progress has been made since the formation of the UN in 1945.

It must be stressed that the use of uncalibrated, non-traceable standards is to be discouraged, especially when the work is likely to spread into international cooperation at high levels of standardization. The uncalibrated instrument, and even worse the instrument that cannot be calibrated, are likely to cause considerable inefficiency and rework of research or of manufactured parts, work that all too often has to be carried out without the reasons for doing so being properly identified and the cause corrected. A balance must, however, be struck between the calibration level and periodicity employed with the task in hand; calibration is not without its cost. Sydenham (1978) provides a summary of the things that a calibration facility operator might need to know about in operating such a facility.

3.5 STANDARDS OF SPECIFICATION

The standards referred to above are concerned with definition of the physical units such as length, mass, time, velocity, pressure, temperature, density, and so on. They exist as physical apparatus and form a class of standard. A second, much larger group of standards—they are generated at around 2000 per month across the world (BSI, 1978); the US alone has around 600 organizations producing them (Slattery, 1971, 1972)—comprises those issued, as documents, to define such matters as terminology, methodology, testing procedures, tolerances, analyses of materials, safe practices, classification schemes for products, product performance, and other characteristics. These are generally termed either *standard specifications, commercial standards, industrial standards* or *technical standards*. The Standards Council of Canada, who operate the National Standards System of Canada, officially defines this form of standard (SCC, 1976), as 'approved rules for an orderly approach to a specific activity'. There does not appear to be a formal generally accepted definition in existence that distinguishes this form of standard from those of the measurement units discussed in Section 3.4.

These pertain, directly and indirectly, to the quality and performance of products and services. They began as standardization for industry spreading later into consumer products. They occur as codes, rules, regulations, and specifications to give control to the supply, erection, use, and operation of articles, materials, and equipment of all kind. They may pertain to raw materials, components, subassemblies, finished products, safety, design, construction, testing, and quality of performance. If it can be measured or related to measurement subjects it could be the subject of a standard specification.

Some of the standards issued contain only subjective measurement statements, they do not necessarily relate to physical measurement of parameters of their subjects. The name *standard* is, however, likely to be confusing. Indeed it has caused considerable confusion with physical standards. The National Standards Laboratory, Sydney, changed its name to the National Measurement Laboratory (NML) largely because of this: its functions were being confused with those of the Standards Association of Australia (SAA) which authorizes the standard specifications.

Standard specifications generally deal with standardization of terms and definitions, creation of standards of design or performance (which helps the tendering process by providing standardized statements for writing into contractual agreements), with the prescription of standards of quality and the associated tests by which to decide the quality, with rationalization of sizes and levels of quality to keep the options down to an efficient minimum number, and with the control of dimensions for components so that they can be reliably interchanged.

England is the acknowledged birthplace of this form of standard—the first such case arose with the standard thread that Whitworth proposed in 1840. The British Standards Institution (BSI) was formed in 1901. Standards were progressively evolved by the work of many interested parties; they are not issued solely by civil service action. They are not in themselves all necessarily legal identities but many are required by relevant authorities, such as electric power supply companies, to be adhered to if a product is to be sold or used in a cooperating situation. Insurance companies, learned institutions, industrial consortia, government agencies, statutory authorities, private firms, and the public at large can each have a hand in establishing the consensus needed for the generation of a standard specification. Specifications may be issued by a national institution and also by professional bodies, the BSI and the American Society of Mechanical Engineers (ASME) being examples of each. The majority of such specifications are adhered to voluntarily, but some are enforced compulsorily.

The first stage in preparation of a standard specification is for a responsible, appropriate, and authorized body to study the case put to it to establish if the need for a specification is worthy. The various organizations usually have established committees or councils responsible for defined areas. The need for a new standard, or for a revision, is generated by a request arising from any source or from the members of the committees who may sense the need from their intimate knowledge of the area they represent.

A conference will then be called of relevant organizations who will use the eventual standard. If this conference agrees to the need it decides the terms of reference and the task is put into the hands of the relevant drafting committee.

The drafting committee acts in a largely autonomous manner proceeding with the task of preparing a *draft standard* or *draft proposal* (DP) for public review and comment. This might be done through subcommittees if the task is large.

The draft document is publically circulated for several months to give knowledgeable persons and bodies a chance to prepare a submission of their views about the draft proposal which they put to the committee. These views are taken into consideration by the drafting committee who finalize the document and then submit it to their authorizing body who scrutinize the whole operation. Finally, if satisfied as the result of a voting procedure, the latter issues the final draft as the declared standard. Creating the international organization standards requires more stages and is usually done by correspondence rather than by face-to-face meetings.

The material contained in the standard can be brought together by many means. It might be adapted from an already existing standard of the two main international organizations, ISO and IEC, or be from another country. If it is identical with the already existing standard it is declared as being *endorsed* by the next user who issues a code number in line with the local numbering system. Another way to obtain the content is from a round robin procedure that decides what is a reasonable and established good practice to write into the standard. Some standards are produced from the work of the committee, the members of which put together their experience together with solicited information. Staff of the national standards body may assist. A trade body may provide a basis for the draft. In short, any competent person or body can participate in this way.

It is important to realize that standard specifications are not statements made for all time. They are always subject to review so that they will remain an accurate statement of contemporary workable practice. Some date faster than others. Threads, for example, have been reviewed at only roughly twenty-year periods whilst electric lamps needed review at some two-yearly intervals in their early years of issue.

A national standards association generally has an official logo as its *mark of conformity*. Only on products and the like that meet the standards, and are tested to do so by an appropriate authority, can the official logo be used.

Identifying products and causing them to comply with standard specifications can be a very valuable method of stabilizing quality and performance. It also greatly reduces contractual statements and engineering specification writing. It leads to efficiency in manufacture and use by aiding stability and rationalization. However, a note of caution must be sounded that the use of standard specifications does have disadvantages. Lack of detailed compliance to a given standard can be used to provide protection for an industry within a country from imports. It also slows up the release of products onto the market. Designs and contract writers must also keep themselves up to date with revisions. Singular, international agreement and endorsement of all standards by all countries will, theoretically, eventually lead to common codes but this seems rather an impossible hope in view of the number of standard specifications that are in existence.

For standard specifications to be effective they must be freely available for consultation and purchase and there must be a suitable organization to handle the administration of sales and updates, advice and standards drafting. Technologically-based countries generally have a government agency to do this. Table 3.4 gives a list of national level standards organizations; note the existence of the two international bodies ISO and IEC. Burton (1976), Mason and Peiser (1971), Ollner (1974), SAA (1977), Sanders (1972), Stewart (1977), and Verman and Kaul (1970) each provide insight into standardization and its procedures.

Specific area standards are mentioned in the respective chapters given in this handbook. There are too many standards on measurement-related subjects for them to be given here: those in the well known standards organizations lists are relatively easy to locate especially if a standards library can be used. A complete set of national standards and some of other organizations is usually to be found in the offices of the national body but it is unrealistic to expect that all standards known can be found in a single location. The BSI library receives as complete a set as can probably be found. The usual approach to the problem is to collect, through purchase, those that relate to one's specific area of endeavour.

Three useful and relevant documents to measuring systems, not in those systems of publications, are a guide to purchase procurement of complex electronic and supervisory systems (CAMA, 1974), an in-house publication comparing the symbols used in electrical systems for in the United Kingdom, United States of America, Canada, Germany, and as stated by the International Electrotechnical Commission (IEC) (Neumüller, 1973), and the Instrument Society of America (ISA) publication of standards and practices for instrumentation (ISA, 1977).

At the international level of standardization there exist two complementary organizations: ISO and IEC. ISO, the International Organization for Standardization, had in 1978, 68 members, 17 corresponding members and it liaised then with over 350 organizations. By 1978 ISO had produced 3750 ISO standards documents, these being the result of the work of 100,000 experts contributing through 1940 technical bodies. A general guide is issued (ISO, 1979a) along with details of the technical programme (ISO, 1979b) and an annual catalogue (ISO, 1979c). Liaisons are officially listed in ISO (1977a). ISO (1978) provides definition of the terms used in processes of standardization and certification. A series entitled 'Bibliography' is published as guides to general material of permanent nature. ISO (1977b) is concerned with standards for documentation and terminology.

With 68 member countries, some of which internally have several hundred standardizing agencies, it is clear that the task of locating the existence and whereabouts of a standard on any particular subject can only be attempted efficiently by computer-based procedures. In 1975 ISO decided to create a

Table 3.4   Standards organizations having membership of ISO (from ISO, 1979a). (Code letters denote organization name not standards code letters, which are given in Table 3.5)

International

International Organization for Standardization (ISO)

1, rue de Varembé
Case postale 56
CH-1211 Genève 20
Switzerland

International Electrotechnical Commission (IEC)

1, rue de Varembé
Case postale 56
CH-1211 Genève 20
Switzerland

National

Albania (BSA)

Byroja e Standarteve
Prane Komisionit te Planit te shtetit
Tiranë

Algeria (INAPI)

Institut algérien de normalisation
et de propriété industrielle
5, rue Abou Hamou Moussa
BP 1021, Centre de tri, Algiers

Australia (SAA)

Standards Association of Australia
Standards House
80–86 Arthur Street
North Sydney, NSW 2060

Austria (ON)

Osterreichisches Normungsinstitut
Leopoldsgasse 4
Postfach 130
A-1021 Vienna 2

Bangladesh (BDSI)

Bangladesh Standards Institution
3-DIT (Extension) Avenue
Motijheel Commercial Area
Dacca 2

Belgium (IBN)

Institut belge de normalisation
Av. de la Brabançonne, 29
B-1040 Brusseles

Brazil (ABNT)

Associação Brasileira de Normas Técnicas
Av. 13 de Maio, n° 13–28° andar
Caixa Postal 1680
CEP: 20. 000, Rio de Janeiro

Bulgaria (DKC)

State Committee for Standardization
at the Council of Ministers
21, 6th September Str.
Sofia

Canada (SCC)

Standards Council of Canada
International Standardization Branch
Meadowvale Corporate Centre
2000 Argentina Road, Suite 2-401
Mississauga, Ontario
L5N 1V8

Chile (INN)

Instituto Nacional de Normalización
Matias Cousino 64–6° piso
Casilla 995, Correo 1
Santiago

China (CAS)
China Association for Standardization
PO Box 820
Peking

Colombia (ICONTEC)
Instituto Colombiano de Normas
    Técnicas
Carrera 37 No. 52–95
PO Box 14237, Bogota

(*continued*)

Table 3.4   (*continued*)

Cuba (NC)	*Cyprus* (CYS)
Comité Estatal de Normalización	Cyprus Organization for Standards
5ta nr 306 entre c y d vedado	and Control of Quality
Zona postal 4	Ministry of Commerce and Industry
Havana	Nicosia
Czechoslovakia (CSN)	*Denmark* (DS)
Úřad pro normalizaci a měření	Dansk Standardiseringsraad
Václavské náměsti 19	Aurehøvej 12
113 47 Prague 1	Postbox 77
	DK-2900 Hellerup
Egypt, Arab Republic of (EOS)	*Ethiopia* (ESI)
Egyptian Organization for Standardization	Ethiopian Standards Institution
2 Latin American Street	PO Box 2310
Garden City	Addis Ababa
Cairo, Egypt	
Finland (SFS)	*France* (AFNOR)
Suomen Standardisoimisliitto r.y.	Association française de normalisation
PO Box 205	Tour Europe
SF-00121 Helsinki 12	Cedex 7
	92080 Paris La Defense
Germany, FR (DIN)	*Ghana* (GSB)
DIN Deutsches Institut für Normung	Ghana Standards Board
Burggrafenstrasse 4-10	PO Box M.245
Postfach 1107	Accra
D-1000 Berlin 30	
Greece (ELOT)	*Hungary* (MSZH)
Hellenic Organization for Standardization	Magyar Szabványügyi Hivatal
Didotou 15	Budapest
Athens 144	Pf. 24
	1450
India (ISI)	*Indonesia* (YDNI)
Indian Standards Institution	Yayasan Dana Normalisasi Indonesia
Manak Bhavan	Indonesian Institute of Sciences
9 Bahadur Shah Zafar Marg	Jalan Teuku Chik Ditiro No. 43
New Delhi 110002	PO Box 250
	Djakarta
Iran (ISIRI)	*Iraq* (IOS)
Institute of Standards and	Iraqi Organization for Standards
Industrial Research of Iran	Planning Board
Ministry of Industries and Mines	PO Box 13032
PO Box 2937	Baghdad
Teheran	

Table 3.4 (*continued*)

Ireland (IIRS)	*Israel* (SII)
Institute for Industrial Research and Standards Ballymun Road Dublin 9	Standards Institution of Israel 42 University Street Tel Aviv 69977
Italy (UNI)	*Ivory Coast* (BIN)
Ente Nazionale Italiano di Unificazione Piazza Armando Diaz 2 1 20123 Milan	Bureau ivoirien de normalisation BP 1318 Abidjan
Jamaica (JBS)	*Japan* (JISC)
Jamaican Bureau of Standards 6 Winchester Road PO Box 113 Kingston 10	Japanese Industrial Standards Committee c/o International Standards Office Standards Department, AIST Ministry of International Trade and Industry 33rd Mori Bldg 3-8-21, Toranomon Minato-ku Tokyo 105
Kenya (KEBS)	*Korea, People's Democratic Republic of* (CSK)
Kenya Bureau of Standards PO Box 54974 NHC House Harambee Avenue Nairobi	Committee for Standardization of the Democratic People's Republic of Korea Committee of the Science and Technology of the State Sosong guyok Ryonmod dong P'yongyang
Korea, Republic of (KBS)	*Lebanon* (LIBNOR)
Bureau of Standards Industrial Advancement Administration Yongdeungpo-Dong Yongdeungpo-Ku Seoul	Institut libanais de normalisation BP 195144 Beyrout
Libyan Arab Jamahiriya (LYSSO)	*Malaysia* (SIRIM)
Libyan Standards and Specifications Office Department of Industrial Organization Secretariat of Industry Tripoli	Standards and Industrial Research Institute of Malaysia Lot 10810, Phase 3, Federal Highway PO Box 35, Shah Alam Selangor

(*continued*)

Table 3.4   (*continued*)

Mexico (DGN)	*Morocco* (SNIMA)

Dirección General de Normas
Tuxpan No. 2
Mexico 7, DF
	Service de normalisation industrielle
	   marocaine
	Direction de l'industrie
	Ministère du Commerce, de l'industrie,
	   des mines et de la marine marchande
	Rabat

Netherlands (NNI)	*New Zealand* (SANZ)
Nederlands Normalisatie-instituut Polakweg 5 PO Box 5810 2280 HV Rijswijk ZH	Standards Association of New Zealand Private Bag Wellington
Nigeria (NSO)	*Norway* (NSF)
Nigerian Standards Organisation Federal Ministry of Industries 11 Kofo Abayomi Road Victoria Island Lagos	Norges Standardiseringsforbund Haakon VII's gate 2 N-Oslo 1
Pakistan (PSI)	*Peru* (ITINTEC)
Pakistan Standards Institution 39 Garden Road, Saddar Karachi-3	Instituto de Investigación Tecnológica Industrial y de Normas Técnicas Jr. Morelli—2da. cuadra Urbanización San Borjá—Surquillo Lima 34
Philippines (PS)	*Poland* (PKNiM)
Philippines Bureau of Standards TML Bldg 100 Quezon Avenue Quezon City, Metro Manila PO Box 3719 Manila	Polski Komitet Normalizacji i Miar U1. Elektoralna 2 00-139 Warsaw
Portugal (DGQ)	*Romania* (IRS)
Direcção-Geral da Qualidade Repartição de Normalização Rua José Estêvão, 83-A Lisbon 1	Institutal Român de Standardizare Căsuţa Poştală 63–87 Bucharest 1
Saudi Arabia (SASO)	*Singapore* (SISIR)
Saudi Arabian Standards Organization Airport Street PO Box 3437 Riyad	Singapore Institute of Standards and Industrial Research 179, River Valley Road PO Box 2611 Singapore 6

Table 3.4 (*continued*)

South Africa, Republic of (SABS)

South African Bureau of Standards
Private Bag X191
Pretoria

Spain (IRANOR)

Instituto Nacional de Racionalización y
Normalización
Serrano 150
Madrid 6

Sri Lanka (BCS)

Bureau of Ceylon Standards
53 Dharmapala Mawatha
Colombo 3

Sudan (SSD)

Standards and Quality Control
  Department
Ministry of Industry
PO Box 2184
Khartoum

Sweden (SĬS)

SIS—Standardiseringskommissionen i
Sverige
Tegnérgatan 11
Box 3 295
S—103 66 Stockholm

Switzerland (SNV)

Association suisse de normalisation
Kirchenweg 4
Postfach
8032 Zurich

Thailand (TISI)

Thai Industrial Standards Institute
Department of Science
Ministry of Industry
Rama VI Street
Bangkok 4

Turkey (TSE)

Türk Standardlari Enstitüsü
Necatibey Cad. 112
Bakanliklar
Ankara

United Kingdom (BSI)

British Standards Institution
2 Park Street
London W1A 2BS

USA (ANSI)

American National Standards Institute
1430 Broadway
New York, NY 10018

USSR (GOST)

USSR State Committee for Standards
Leninsky Prospekt 9
Moscow 117049

Venezuela (COVENIN)

Comisión Venezolana de Normas
  Industriales
Av. Boyacà (Cota Mil)
Edf. Fundación La Salle, 5° piso
Caracas 105

Vietnam Socialist Republic of (TCVN)

Départment de normalisation
Comité d'État des sciences et techniques
39, rue Tràn Hung Dao
Ho Chi Min City

Yugoslavia (JZS)

Jugoslovenski zavod za Standardizaciju
Slobodana Penezića-Krcuna br. 35
Pošt. Pregr. 933
11000 Belgrade

Table 3.5   Key to country of origin for some standards
codes (BSI, 1978) (Letters are those that prefix standards code, e.g. BS . . .)

ABNT	Brazil
AISI	USA
ANSI	USA
API	USA
AS	Australia
ASME	USA
AWS	USA
BDS	Bulgaria
BDSS	Bangladesh
BNS	Barbados
BS	United Kingdom
BST	Sweden
CAN	Canada
CAS	Central Africa
CEE	International
CEI	Italy
CEMA	Canada
CGA	Canada
CGSB	Canada
CISPR	International
CKS	South Africa
CNS	China
CODELECTRA	Venezuela
COPANT	Pan America
COVENIN	Venezuela
CRS	Costa Rica
CSN	Czechoslovakia
CUNA	Italy
CYS	Cyprus
DEMKO	Denmark
DGN	Mexico
DGNT	Bolivia
DIN	Germany FR
DS	Denmark
ELOT	Greece
ES	Ethiopia
E.S.	Arab Republic of Egypt
EURONORM	European
Fed.	USA
GOST	USSR
GS	Ghana
I	Portugal
ICAITI	Central America
ICONTEC	Colombia
IEC	International
IEEE	USA
INANTIC	Peru

Table 3.5 (*continued*)

INDITECNOR	Chile
INEN	Ecuador
IOS	Iraq
IRAM	Argentina
IRS	India
IS	India
I.S.	Eire
ISIRI	Iran
ISO	International
IST	Iceland
JIS	Japan
JS	Jamaica
JSS	Jordan
JUS	Yugoslavia
KEMA	Netherlands
KS	South Korea
K.S.	Kenya
KSS	Kuwait
LS	Lebanon
LSS	Libya
MBS	Malawi
MI	Hungary
MNC	Sweden
MS	Malaysia
MSZ	Hungary
NBN	Belgium
MBS	USA
NC	Cuba
NEMA	USA
NEMKO	Norway
NEN	Netherlands
NF	France
NI	Indonesia
NIS	Nigeria
NM	Morocco
NORVEN	Venezuela
NP	Portugal
NPR	Netherlands
NS	Norway
NSRDS	USA
NVS	Norway
NZS	New Zealand
ONORM	Austria
OS	Oman
OVE	Austria
PN	Poland
PS	Philippines
P.S.	Pakistan

(*continued*)

Table 3.5   (*continued*)

SAA	Australia
SABS	South Africa
SAE	USA
SAS	Saudi Arabia
SEMKO	Sweden
SEV	Switzerland
SFS	Finland
SI	Israel
SLS	Sri Lanka
SN	Switzerland
SS	Sweden
S.S.	Sudan
SSS	Syria
STAS	Rumania
TGL	German Democratic Republic
TIS	Thailand
TS	Turkey
TTS	Trinidad & Tobago
UL	USA
ULC	Canada
UNE	Spain
UNEL	Italy
UNI	Italy
UNIT	Uruguay
UTE	France
VDE	Germany FR
VIS	Sweden
VSM	Switzerland
ZS	Zambia

network (ISONET) that would work towards a common data control method for at least the international standardizing bodies. ISO (1977c) lists those bodies in ISONET at that time.

IEC, the International Electro-technical Commission, has a formal agreement with ISO to cover only electrical and electronic activities, leaving all others to ISO. By 1978 the IEC had published more than 1400 standards documents: IEC (1978a) is the list of these. An annual report is published (e.g. IEC, 1978b), which lists panels, officials, new issues plus drafts in consideration. Another document (IEC, 1979) provides the official guide to activities.

When seeking the existence of a suitable standard to use it would be of value to be able to consult a master global index. However, this does not appear to exist, presumably because of the immense number of standards already issued before electronic data base methods were introduced. The usual procedure would be to consult the most recent annual index volume for selected agencies (e.g. BSI, 1979; ANSI, 1977; SAE, 1977). The BSI publishes a monthly bibliography of standards received into its library (BSI, 1978). Chumas (1974) may be valuable in such a search. Table 3.5 provides a finder of the country of origin for given standards letters.

Attempts have been made in the USA to index internal voluntary engineering standards: Slattery (1971, 1972) indexes over 25,000 standards issued by hundreds of US agencies; GSA (1978) adds to these by listing those issued by US Federal Departments for their purchasing requirements; and Chumas (1975) lists standardization activities of 580 internal organizations. The value of contact with a standards librarian cannot be overstressed as an efficient means to make a standards search. IEC and ISO are not the only international standardizing bodies. Because measurement technique applies to all endeavour it may be necessary to investigate the standards issued by other bodies. ISO (1977c) lists these bodies giving a brief description of their operation and scope.

Somewhat surprisingly, because all practising technologists make extensive use of standard specifications, with the possible exception of within the USSR, the subject is not formally taught in its own right. Training is more likely to be provided on a rather *ad hoc* basis with short courses appearing as a new demand appears.

3.6 NATIONAL MEASUREMENT SYSTEMS

3.6.1 The Concept of a National Measurement System

A complex system of measuring capability and application exists in technologically advanced countries. Until recent times this system has not been seen as an identifiable entity nor has it evolved with guidance. In the late 1960's the concept of the *national measurement system* (NMS) came into prominence in the USA. It is 'that system of activity that can be given the credit of enabling manufacture, commerce, trade and communication to develop with some degree of compatibility between the different sectors of a nation's economy and in international arrangements'.

The NMS of the USA has been studied formally at depth (see Huntoon, 1967; Compton, 1973), data having been published by Sangster (1976a, 1976b) in the form of 'impact matrices' showing how measurement of a specific kind, such as, say, temperature or electricity, relate to specific sectors of the national effort.

The value of the NMS of a particular country can be seen simplistically by the value of measurements made in a year in that country. Measurements are very diffuse and extensive in demand: they often tend to go unnoticed. Huntoon (1967) suggests that in the USA (those figures are quoted as they appear to be the only set available) 20,000,000,000 measurements are performed each day. In 1965, the USA industrial sector invested around 3 % GNP into measurements, this figure increasing at around 1 % per annum. Data on properties of materials,

a small part of the total measurement needs, consumed 5% GNP, also increasing each year. Huntoon stated that those industries that heavily invested in
measurements usually showed the greatest productivity. Many products simply
could not be made without automatic measurements to feed the control loops.
In that study the US user appeared to be willing to pay 3% GNP for measurements. Finkelstein (1976) provides some data on the extent of measurements in
the UK.

Published figures having the same degree of research expended on them do
not appear to have been compiled for other countries, but several speakers have
occasionally declared that their own countries ratios are similar to those given
above.

The USA NMS study arose because of three needs for quantified data relating
to effort expended on measurements. The first was to obtain better understanding of basic measurements and their standards, the second was related to
data and standards on materials, and the third reason was for information on
the technological standards and measurements. Norden (1975) gives the
following reasons:

(a)   to develop a structural organization for the system under consideration;
(b)   to identify and quantify the importance of the technologies which use the
      NMS;
(c)   to study the second- and third-order effects of the NMS, i.e. on politics,
      society, economics, environment, etc.;
(d)   to identify potential measurement problems which may arise in tech-
      nologies within the NMS.

Having been shown the extent and cost of the NMS and how it impacts on so
much of a nation's activity it might eventually be expected that more systematic
effort will be devoted to measurements and their effective application.

3.6.2   Traceability, Calibration, and Evaluation

The classic instance that highlighted the need for traceable and standardized
measurements comes from American history (Cochrane, 1966). In 1904 a junior
night watchman, in a building of the National Bureau of Standards, Washington
City, attempted to put out a fire but was unable to fit the hose end to the hydrant
because the threads were not of the same size. Soon after this event a disastrous
fire at Baltimore again demonstrated the need for standardization when out-of-
town fire appliances could not use their equipment for the same reason. A
subsequent study showed that at that time there were not less than 600 different
hose sizes in current use.

Although the concept of traceability of calibrations for measuring instruments
is important it seems that comparatively little has been written about it; textbooks on metrology and instrumentation rarely mention it. Stein (1970) is an

account of the practical pitfalls about traceability that arise when it is assumed that the apparently traceable measuring instrument is regarded as infallible. Daneman (1975) is a good general statement; it also includes a useful bibliography. Julie (1965) describes traceability to the NBS in the USA and Tagaya (1977) covers the situation in Japan.

Traceability is very much a feature of the accounting aspect of an instrument as it involves the certifications of calibration values used; values which are not automatically a feature of an instrument but must be assigned by a calibration procedure.

A calibration is said to be traceable if it can validly be traced back along a defined line of increasingly more certain calibrations to the primary standard, or standards, used in the SI system. This means that any instrument that is traceably calibrated and that is at the same level will have the same assigned values—within the uncertainty allocated to it.

The very best instrument has no traceable validity if it cannot be proved at any time, especially after a failure, that its measurement values indicated are in the official traceable line. If they are then any consequent malfunction or damage to the instrument will not upset the validity and use of past values taken with it and the instrument can be replaced or readjusted to be the same. Ensuring traceability for an instrument is akin to taking out insurance before the disaster occurs. Loss of operation of an uncalibrated and non-traceable instrument means that data previously taken with it cannot be accepted for there is no longer any means to re-establish the instrument calibration after repair or replacement to give the same readings. Discrimination and repeatability will be regained but accuracy is lost.

An essential step to obtaining traceable calibration is the existence of a network of laboratories that can perform the service. This may begin with the national physical standards authority, that holding the primary standards, performing all of the necessary calibrations but the need soon expands beyond that laboratory's capability. At that stage an organization is established to control other capable laboratories. The procedure of authorizing other laboratories having capability in stated areas to make these traceable calibrations is generally known as *accreditation*. As Volume 2 deals with this topic it need not be discussed further here other than to put it in perspective.

The instrument that has been calibrated under traceable conditions is not, however, always a firm point of measurement certainty for it was calibrated under certain stated conditions where the influence quantities and other perturbing parameters were controlled. It must not be forgotten that it may not be providing the same calibration values when placed under service conditions.

The value of a calibrated instrument under different conditions to that of the calibration has not been given the degree of study that it deserves. Indeed, it is common practice to determine the effect on the instrument of influence quantities at the time of the test but it is rare for the instrument user to explore thoroughly and adequately what influence properties are present at the time of subsequent use. Once certain forms of noise data are present in the output signal they cannot be eliminated by post-detection processing. They must be eliminated at the sensing interface stage.

Effective calibration is sometimes better performed on site using mobile laboratories (Daneman, 1978; Morton, 1978) or a temporary calibration laboratory.

In order to retain controlled uncertainty throughout the traceable chain each stage needs to have less uncertainty than that below it: there was once a rule in vogue to use stages ten or more times better at each level, but this rule has come under fire in recent times as being too optimistic. Often an external agency will develop a piece of apparatus that appears to be as good, if not better than, the official standard. It cannot be given a traceable calibration to the ultimate of its capability, nor can it be used to replace the official standard. Legal and experimental processes will be initiated and eventually the apparatus might be adopted as the new standard, but this takes time. The process usually begins when a measurement gap develops between users and the standardizing groups and then, as industry and science use their own resources to solve the problem, the situation changes to that of a measurement pinch (see Cochrane, 1966). Standards laboratories are unable to keep far enough ahead at all times for all required variables due to their limited resources and the time constants of the development processes involved.

Calibration proves the validity of an instrument to provide an accurate indication of the variable it is made to quantify. Such factors as the ability of the instrument to perform for long periods of operation, to withstand reasonable shock loads, temperature excursions, relative humidity changes, and more factors are not strictly matters of calibration. In recent decades there has arisen the concept of *evaluating* instruments, along with their manuals, servicing features, and any parameters of importance.

It is a costly matter to evaluate an instrument but the cost can be very worthwhile when compared with the costs that use of the instrument carries with it. Figure 3.3 shows the bulk of the cost sources that are associated with an instrument during its life cycle. Evaluation will help to minimize many of the areas of cost shown. The place of evaluation is in situations where the instrument plays a vital role or where instruments are to be used in situations where eventual contractual disputes may arise. An evaluated instrument design is more likely to bear post-event scrutiny than one not tested in this way.

It is an unfortunate fact that the general quality of much of manufactured instrumentation is not up to the expectations of the customer at the time of delivery. Moss (1978) has studied the quality of contracted instrumentation delivered to a major USA agency. In the study period around 25 % of delivered instrumentation was rejected at the acceptance test stage and one out of ten items needed attention to obtain proper operation. This is for contracted

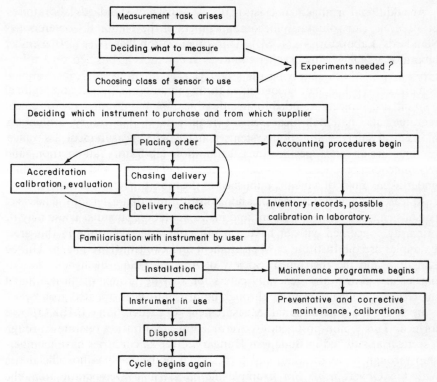

Figure 3.3 Chart showing sources of cost arising through use of an instrument

products which would be expected to be better than 'off-the-shelf lines' sold on the open market. Evaluation is part of the quality control of instrumentation and it is to be expected that more will be conducted as time passes. Evaluation is described in greater detail in Volume 2.

3.7 RELEVANT INSTITUTIONS AND ACTIVITIES

To complete this general introduction to the practice and administration of measurements it is of value to consider the kinds of institutions and organized activities that relate closely to some factor of measurement as a distinct entity. It is not possible to list all activities here since they number too many.

Mention has already been made of the group of national- and international-level laboratories that maintain the primary standards and of their sister organizations that organize and issue standards specifications.

A third group that is steadily growing in size is that formed by those national organizations that operate accreditation schemes. The first in existence, at national level, was the National Association of Testing Authorities (NATA) in Australia—for more about accreditation see Volume 2.

An additional arrangement exists in the USA in which standards laboratories, at all levels, can voluntarily become members of the National Conference of Standards Laboratories (NCSL). This organization has over 350 member laboratories, a few being in other countries. It conducts its business on a voluntary basis, the NBS supplying the secretariat, and arranges a regular annual conference. Work is mainly conducted through numerous subject and regional committees. Subjects attended to include education and training, measurement requirements, national measurement requirements, laboratory evaluation, biomedical safety, calibration systems management, measurement assurance, product design and specifications, calibration laboratory automation, and recommended practices. A considerable amount of research and investigation is undertaken, much of which is published in the monthly *NCSL Newsletter*.

At least 40 countries have professional institutions that cater for the needs of persons engaged in measurement and instrumentation pursuits. These may be primarily concerned with or have a major committee operating in this area. Examples are the Institute of Measurement and Control (IMC) in the United Kingdom and the Instrument Society of America (ISA). Each body has its particular features and generally caters for varying groups of measurement interest ranging from scientific through industrial to sales oriented groups.

In 1965 the International Measurement Confederation (IMEKO) was formed. This, a non-profit, non-governmental organization operates through a secretariat situated in Budapest, Hungary. Over 25 countries have membership through a suitable national professional body that has no official ties with the Government. For example in the German Democratic Republic membership is held by the *Gesellschaft fur Mess- und Automatisierungstechnik* and in China by the Chinese Scientific Society for Measurements and Instruments. Over the years since 1965 membership has spread from the original countries (Hungary, Poland, and USSR) to include firstly many northern-hemisphere nations and then in the 1970's developing countries in the India–Asia region and those of Australasia. However, the African and South American continents are poorly represented in the list of members.

IMEKO conducts its business by correspondence, through annual meetings arranged by the various working technical committees, and through regular international congresses. IMEKO is one of the five sister members of the FIACC group, two others being the International Federation of Automatic Control (IFAC) and the International Federation of Information Processing (IFIP).

Each of the above institutional arrangements is based on, and controlled by, a widespread membership forming some kind of basically capital-absent enterprise. They can provide interaction between people but are not easily able to conduct research and development undertakings.

Additional to the government institutions already mentioned, that possess established laboratories for advancing the science and practice of measurement, there also exist several other unique laboratories. In India there is a major facility, the Central Scientific Instrument Organization (CSIO), working for the development and transfer into application via industrial marketing, of scientific and industrial instrumentation. CSIO, a branch of the national government scientific service, has around 1000 staff who are committed to producing instrument designs that are vital to the development of the country. Over one hundred designs have been developed to the well-engineered prototype stage ready for a manufacturer to take up for manufacture and marketing. CSIO also operates several regional instrument repair centres throughout India as well as training schools and international aid programmes.

At CSIO emphasis is on the development of more conventional, already-existing instruments. In the UK the Sira Institute formerly the British Scientific Instrument Research Association, set up through partial funding from government plus subscriptions from subscriber trade organizations, has a brief to advance instrumentation in all of its phases. Efforts over the past decades include design and research on new instruments, services to the general public, evaluation of instruments, provision of training, advice to requests, and supply of testing and calibration services.

The work of a very small number of private consultants, who have become expert in such matters as international laboratory design, instrument design, marketing, survey work, and training, must also be recorded. These persons have accumulated experience that places them in a unique position in matters of measurement and instrumentation.

This brief account would not be complete without mention of the value of trade fairs and exhibitions organized for the trade to display its wares. Instrumentation does occasionally have events specifically devoted to it but more generally it would find a display outlet in either control, optics or fine-mechanism oriented events, or within exhibitions concerned with applications that make use of measurement. Many such events combine a conference with their exhibition.

REFERENCES

ANSI (1977). *Catalog of American National Standards*, American National Standards Institute, New York (plus many supplements).

Bailey, A. E. (1977). 'The measures of man', *Proc. IEE*, **124**, No. 1, 83–8.

Beckwith, T. G., and Buck, N. L. (1969). *Mechanical Measurements*, Addison-Wesley, Reading, USA.

Bell, G. A., and Clarke, A. L. (1975). 'The realization of a standard of density', in *46th ANZAAS Congress, Canberra, Australia, January 1975*.

Bell and Howell (1974). *The Bell and Howell Pressure Transducer Handbook*, CEC/Instruments Div., Bell and Howell, Pasadena, USA.

Bendick, J. (1947). *How Much and How Many: the Story of Weights and Measures*, McGraw-Hill, New York.

Bradshaw, M. F. (1975). *Decimal and Metric Reference Book*, E. J. Arnold, Leeds, UK.
BSI (1978). *Worldwide List of Published Standards*, British Standards Institution, London (continuing monthly bibliography, November issue consulted).
BSI (1979). *BS Yearbook 1979—complete to 30 September 1978*, British Standards Institution, London.
Burton, W. K. (1976). *Measuring Systems and Standards Organizations—SRIO*, American National Standards Institute, New York.
CAMA (1974). *A Guide to the Procurement of Complex Electronic Control and Supervisory Systems*, Control and Automation Manufacturers Association, London.
Chiswell, B., and Grigg, E. C. M. (1971). *SI Units*, Wiley, London.
Chumas, S. J. (1974). *Index of International Standards: NBS Spec. Publ. 390*, National Bureau of Standards, Washington DC.
Chumas, S. J. (1975). *Directory of United States Standardization Activities: NBS Spec. Publ. 417*, National Bureau of Standards, Washington DC.
Clason, W. E. (1977). *Dictionary of Measurement and Control*, North-Holland, Amsterdam (in English, American, French, Spanish, Italian, Dutch, German).
Cochrane, R. C. (1966). *Measures for Progress—a History of the National Bureau of Standards*, National Bureau of Standards, US Dept. of Commerce, Washington DC.
Compton, P. R. (1973). 'National measurement system study', *NCSL Newsletter*, **13**, No. 3, 10–14.
Daneman, H. (1975). *Some Aspects of Traceability*, Joint Industrial Measurements and Standards Committee, Japan Industrial Technology Association.
Daneman, H. (1978). 'Mobile calibration vans', *NCSL Newsletter*, **18**, No. 2, 43–5.
Dellow, E. L. (1970). *Measuring and Testing*, David and Charles, Newton Abbot, UK.
Dietrich, C. F. (1973). *Uncertainty, Calibration and Probability: the Statistics of Scientific and Industrial Measurement*, Adam Hilger, Bristol.
Ellis, K. (1973). *Man and Measurement*, Priory Press, London.
Esnault-Pelterie, R. (1950). *Dimensional Analysis and Metrology (the Giorgi System)*, Rouge, Lausanne.
Feinstein, A. R. (1971). 'On exorcising the ghost of Gauss and the curse of Kelvin', *Clin. Pharmacol. Therap.*, **12**, 1003–16.
Feirer, J. L. (1977). *SI Metric Handbook*, Scribners, New York.
Finkelstein, L. (1976). 'Preliminary study of the economic significance of industrial measurement and instrumentation technology for the U.K.', *Intern. Rep.* DSS/LF/110, *City University, London*.
Foxboro-Yoxall (1972). *Process Control Instrumentation—with Explanatory Notes: Publ. 105E*, Foxboro-Yoxall, Reading, UK.
GSA (1978). *Index of Federal Specifications and Standards*, General Services Administration Federal Supply Service, Washington DC (with cumulative bimonthly supplement).
Herceg, E. E. (1972). *Handbook of Measurement and Control*, HB-72, Schaevitz Engineering, Pennsauken, USA.
HMSO (1972). *Bases of Measurement—Definition and Realisation at the National Physical Laboratory, Ref. 56–6946, 5/72*, HMSO, London.
Huntoon, R. D. (1967). 'Concept of national measurement system', *Science*, **158**, 67–71.
IEC (1978a). *Catalogue of Publications*, International Electrotechnical Commission, Geneva (annual catalogue).
IEC (1978b). *Report on Activities for 1977*, International Electrotechnical Commission, Geneva (annual report).
IEC (1979). *Annuaire 1979 Handbook*, International Electrotechnical Commission, Geneva (supplement also issued).
IMEKO (1980). *Bibliography of Books Published on Measurement Science and Technology—with Emphasis on the Physical Sciences*, P. H. Sydenham (Ed.), TC-1 Committee, IMEKO, available Department of Applied Physics, Technical University, Delft, The Netherlands.

Institute of Measurement and Control (1978). *Measurement and Control*, **11** (monthly issues contain individual articles on calibration and standards).

ISA (1977). *Standards and Practices for Instrumentation*, 5th Edn, Instrument Society of America, Pittsburgh, USA.

ISO (1977a). *ISO Liaisons*, International Organization for Standardization, Geneva.

ISO (1977b). *International Standards for Documentation and Terminology: ISO Biblio*. 7, July, International Organization for Standardization, Geneva.

ISO (1977c). *Directory of International Standardizing Bodies*, International Organization for Standardization, Geneva.

ISO (1978). *General Terms and their Definitions Concerning Standardization and Certification: ISO Guide* Z-1978(E), International Organization for Standardization, Geneva.

ISO (1979a). *Memento 1979*, International Organization for Standardization, Geneva.

ISO (1979b). *Technical Programme 1979*, International Organization for Standardization, Geneva.

ISO (1979c). *ISO Catalogue*, International Organization for Standardization, Geneva (annual issue).

Julie, L. (1965). 'NBS traceability', *Electron. Instrum. Dig.*, Nov./Dec.

Kisch, B. (1965). *Scales and Weights: an Historical Outline*, Yale University Press, New Haven, Conn., USA.

Klein, H. A. (1975). *The World of Measurements*, Allen and Unwin, London.

Lowe, D. A. (1975). *A Guide to International Recommendations on Names and Symbols for Quantities and on Units of Measurement*, World Health Organisation, Geneva.

Mason, H. L., and Peiser, H. S. (1971). *Proc. Sem. on Metrology and Standardisation in Less Developed Countries: the Role of National Capability for Industrializing Economics*, National Bureau of Standards, Washington DC.

Massey, B. S. (1971). *Units, Dimensional Analysis and Physical Similarity*, Van Nostrand Reinhold, New York.

McNish, A. G. (1958). 'Classification and nomenclature for standards of measurement', *IRE Trans. Instrum.*, **I-7**, 371.

Metrication Board, UK (1971). *Metric Booklist, Post-School Textbooks and Reference Books using SI Units*, HMSO, London.

Metric Conversion Board, Australia (1972). *Metric Handbook:* SAAMH1-1972, Standards Association of Australia, Sydney.

Morton, J. (1978). 'Rockwell International, Avionics and Missiles Group metrology standards laboratory', *NCSL Newsletter*, **18**, No. 2, 38–42.

Moss, C. (1978). 'Quality of precision measurement equipment', *NCSL Newsletter*, **18**, No. 1, 23–4.

NBS (1976). *Brief History of Measurement Systems—with a Chart of the Modernized Metric System: NBS Spec. Publ.* 304A, *Revised August 1976*, National Bureau of Standards, Washington DC (wall chart also available).

NBS (1977). *Index of Technical Activities*, US Dept. of Commerce, Washington DC.

Neumüller, R. (1973). *How to Read German Schematic Diagrams*, Siemens Aktiengesellschaft, West Germany.

NML (1975). *Tests and Measurements*, CSIRO, Melbourne.

Norden, B. N. (1975). 'National measurement system—length and related dimensional measurements—a micro study', *NCSL Newsletter*, **15**, No. 2, 21–6.

NS Corp. (1977). *The Pressure Transducer Handbook*, National Semiconductor Corp., Santa Clara, USA.

O'Higgins, P. J. (1966). *Basic Instrumentation—Industrial Measurement*, McGraw-Hill, New York.

Ollner, J. (1974). *The Company and Standardization*, Swedish Standards Institution, Stockholm.

O'Neill, P. J. (1976). *The Wiley Metric Guide*, Wiley, Sydney.

Page, C. H., and Vigoureux, P. (1975). *The International Bureau of Weights and Measures 1875–1975*, Transl. BIPM Centenn. Vol., *NBS Spec. Publ.* 420, National Bureau of Standards, Washington DC.

Pankhurst, R. C. (1964). *Dimensional and Scale Factors*, Chapman and Hall, London.

Royal Society (1975). *Quantities, Units and Symbols*, Royal Society, London.

SAA (1977). *International Standardization and National Standards; a survey of the objectives and structures of ISO and IEC, and of the procedures for Australian participation therein*, Standards Association of Australia, Sydney.

SAE (1977). *SAE Handbook*, Society of Automotive Engineers, Warrendale, USA.

Sanders, T. R. B. (1972). *The Aims and Principles of Standardisation*, International Organization for Standardization, Geneva.

Sangster, R. C. (1976a). *Transactions Matrix Description of the National System of Physical Measurements*, NBSIR 75-943, National Bureau of Standards, Washington DC.

Sangster, R. C. (1976b). *Collected Executive Summaries Studies of the National Measurement System 1972–75*, NBSIR 75-947, National Bureau of Standards, Washington DC.

SCC (1976). *The National Standards System of Canada—the Second Five Years 1976/1980*, Standards Council of Canada, Ottawa (French/English combined).

Simone, D. V. De (1971). *A Metric America—a Decision Whose Time has Come*, NBS *Spec. Publ.* 345, National Bureau of Standards, Washington DC.

Simone, D. V. De, and Treat, C. F. (1971). *A History of the Metric System Controversy in the United States*, *NBS Spec. Publ.* 345-10, National Bureau of Standards, Washington DC.

Slattery, W. J. (1971). *An Index of US Voluntary Engineering Standards*, *NBS Spec. Publ.* 329, National Bureau of Standards, Washington DC.

Slattery, W. J. (1972). *An Index of US Voluntary Engineering Standards, Suppl.* 1, *NBS Spec. Publ.* 329-1, *Suppl.* 2 *Spec. Publ.* 329-2, 1975. National Bureau of Standards, Washington DC.

Stata, R. (1969). *A Selection Handbook and Catalog Guide to Operational Amplifiers*, Analog Devices Inc., Cambridge, USA.

Stein, P. K. (1970). 'Traceability—the golden calf', *Measurements and Data*, **2**, No. 4, 97–105.

Stewart, W. I. (1977). *Standardization in Australia—Some Papers and Comments*, Standards Association of Australia, Sydney.

Sydenham, P. H. (1974). *Stathmology*, Dept. of Continuing Education, Univ. New England, NSW, Australia.

Sydenham, P. H. (1978) 'Guide for calibration and testing facility operators', in *Accredited Testing Laboratories Conf., September 1978, Sydney*, National Association of Testing Authorities and New South Wales Science and Technology Council.

Sydenham, P. H. (1979). *Measuring Instruments—Tools of Knowledge and Control*, Peter Peregrinus, London.

Tagaya, H. (1977). 'Approach to a national traceable system in Japan', *NCSL Newsletter*, **17**, No. 3, 44–77.

Verman, L. C. (1973). *Standardization; a New Discipline*, Archon Books, Hamden, Conn., USA.

Verman, L. C., and Kaul, J. (1970). *Metric Change in India*, Indian Standards Institution, New Delhi.

Handbook of Measurement Science, Vol. 1
Edited by P. H. Sydenham
© 1982 John Wiley & Sons Ltd.

Chapter

4

E.-G. WOSCHNI

Signals and Systems in the Time and Frequency Domains

Editorial introduction

Measurement is the procedure by which information is extracted about parameters of a system, mapping the basic information entity into a meaningful knowledge statement. The measurement information is conveyed from the system to the point of observation in the form of signals. A considerable body of knowledge now exists for processing signals: this chapter provides a condensed summary of that knowledge. It embodies elements taken from many subject groups—namely, signal theory, information theory, communication theory, transfer function theory, and more. The boundary of these topics is such that they overlap each other to a great extent.

The Chapter is intended to promote awareness of the many mathematical techniques that can be brought to bear at one or many stages of a measuring instrument data handling chain. It is provided to act very much as a point of mathematical reference for the rest of the handbook. As such the reader will find many of the topics are used in subsequent chapters that often extend this material into the special area of interest to the subject in discussion.

The symbols and terms of mathematics are generally international and thus are, to a large degree, uniformly used. Different usage does, however, occur from writer to writer. In the light of there being no single absolute nomenclature standard to work to, it was decided that authors use that with which they are familiar provided adequate definition is provided in their contribution. In general, each chapter stands alone as a contribution containing sufficient preamble to introduce the topic.

4.1 INTRODUCTION

As represented in Figure 4.1, the measurement-technological task is to establish the characteristics of actually occurring input signals x_r from known (measured) output signals y_r. Signals having several components, x_r, y_r for example, can be summarized to form signal vectors $\mathbf{x}, \mathbf{y}, \mathbf{z}$, in the same way as can the disturbances

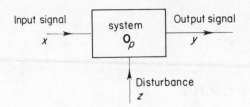

Figure 4.1   Task of measurement

z_r. The output quantities are a function of the input quantities, and the relationship between them is given by the behaviour of the measuring system. The aim is to have a certain behaviour, which is represented by an ideal mathematical operation $\mathbf{O}_{id}$:

$$\mathbf{y}_{id} = \mathbf{O}_{id}\{\mathbf{x}\} \tag{4.1a}$$

In the individual case, this operation may be a constant with one dimension, to give an example, but it may also be a differentiation or integration, as for instance in the case of measuring devices for averaging.

The practical measuring system with the connection $\mathbf{O}_{real}$ between its output and input quantities also includes real, falsified output quantities which depend on the disturbances $\mathbf{z}$:

$$\mathbf{y}_{real} = \mathbf{O}_{real}\{\mathbf{x}; \mathbf{y}\} \tag{4.1b}$$

Hence, an error ε occurs given by

$$\varepsilon = \mathbf{y}_{id} - \mathbf{y}_{real} \tag{4.1c}$$

In this chapter, these relationships will be dealt with in detail, in particular those with time-varying signals. The present discussion has the following four objectives:

(a)   description of the signals by characteristic values and functions;
(b)   description of measuring systems by means of characteristic values and functions;
(c)   description of the errors and deduction of quality criteria;
(d)   means for optimizing the system, that is, for minimizing errors.

Objective (a) is discussed in Section 4.2, the other objectives, which are based upon the first, are dealt with in Sections 4.3 and 4.4. Reference is made in each case to descriptions in the time and frequency domains. Figure 4.2 shows the objectives which have been outlined above and which are tailored especially to the following two fundamental problems of measurement technology.

Signal identification. In Figure 4.2a the aim is to determine the signal parameters of the input signals in terms of known (measured) output signals $\mathbf{y}$, where, if possible, disturbance variables should have no influence whatsoever.

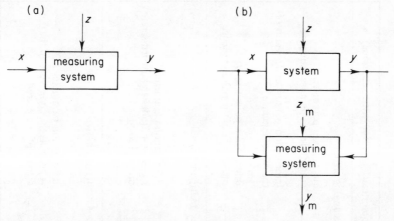

Figure 4.2   (a) Identification of signals; (b) identification of systems

System identification. Figure 4.2b, on the other hand, depicts measurement of the parameters of a system. In that case input test signals **x** and associated output signals **y** are both fed to the measuring system. The parameters which are characteristic for the system have to be determined from the output quantities of this measuring system $\mathbf{y_M}$.

4.2   SIGNALS

4.2.1   Classification of Signals

In addition to the useful signal **x**, interfering signals **z** occur. Both types of signal are carriers of information: wanted information in the case of the useful signal and unwanted in the case of the interfering signal. Both signals can be treated with the same methods.

For signals to be carriers of information, is necessary that certain parameters a_r of these signals be allowed to change:

in a signal $\mathbf{x}(a_r)$ it must be possible, for the purpose of information transmission, to change these information parameters. Information parameters a_r are those parameters of the signal upon which the behaviour of the information to be transmitted is mapped.   (4.2)

Table 4.1 gives a survey of signal designations: A distinction is made between *analog* signals, which are signals having no quantization of the information parameter, and *discrete* signals in which, because of quantization, the information can assume a finite number of values. An important special case is that involving binary signals in which the information parameter can take only two discrete values: 0 and 1.

Table 4.1   Classification of signals

Characteristic			Without quantization of time (continuous)	With quantization of time (discontinuous)
Without quantization of the information-parameter	analog		**Continuous analog signal** information parameter I: signal amplitude — information parameter I: phase shift or frequency	**Discontinuous analog signal** information parameter I: amplitude of the pulse sequence — information parameter I: width of the pulse sequence — information parameter I: phase shift of the pulse sequence
With quantization of the information parameters	discrete	multipoint-signal	**Continuous multipoint signal** information parameter I signal amplitude can have n discrete quantities	**Discontinuous multipoint signal** information parameter I signal amplitude can have n discrete quantities
		binary	**Continuous binary signal** information parameter I can have only quantities 0 and 1	**Discontinuous binary signal** information parameter I is given as the code word of the 0, 1 signals of one period

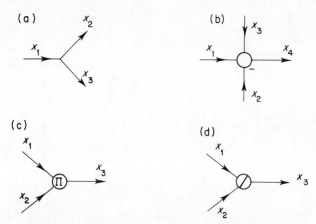

Figure 4.3 (a) Branching: $x_2 = x_3 = x_1$; (b) adding or subtraction: $x_4 = x_1 + x_3 - x_2$; (c) multiplication: $x_3 = x_1 \cdot x_2$; (d) division: $x_3 = x_1/x_2$

In time-dependent signals—most signals are time dependent or are converted to become time-dependent signals by scanning, as is the case in television—the information parameter can either change at any time (continuous signals), or changes are possible at given cycle times only, due to time quantization.

If the entire behaviour of the signal, including future behaviour, is known, the signal is said to be a *determined* one. The transmission or measurement of this signal, naturally, does not produce any gain of information. This type of signal plays a major role as test signals (e.g. impulse function, step function). Contrary to this, a signal to be measured has little *a priori* information. Signals with an unknown flow are called *non-determined* signals. If they are described by a probability distribution, they may also be called *stochastic* signals.

To trace and represent the flow of the signal, signal flow graphs are used. Figure 4.3 gives a survey of the graphical representation of the branching, addition, subtraction, multiplication or division of signals. Signals may be represented in the time and frequency (spectral) domains. For the formation of statistical characteristic values, time or statistical means are used. Furthermore, signal representations are applied by making use of geometrical relationships.

4.2.2 Signals in the Frequency Domain

4.2.2.1 *Periodic signals: Fourier spectrum*

A very important and determined fundamental signal, which is also of major importance as a test signal, is the harmonic oscillation

$$x = \hat{X} \sin(\omega t + \varphi) \tag{4.3a}$$

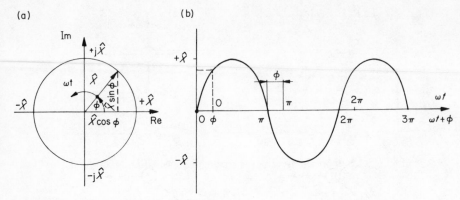

Figure 4.4   Harmonic oscillation: (a) indicator representation; (b) time representation

where $\hat{X}$ is the amplitude, $\omega(=2\pi f)$ is the angular frequency, $T(=1/f)$ is the oscillation period and φ is the phase angle (often zero). Representation in the complex plane, as shown in Figure 4.4, yields, according to the so-called 'symbolic method', complex and oriented indicators x and, consequently, the relationship

$$x = \hat{X} \exp[j(\omega t + \varphi)] = \hat{X} \cos(\omega t + \varphi) + j\hat{X} \sin(\omega t + \varphi) = \hat{X} e^{j\omega t} \quad (4.3b)$$

which is equivalent to equation (4.3a) and which has the oriented complex amplitude

$$\hat{X} = \hat{X} e^{j\varphi}. \quad (4.3c)$$

According to Figure 4.4, the directed quantity $\hat{X}$ can be split into real and imaginary parts:

$$\hat{X} = A + jB \,|\hat{X}| = \hat{X} = \sqrt{(A^2 + B^2)} \qquad \varphi = \arctan(B/A) \quad (4.3d)$$

The advantage of the symbolic method consists, above all, in the possibility of having a simple and easily understandable addition of several partial oscillations having the same frequency (Woschni, 1981). Periodic signals are particularly useful as test signals, because the same signal behaviour is obtained after each cycle period T and can be observed on an oscilloscope synchronized with T.

According to Fourier, it is possible to represent this type of signal having a cyclic time behaviour $x(t)$ by a series of sinusoidal and cosinusoidal oscillations with frequencies which are multiples of the fundamental frequency ω_0 given by

$$\omega_0 = 2\pi f_0 = 2\pi/T \quad (4.4a)$$

as shown in Figure 4.5, which depicts two sinusoids (of many) that form the cyclic rectangular oscillation

$$x(t) = \tfrac{1}{2}A_0 + \sum_{n=1}^{\infty} [A_n \cos(n\omega_0 t) + B_n \sin(n\omega_0 t)] \quad (4.4b)$$

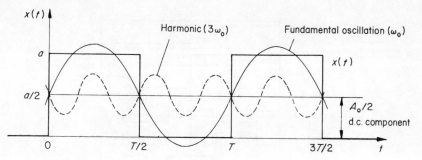

Figure 4.5   Representation of the rectangular oscillation by sinusoids

with the amplitude spectrum

$$C_n = \sqrt{(A_n^2 + B_n^2)} \qquad (4.4c)$$

The Fourier coefficients can be calculated from the relationships

$$A_n = \frac{2}{T} \int_{-T/2}^{+T/2} x(t)\cos(n\omega_0 t)\, dt \qquad (4.5a)$$

$$B_n = \frac{2}{T} \int_{-T/2}^{+T/2} x(t)\sin(n\omega_0 t)\, dt \qquad (4.5b)$$

From equations (4.5a, b), one can see that there are only cosine terms for even time functions $x(t) = x(-t)$, and only sine terms for odd functions $x(t) = -x(-t)$.

Transformation with the aid of Euler's theorem, leads to the complex Fourier series

$$x(t) = \frac{1}{T} \sum_{n=-\infty}^{+\infty} \hat{X}(jn\omega_0)e^{jn\omega_0 t} \qquad (4.6a)$$

with the complex coefficient

$$\hat{X}(jn\omega_0) = \int_{-T/2}^{+T/2} x(t)e^{-jn\omega_0 t}\, dt \qquad (4.6b)$$

and the amplitude spectrum $|C_{nk}|$ given by

$$|C_{nk}| = \left| \frac{1}{T} \hat{X}(jn\omega_0) \right| \qquad (4.6c)$$

In addition to the system of orthogonal functions based upon sine and cosine functions other orthogonal systems have recently been introduced, in particular the Walsh functions (Harmuth, 1970) which are shown up to eighth order in

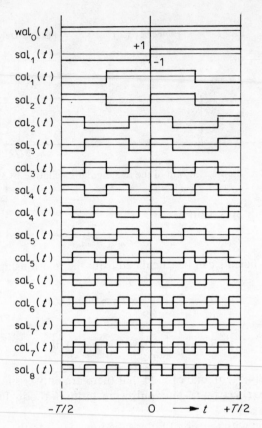

Figure 4.6 Walsh-functions up to eighth order

Figure 4.6. From equations (4.4) and (4.5), respectively, the corresponding relationships are developed:

$$x(t) = W_0 + \sum_{n=1}^{\infty} [W_{c_n} \mathrm{cal}_n(t) + W_{s_n} \mathrm{sal}_n(t)] \tag{4.7a}$$

with the Walsh coefficients W_{c_n} and W_{s_n}:

$$W_{c_n} = \frac{1}{T} \int_{-T/2}^{+T/2} x(t)\, \mathrm{cal}_n(t)\, \mathrm{d}t \tag{4.7b}$$

$$W_{s_n} = \frac{1}{T} \int_{-T/2}^{+T/2} x(t)\, \mathrm{sal}_n(t)\, \mathrm{d}t \tag{4.7c}$$

As one can see, the Walsh spectra, which are also called *sequency spectra*, are superior to the Fourier spectra, in that the multiplication with sine and cosine

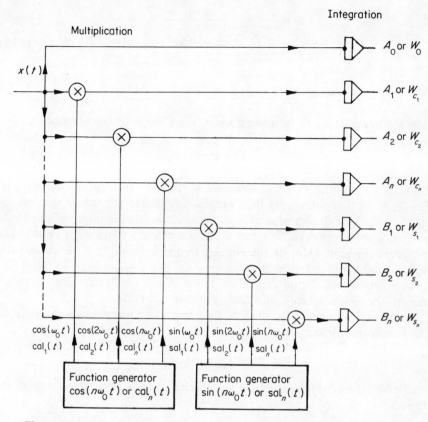

Figure 4.7   Experimental determination of the Fourier and Walsh coefficients

functions, respectively, is obviated by a simple reversal of signs. Therefore, these spectra can be determined more easily by experiment than the Fourier spectra (Figure 4.7).

4.2.2.2   *Non-periodic signals, spectral amplitude density, Fourier transform*

Non-periodic functions are of a great importance both as determined signals, i.e. test signals (step function, impulse function) and as non-determined signals (unknown signals which are to be measured). A discrete Fourier spectrum exists for periodic signals, whereas a continuous spectrum develops for non-periodic signals. It is obtained from the Fourier series (equation (4.6)) by passing to the limit:

$$T \to \infty \qquad \omega_0 = 2\pi/T \to d\omega \qquad 1/T \to d\omega/2\pi \qquad n\omega_0 \to \omega \qquad (4.8a)$$

In this case

$$\hat{X}(j\omega) = \int_{-\infty}^{+\infty} x(t)e^{-j\omega t}\,dt = F\{x(t)\} \tag{4.8b}$$

$$x(t) = \frac{1}{2\pi}\int_{-\infty}^{+\infty} \hat{X}(j\omega)e^{j\omega t}\,d\omega = F^{-1}\{\hat{X}(j\omega)\} \tag{4.8c}$$

Here the integrals are to be understood as the Cauchy principal value

$$\lim_{c\to\infty}\int_{-c}^{+c}\cdots$$

F and F^{-1} are the Fourier and inverse Fourier transforms respectively. Physically, $\hat{X}(j\omega)$ represents the complex amplitude related to $d\omega$, and is therefore also called *spectral amplitude density* having the dimension of amplitude per frequency interval, i.e. $V \cdot s$ and V/Hz, respectively. Depending on whether the frequency scale [Hz] or the angular frequency scale $[s^{-1}]$ is chosen the values will differ by a factor 2π.

An identical calculation can be made for Walsh functions. This, however, leads to the *sequential amplitude density* (Harmuth, 1970).

The Fourier transform has the following important properties and theorems. The transform is linear, i.e.

$$\sum_{v} a_v x_v(t) \circ\!\!- \sum_{v} a_v \hat{X}_v(j\omega) \tag{4.9a}$$

(where $\circ\!\!-$ is the sign for 'assignment'). For a change of the time scale, the relationship is

$$x(at) \circ\!\!- \frac{1}{|a|}\hat{X}\left(\frac{j\omega}{a}\right) \tag{4.9b}$$

Particular importance should also be attributed to the displacement theorems, namely the time displacement

$$x(t - t_0) \circ\!\!- \hat{X}(j\omega)\exp(-j\omega t_0) \tag{4.9c}$$

and the frequency shift

$$x(t)\exp(j\omega t_0) \circ\!\!- \hat{X}[j(\omega - \omega_0)] \tag{4.9d}$$

For differentiation, one obtains:

$$\frac{d^n x(t)}{dt^n} \circ\!\!- (j\omega)^n \hat{X}(j\omega) \tag{4.9e}$$

and for convolution:

$$\int_{-\infty}^{+\infty} x_1(\tau)x_2(t - \tau)\,d\tau \circ\!\!- \hat{X}_1(j\omega)\hat{X}_2(j\omega) \tag{4.9f}$$

4.2.2.3   Spectral power density

For non-deterministic signals, let us assume in the following that they are stationary, that is, that their time averages are not time-dependent quantities. To identify such signals $x(t)$, the *spectral power density* $S_{xx}(\omega)$ is used. It is defined as the part of the power ΔP, which falls into a differentially small frequency range $\Delta\omega$, that is,

$$S_{xx}(\omega) = \lim_{\Delta\omega \to 0} \frac{\Delta P}{\Delta\omega} = \frac{dP}{d\omega} \tag{4.10a}$$

In contrast to the spectral density, which cannot be determined in the case of random signals, the spectral power density is a real-valued function of the frequency ω. It does not contain any phase information. The latter is lost in the calculation of the average value, which is necessary for the formation of the power. This can also be seen from the relationship existing on the basis of Parseval's equation, by averaging over a time domain T (Zadeh and Desoer, 1963; Woschni, 1973):

$$S_{xx}(\omega) = \frac{1}{2\pi} \lim_{T \to \infty} \frac{|\hat{X}(j\omega)|^2}{2T} \tag{4.10b}$$

Consequently, the spectral power density is real and always positive, and it is an even function for which $S_{xx}(\omega) = S_{xx}(-\omega)$. Since the phase angle is missing, $S_{xx}(\omega)$ does not contain the full information about $x(t)$; a reverse calculation is not possible. The power P of the entire signal existing in the whole frequency domain, can be calculated on the basis of Parseval's equation, for the energy W

$$W = \int_{-\infty}^{+\infty} x^2(t)\, dt = \frac{1}{2\pi} \int_{-\infty}^{+\infty} |\hat{X}(j\omega)|^2\, d\omega$$

(Zadeh and Desoer, 1963; Woschni, 1981):

$$P = \overline{x^2(t)} = \lim_{T \to \infty} \frac{1}{2\pi} \int_{-T}^{+T} x^2(t)\, dt = \int_{-\infty}^{+\infty} \lim_{T \to \infty} \frac{1}{2\pi} \frac{|\hat{X}(j\omega)|^2}{2T}\, d\omega = \int_{-\infty}^{+\infty} S_{xx}\, d\omega \tag{4.10c}$$

From equations (4.10a,b) it can be concluded that a random phenomenon contains a periodic component of frequency ω_0 with amplitude $\hat{X}$. Dirac delta functions will develop in $S_{xx}(\omega)$ at the frequencies $\pm\omega_0$ (Woschni and Krauss, 1976):

$$S_{xx}(\omega)|_{\omega_0} = \tfrac{1}{2}\hat{X}^2\delta(|\omega| - \omega_0) \tag{4.10d}$$

Furthermore, from equation (4.10c) it follows that the power density $S_{xx}(\omega)$ must decrease rapidly from a certain critical frequency and must vanish at higher frequencies because of the requirement of boundedness of the power P. Depending on the critical frequency ω_c, a distinction is made between narrow-band and wide-band signals.

4.2.2.4 *Practical investigations*

The foundation for investigations in practice is the possibility of determining the characteristic functions and values by experiment. For the registration of spectral amplitude density and power density, respectively, one applies the same principles. In the case of filtering, use is made of several filters of bandwidth $\Delta\omega$ which are staggered in the frequency and whose outputs will be connected one after the other to a display unit for the voltage which is proportional to $|\hat{X}(j\omega)|$. According to Figure 4.8a, it is likewise possible to synchronize the switch with

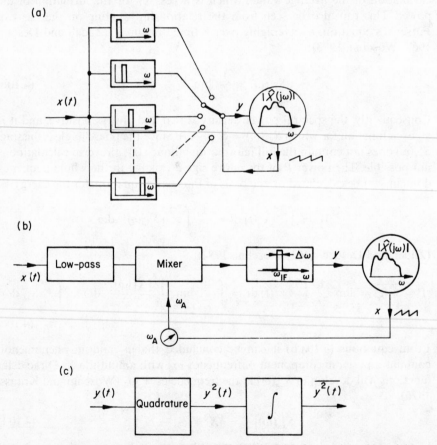

Figure 4.8   Spectral analyser: (a) switched-filter principle of operation; (b) variable centre frequency of single filter method; (c) formation of the power density

a sweep voltage that deflects the beam of an oscilloscope in the x-direction, proportionally to ω; consequently, the spectrum $|\hat{X}(j\omega)| = f(\omega)$ can be recorded.

Figure 4.8b shows another method which is characterized by the fact that only one filter of bandwidth $\Delta\omega$ and of centre frequency ω_{IF} (which is also called intermediate frequency is required. Tuning is carried out by mixing with the continuously tunable auxiliary frequency ω_A, where the following frequencies will be allowed to pass:

$$\omega = \omega_{IF} - \omega_A; \quad \omega^* = \omega_{IF} + \omega_A \qquad (4.11a,b)$$

The low-pass filter at the input eliminates the image frequency ω^*, so that one can cover the entire frequency domain required by tuning ω_A. Synchronizing the sweep signal with the x-axis sweep will generate the amplitude spectrum on the oscilloscope display. It is important to ensure that the filter has enough time to respond to the starting surge. The transient time t_{tr} of the filter needed is, according to the Shannon sampling theorem (Section 4.3.7):

$$t_{tr} = 1/2\Delta f = \pi/\Delta\omega \qquad (4.11c)$$

The sampling of the spectrum must, therefore, be carried out relatively slowly (low sweep frequency of the oscilloscope), for which reason long-persistence cathode ray oscilloscope methods are required. During the scanning run, the spectrum must be practically stable; it must be a steady signal in the statistical sense.

In addition to filtering and variable acoustic frequency methods, other techniques are used which make use of computers to implement the Fourier transform according to equation (4.8b) directly. This will be discussed in detail in Section 4.2.4, where specially adapted techniques for the fast Fourier transform will be dealt with. The computing advantages which arise when the amplitude sequential spectrum defined by the Walsh functions is used instead of the amplitude frequency spectrum, have already been pointed out in connection with Figure 4.7. While the techniques described in Figure 4.8 supply the amplitude response of the spectrum $|\hat{X}(j\omega)|$ only, it is possible to obtain additional phase information from the relationship $\varphi = \arg[\hat{X}(j\omega)]$ using computing methods.

To display the power spectrum the formulation $\Delta P/\Delta\omega = \overline{\Delta x^2(t)}/\Delta\omega$ is implemented, as outlined in Figure 4.8c, between the filter output and the input of the display unit and the oscilloscope.

To give some examples, consider some more signals which are used as test signals as well as for the approximate representation of measurement signals (refer to Section 4.2.7).

For a periodic sequence of rectangular pulses having the pulse width: repetition ratio $\tau = \Delta T/T$ (see Figure 4.9a), after substitution into equations

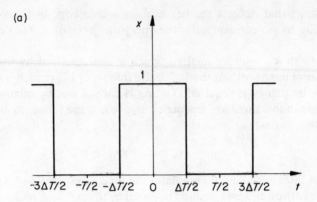

(a)

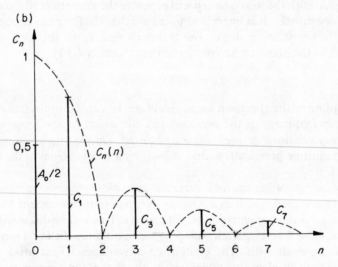

(b)

Figure 4.9   (a) Sequence of pulses; (b) spectrum for $\tau = 0.5$

(4.5a,b) and (4.4c) or (4.6b,c) and after an elementary calculation, one obtains pure cosinusoidal oscillations having the amplitudes

$$C_n = A_n = 2\tau \frac{\sin(n\pi\tau)}{n\pi\tau} \tag{4.12a}$$

Figure 4.9b shows the spectrum as well as the envelope curve for $\tau = 0.5$. The relationship with Figure 4.5 becomes immediately evident: the smaller the pulse width, i.e. $t_{tr} = \Delta T$. The pulse height is still correctly indicated, whereas the region up to the first zero of the envelope curve. For $\tau \to 0$, a constant spectrum results because the first zero shifts towards $\omega = \infty$. This case is significant as a test signal. For $\tau \to 0$, the unit impulse becomes the Dirac function $\delta(t)$ with the normalization

$$\int_{-\infty}^{+\infty} \delta(t)\, dt = \int_{-0}^{+0} \delta(t)\, dt = 1$$

According to equation (4.6b), this has a spectral amplitude density $\hat{X}(j\omega) = 1$, with only cosine oscillations of constant amplitude occurring.

Another important test signal is the unit step $w(t)$, which will be explained in detail in Section 4.2.7. For this, because

$$x = \begin{cases} 0 & t < 0 \\ 1 & t > 0 \end{cases}$$

(from equation (4.6b) for the spectral power density), $\hat{X}(j\omega) = 1/j\omega$ showing that only sinusoidal oscillations occur with an amplitude which declines hyperbolically with $1/\omega$.

The power spectrum declines with $|\hat{X}(j\omega)|^2$ according to equation (4.10b). This means that, in the case of a periodic rectangular oscillation, the major part of the power lies within the frequencies up to the first zero, and that $S_{xx}(\omega)$ is also constant for the unit impulse. The spectral power density for the unit step decreases very rapidly (as $1/\omega^2$) with the rising frequency. Another very important signal is resistance noise (Johnson noise), the power spectrum for which is constant up to very high frequencies (more than 10^{12} Hz) and assumes the value of $S(\omega) = kTR/\pi$ ($k = 1.37 \times 10^{-23}$ W s K^{-1}), T = absolute temperature (in K) and R = resistance (in Ω). Consequently, the noise voltage for a frequency bandwidth of $\Delta\omega$ is:

$$U_{eff} = \left(\frac{4}{2\pi} kTR\Delta\omega\right)^{1/2} \tag{4.12b}$$

For more details and further typical signals, see Section 4.2.7.

4.2.3 Signals in the Time Domain

4.2.3.1 Mean values

A time function $x(t)$ can be characterized by time averages of the nth order, which are also called moments of the nth order:

$$\overline{x^n(t)} = \frac{1}{2T} \int_{-T}^{+T} x^n(t)\, dt \tag{4.13a}$$

or, for non-periodic signals:

$$\overline{x^n(t)} = \lim_{T \to \infty} \frac{1}{2T} \int_{-T}^{+T} x^n(t)\, dt \tag{4.13b}$$

For $n = 1$, one obtains the linear (arithmetic) average which, from the physical point of view, can be interpreted to be the zero-frequency component of the signal or, according to Section 4.2.2.1, the Fourier coefficient $A_0/2$.

Of particular importance is the average value for $n = 2$, which, as a mean square value according to equation (4.10c), represents a measure for the power. The root from the mean square value is the effective value X_{eff} given by

$$X_{eff} = \sqrt{[\overline{x^2(t)}]} \tag{4.13c}$$

The following relationship exists between the component $\overline{x(t)}$, the component $\underset{\sim}{x}(t)$, and the mean square value X_{eff}:

$$X_{eff} = \overline{x^2(t)} = [\overline{x(t)}]^2 + \overline{\underset{\sim}{x}^2(t)} \tag{4.13d}$$

For a harmonic oscillation, one obtains $X_{eff} = \hat{X}/\sqrt{2}$ for the effective value.

4.2.3.2 Correlation function

The correlation function $\psi(\tau)$ represents a generalized mean square value, where a function is multiplied by the function displaced by time τ, and where the mean value is formed. If this function is the same function x, we call it an *autocorrelation* function $\psi_{xx}(\tau)$, where

$$\psi_{xx}(\tau) = \lim_{T \to \infty} \frac{1}{2T} \int_{-T}^{+T} x(t)x(t + \tau)\, dt = \overline{x(t)x(t + \tau)} \tag{4.14a}$$

It is suitable for making statistical statements on the internal relationships between function sections, as is now shown in a survey of its typical properties:

(a) In averaging, the phase information is lost, as is the case for the spectral power density (Section 4.2.2.2). Therefore, there are also direct relationships between the autocorrelation function $\psi_{xx}(\tau)$ and the spectral power density $S_{xx}(\omega)$, which will be discussed in detail in Section 4.2.3.3. However, periodic components in the signal $x(t)$ will be maintained without giving consideration to the phase position, because the following expression applies to the autocorrelation function of the harmonic oscillation, independently of the phase position:

$$\psi_{xx}(\tau) = \tfrac{1}{2}\hat{X}^2 \cos(\omega\tau) = X_{eff}^2 \cos(\omega\tau) \tag{4.14b}$$

(b) The value for $\tau = 0$ represents, according to (4.14a), the mean square value and is the maximum value of the autocorrelation function

$$\psi_{xx}(0) = \overline{x^2(t)} = X_{eff}^2 \tag{4.14c}$$

The other threshold value for $\tau \to \infty$ is the square of the linear mean value:

$$\lim_{\tau \to \infty} \psi_{xx}(\tau) = [\overline{x(t)}]^2 \tag{4.14d}$$

(c)   Since it is of no significance whether the function $x(t)$ in equation (4.14a) is displaced towards positive or negative times, the autocorrelation function is an even function:

$$\psi_{xx}(\tau) = \psi_{xx}(-\tau) = \overline{x(t)x(t+\tau)} = \overline{x(t)x(t-\tau)} \qquad (4.14e)$$

If two different signals $x(t)$, $y(t)$ are being compared one with the other, the measure used for the statistical relationship between them is the *cross-correlation* function $\psi_{xy}(\tau)$ according to the definition:

$$\psi_{xy}(\tau) = \lim_{T \to \infty} \frac{1}{2T} \int_{-T}^{+T} x(t)y(t+\tau)\,\mathrm{d}t = \overline{x(t)y(t+\tau)} \qquad (4.15a)$$

In measurement technology, the cross-correlation function plays a major role in solving system identification tasks during normal operation by means of the disturbances (see Section 4.3.8). The solution of this measurement problem is the foundation for adaptive systems (Davies, 1970).

In contrast to the autocorrelation function, the cross-correlation function has the following features:

(a)   It has no even functions, but the following relationship holds:

$$\psi_{xy}(\tau) = \psi_{yx}(-\tau) \qquad (4.15b)$$

(b)   It contains relative phase information concerning the two events $x(t)$, $y(t)$. In particular, the cross-correlation function of two harmonic signals with the same frequency disappears if the phase shift is $\pm \pi/2$, as can be seen following substitution into equation (4.15a). Likewise, the cross-correlation of two harmonic oscillations is zero if the frequencies are unequal.

(c)   The limiting cases are:

$$\psi_{xy}(0) = \psi_{yx}(0) = \overline{x(t)y(t)} \qquad (4.15c)$$

$$\lim_{\tau \to \infty} \psi_{xy}(\tau) = \overline{x(t)} \cdot \overline{y(t)} \qquad (4.15d)$$

The experimental registration of the correlation function will be discussed in Section 4.2.3.4.

4.2.3.3   *Relations to spectral power density: Wiener–Chinchine theorem*

The autocorrelation function, like the spectral power density, contains no phase information; this is lost in both cases because of the averaging operation. There is a relationship between both functions, as is the case between the time behaviour of the signal $x(t)$ and the corresponding spectral amplitude density $X(j\omega)$, via the Fourier transform; this relationship is known as the Wiener–Chinchine theorem (Woschni, 1981; Davies, 1970):

$$S_{xx}(\omega) = \frac{1}{2\pi} \int_{-\infty}^{+\infty} \psi_{xx}(\tau) e^{-j\omega\tau} \, d\tau = \frac{1}{2\pi} F\{\psi_{xx}(\tau)\} \qquad (4.16a)$$

$$\psi_{xx}(\tau) = \int_{-\infty}^{+\infty} S_{xx}(\omega) e^{j\omega\tau} \, d\omega = 2\pi F^{-1}\{S_{xx}(\omega)\} \qquad (4.16b)$$

Since the autocorrelation function is an even function (Section 4.2.3.2), cosinusoidal oscillations only occur. Consequently, equations (4.16a,b) can be rewritten

$$S_{xx}(\omega) = \frac{1}{\pi} \int_{0}^{\infty} \psi_{xx}(\tau) \cos(\omega\tau) \, d\tau \qquad (4.16c)$$

$$\psi_{xx}(\tau) = 2 \int_{0}^{\infty} S_{xx} \cos(\omega\tau) \, d\omega \qquad (4.16d)$$

If the autocorrelation function in equations (4.16a,b) is substituted by the cross-correlation function $\psi_{xy}(\tau)$, then the corresponding relationships with the cross-power density will be obtained:

$$S_{xy}(j\omega) = \frac{1}{2\pi} \int_{-\infty}^{+\infty} \psi_{xy}(\tau) e^{-j\omega\tau} \, d\tau = \frac{1}{2\pi} F\{\psi_{xy}(\tau)\} \qquad (4.17a)$$

$$\psi_{xy}(\tau) = \int_{-\infty}^{+\infty} S_{xy}(j\omega) e^{j\omega\tau} \, d\omega = 2\pi F^{-1}\{S_{xy}(j\omega)\} \qquad (4.17b)$$

4.2.3.4 Practical investigations

The characteristic functions and values which have been introduced can be found by experiment.

For the measurement of the linear mean value $\overline{x(t)}$, use is made of either moving-coil instruments or transistor voltmeters having a series-connected integration link; compare Figure 4.10a, with Figure 4.10b showing a simple analog circuit for finding the average value according to the equation

$$u_o = \frac{1}{C} \int i_C \, dt \approx \frac{1}{RC} \int u_i \, dt \qquad (4.18)$$

using the assumption $u_o \ll u_i$, i.e. $R \gg 1/\omega C$.

The mean square value can be found by means of the principle shown in Figure 4.11. To obtain the square of a value use is made of either electronic circuits with a square-law characteristic, e.g. diodes or transistors (as in the case

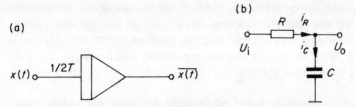

Figure 4.10   Measurement of the linear mean value: (a) basic circuit;
(b) simple realization

of transistor voltmeters), or a measuring device having a square-law characteristic, such as a soft-iron or hot-wire instrument (heating $\approx P \approx I^2R$).

Figure 4.12 shows the basic system for determining the autocorrelation function of the cross-correlation function. Delay due to a delay section, multiplication, and averaging yield, according to equation (4.14a), the autocorrelation
function (switch in position A) or, according to equation (4.15a), the cross-
correlation function (switch in position B). Since the integration time T cannot
be chosen to be infinite, only the short-term correlation function

$$\psi_k(\tau) = \frac{1}{2T} \int_{-T}^{+T} \cdots \mathrm{d}t \qquad (4.19)$$

will be measured in practice; under certain circumstances it reflects the actual
behaviour of $\psi(\tau)$ with sufficient accuracy (Section 4.2.4). In addition to the
analog methods discussed, used to ascertain experimentally the characteristic

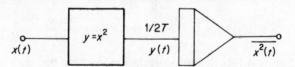

Figure 4.11   Measurement of the mean square value

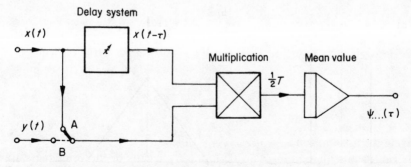

Figure 4.12   Experimental determination of the correlation function: (a) autocorrelation function $\psi_{xx}(\tau)$; (b) cross-correlation function $\psi_{xy}(\tau)$

functions and values in the time domain, an increasing use of digital methods is evident. In this case, the function $x(t)$ is split up into various values (support values) at different times t, the sampling theorem (equation 4.11c)) having to be observed. With the aid of the corresponding relationships, it is possible, for instance, to calculate the correlation function point by point from the support values.

Further methods are based on the application of the Fourier transform, i.e. calculation of the correlation function from the power spectrum which has been found by experiment (see Section 4.2.4).

Consider now some typical cases which are to stand as examples (see also Section 4.2.7). For the autocorrelation function of a rectangular pulse according to Figure 4.13a,

$$\psi_{xx}(\tau) = \frac{1}{2T_1} \int_{-T_1+\tau}^{+T_1} \hat{X} \cdot \hat{X} \, dt = \hat{X}^2 \left(1 - \frac{|\tau|}{2T_1}\right)$$

The result shown in Figure 4.13b reveals that if the pulse width $2T_1$ decreases in the limiting case $T_1 \to 0$, that is for the unit impulse $\delta(t)$, the autocorrelation function is also a delta function. This also follows immediately from the calculation of the autocorrelation function of white noise having a constant power spectrum $S_{xx}(\omega) = $ constant, as is the case for unit impulse. For this calculation equations (4.16b) and (4.16d) are used.

The relationship between the width of the power spectrum and the corresponding autocorrelation function can also be seen if the autocorrelation function for narrow-band noise is calculated in accordance with Figure 4.14a:

$$\psi_{xx}(\tau) = a \int_{-\omega_g}^{+\omega_g} e^{-j\omega t} \, d\omega = 2a\omega_g \frac{\sin(\omega_g \tau)}{\omega_g \tau}$$

As shown in Figure 4.14b, the autocorrelation function becomes smaller as the noise bandwidth increases, degenerating into a delta function for white noise.

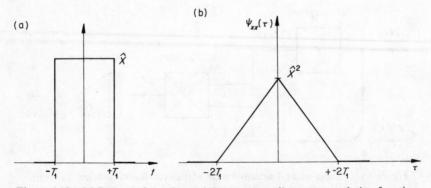

Figure 4.13   (a) Rectangular pulse and (b) corresponding autocorrelation function

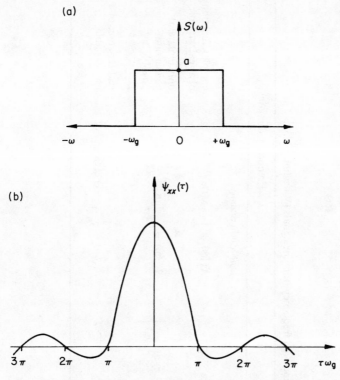

Figure 4.14   (a) Narrow-band noise and (b) corresponding autocorrelation function

This shows clearly that statistical relationships cease to exist if two noise signals
are slightly displaced relative to each other.

4.2.4   Relevance of Time and Frequency Ranges and Transforms Between them

In the following, a summary will be given of the relationships established so far
between the various signal representations in both the time domain and the
frequency domain. Also consideration will be given to the possibilities of conversion indicated above. Table 4.2 contains a survey of such relationships.

While the time function $x(t)$ and the spectral amplitude density $\hat{X}(j\omega)$ contain
the full information concerning the signal, this is not the case for the functions
resulting from averaging (autocorrelation function and power density). Phase
information is lost due to averaging. For this reason, conversions are possible
only in the direction indicated by the arrow.

Conversions are possible, via the Fourier transform, between functions in the
time domain and corresponding functions in the frequency domain. Therefore,

Table 4.2 Relationships between time- and frequency-domain signals (Woschni, 1973)

Time domain		Frequency domain	
function of time		amplitude density	
$x(t) = \dfrac{1}{2\pi} \displaystyle\int_{-\infty}^{+\infty} \hat{x}(j\omega)e^{j\omega t}\, d\omega$	$\Leftarrow F \Rightarrow$	$\hat{x}(j\omega) = \displaystyle\int_{-\infty}^{+\infty} x(t)e^{-j\omega t}\, dt$	complex function of a real variable
real function of a real variable	$\longrightarrow$	$S_{xx}(\omega) = \dfrac{1}{2\pi}\lim_{T\to\infty}\dfrac{|\hat{x}(j\omega)|^2}{2T}$	only unilateral conversion possible
mean value operation; phase information is lost			
autocorrelation function			
$\psi_{xx}(\tau) = \lim_{t\to\infty}\dfrac{1}{2T}\displaystyle\int_{-T}^{+T} x(t)x(t+\tau)\, dt$			
$\psi_{xx}(\tau) = \displaystyle\int_{-\infty}^{+\infty} S_{xx}(\omega)e^{j\omega t}\, d\omega$	$\Leftarrow F \Rightarrow$	$S_{xx}(\omega) = \dfrac{1}{2\pi}\displaystyle\int_{-\infty}^{+\infty} \psi_{xx}(\tau)e^{-j\omega\tau}\, dt$	real function of a real variable
real function of a real variable			

it does not matter, as far as the significance of the statement is concerned, which of these two functions is measured, rather it is a question of convenience. For instance, it is useful to carry out measurements in optical communications predominantly in the time domain. In vibration measuring technology, however, it is preferable to carry out the measurements in the frequency domain.

To accomplish the Fourier transform, digital computers are used at the present time. The basis for this is the *fast Fourier transform* (FFT). Using a transform specially tailored to fit the treatment in the digital computer, discrete Fourier transform programs have been established which save computing time, and where at least 1000 support points are quite usual. The calculation supplies the Fourier coefficients. As the number of support points is limited, the short-term correlation function is determined, which, however, is practically identical with the correlation function, provided the correlation time $\tau_k \leqslant T$ (refer to equation (4.19)). For more detail see Birgham (1974).

The importance of the conversions has already been discussed. Mention should again be made of the fact that $x(t)$ and $\hat{X}(j\omega)$ are related in the same way as are $\psi_{xx}(\tau)$ and $S_{xx}(\omega)$. For instance, a constant amplitude density has the delta function, which is a time function, as an autocorrelation function, just as the constant power density in white noise has.

4.2.5 Characteristics of Signals: Using Probability Functions

4.2.5.1 *Probability function and probability density*

To describe randomly fluctuating events $\xi(t)$, use is made of characteristic functions which are based on the theory of probabilities.

The probability distribution $W(x)$, which is also called the first-order distribution function, indicates the probability p that the signal $\xi(t)$ remains smaller than a barrier x, i.e.

$$W(x) = p[\xi(t) < x] \tag{4.20a}$$

The limiting values of the probability distribution

$$\lim_{x \to -\infty} W(x) = 0 \tag{4.20b}$$

$$\lim_{x \to +\infty} W(x) = 1 \tag{4.20c}$$

are also clearly understandable, because they mean the impossibility of a value smaller than $-\infty$ as well as the certainty of the occurrence of any signal value $\xi(t)$. For continuous functions $\xi(t)$, the probability distribution $W(x)$ is a monotonically increasing function.

The probability density $w(x)$ is the probability related to Δx for the fact that the values of the event $\xi(t)$ are within a narrow region near the value x, that is

$$w(x) = \frac{1}{\Delta x} p[x \leqslant \xi(t) < x + \Delta x]; \quad \Delta x \to dx \qquad (4.21a)$$

When comparing the Figures 4.15a, and 4.15b it can be seen that

$$\int_{-\infty}^{x} w(u)\, du = W(x) \qquad (4.21b)$$

and

$$dW(x)/dx = w(x) \qquad (4.21c)$$

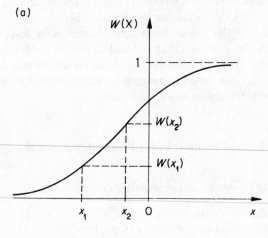

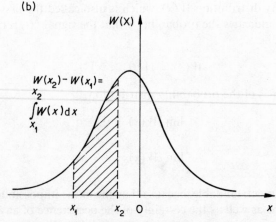

Figure 4.15   (a) Probability distribution and (b) corresponding probability density

Taking into consideration the limit (equation (4.20c)) the normalization is

$$\int_{-\infty}^{+\infty} w(x)\, dx = 1 \tag{4.21d}$$

As shown in Figure 4.15, the probability that $\xi(t)$ lies within the interval x_2 to x_1 is calculated by:

$$p[x_1 \leqslant \xi < x_2] = W(x_2) - W(x_1) = \int_{x_1}^{x} w(x)\, dx \tag{4.21e}$$

For multidimensional distributions $x_1, x_2, \ldots, x_n$, it is possible to introduce compound probability distributions $W(x_1, x_2, \ldots, x_n)$ and compound probability distribution densities $w(x_1, x_2, \ldots, x_n)$:

$$W(x_1, x_2, \ldots, x_n) = p[\xi_1(t) < x_1, \xi_2(t) < x_2, \ldots, \xi_n(t) < x_n] \tag{4.22a}$$

$$w(x_1, x_2, \ldots, x_n) = \frac{\partial^n}{\partial x_1 \partial x_2, \ldots, \partial x_n} W(x_1, x_2, \ldots, x_n) \tag{4.22b}$$

Furthermore, conditional probability distributions $W(x_1/x_2)$ and conditional probability distribution densities $w(x_1/x_2)$ are defined. They indicate the probability that the value x_1 occurs on condition that the value x_2 already exists. The following relationships hold for the compound probability density:

$$w(x, y) = w(x|y) \cdot w(y) = w(y|x)w(x) \tag{4.22c}$$

Of utmost importance in practice is the Gaussian distribution density:

$$w(x) = \frac{1}{\sqrt{(2\pi)}\sigma} \exp\left(\frac{-(x-a)^2}{2\sigma^2}\right), \tag{4.23a}$$

where $a = \overline{x(t)}$ is the linear mean value and σ the standard deviation, related to the square mean value, $\overline{x^2(t)}$, by

$$\sigma = \sqrt{[\overline{x^2(t)} - a^2]} \tag{4.23b}$$

Figure 4.16 shows the Gaussian distribution density for $a = 0$.

4.2.5.2 *Relations to the mean values: ergodic theorem*

The expectation value E of a function $f(x)$ is defined as follows:

$$E\{f(x)\} = \int_{-\infty}^{+\infty} f(x)w(x)\, dx \tag{4.24a}$$

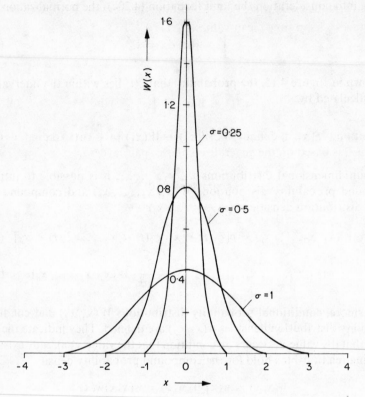

Figure 4.16   Gaussian distribution density

For $f(x) = x^n$, the moment M_n of nth order is obtained. Of particular importance is the moment of first order, denoted as the linear phase space average $\tilde{x}$:

$$\tilde{x} = M_1 = E\{x\} = \int_{-\infty}^{+\infty} xw(x)\,\mathrm{d}x \tag{4.24b}$$

and the square phase space average:

$$\tilde{x}^2 = M_2 = E\{x^2\} = \int_{-\infty}^{+\infty} x^2 w(x)\,\mathrm{d}x \tag{4.24c}$$

Accordingly, the expectation value for the simultaneous occurrence of $x_1(t)$ and $x_2(t + \tau)$ is obtained:

$$x_1(t)x_2(t + \tau) = E\{x_1(t), x_2(t + \tau)\}$$

$$= \int_{-\infty}^{+\infty} \int_{-\infty}^{+\infty} x_1(t)x_2(t + \tau)w[x_1(t), x(t + \tau)]\,\mathrm{d}x_1\,\mathrm{d}x_2 \tag{4.24d}$$

If this is an ergodic event, that is, if the ergodic theorem is satisfied, the phase space averages $\widetilde{x^n}$ and time mean values $\overline{x^n(t)}$ are

$$E\{x^n\} = \tilde{x}^n = \int_{-\infty}^{+\infty} x^n w(x)\, dx$$

$$= \overline{x^n(t)} = \lim_{T \to \infty} \frac{1}{2T} \int_{-T}^{+T} x^n(t)\, dt \tag{4.25a}$$

With equation (4.24d), a definition can be obtained of the correlation function $\psi_{xy}(\tau)$, which is based on the generalized phase space average:

$$\psi_{xy}(\tau) = \lim_{T \to \infty} \frac{1}{2T} \int_{-T}^{+T} x(t)y(t + \tau)\, dt$$

$$= \overline{x(t)y(t + \tau)}$$

$$= \widetilde{x(t)y(t + \tau)}$$

$$= \int_{-\infty}^{+\infty} \int_{-\infty}^{+\infty} x(t)y(t + \tau)w[x(t), y(t + \tau)]\, dx\, dy \tag{4.25b}$$

For the particularly important Gaussian distribution density according to equation (4.23a), one calculates:

$$M_1 = \tilde{x} = \overline{x(t)} = \int_{-\infty}^{+\infty} \frac{x}{\sqrt{2\pi}\,\sigma} \exp\left(\frac{-(x - a)^2}{2\sigma^2}\right) dx = a \tag{4.25c}$$

$$M_2 = \tilde{x}^2 = \overline{x^2(t)} = \int_{-\infty}^{+\infty} \frac{x^2}{\sqrt{2\pi}\,\sigma} \exp\left(\frac{-(x - a)^2}{2\sigma^2}\right) dx = a^2 + \sigma^2 \tag{4.25d}$$

4.2.5.3   Practical investigations

In order to register the probability distribution and density electronic majority decision elements having an adjustable threshold value x are used, as outlined in Figure 4.17. This can be done by either a triggering circuit or a voltage divider having a biased diode. With this, a normalization is to be carried out such that the corresponding conditions, equations (4.20c) and (4.21d), are observed. The arrangement is also suitable for recording on the oscilloscope screen, provided the voltage of the sweep generator for the x-deflection of the oscilloscope is used to control the threshold value x. The sweep frequency must be slow enough to ensure adequate averaging occurs. Oscilloscopes with long screen persistence times are used.

By coupling several installations in accordance with Figure 4.17b, compound probability distributions $W(x, y)$ can also be recorded. For this purpose, the trigger outputs of one arrangement for each event x, y will be connected with an AND element and further processed as shown in Figure 4.17b (Woschni, 1968).

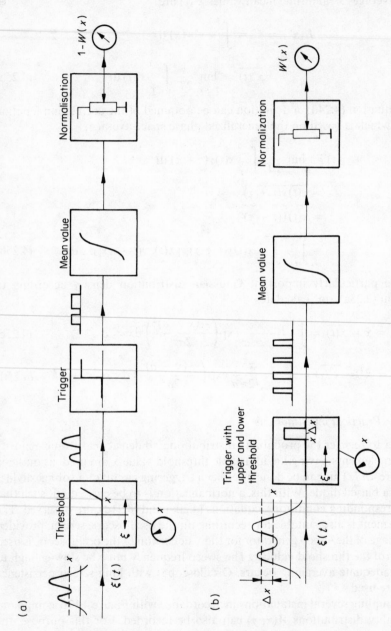

Figure 4.17 (a) Registration of the probability distribution and (b) probability density

In practice it is possible to describe many events, at least approximately, by the Gaussian distribution of equation (4.23a). The probability for the fluctuation process to lie within the range $-x \leqslant \xi(t) < +x$, or $a - x \leqslant \xi(t) < a + x$, where a is a constant, is:

$$p[-x \leqslant \xi(t) < +x] = \frac{1}{\sqrt{2\pi}\,\sigma} \int_{-x}^{+x} \exp\left(\frac{-\xi^2}{2\sigma^2}\right) d\xi \qquad (4.26a)$$

To evaluate this the probability integral is used which is tabulated in the following form (Jahnke, 1960):

$$\phi(x) = \frac{2}{\sqrt{\pi}} \int_0^x \exp(-u^2)\, du \qquad (4.26b)$$

Extracts are shown in Table 4.3.

It may have been found by measurement, for example, that the length of workpieces having an average value of $a = 10$ cm satisfies a Gaussian distribution and shows a standard deviation of $\sigma = 3$ mm. What matters then might be the number of workpieces that lie within an admissible tolerance range of 10 mm $\pm$ 4 mm. Evaluation according to equation (4.26) shows that 82% of the pieces are within the tolerance range, and that the remaining pieces lie outside this range.

4.2.6 Geometrical Signal Representations

4.2.6.1 *In Euclidean signal-space*

Signals with n components $x_1, \ldots, x_n$ can be represented by a signal vector $\mathbf{x}$ in the n-dimensional space:

$$\mathbf{x} = (x_1, x_2, x_3, \ldots, x_n) \qquad (4.27a)$$

where the end point of the vector determines the corresponding signal.

Table 4.3 Gaussian probability integral $\phi(x)$ evaluated for a range of x

x	0	2	4	6	8
0.0	0.0000	0.0226	0.0451	0.0676	0.0901
0.1	0.1125	0.1348	0.1569	0.1790	0.2009
0.2	0.2227	0.2443	0.2657	0.2869	0.3079
0.3	0.3286	0.3491	0.3694	0.3893	0.4090
0.5	0.5205	0.5379	0.5549	0.5716	0.5879
0	0.0000	0.2227	0.4284	0.6039	0.7421
1	0.8427	0.9103	0.9523	0.9763	0.9891
2	0.9953	0.9981	0.9993	0.9998	0.9999

For analog signals, use is made of Euclidean space, where the Pythagorean theorem applies. Thus, for the magnitude of the vector, also called 'norm' in geometry, one obtains:

$$\|\mathbf{x}\| = \left(\sum_{r=1}^{n} x_r^2 \right)^{1/2} \tag{4.27b}$$

The various components are then the projections onto the different axes of the n-dimensional space, which is called signal space.

The notion of the distance d between two or more signals is of great practical importance. This distance results as a norm of the difference between two signal vectors $\mathbf{x}$ and $\mathbf{y}$ as follows:

$$d(\mathbf{x}, \mathbf{y}) = \|\mathbf{x} - \mathbf{y}\| = \left(\sum_{v=1}^{n} |x_v - y_v|^2 \right)^{1/2} \tag{4.27c}$$

The scalar product

$$\mathbf{x} \cdot \mathbf{y} = \sum_{v=1}^{n} x_v y_v \tag{4.28a}$$

can be used to write the angle α between the two vectors in the following way:

$$\cos \alpha = \frac{\mathbf{x} \cdot \mathbf{y}}{\|\mathbf{x}\| \, \|\mathbf{y}\|} \tag{4.28b}$$

For continuous analog signals $x(t)$, which are defined in the range $a \leqslant t < b$, one can accordingly indicate a norm:

$$\|\mathbf{x}\| = \left(\int_a^b x^2(t) \, dt \right)^{1/2} \tag{4.29a}$$

Physically, it represents the square root of the energy of the signal. A special Hilbert space is thus defined (Blumenthal, 1961). The distance between two signals corresponds to the root mean square error:

$$d(\mathbf{x}, \mathbf{y}) = \|\mathbf{x} - \mathbf{y}\| = \left(\int_a^b |x(t) - y(t)|^2 \, dt \right)^{1/2} \tag{4.29b}$$

It is often used in measurement technology and cybernetics as a measure for the error (see Section 4.3.10).

4.2.6.2   In Non-Euclidean space: Hamming distance

The above representation in the Euclidean space is suitable for analog signals. Use is made of a representation in non-Euclidean signal space for discrete signals whose importance is constantly increasing. In this space, the distance between two vectors $\mathbf{x}$, $\mathbf{y}$ is defined to be the sum of the differences of the individual components (Blumenthal, 1961)

$$d(\mathbf{x}, \mathbf{y}) = \|\mathbf{x} - \mathbf{y}\| = \sum_{v=1}^{n} |x_v - y_v| \qquad (4.30a)$$

according to the norm of this space:

$$\|\mathbf{x}\| = \sum_{v=1}^{n} x_v \qquad (4.30b)$$

The most important discrete signals are binary signals, that is signals in which the individual components can assume the values 1 and 0 only. For such signals, the signal space constitutes a n-dimensional hypercube having the edge length 1, where the various edges are occupied by possible signals only.

Figure 4.18 shows such signal words with one, two, and three bits. Obviously, the representation in non-Euclidean space with the norm according to equation (4.30b) has the advantage that the distance d indicates the number of digits by which two signal words differ from each other. The similarly defined minimum distance in a signal alphabet is called the Hamming distance and constitutes an important characteristic value for the investigation into a system's insensitivity to noise (Peterson, 1962). To investigate distances between signals, use is also made of distance matrices. In the analog-to-digital conversion of signals, a transform between the corresponding signal spaces takes place.

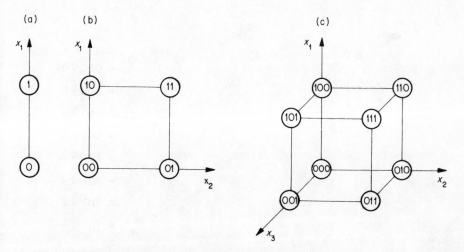

Figure 4.18   Representation of a binary signal in the signal space: (a) one bit; (b) two bits; (c) three bits

4.2.6.3 *Representation as a codegraph*

For an easily understandable explanation of the signal structure, the *codegraph* may be used. Based on the results of the graph theory, the *m* possible states (symbols) of a signal are shown as an assembly according to Figure 4.19. Figure 4.19b shows the application of this type of representation to a binary signal having three digits. Where the individual code words are of the same length, they can be separated one from the other by counting. No characters are required for the separation of the individual words; the code is irreducible. Obviously, this is the case if the end points of the codegraph are occupied by code words only.

4.2.7 Typical Signals

The essential properties of important signals are summarized in Table 4.4. Individual signals have already been presented as examples in the relevant sections. The harmonic oscillation, the unit impulse, and unit step are typical test signals for system identification. White noise (wide-band noise) has been included in the summary as the most important interfering signal. For this signal, there is no amplitude density $\hat{X}(j\omega)$, as has already been discussed. The relationships which have already been pointed out between time and frequency functions as Fourier transforms become evident: the narrower the time signals, the wider the band of the frequency signals with the extreme assignments between the constant behaviour in the frequency domain and the function behaviour in the time domain.

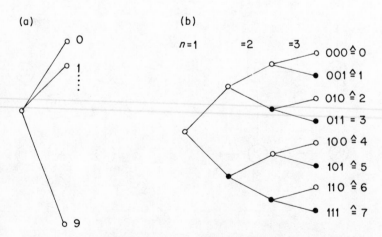

Figure 4.19   Representation as a codegraph: (a) signal with $m = 10$ possible states; (b) binary signal with the length $n = 3$

Table 4.4 Properties of important signals

Characteristic	Time function $x(t)$	Amplitude density $|\hat{X}(j\omega)|$	Power density $S_{xx}(\omega)$	Autocorrelation function $\psi_{xx}(\tau)$	Remarks
Harmonic vibration					Test signal
Unit step $w(t)$					Test signal
Unit pulse $\delta(t)$					Test signal
Pulse					Important for approximations
Ramp function					Important for approximations
White noise					Typical disturbance

Of great importance for measurement technology in particular are estimates and approximate considerations. In contrast to other fields of information engineering, for instance control engineering, where input signals are given and output signals are to be calculated, in measuring technology the opposite has to be done. For this reason, before choosing a suitable measuring device, it is necessary to make an assessment of the signal behaviour to be expected, on the basis of *a priori* information, and then choose the device. To this end, it is usual to approximate the quantity to be measured by either a pulse-shaped or ramp-shaped signal. Therefore, these two signals have also been included in Table 4.4. The pulse duration and the duration of the ramp function, respectively, can be taken from data on the technological process to be investigated, for instance the speed in rotating machines. According to the sampling theorem, a threshold frequency ω_g corresponds to signal duration $2T_1$ (refer to equation (4.11c) of Section 4.2.2.4):

$$\omega_g = \pi/2T_1 \tag{4.30c}$$

Above this frequency there are only spectral oscillations having a relatively low amplitude; the amplitude or power spectrum mainly lies below this threshold frequency (called band-limited signals). To avoid major measurement errors, it is therefore sufficient for the measuring device to cover this frequency domain, i.e. $0 \geqslant \omega > \omega_g$ (see also Section 4.3.7).

4.3   SYSTEMS

4.3.1   Classification of Systems

According to Figure 4.20 a system can be interpreted as a 'black box' with a family of input variables x_r, which can be regarded as a signal vector **x**, and a family of corresponding output variables y_r, forming a vector **y**. The interior of the 'black box' may consist of several elements either electrical, hydraulic, pneumatic or of other energy nature. The overall behaviour of these systems is given by a mathematical equation of the following form:

$$\mathbf{y} = \mathbf{O}\{\mathbf{x}\} \tag{4.31}$$

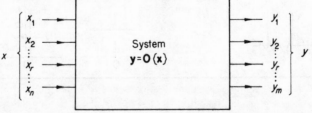

Figure 4.20   Definition of a system

Different systems may have the same overall behaviour, that means the same relation (4.31). This means that a given system may be substituted by another one with the same mathematical equation, for example by a computer of either analog or digital form (an aspect of modelling). Because of the convenient adaptability of computers to the behaviour of a given system by means of programming such modelling methods play an important role in the field of cybernetics. Measurement as a part of cybernetics can be comprehended as a mapping of the input signals on the space of the output signals (Finkelstein, 1976).

Technical systems are characterized as those with active or only passive elements (so called 'active' or 'passive systems') or they are described by the number of ports (two-port, three-port, etc.).

Table 4.5 gives a survey of the classification of technical systems. With respect to the difficulties for the treatment of a system it is of great importance to know if the system is linear or not, because in the linear case the superposition law is valid. In this chapter normally it is assumed that the system is a linear one. Methods of linearization will be treated in Section 4.3.2. Another typical characteristic of a system is whether the parameters describing the behaviour are functions of time or not. Furthermore most of the systems used in cybernetics are so called unidirectional systems, that means that the parameters of the system are independent of those of the following system. We will only deal with systems that fulfill this assumption. Otherwise the results of four-pole theory must be applied (Feldtkeller, 1962).

Table 4.5 Classification of technical systems

System	Linear system	Non-linear system
Mark	Parameters are constant, independent of amplitude. Superposition law is valid	Parameters are a function of amplitude. Superposition law is not valid
Parameters are time invariant	Examples: most of measurement systems with small input levels: amplifiers; filters; transducers	Examples: systems with large input levels: output amplifiers; driver stages. Often linearization is possible
Parameters are functions of time	Examples: controlled amplifiers with multiplicative properties and small input levels: modulators; frequency multipliers; parametric amplifiers	Examples: controlled amplifiers with multiplicative properties and large input levels: modulators, frequency modulators; frequency multipliers

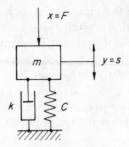

Figure 4.21 Spring–
mass–damper system

As an example of great practical importance in the field of measurement the spring–mass–damper system, as illustrated in Figure 4.21, will be considered. This system is used in numerous sensors, for, among others, the measurement of length or force. If force F is the input variable x, and length s the output variable y, the following special equation corresponding to the general relation (4.31) is obtained:

$$m\ddot{y} + k\dot{y} + cy = x \qquad (4.32)$$

This second-order differential equation describes the system's dynamic behaviour.

4.3.2 Modelling and Linearization

4.3.2.1 *General remarks*

As mentioned above, systems with different interior elements or form of energy can have the same mathematical relationship between output and input variables. From this fact it follows that a given system can be represented by another system having the same overall behaviour. This modelling has the advantage that with the model system the parameters and structures may be changed easily by programming a computer. Furthermore it is possible to observe the input and output quantities in a convenient way by means of oscilloscopes or plotters and to change the scale of coordinates or time axes. Important methods of modelling are analogies, application of block diagrams, and linearization.

4.3.2.2 *Analogies*

Of great importance is the fact that mechanical systems, in the same way as pneumatic or hydraulic and other systems, can be presented by electrical

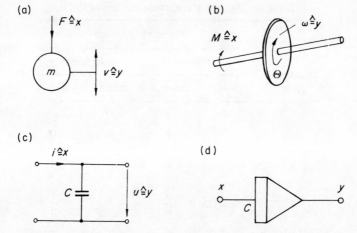

Figure 4.22 Examples from mechanical and electrical systems:
(a) Translation; (b) rotation; (c) capacity; (d) analog computer

systems, as shown in Figure 4.22. In the cases illustrated in this arrangement the
following equations are valid:

(a) In the mechanical example (mass m, velocity v, force F)

$$v = \frac{1}{m} \int F \, dt \qquad (4.33a)$$

(b) In the case of rotatory motion (moment of inertia Θ, angular velocity ω,
torque M)

$$\omega = \frac{1}{\Theta} \int M \, dt \qquad (4.33b)$$

(c) In the electrical system (voltage u, capacitance C, current i)

$$u = \frac{1}{C} \int i \, dt \qquad (4.33c)$$

(d) In the general case of an analog computer (constant c)

$$y = c \int x \, dt \qquad (4.33d)$$

Generalizing the dependences illustrated in Figure 4.22 realizes the survey of
analogies between electrical and mechanical systems shown in Table 4.6. For
more detail refer to Olson (1943), Koenig and Blackwell (1961), and Reichardt
(1960). For the spring–mass–damper system with differential equation (4.32)

Table 4.6 Survey of electromechanical analogies

| Electrical system | Mechanical system | | | |
| | Translation | | Rotation | |
	direct	indirect	direct	indirect
i	F	v	M	ω
u	v	F	ω	M
R	$\dfrac{1}{k}$	k	$\dfrac{1}{k}$	k
C	m	$\dfrac{1}{m}$	θ	$\dfrac{1}{\theta}$
L	$\dfrac{1}{C}$	C	$\dfrac{1}{C}$	C

and the representation of Figure 4.21 the electrical models shown in Figure 4.23 are obtained. Direct analogy (see Table 4.6) yields the parallel circuit Figure 4.23a with the equation

$$C\frac{du}{dt} + \frac{1}{R}u + \frac{1}{L}\int u\, dt = i \tag{4.34a}$$

while indirect analogy provides the series circuit given in Figure 4.23b:

$$\frac{1}{C}\int i\, dt + Ri + L\frac{di}{dt} = u \tag{4.34b}$$

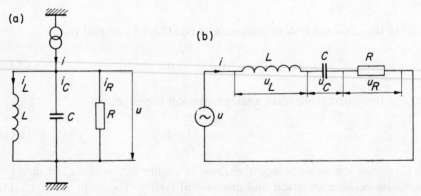

Figure 4.23 Electrical model of the spring–mass–damper system of Figure 4.21:
(a) parallel circuit, (b) series circuit

Table 4.7 Symbols for block diagrams

Function	Branch	Summation	Subtraction	Sign inversion	General system
Symbol					
Equation	$y_1 = y_2 = x$	$y = x_1 + x_2$	$y = x_1 - x_2$	$y = -x$	$y = f(x, t)$

Function	Constant factor	Integrator	Non-linear system	Root calculation	Multiplier
Symbol					
Equation	$y = kx$	$y = \int x \, dt$	$y = f(x)$	$y = \sqrt[n]{x}$	$y = x_1 \cdot x_2$

As can be seen from a comparison of Figures 4.21 and 4.23 the indirect analogy is more convenient because a mechanical parallel circuit is modelled as an electrical parallel circuit.

4.3.2.3 Block diagrams

Measurement systems are built up from particular subsystems. Therefore it is suitable to represent each subsystem by a block including a symbol indicating the operation the subsystem has to realize. Table 4.7 contains some of these symbols and signs used for the demonstration of the interconnections between the systems. (It must be pointed out that many standards for such are in use.) It is supposed that the systems are unidirectional. Otherwise it is possible to describe the behaviour of such a system by means of an interconnection between several unidirectional systems. Figure 4.24 illustrates three typical methods of interconnection of systems. The output–input relation of a system is given by the equation

$$y = Gx \tag{4.35}$$

The frequency response G_e for an equivalent system having the same overall behaviour to that of the interconnection of the subsystems shown is given by:

(a) series circuit (Figure 4.24a)

$$G_e = \prod_{r=1}^{n} G_r \tag{4.36a}$$

(b)   parallel circuit (Figure 4.24b)

$$G_e = \sum_{r=1}^{n} G_r \qquad (4.36b)$$

(c)   in opposition (Figure 4.24c)

$$G_e = \begin{cases} \dfrac{G_1}{1 - G_1 G_2} & \text{feedforward, oscillator} \qquad (4.36c) \\[2ex] \dfrac{G_1}{1 + G_1 G_2} & \text{negative feedback, control} \qquad (4.36d) \end{cases}$$

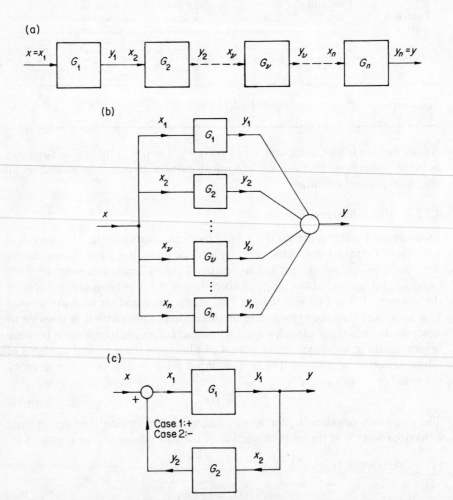

Figure 4.24   Typical system connections: (a) series circuit; (b) parallel circuit;
(c) connection in opposition

The product $G_1 G_2$ in equations (4.36c,d) describes the frequency response of the open control loop.

4.3.2.4 Linearization

Linear systems are distinguished by the validity of the superposition law. Non-linear systems are often linearized to enable the advantages of linear systems to be used. The following preliminary conditions must be fulfilled in such a strategy:

(a) only small deviations of the characteristics from the linear course can be used;
(b) only a relatively small drive range of the non-linear characteristic can be tolerated.

For the linearization the Taylor series expansion of the non-linear characteristic $y = f(x)$ at the working point $y_0 = f(x_0)$ is employed. Writing only the deviations from the working point Δx, Δy yields

$$y - y_0 = \Delta y = \left.\frac{\partial f}{\partial x}\right|_{x_0} \Delta x + \frac{1}{2}\left.\frac{\partial^2 f}{\partial x^2}\right|_{x_0} \Delta x^2 + \frac{1}{6}\left.\frac{\partial^3 f}{\partial x^3}\right|_{x_0} \Delta x^3 + \cdots. \qquad (4.37a)$$

In practical measurement technique the input variable is often the sinusoidal function $x = \hat{X} \sin(\omega t)$. The proportion of the dominant wave with frequency ω at the output $\hat{Y}_\omega$, expressed as a ratio with the amplitude of the input $\hat{X}$, gives

$$\frac{\hat{Y}_\omega}{\hat{X}} = \left.\frac{\partial f}{\partial x}\right|_{x_0} \left(1 + \frac{1}{8}\frac{\partial^3 f/\partial x^3|_{x_0}}{\partial f/\partial x|_{x_0}} \hat{X}^2\right) \qquad (4.37b)$$

This function is the describing function; it expresses the frequency response of a non-linear system. Depending on the sign of the third-order differential coefficient, the describing function either increases or decreases with the square of the amplitude of the input as Figure 4.25 shows. In practice the case $\partial^3 f/\partial x^3|_{x_0} > 0$ can be troublesome because the amplification factor is increasing with increasing amplitude leading to an unstable oscillating régime (Woschni, 1973).

Important in the field of measurement is the rectification effect using a square-wave characteristic

$$\Delta y_0 = \frac{1}{4}\left.\frac{\partial^2 f}{\partial x^2}\right|_{x_0} \hat{X}^2 \qquad (4.37c)$$

In this case distortion appears that is described by the distortion factors

$$k_2 = \frac{\hat{Y}_{2\omega}}{\hat{Y}_\omega} = \frac{1}{4}\left(\frac{\partial^2 f/\partial x^2|_{x_0}}{\partial f/\partial x|_{x_0}}\right)\hat{X}, \qquad (4.37d)$$

$$k_3 = \frac{\hat{Y}_{3\omega}}{\hat{Y}_\omega} = \frac{1}{24}\left(\frac{\partial^3 f/\partial x^3|_{x_0}}{\partial f/\partial x|_{x_0}}\right)\hat{X}^2 \qquad (4.37e)$$

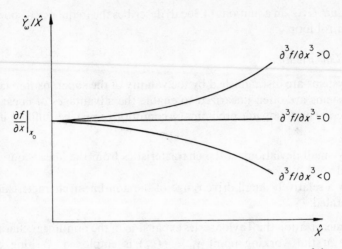

Figure 4.25   Ratio of the amplitude of the dominant wave of the output $\hat{Y}_\omega$ to the amplitude of the input $\hat{X}$ (describing function)

4.3.3   Systems in the Time Domain

4.3.3.1   *Description by differential equations*

The oldest method used in solving problems of system analysis is the method of differential equations. A linear system is described by a linear differential equation of the following form:

$$a_n y^{(n)} + a_{n-1} y^{(n-1)} + \cdots + a_r y^{(r)} + \cdots + a_2 \ddot{y} + a_1 \dot{y} + a_0 y$$
$$= b_0 x + b_1 \dot{x} + \cdots + b_m x^{(m)} \tag{4.38a}$$

This equation may be written

$$T_n^n y^{(n)} + T_{n-1}^{n-1} y^{(n-1)} + \cdots + T_r^r y^{(r)} + \cdots + T_2^2 \ddot{y} + T_1 \dot{y} + y$$
$$= G_0 x + \frac{b_1}{a_0} \dot{x} + \cdots + \frac{b_m}{a_0} x^{(m)} \tag{4.38b}$$

where G_0 is the static sensitivity:

$$G_0 = \frac{b_0}{a_0} = \frac{\Delta y}{\Delta x} \tag{4.38c}$$

G_0 can be measured as an amplification factor by means of a small alteration Δx as stated in equation (4.38c). The coefficients

$$T_r = \sqrt[r]{(a_r/a_0)} \tag{4.38d}$$

are time constants. For a differential equation of nth order n time constants are defined, the greatest of which is used to estimate the duration of the transient function. The solution of the differential equation consists of two additive components, a stationary and a dynamic portion:

$$y(t) = y_{st} + y_d(t) \tag{4.39a}$$

To solve the homogeneous differential equation leading to the dynamic portion $y_d(t)$, the assumption

$$y_d = C e^{pt} \tag{4.39b}$$

is used. The zero points of the characteristic equation

$$T_n^n p^n + T_{n-1}^{n-1} p^{p-1} + \cdots + T_2^2 p^2 + T_1 p + 1 = 0 \tag{4.39c}$$

signify whether the solution is stable or not. In particular these so called eigenvalues p_r prove that

(a)   if the real part is less than zero, i.e. $Re(p_r) < 0$, a stable solution exists;
(b)   if the real part is greater than zero, i.e. $Re(p_r) > 0$, an unstable solution exists;
(c)   if the eigenvalues are complex, oscillations with decreasing or increasing amplitudes exist.

With the eigenvalues p_r the dynamic solution yields (Coddington and Levinson, 1955)

$$y_d = \sum_{r=1}^{n} C_r \exp(p_r t) \tag{4.39d}$$

If a double root p_0 arises

$$y_d = (C_1 + C_2 t) \exp(p_0 t) \tag{4.39e}$$

The stationary solution y_{st} is to be found by means of suitable terms satisfying the inhomogeneous differential equation (4.38b).

As may be seen from equation (4.39d) the eigenvalue p_r corresponds to a time constant $T_r = 1/p_r$, the greatest value of which (T_{max}) is responsible for the duration of the transient process. Because $e^{-3} = \frac{1}{20} \triangleq 5\%$ the transient process approximately continues and

$$t_{tr} = 3T_{max} \tag{4.39f}$$

where t_{tr} is the transient time.

4.3.3.2   Basics of state space description

The basis of the state space description is the classical differential equation discussed above. The state of a system is described by a set of state variables y_r. The number of these state variables agrees with the degree of the differential equation. Today the method is of growing importance because of its aptitude for computer simulation of systems. The nth-order differential equation given for computer simulation of systems.

The nth-order differential equation given by (4.38b) may be transformed into a system of n differential equations of first order:

$$
\begin{aligned}
y &= y_1 \\
\dot{y} &= y_2 = \dot{y}_1 \\
\ddot{y} &= y_3 = \dot{y}_2 \\
&\;\;\vdots \\
y^{(n-1)} &= y_n = \dot{y}_{n-1}
\end{aligned}
\tag{4.40a}
$$

$$
y_n = -\frac{1}{T_n^n} y_1 - \frac{T_1}{T_n^n} y_2 - \frac{T_2^2}{T_n^n} y_3 - \cdots - \frac{T_{n-1}^{n-1}}{T_n^n} y_n + \frac{1}{T_n^n} x
$$

This system of equations can be written in the form of a vector differential equation (Zadeh and Desoer, 1963):

$$
\frac{d}{dt}
\begin{bmatrix} y_1 \\ y_2 \\ y_3 \\ \vdots \\ y_n \end{bmatrix}
=
\begin{bmatrix} \dot{y}_1 \\ \dot{y}_2 \\ \dot{y}_3 \\ \vdots \\ \dot{y}_n \end{bmatrix}
=
\begin{bmatrix}
0 & 1 & 0 & 0 & \cdots & 0 \\
0 & 0 & 1 & 0 & \cdots & 0 \\
0 & 0 & 0 & 1 & \cdots & 0 \\
\vdots & \vdots & \vdots & \vdots & \cdots & \vdots \\
-\dfrac{1}{T_n^n} & -\dfrac{T_1}{T_n^n} & -\dfrac{T_2^2}{T_n^n} & -\dfrac{T_3^3}{T_n^n} \cdots & -\dfrac{T_{n-1}^{n-1}}{T_n^n}
\end{bmatrix}
\begin{bmatrix} y_1 \\ y_2 \\ y_3 \\ \vdots \\ y_n \end{bmatrix}
+
\begin{bmatrix} 0 \\ 0 \\ 0 \\ \vdots \\ \dfrac{1}{T_n^n} \end{bmatrix} x
\tag{4.40b}
$$

or with the abridgements $\mathbf{y}$, $\mathbf{T}$, $\mathbf{b}$, and $\mathbf{x}$:

$$
\dot{\mathbf{y}} = \mathbf{T}\mathbf{y} + \mathbf{b}\mathbf{x}
\tag{4.40c}
$$

Consider as an example the spring–mass–damper system of Figure 4.21 with the differential equation (4.32). The vector equation is given by

$$
\begin{bmatrix} \dot{y}_1 \\ \dot{y}_2 \end{bmatrix}
=
\begin{bmatrix} 0 & 1 \\ -\dfrac{c}{m} & -\dfrac{k}{m} \end{bmatrix}
\cdot
\begin{bmatrix} y_1 \\ y_2 \end{bmatrix}
+
\begin{bmatrix} 0 \\ \dfrac{1}{m} \end{bmatrix} x
\tag{4.40d}
$$

The state variables are the displacement $y = y_1$ of the mass and the velocity $\dot{y} = y_2$. Figures 4.26 and 4.27 respectively show the programming of analog and digital computers for modelling this system. The relationships between programming and state space description are easy to recognize.

4.3.3.3 Step response function, pulse response function

For system description and identification certain response (output, answer) functions for test signals at the input are generally used. These are presented

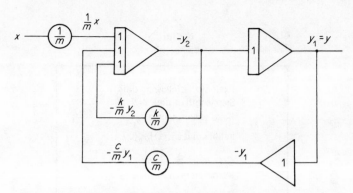

Figure 4.26   Direct programming of a spring–mass–damper system
on an analog computer

here as time-dependent functions but, as is to be seen in Section 6.3.2, they can
be used with any general variable x, spatial distribution being one used in optical
systems.

(a)   The response to a step function with a step amplitude of 1 (unit step func-
      tion, $w(t)$) is the unit step response or transient response $h(t)$ as illustrated
      in Figure 4.28a.
(b)   The response to a pulse function with an integral value of 1 (Dirac delta
      function, $\delta(t)$) yields the unit pulse response or weighting function $g(t)$
      shown in Figure 4.28b.

Because of the linearity of the system the response function $y(t)$ is to be divided
by the step amplitude or the integral value respectively to get the normalized
function. In the case of system identification if the input function generated by a
signal generator is not the ideal function but a function with a rise time $\Delta\tau$ as
signified in Figure 4.28, the condition must be fulfilled that the rise time $\Delta\tau$, or
the pulse width $\Delta\tau$, are very much smaller than the transient time t_{tr} of the system
(Woschni, 1973).

   As can be seen from the comparison of parts of Figure 4.28 the pulse function
is connected with the step function by means of a differentiation, that is, in the
sense of distribution (Gelfand and Schilow, 1960),

$$\mathrm{d}\omega(t)/\mathrm{d}t = \delta(t) \tag{4.41a}$$

For linear systems it is immaterial whether the differentiation is realized at the
input or output side of the system, that means

$$h(t) = \int g(t)\,\mathrm{d}t \tag{4.41b}$$

It is only a matter of suitability, whether the transient or weighting functions are
used for identification. For a system with first-order delay, as is used for the

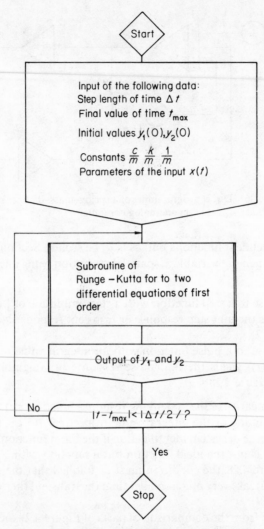

Figure 4.27   Flow chart for the programming
of a spring–mass–damper system on a digitial
computer.

approximation of several systems in measurement technology, we gain the
differential equation

$$T_1 \dot{y} + y = x \qquad (4.41c)$$

with eigenvalue

$$p_1 = -1/T_1 \qquad (4.41d)$$

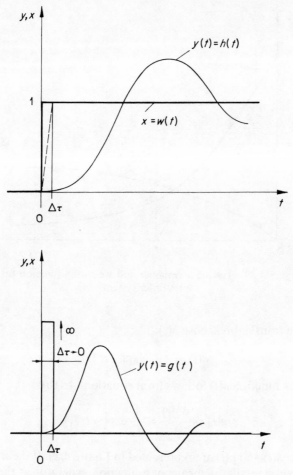

Figure 4.28 (a) Unit step $w(t)$ and transient response $h(t)$;
(b) delta function $\delta(t)$ and weighting function $g(t)$

The stationary solution may be found by means of the assumption

$$y_{\text{st}} = c_2$$

and the total solution yields

$$y = y_{\text{st}} + y_{\text{d}} = c_2 + c_1 \exp(-t/T_1)$$

Using the boundary conditions

$$y|_{t=-\infty} = 0 \quad y|_{t=\infty} = 1$$

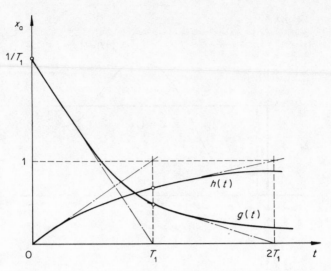

Figure 4.29   Transient response and weighting function for a
first-order system

we get for the transient response $h(t)$:

$$h(t) = 1 - \exp(-t/T_1) \tag{4.41e}$$

The weighting function $g(t)$ follows from equation (4.41b):

$$g(t) = \frac{dh(t)}{dt} = \frac{1}{T_1} \exp(-t/T_1) \tag{4.41f}$$

The functions $h(t)$ and $g(t)$ are represented in Figure 4.29. If these functions are obtained experimentally by means of a function generator at the input of the system examined the time constant T_1 is given by the length of the subtangent (Figure 4.29). Furthermore the figure shows the transient time t_{tr} to be nominally three times the time constant T, for $e^{-3} = \frac{1}{20}$. For more details referring to testing of systems and important examples see Sections 4.3.8 and 4.3.9.

4.3.3.4   *Generalized response functions: convolution*

Transient response and weighting functions are response functions to special input signals. In the general case the input function is broken down into a series of weighted Dirac delta functions, which are time-delayed as represented in Figure 4.30. The pulse at the time τ_1 yields the output

$$x(\tau_1)\Delta\tau \cdot g(t_1 - \tau_1)$$

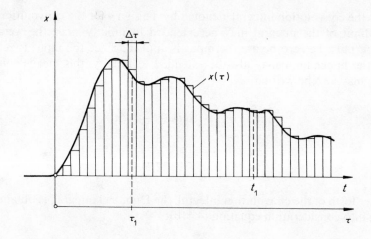

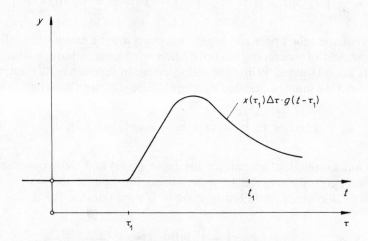

Figure 4.30   Explanation of the convolution integral

Because of the linearity of the system the superposition law is valid, that means the entire output is the sum (integral) of all inputs at the time $t - \tau_1 > 0$:

$$y(t) = \int_0^t x(\tau)g(t - \tau)\, d\tau$$

$$= \int_0^t x(t - \tau)g(\tau)\, d\tau$$

$$= x(t) * g(t) \qquad (4.42a)$$

This is the convolution integral, denoted by the sign $*$ for the convolution. The lower limit of the integral may be extended to infinity since the weighting function must be zero before the input is applied. Since the input is zero for $t < 0$ the upper limit may also be extended to infinity, this means equation (4.42a) may also be written

$$y(t) = \int_{-\infty}^{+\infty} x(\tau)g(t - \tau)\, d\tau$$

$$= \int_{-\infty}^{+\infty} x(t - \tau)g(\tau)\, d\tau \qquad (4.42b)$$

Another form of the convolution integral, the Duhamel integral, is obtained by taking into consideration equation (4.41b):

$$y(t) = \frac{d}{dt} \int_0^t x(\tau)h(t - \tau)\, d\tau = \frac{d}{dt} \int_0^t x(t - \tau)h(\tau)\, d\tau = \frac{d}{dt}[x(t) * h(t)] \quad (4.42c)$$

In this equation the upper and lower limits may also be extended to infinity.

In the field of measurement convolution is of great importance for system identification (Davies, 1970). The autocorrelation function at the output of a system and the input are related by a double convolution (Woschni, 1981)

$$\psi_{yy}(\tau) = \int_0^{\infty} \int_0^{\infty} \psi_{xx}(\tau + \tau_1 - \tau_2)g(\tau_1)g(\tau_2)\, d\tau_1\, d\tau_2 \qquad (4.43a)$$

If the autocorrelation function of the input $\psi_{xx}(\tau)$ and the cross-correlation function $\tau_{xy}(\tau)$ are measured the weighting function of the system can be calculated, for which the following relation is valid (Davies, 1970):

$$\psi_{xy}(\tau) = \int_0^{\infty} g(t)\psi_{xx}(\tau - t)\, dt \qquad (4.43b)$$

Deconvoluting equation (4.43b) gives the required pulse response function $g(t)$; the system may be regarded as being identified. Direct deconvolution techniques using equation (4.43b) can be difficult. Therefore methods in the frequency domain were developed as shown in the following sections. In the special case of white noise with an autocorrelation function $\psi_{xx}(\tau - t) = 2\pi S_0 \delta(\tau - t)$ the deconvolution degenerates to the equation

$$\psi_{xy}(\tau) = 2\pi S_0 g(\tau) \qquad (4.43c)$$

This means that the weighting function corresponds directly to the cross-correlation function (Woschni, 1973).

4.3.4 Systems in the Frequency Domain

4.3.4.1 *Frequency response: logarithmic characteristics*

Using a sinusoidal input and taking up both amplitude of the output $\hat{Y}$ normalized to the input $\hat{X}$ and phase φ as a function of frequency ω provides the frequency response for which

$$\text{amplitude characteristic is } |G(j\omega)| = \hat{Y}/\hat{X} \qquad (4.44a)$$

$$\text{phase characteristic is } \varphi(\omega) \measuredangle y, x \qquad (4.44b)$$

In the complex presentation the input $x = \hat{X}\,e^{j\omega t}$ yields the output $y = \hat{Y}\,e^{j(\omega t + \varphi)}$ $= \hat{Y}\,e^{j\varphi}\,e^{j\omega t}$, that means the complex frequency response is

$$\frac{\hat{Y}}{\hat{X}}\,e^{j\varphi} = G(j\omega) = |G(j\omega)|e^{j\varphi(\omega)} \qquad (4.44c)$$

This function is represented in the complex plane as a locus diagram. The complex frequency response may be split into a real and an imaginary part:

$$G(j\omega) = P(\omega) + jQ(\omega) \qquad (4.44d)$$

with the relations

$$|G(j\omega)| = \sqrt{[P^2(\omega) + Q^2(\omega)]} \qquad (4.44e)$$

$$\varphi(\omega) = \arctan\left(\frac{Q(\omega)}{P(\omega)}\right) \qquad (4.44f)$$

Figure 4.31 explains the representation of the frequency characteristics. If the differential equation is given it is very convenient to obtain the frequency response by means of the terms

$$x^{(\mu)} = (j\omega)^{\mu}\hat{X}\,e^{j\omega t}; \quad y^{(r)} = (j\omega)^r\hat{Y}\,e^{j\varphi}\,e^{j\omega t} \qquad (4.45a)$$

Substituting the differential equation (4.38b) and solving the output–input relation yields the complex frequency response

$$G(j\omega) = \frac{\hat{y}\,e^{j\varphi}}{\hat{X}} = \frac{G_0 + j\omega(b_1/a_0) + \cdots + (j\omega)^m(b_m/a_0)}{1 + j\omega T_1 + \cdots + (j\omega)^n T_n^n} \qquad (4.45b)$$

This implies that substitution of nth-order differentiation by $(j\omega)^n$ and nth-order integration by $(1/j\omega)^n$ will yield the output–input relation.

The frequency response can be measured, by way of equations (4.44a,b), using a sinusoidal test signal. If the input is a stochastic signal with power density $S_{xx}(\omega)$ the output power density $S_{yy}(\omega)$ may be calculated according to (Davies, 1970)

$$S_{yy}(\omega) = |G(j\omega)|^2 S_{xx}(\omega) \qquad (4.45c)$$

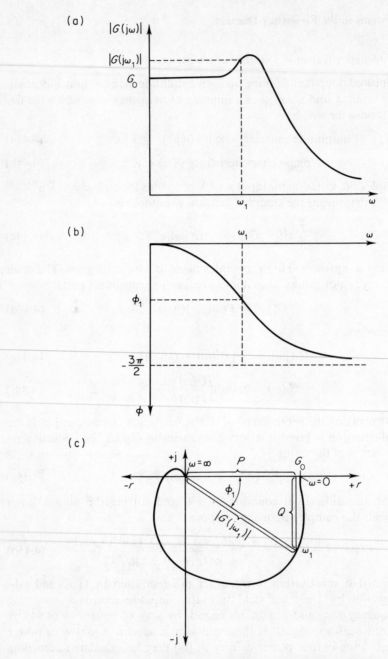

Figure 4.31 Graphical representation of frequency characteristics: (a) amplitude characteristic; (b) phase characteristic; (c) locus diagram

The double convolution (equation (4.43a)) in the time domain corresponds with a multiplication with $|G(j\omega)|^2$ in the frequency domain (Woschni, 1973).

Of great practical importance is the plotting of the amplitude characteristic in a double-logarithmic calibration graph and the phase characteristic with a linear φ-axis and a logarithmic ω-axis, known as logarithmic frequency characteristics. The amplitude is generally measured in decibel (dB) units,

$$20 \log(\hat{Y}/\hat{X}) = 20 \log|G(j\omega)| \tag{4.45d}$$

in order that a linear scale can be used for the y-axis (Bode diagram). The advantage of this method is illustrated in Figure 4.32. As treated in Section 4.3.2.3 the overall behaviour of series-connected systems is given by the multiplication of the frequency responses of these systems (equation (4.36a)). Because of the logarithmic representation the multiplication is simplified to a summation, which can easily be realized graphically.

For a system with first-order delay, used for approximation of more complicated systems, from the differential equation (4.41c) of Section 4.3.3.3 we get

$$G(j\omega) = 1/(1 + j\omega T_1) \tag{4.46a}$$

$$|G(j\omega)| = 1/\sqrt{1 + \omega^2 T_1^2} \tag{4.46b}$$

$$\varphi(\omega) = -\arctan(\omega T_1) \tag{4.46c}$$

The frequency response functions are featured in Figure 4.33. From Figure 4.33b results a critical frequency f_c (or ω_c) given by

$$2\pi f_c = \omega_c = 1/T_1 \tag{4.46d}$$

used for approximations (Woschni, 1973).

Important examples of measurement systems are discussed in Section 4.3.9.

4.3.4.2 Transfer function

A generalization of the complex frequency response arises if the frequency $j\omega$ is extended to the complex frequency

$$p = j\omega + \delta \tag{4.47a}$$

with the increase constant δ.

This complex frequency p is the same as the variable p (or s in the mathematical literature) used with Laplace transformation (see Section 4.3.5). In the physical sense p means a harmonic oscillation with an exponential increasing or decreasing amplitude (Woschni, 1973):

$$x = \hat{X} e^{pt} = \hat{X} e^{\delta t} e^{j\omega t} \tag{4.47b}$$

If p is represented in the p-plane the left-hand side of this plane signifies stable solutions ($\delta < 0$) while the right-hand side leads to unstable solutions ($\delta > 0$).

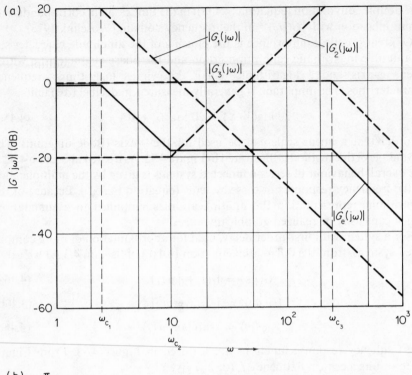

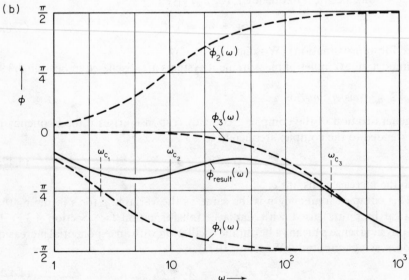

Figure 4.32   Graphical determination, through addition, of the frequency responses
of series-connected systems (Woschni, 1981)

(a)

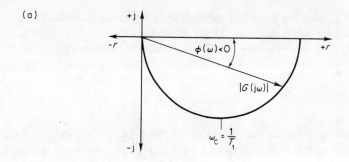

(b)

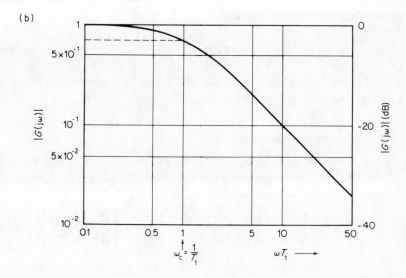

(c)

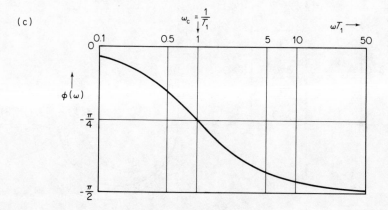

Figure 4.33   (a) Locus diagram; (b) amplitude characteristic and (c) phase characteristic for a first-order system

This criterion is used for assessing the stability of systems as treated in Section 4.3.6. The complex frequency p used instead of ω results in the transfer function $G(p)$ for the system described by equations (4.38b) or (4.45b):

$$G(p) = \frac{\dfrac{b_0}{a_0} + p\dfrac{b_1}{a_0} + \cdots + p^m\dfrac{b_m}{a_0}}{1 + pT_1 + \cdots + p^nT_n^n} = \frac{\sum_{r=0}^{m}\dfrac{b_r}{a_0}p^r}{\sum_{\mu=0}^{n} T_\mu^\mu p^\mu} \qquad (4.47c)$$

The $G(p)$-plane is a conformal mapping of the p-plane (Zadeh and Desoer, 1963), that means the side-directions of the curves remain valid, as the next example shows. The system with first-order delay with the frequency responses (4.46a,b,c) has the transfer function

$$G(p) = \frac{1}{1 + pT_1} \qquad (4.47d)$$

given in Figure 4.34 with several values of δ. It can be observed that Figure 4.33 is a special case included in Figure 4.34.

4.3.4.3 Pole–zero configuration

By means of searching the zero points of both the numerator p_r^* and the divisor p_μ of the fraction (4.47c) one gains the equivalent product representation, the polynomial equation alternative of equation (4.47c)

$$G(p) = c\,\frac{(p - p_1^*)(p - p_2^*)\cdots(p - p_m^*)}{(p - p_1)(p - p_2)\cdots(p - p_n)} = c\,\frac{\prod_{r=1}^{m}(p - p_r^*)}{\prod_{\mu=1}^{n}(p - p_\mu)} \qquad (4.48a)$$

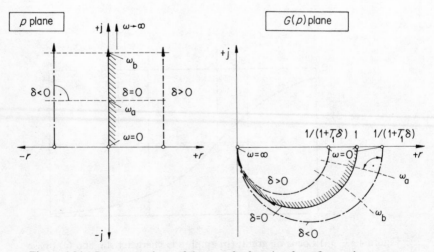

Figure 4.34   Representations of the transfer function for a first-order system

having the poles p_μ and the zeros p_r^* of the transfer function. Therefore the properties of a system are described completely, up to a constant c, by the position of poles and zeros, represented in the pole–zero plane (Figure 4.35). The poles agree with the eigenvalues of the differential equation.

The frequency response can be calculated from the pole–zero plane by

$$G(j\omega) = c \frac{\prod_{r=1}^{m} (j\omega - p_r^*)}{\prod_{\mu=1}^{n} (j\omega - p_\mu)} \tag{4.48b}$$

Figure 4.35 shows how, for series connections, the pole–zero representation of a complicated system can be split into a sum of simpler systems. Poles and zeros at the same point cancel each other; this feature is used for the correction of systems by means of additional series-connected correcting elements (Section 4.3.10). The position of the poles contains information as to whether the system is stable or not: existence of poles in the right half-plane signifies instability

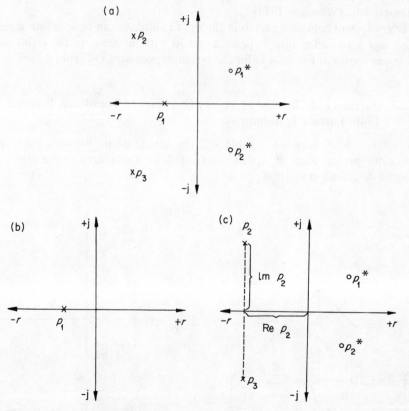

Figure 4.35   (a) Splitting up the pole–zero plane of a system; (b, c) into series-connected subsystems ($\times$ poles; $\bigcirc$ zeros)

(Section 4.3.6). In measurement the phase shift bridge (Figure 4.36a) is often used. It has the transfer function

$$G(p) = \frac{1}{2} \cdot \frac{1 - pCR}{1 + pCR}$$

and the pole and zero at

$$p_0 = -\frac{1}{RC}; \quad p_0^* = \frac{1}{RC}.$$

The pole–zero plane (Figure 4.36b) shows a symmetrical position of pole and zero referred to the imaginary axis. This configuration of poles and zeros is typically for all-pass circuits having a constant amplitude characteristic and a frequency-dependent phase characteristic as follows directly from equation (4.48b). All-pass systems play an important role for the correction of the phase characteristic (Woschni, 1981).

Every system containing zeros in the right half-plane can be split into an all-pass and a so called minimal-phase system without zeros in the right half-plane, as shown in Figure 4.37 for the system represented in Figure 4.35c.

4.3.5 Relevance of Time and Frequency Domains and Transforms Between Both: Laplace Transform

In Section 4.2.4, especially Table 4.2, the relationships between time and frequency presentation of signals are treated. In a similar way the relations may be described in systems.

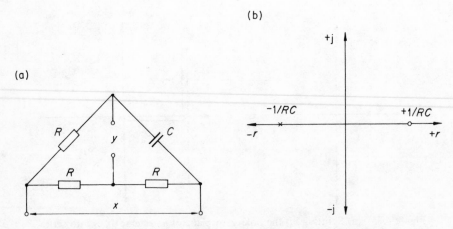

Figure 4.36   Phase shift bridge: (a) circuit; (b) pole–zero plane diagram

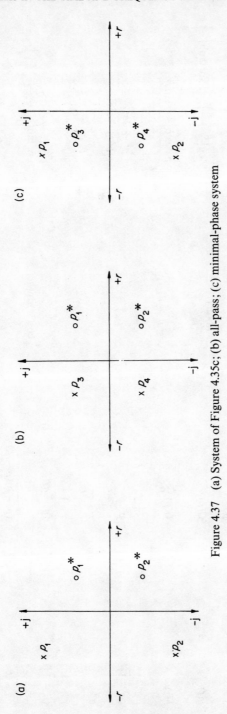

Figure 4.37 (a) System of Figure 4.35c; (b) all-pass; (c) minimal-phase system

Table 4.8   Laplace transforms of time-dependent functions (Woschni and Kraus, 1976)

Time function, $f(t)$	Laplace transform, $F(p)$	Time function, $f(t)$	Laplace transform, $F(p)$
0	0	$\dfrac{t}{2}\sinh(at)$	$\dfrac{ap}{(p^2-a^2)^2}$
$\delta(t)$	1	$\dfrac{\sin(at)}{t}$	$\arctan\left(\dfrac{a}{p}\right)$
$w(t)$	$\dfrac{1}{p}$	$\dfrac{2\sinh(at)}{t}$	$\log\left(\dfrac{p+a}{p-a}\right)=2\,\mathrm{arc\,tanh}\left(\dfrac{a}{p}\right)$
$g(t)$	$G(p)$	$\cos(at+b)$	$\dfrac{p\cos b - a\sin b}{p^2+a^2}$
$h(t)$	$G(p)\dfrac{1}{p}$	$\cos^2(at)$	$\dfrac{p^2+2a^2}{p(p^2+4a^2)}$
$\dfrac{t^{n-1}}{(n-1)!}$	$\dfrac{1}{p^n}$	$\cosh^2(at)$	$\dfrac{p^2+2a^2}{p(p^2-4a^2)}$
e^{at}	$\dfrac{1}{p-a}$	$\dfrac{t}{2}\cos(bt)\sin(at)$	$\arctan\left(\dfrac{2ap}{p^2-a^2+b^2}\right)$
$\dfrac{t^{n-1}}{(n-1)!}\,\mathrm{e}^{at}$	$\dfrac{1}{(p-a)^n}$	$2\,\dfrac{\cos(bt)-\cos(at)}{t}$	$\log\left(\dfrac{p^2+a^2}{p^2+b^2}\right)$
$\dfrac{\mathrm{e}^{at}-\mathrm{e}^{bt}}{a-b}$	$\dfrac{1}{(p-b)(p-a)}$	$2\,\dfrac{\cosh(bt)-\cosh(at)}{t}$	$\log\left(\dfrac{p^2-a^2}{p^2-b^2}\right)$
$\dfrac{a\mathrm{e}^{at}-b\mathrm{e}^{bt}}{a-b}$	$\dfrac{p}{(p-b)(p-a)}$	$\dfrac{a\sin(bt)-b\sin(at)}{ab(a^2-b^2)}$	$\dfrac{1}{(p^2+a^2)(p^2+b^2)}$
$\sin(at)$	$\dfrac{a}{p^2+a^2}$		
$t\sin(at)$	$\dfrac{2ap}{(p^2+a^2)^2}$		

$f(t)$	$F(p)$
$e^{-\lambda t}\sin(at)$	$\dfrac{a}{(p+\lambda)^2 + a^2}$
$\sinh(at)$	$\dfrac{a}{p^2 - a^2}$
$e^{\lambda t}\sinh(at)$	$\dfrac{a}{(p-\lambda)^2 - a^2}$
$\cos(at)$	$\dfrac{p}{p^2 + a^2}$
$t\cos(at)$	$\dfrac{p^2 - a^2}{(p^2 + a^2)^2}$
$e^{-\lambda t}\cos(at)$	$\dfrac{p+\lambda}{(p+\lambda)^2 + a^2}$
$\cosh(at)$	$\dfrac{p}{p^2 - a^2}$
$e^{\lambda t}\cosh(at)$	$\dfrac{p-\lambda}{(p-\lambda)^2 - a^2}$
$\sin(at + b)$	$\dfrac{a\cos b + p\sin b}{p^2 + a^2}$
$\sin^2(at)$	$\dfrac{2a^2}{p(p^2 + 4a^2)}$
$\sinh^2(at)$	$\dfrac{2a^2}{p(p^2 - 4a^2)}$
$\dfrac{t}{2}\sin(at)$	$\dfrac{ap}{(p^2 + a^2)^2}$

$f(t)$	$F(p)$
$\dfrac{\cos(bt) - \cos(at)}{a^2 - b^2}$	$\dfrac{p}{(p^2 + a^2)(p^2 + b^2)}$
$\sin(at)\sinh(at)$	$\dfrac{2a^2 p}{p^4 + 4a^4}$
$\sin(at)\cosh(at)$	$\dfrac{a(p^2 + 2a^2)}{p^4 + 4a^4}$
$\cos(at)\sinh(at)$	$\dfrac{a(p^2 - 2a^2)}{p^4 + 4a^4}$
$\cos(at)\cosh(at)$	$\dfrac{p^3}{p^4 + 4a^4}$
$\dfrac{b\sinh(at) - a\sinh(bt)}{a^2 - b^2}$	$\dfrac{ab}{(p^2 - a^2)(p^2 - b^2)}$
$\dfrac{\cosh(at) - \cosh(bt)}{a^2 - b^2}$	$\dfrac{p}{(p^2 - a^2)(p^2 - b^2)}$
$\dfrac{1}{\sqrt{(\pi t)}}$	$\dfrac{1}{\sqrt{p}}$
$2\sqrt{(t/\pi)}$	$\dfrac{1}{p\sqrt{p}}$
e^{t-a}	$e^{-ap}\dfrac{1}{p-1}$
$\dfrac{1}{\sqrt{(\pi t)}}e^{-at}$	$\dfrac{1}{\sqrt{(p+a)}}$
$\dfrac{e^{-at}}{\sqrt{(b^2 - a^2)}}\sin[\sqrt{(b^2 - a^2 t)}]$	$\dfrac{1}{p^2 + 2ap + b^2}$
$\dfrac{1}{t}(e^{bt} - e^{at})$	$\log\!\left(\dfrac{p - a}{p - b}\right)$

The basic idea of the calculation of the time functions, i.e., $g(t)$ or $h(t)$, is as follows. Both the spectral function $\hat{X}(j\omega)$ of the input and the frequency response of the system $G(j\omega)$ or generalized) $G(p)$ are given. It is then possible to derive the output caused by any of the several sinusoidal input spectral oscillations by multiplying the complex spectral density $\hat{X}(j\omega)$ with the frequency response then applying summation (integration) of all frequency components. This is valid because the superposition law applies. In system theory the Laplace transform is preferred to the Fourier transform because it converges more quickly. It can be derived by substituting $j\omega \rightarrow p$ in the Fourier transform equation (4.28)

$$F(p) = \int_0^\infty f(t)e^{-pt}\,dt = L\{f(t)\} \tag{4.49a}$$

$$f(t) = \frac{1}{2\pi j}\int_{c-j\infty}^{c+j\infty} F(p)e^{pt}\,dp = L^{-1}\{F(p)\} \tag{4.49b}$$

In system theory this so called one-sided Laplace transform is made use of, for only the region $t \geq 0$ is interesting. For solving optical problems the two-dimensional Laplace transform or Fourier transform is applied (Goodman, 1968). The convergence abscissa c in equation (4.49b) is chosen in such a way that the poles remain to the left of this abscissa. Due to the residue theorem (Kaplan, 1962),

$$\int_{c-j\infty}^{c+j\infty} F(p)e^{pt}\,dp = 2\pi j \sum \text{Res}(p_r) \tag{4.50a}$$

and for a pole of nth order

$$\text{Res}(p_0) = \frac{1}{(n-1)!}\lim_{p\to p_0}\frac{d^{n-1}}{dp^{n-1}}[F(p)e^{pt}(p-p_0)^n] \tag{4.50b}$$

Table 4.8 provides (and later in Figure 6.2) Laplace transforms for commonly met time functions. Table 4.9 gives the Laplace transforms of test signals (Woschni and Kraus, 1976). Table 4.10 surveys theorems of the Laplace transform. Physical considerations lead to the following relationships between output and input (Woschni, 1973):

$$L\{y(t)\} = L\{x(t)\}G(p) \qquad y(t) = L^{-1}\{L\{x(t)\}G(p)\} \tag{4.51a}$$

This equation yields for the transfer function $G(p)$:

$$G(p) = \frac{L\{y(t)\}}{L\{x(t)\}} = \frac{\hat{Y}(p)}{X(p)} \tag{4.51b}$$

Table 4.9 Laplace transforms of test signals (Woschni and Kraus, 1976)

Aperiodic test signal function	Equation	Laplace transform
	(Dirac) pulse with integral value $F_\perp$ $x(t) = F_\perp \delta(t); \; \delta(t) = \begin{cases} \infty & \text{for } t = 0 \\ 0 & \text{for } t \neq 0 \end{cases}$	$F_\perp(p) = F_\perp$
	step function $x(t) = Aw(t); \; w(t) = \begin{cases} 0 & \text{for } t < 0 \\ 1 & \text{for } t > 0 \end{cases}$	$F_\Gamma(p) = \dfrac{A}{p}$
	ramp function $x(t) = \begin{cases} 0 & \text{for } t \leqslant 0 \\ At & \text{for } t \geqslant 0 \end{cases}$	$E_{\diagup}(p) = \dfrac{A}{p^2}$
	rectangular pulse $x(t) = \begin{cases} 0 & \text{for } t < 0, t > T \\ A & \text{for } 0 < t < T \end{cases}$	$E_{\sqcap}(p) = A\,\dfrac{1 - e^{-pT}}{p}$

(continued)

Table 4.9   (continued)

Aperiodic test signal function	Equation	Laplace transform
triangle pulse	$$x(t) = \begin{cases} 0 & \text{for } t \leq 0,\, t \geq T \\[2mm] \dfrac{2A}{T}\,t & \text{for } 0 \leq t \leq \dfrac{T}{2} \\[2mm] 2A\left(1 - \dfrac{t}{T}\right) & \text{for } \dfrac{T}{2} \leq t \leq T \end{cases}$$	$$E_\wedge(p) = \frac{2A}{p}\left(\frac{\left(1 - e^{-p(T/2)}\right)^2}{p}\right)$$
trapez. pulse	$$x(t) = \begin{cases} 0 & \text{for } t \leq 0,\, t \geq T \\[2mm] \dfrac{A}{aT}\,t & \text{for } 0 \leq t \leq aT \\[2mm] A & \text{for } aT \leq t \leq (1-a)T \\[2mm] \dfrac{A}{a}\left(1 - \dfrac{t}{T}\right) & \text{for } (1-a)T \leq t \leq T \end{cases}$$	$$E_\cap(p) = \frac{A}{aT}\,\frac{\left(1 - e^{-aTp}\right)\left(1 - e^{-(1-a)Tp}\right)}{p^2}$$
half-sine pulse	$$x(t) = \begin{cases} 0 & \text{for } t \leq 0,\, t \geq T \\[2mm] A\,\sin\left(\dfrac{\pi}{T}\,t\right) & \text{for } 0 \leq t \leq T \end{cases}$$	$$E_\Omega(p) = \frac{A\pi}{p}\,\frac{1 - e^{-pT}}{p^2 + (\pi/p)^2}$$

Table 4.10   Theorems of Laplace transform

(1)	Addition theorem

$$L\{f_1(t) + f_2(t)\} = L\{f_1(t)\} + L\{f_2(t)\}$$

(2)	Multiplication theorem

$$L\{af(t)\} = aL\{f(t)\}$$

(3)	Shifting theorem

$$\text{for } a > 0 \qquad L\{f(t - a)\} = e^{-pa}F(p)$$

$$\text{or } L\{f(t + a)\} = e^{pa}\left(F(p) - \int_0^a e^{-pt}f(t)\,dt\right)$$

(4)	Likeness theorem

$$L\{f(at)\} = \frac{1}{a}F\left(\frac{p}{a}\right) \quad \text{if } a > 0$$

(5)	Attenuation theorem

$$L\{e^{-at}f(t)\} = F(p + a) \quad \text{if } \operatorname{Re}(p + a) \geqslant p_0 > 0$$

(6)	Limit theorem

$$\lim_{t \to \infty} f(t) = \lim_{p \to 0} pF(p); \; \lim_{t \to 0} f(t) = \lim_{p \to \infty} pF(p)$$

(7)	Integration theorem

$$L\left\{\int_0^t f(\tau)\,d\tau\right\} = \frac{1}{p}L\{f(t)\} \quad \text{if } \operatorname{Re}(p) > 0$$

(8)	Differentiation theorem

$$L\{f^{(n)}(t)\} = p^n L\{f(t)\} - p^{n-1}f(+0) - \cdots - f^{(n-1)}(+0),$$

if the limits

$$\lim_{t \to 0} f(t) = f(+0); \; \lim_{t \to 0} \dot{f}(t) = \dot{f}(+0); \ldots;$$

$$\lim_{t \to 0} f^{(n-1)}(t) = f^{(n-1)}(+0), \text{ exist}$$

(9)–(10)	Convolution theorem

If the integrals $\int e^{-pt}f_1(t)\,dt$ and $\int e^{-pt}f_2(t)\,dt$ both are absolutely convergent or at most one absolutely and the other conditionally convergent yields

$$L\{f_1(t)\}L\{f_2(t)\} = L\{f_1(t) * f_2(t)\}$$

with

$$f_1(t) * f_2(t) = \int_0^t f_1(\tau)f_2(t - \tau)\,d\tau = \int_0^t f_1(t - \tau)f_2(\tau)\,d\tau$$

Using the Laplace transforms of the unit step and the Dirac function (Tables 4.8 and 4.9) provides, instead of equation (4.51b),

$$G(p) = \frac{L\{g(t)\}}{L\{\delta(t)\}} = L\{g(t)\} = \int_0^\infty g(t)e^{-pt}\,dt \qquad (4.51c)$$

or

$$G(p) = \frac{L\{h(t)\}}{L\{w(t)\}} = pL\{h(t)\} = p\int_0^\infty h(t)e^{-pt}\,dt \qquad (4.51d)$$

and for the calculation of the time functions

$$g(t) = L^{-1}\{G(p)\} = \frac{1}{2\pi j}\int_{c-j\infty}^{c+j\infty} G(p)e^{pt}\,dp; \qquad (4.51e)$$

$$h(t) = L^{-1}\left\{\frac{G(p)}{p}\right\} = \frac{1}{2\pi j}\int_{c-j\infty}^{c+j\infty} \frac{G(p)}{p}\,e^{pt}\,dp \qquad (4.51f)$$

From equation (4.51a) it follows that

$$y(t) = L^{-1}\{L\{x(t)\}L\{g(t)\}\} = \int_0^t x(\tau)g(t-\tau)\,d\tau = x(t) * g(t) \quad (4.51g)$$

or

$$y(t) = L^{-1}\{L\{x(t)\}pL\{h(t)\}\} = \frac{d}{dt}\int_0^t x(\tau)h(t-\tau)\,d\tau = \frac{d}{dt}[x(t) * h(t)]$$
$$(4.51h)$$

giving the convolution theorem of the Laplace transform (Table 4.10). Figure 4.38 demonstrates this fact. By means of these relations the deconvolution problem may be solved (Davies, 1970):

$$x(t) = L^{-1}\left\{L\{y(t)\}\frac{1}{G(p)}\right\} = \int_0^t y(\tau)g^*(t-\tau)\,d\tau = x(t) * g^*(t) \quad (4.51i)$$

with

$$g^*(t) = L^{-1}\left\{\frac{1}{G(p)}\right\}$$

The deconvolution often becomes extremely difficult because it is generally not possible to realize the inverse system functions. An alternative approach to the solution, by means of Fourier transforms, is often used (Davies, 1970, see also Section 4.3.3.4).

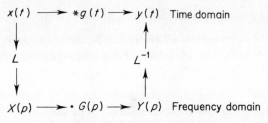

Figure 4.38   Convolution relationships

4.3.6   Stability

A system is stable if one of the following conditions is met:

(i)   The eigenvalues of the differential equation p_r exhibit a real part less than zero, i.e.

$$\text{Re}(p_r) = \delta_r < 0 \tag{4.52a}$$

(ii)   In the right-hand side of the pole–zero plot there are

$$\text{no poles of the transfer function;} \tag{4.52b}$$

(iii)   Decreasing eigenfunctions occur; this means that the weighting function $g(t)$ fulfils the condition

$$\lim_{c \to \infty} \int_0^c |g(t)| \, dt \leqslant M < \infty \tag{4.52c}$$

For the examination of these stability conditions certain stability criteria can be applied to test the situation:

(a)   If the following differential equation is given

$$a_n y^{(n)} + a_{n-1} y^{(n-1)} + \cdots + a_0 y = b_0 x + \cdots + b_m x^{(m)}$$

then the Hurwitz–Routh criterion can be used to test the stability. All coefficients a_r, and the determinants

$$\Delta_\mu = \begin{vmatrix} a_1 & a_0 & 0 & \cdots & 0 \\ a_3 & a_2 & a_1 & a_0 & 0 \\ \vdots & \vdots & \vdots & \vdots & \vdots \\ a_{2\mu-1} & a_{2\mu-2} & & \cdots & a_\mu \end{vmatrix} \tag{4.52d}$$

have to be positive, i.e. $a_r > 0$, $\Delta_\mu > 0$.

(b)   For practical applications graphical methods based on the locus diagram representations are of value. The transfer function $G(p)$ consists of polynomials in the numerator $N(p)$ and the divisor $D(p)$:

$$G(p) = N(p)/D(p) \tag{4.53a}$$

Application of the stability condition (4.52b) shows that

the polynomial of the divisor $D(p)$ is not permitted to have zeros
in the right half-plane including those lying upon the imaginary axis   (4.53b)
(stablility limit).

For testing the stability of a given system the divisor of the frequency response,
i.e. $D(j\omega)$, is drawn on the complex plane (locus diagram) forming a closed
curve by inclusion of negative frequencies. Because of the conformal mapping
the unstable field is always that lying on the right-hand side of the locus curve
drawn from $\omega = -\infty$ to $\omega = +\infty$ (see Figure 4.39). To test for stability it is
necessary to verify where the zero point is situated:

If the zero point of the $D(p)$-plane remains left of the locus curve
$D(j\omega)$, passed through from $\omega = -\infty$ to $\omega = +\infty$, the system is   (4.53b)
stable. If not it is unstable (Figure 4.40).

The procedure is also suitable for the statement of the stability margin (see
Woschni, 1973 or Zadeh and Desoer, 1963).

As an example of a commonly met system in instrument systems consider the
feedback system shown in Figure 4.24c. For negative feedback (control) the
frequency response is given by equation (4.36d) of Section 4.3.2.3.

$$G_e(j\omega) = \frac{G_1(j\omega)}{1 + G_1(j\omega)G_2(j\omega)}$$

The $D(j\omega)$ function is

$$1 + G_1(j\omega)G_2(j\omega) \qquad\qquad (4.53c)$$

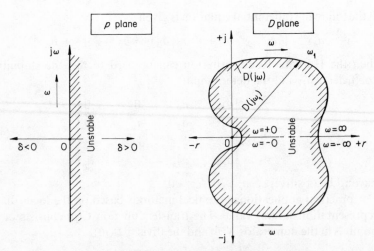

Figure 4.39   Locus diagram of the divisor $D(p)$

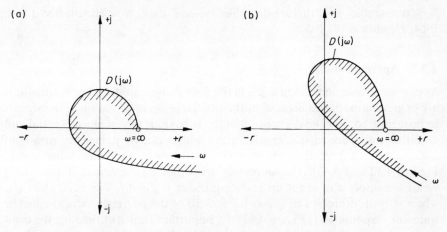

Figure 4.40 (a) Unstable system; (b) stable system

Instead of testing this function it is more convenient to verify that the open-loop frequency response

$$G_1(j\omega)G_2(j\omega) \tag{4.53d}$$

fulfils the condition related to the point '-1' as shown in Figure 4.41 (Nyquist diagram). The diagram illustrates that an increasing amplification will take the system from a stable to an unstable state. This fact may be caused by a non-linearity and may lead to unstable oscillatory behaviour, (see Section 4.3.2.4).

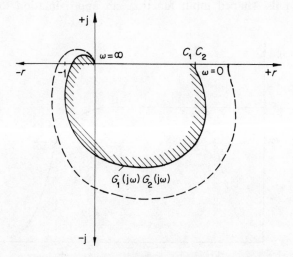

Figure 4.41 Testing stability of closed-loop systems
(control)

Structure-stable or structure-unstable systems may be distinguished (Tou, 1964; Newton *et al.*, 1957).

4.3.7 Approximations

As proved already in Section 4.2.7, in the field of measurement approximations are of great importance because of the small *a priori* information of the signals to be measured. For the description of the behaviour of systems it is usual and useful to introduce characteristic values which can be gained by means of approximations from the functions in the frequency and time domains (see Figures 4.42 and 4.43). The amplitude–frequency characteristic (Figure 4.42) yields the upper and lower critical frequencies $f_{c,u}$ and $f_{c,1}$ or $\omega_{c,u}$ and $\omega_{c,1}$, where $|G(j\omega)|$ decreases to $\sqrt{2} = 0.7 \triangleq 3$ dB of the reference value. From the transient response $h(t)$ (Figure 4.43) the important transient time t_{tr}, the dead time t_d, the delay time t_1, the compensation time t_c, and the maximum overshoot Δx_0 are obtained. The transient time is approximately (see Section 4.3.3.1)

$$t_{tr} = 3T_{max} \tag{4.54a}$$

and from the sampling theorem the transient time and the critical frequency are related by

$$f_{c,u} = 1/2t_{tr} \tag{4.54b}$$

In measurement systems engineering these approximate considerations are of great importance, for example as in the selection of suitable measuring instruments. Consider the following example. Figure 4.44 illustrates the problem of measuring a pulse-shaped input function, an approximation for numerous

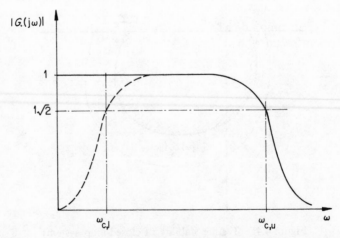

Figure 4.42   Definition of the critical frequencies

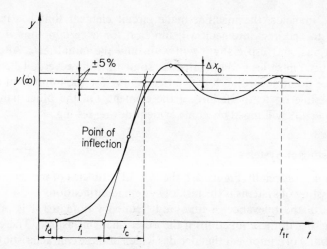

Figure 4.43   Definition of the characteristic values in the time
domain

measuring tasks (Woschni, 1972a). In this figure three cases are presented. The
long dashes represent the example in which the transient time is equal to the
pulse width, i.e. $t_{tr} = \Delta T$. The pulse height is still correctly indicated, whereas the
pulse shape is strongly distorted. In contrast to this, considerable error is introduced when measuring the pulse height if the transient time is too long (short
dashes). To permit proper determination of the pulse shape, the transient time
has to be substantially shorter than the pulse width (chain curve). The same
considerations may also be useful for analysing measuring errors. If one expects,
for instance, a pulse-shaped behaviour of the output and receives, as the response, an output variable with a heavily prolonged trailing edge, then an error
will be present (short dashes in Figure 4.44). It is noteworthy that minor and
medium errors in amplitude measurement frequently produce more detri-
mental effects in practice than the very large measuring errors. For errors of

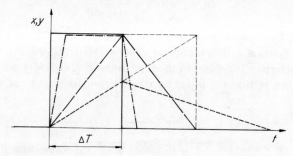

Figure 4.44   Approximation of the output for the case of
a pulse-shaped input

some 10% in fields, the machine-made circuit element that has been sized according to this measurement will function for a certain time due to the safety margins, and especially it will withstand the initial tests. After having been produced in series and operated for some time, however, all elements fail at the same point according to the fatigue curve for the number of stress reversals possible up to the failure of the element. On the other hand, major measuring errors will mostly become evident during testing.

4.3.8   Testing of Systems

In Section 4.1, especially Figure 4.2, the typical problems of measurements are treated. Testing of systems is the task of system identification.

Because of the relevance of time and frequency domains it is possible to calculate either of these functions if the other one is measured. Therefore, it is the availability of equipment that decides which of these characteristic functions is to be found.

Testing of systems is performed by means of test signals produced by a test generator, recording the corresponding output signals as illustrated in Figure 4.45. Table 4.11 gives a survey of the test signals used, the output signals, and the characteristic values used for approximations. Today the process of obtaining the characteristic functions is often automated, making use of programmable function generators that are controlled by microcomputers, to form the appropriate input signals.

In particular the frequency response $G(j\omega)$ is obtained by measuring both phase angle and proportion of the amplitudes of output to a given sinusoidal input. Before taking the true values it is necessary to wait until the steady-state solution appears, i.e. $t > t_{tr} = 1/(2f_c)$. If the measuring device itself has a non-ideal frequency response $G_x(j\omega)$, $G_y(j\omega)$ the real frequency response $G(j\omega)$ may be calculated from the wrong value $G^*(j\omega)$ by means of the relation (Woschni, 1972a)

$$G(j\omega) = G^*(j\omega)\, \frac{G_x(j\omega)}{G_y(j\omega)} \qquad (4.55a)$$

Othere principles of calibration make use of comparison systems or reciprocity principles for systems with reversible operation (Woschni, 1972a).

If the characteristic functions in the time domain, $g(t)$ of $h(t)$, are to be found a problem arises in that the input signals are not the ideal ones, as shown in

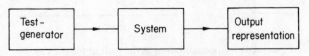

Figure 4.45   Testing of systems

Table 4.11 Survey of test signals

Characteristic function	Input test signal	Output function	Characteristic values
Differential equation	Not specified	Not specified	Time constants T_r
In the frequency domain	Harmonic oscillation $x = \hat{X}e^{j\omega t}$	Frequency response $G(j\omega)$; Transfer function $G(p)$	Critical frequency f_c; ω_c
In the time domain	Unit step function $w(t)$ Unit pulse = Dirac delta function $\delta(t)$	Transient response $h(t)$ Weighting function $g(t)$	Transient time t_r dead time t_d delay time t_l compensating time t_c overshoot Δx_0
Stochastic functions	White noise $S(\omega)$ = constant	Cross-correlation function $\psi_{xy}(\tau) \triangleq g(\tau)$	Transient time t_{tr} Correlation time t_{corr}

Figure 4.46 for a non-ideal step function $w^*(t)$. Instead of the real transient response $h(t)$, $h^*(t)$ is used:

$$h^*(t) = \frac{d}{dt} \int_0^t w^*(\tau)h(t - \tau)\, d\tau \qquad (4.55b)$$

By means of a deconvolution it is possible to gain the real transient response where $w^*(t)$ is known (Woschni, 1972a). In practice it is usually sufficient that the rise time of the step function is smaller than one magnitude of the transient time of the system to be tested.

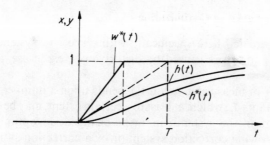

Figure 4.46 Testing with a non-ideal step function $w^*(t)$

Of great practical importance, especially for self-adaptive systems, are those methods of identification which use the noise at the input of a system (Davies, 1970). As shown in Section 4.3.3.4 the weighting function $g(t)$ may be calculated by deconvoluting the relation

$$\psi_{xy}(\tau) = \int_0^\infty g(t)\psi_{xx}(t - \tau)\,\mathrm{d}t \tag{4.55c}$$

Because deconvolution technique becomes difficult, methods in the frequency domain or special noise sources are used (Davies, 1970). Use of white noise yields

$$\psi_{xy}(\tau) = \text{constant} \cdot g(\tau) \tag{4.55d}$$

that is, the weighting function becomes a constant multiplied by the cross-correlation function (Woschni, 1981). For more detail, including errors arising from use of a non-ideal correlation function, and the reduction of errors see Davies (1970).

4.3.9 Typical Systems

Table 4.12 gives a survey of the most important measurement systems and characteristic functions in both the time and frequency domains (Woschni, 1981). The results of this table, in principle at least, may be used to find optimal parameters of a system. For example, the best damping of the spring–mass–damping system, typical of a great number of measuring systems, can be read off to be approximately 1 (precise value 0.7). The transfer functions of typical systems as illustrated in Table 4.12 allow the user to gain a survey of typical curve distortions and their causes using the methods for approximations of Section 4.3.9. These considerations lead to the results summarized in Table 4.13. As input function a pulse-shaped curve is assumed. Errors are dealt with in more detail in Chapter 6 of this handbook.

4.3.10 Some Remarks on Optimization

Optimization is treated here specifically with respect to system theory. Both dynamic errors and errors caused by disturbances, e.g. noise, are taken into consideration.

The behaviour of measuring systems depends upon a number of parameters k_i, e.g. time constants T_r, which, at least to a certain extent, may be freely selected. If the parameters of the system cannot be changed, an optimization can be effected by a following correction system or by a correction system connected in opposition, the parameters of such a system being adjustable. With the advance of microprocessors, computers of that kind will be employed for this purpose, the behaviour of the computers necessary for the optimum behaviour of the entire system being ensured by suitable programming.

The fundamental idea is shown for one-dimensional problems in Figure 4.47. The difference between the output variables of the ideal optimum system (model) and the real system, with a following correction system, is formed and assessed according to the computing instruction defined by the optimization criterion. The parameters k_i of the correction system are set so that the real system resembles the optimum system as far as possible. Principally, we distinguish between static and dynamic optimization (Bellman, 1961) according to the definition of the performance criterion. If the system is dimensioned so that a performance function that depends directly on the parameters k_i becomes optimum, we have a static optimization. In dynamic optimization, on the other hand, a process $x(t)$ is sought such that a performance function depending on this process—a functional—assumes an extremal value (Ventcelj, 1964).

The previous considerations imply that the central problem of optimization is the creation of a suitable optimization criterion.

In most cases the criterion of the mean square error, as shown in Figure 4.48, is used. If $G_{real}(j\omega)G_{corr}(j\omega)$ is the frequency response of the system corrected by a following correction system $G_{corr}(j\omega)$ as shown in Figure 4.47, we obtain the mean square error (Woschni and Kraus, 1975):

$$\overline{\varepsilon^2} = \int_0^\infty S_{xx}(\omega)|G_{id}(j\omega) - G_{real}(j\omega)G_{corr}(j\omega)|^2 \, d\omega$$

$$+ \int_0^\infty S_{zz}(\omega)|G_{corr}(j\omega)|^2 \, d\omega$$

$$= \overline{\rho^2} + P_z \tag{4.56a}$$

with the assumption having been made that the disturbances occur prior to the correction computer.

The total error according to equation (4.56a) is composed of two components: the error due to insufficient dynamic behaviour $\overline{\rho^2}$ and the error due to the disturbance P_z. If there is no correlation between signal source and disturbance source, as has been assumed here, the two error components must be added to arrive at the total error. When the error components and total error are plotted against the correction degree 'a', relationships will always result as presented in Figure 4.49. The dynamic error component $\overline{\rho^2}$ decreases as 'a' increases and disappears for 'a' $\to \infty$ (ideal correction), whereas the error component due to the disturbance increases with 'a'. Figure 4.50 provides the physical demonstrative explanation. In this figure the amplitude responses $|G_{real}(j\omega)|$ of the uncorrected system as well as of the correction system $|G_{corr}(j\omega)|$ are plotted. To compensate for the decrease of the uncorrected system at frequencies above the limiting frequency $\omega_{c, real}$ the correction system must raise these frequencies

Table 4.12 Survey of typical systems and their characteristic functions

Mathematical formulation	Transfer locus	Amplitude phase curve
k	$+j$, k, $-j$, $+r$	$\lvert G(j\omega)\rvert$, k, ω ; $\phi(\omega)$
$\dfrac{k}{1+T_1 p}$	$+j$, $-\pi/4\ k$, $+r$, ω, $-j$, $\omega_g/\sqrt{2}=1/T_1$	$G(j\omega)$, k, $k/\sqrt{2}$, $\omega_g=1/T_1$, ω ; $\omega_g=1/T_1$, $-\pi/4$, $-\pi/2$, $\phi(\omega)$
$\dfrac{k}{1+T_1 p+T_2^2 p^2}$	$+j$, $-r$, k, $+r$, $\dfrac{kT_2}{T_1}$, ω, $1/T_2=\omega_o$, $-j$	$\lvert G(j\omega)\rvert$, kT_2/T_1, $T_1/2T_2=D<1$, k, $D>1$ $1/T_2=\omega_o$, ω ; $1/T_2=\omega_o$, $D>1$, $D<$, $-\pi$, $\phi(\omega)$
$\dfrac{-kp^2}{1+T_1 p+T_2^2 p^2}$	k/T_2^2 $+j$, $-r$, T_2, $+r$, ω, $-j$ $1/T_2=\omega_o$	$\lvert G(j\omega)\rvert$, $D<1$, k/T_2, $D>1$, $1/T_2=\omega_o$, ω ; $1/T_2=\omega_o$, $D>1$, $-\pi$, $D>$, $\phi(\omega)$
$e^{-t_t p}$	$\omega=n\pi/t_t$ $+j$, $-r$, ω, $+r$, $-k$ $+k$, $-j$ $\omega=n\pi/2t_t$	$\lvert G(j\omega)\rvert$, k, ω ; $\pi/2t_t$, π/t_t, $-\pi/2$, $-\pi$, $\phi(\omega)$
$\dfrac{ke^{-t_t p}}{1+T_1 p}$	$+j$, $-r$, k, $+r$, ω, $-j$	$\lvert G(j\omega)\rvert$, t, $k/\sqrt{2}$, $\omega_g=1/T_1$, ω ; $\phi(\omega)$
$\dfrac{k}{p}\,\dfrac{1}{1+T_1 p}$	$-r$, $+r$, ω, $-j$	$\lvert G(j\omega)\rvert$, ω ; $-\pi/2$, π, $\phi(\omega)$
$\dfrac{kp}{1+T_1 p}$	$+j$, ω, k/T_1, $-r$, $+r$, $-j$	$\lvert G(j\omega)\rvert$, kT_1, $\omega_g=1/T_1$, ω ; $\phi(\omega)$, $\pi/2$, $\pi/4$, $\omega_g=1/T_1$, ω

poles ($\times$) and zeros (0)	Function of time		Examples for systems with such a behaviour
	transfer function $x_a/1 = h(t)$	weighting function $a_a/1 = g(t)$	
None	k (step at $t=0$)	$x_{a_0} \to \infty$, $\int x_{a_0}\, dt = k$, $\Delta\tau \to 0$	systems with proportional behaviour (idealised)
$-1/T_1$	k	k/T_1	temperature-measuring arrangements without protective tube; very heavily damped systems capable of vibrating with proportional behaviour (idealized); system with delay of the first order and compensation
ω_0, $\cos^{-1}D$, $1/T_2 = \omega_0$	k		spring–mass damping system, $D \leq 1$, temperature-measuring arrangements with protective tube; $D < 1$, approximation behaviour for vibration systems
Double zero, $1/T_2 = \omega_0$			spring–mass damping system without fixed point $D \leq 1$
Cannot be exactly represented	k	∞, $\Delta\tau \to 0$	system with pure dead time (idealization), caused, e.g., by pipeline, transport path, etc.
$-1/T_1$; factor $e^{-t_1 p}$ cannot be exactly represented.	k	k/T_1	real systems with dead time, e.g., temperature-measuring arrangement with heat conductivity feed to transducer
$-1/T_1$	k	k	real integrator
$-1/T_1$	k/T_1	∞, $\Delta\tau \to 0$, T_1	real differentiator ($x_a \gtrapprox \dot{x}_e$)

Table 4.13   Typical wave-shape distortions and their causes

— Output variable $x_a(t)$ ----- Input variable $x_e(t)$ Curve shape	Typical features	Causes: amplitude curve, phase curve $|G(j\omega)|$; $\varphi(\omega)$	Typical features	Remarks
1	sharp corners steel edges straight top		ideal system, i.e., $f_{gu} = 0$; $f_{go} = \infty$ ideal phase curve	practically not fully satisfied; only approximately for slow process (T great)
2	blurred corners oblique edges straight top		upper limiting frequency f_{go} too low; lower limiting frequency $f_{gu} = 0$ ideal phase curve	short peaks of the input function are not reproduced or are reproduced with wrong height; remedy: to increase upper limiting frequency, e.g., by reducing masses (idealized system)
3	sharp corners steep edges inclined top		lower limiting frequency f_{gu} too high; upper limiting frequency $f_{go} = \infty$	short peaks are correctly indicated; the effect of the error is the stronger the slower the process takes place; pure static calibration end measurement not possible (e.g., piezoelectric measuring device)

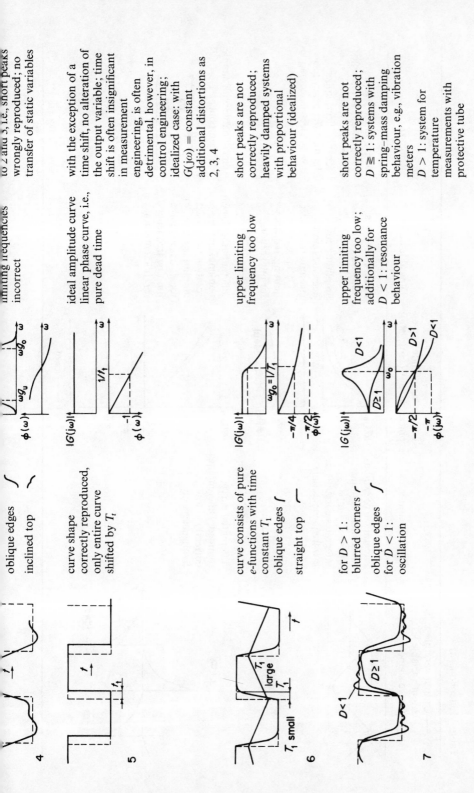

		limiting frequencies incorrect		to 2 and 3, i.e., short peaks wrongly reproduced; no transfer of static variables
4	oblique edges inclined top			
5	curve shape correctly reproduced, only entire curve shifted by T_t	ideal amplitude curve linear phase curve, i.e., pure dead time		with the exception of a time shift, no alteration of the output variable; time shift is often insignificant in measurement engineering, is often detrimental, however, in control engineering; idealized case: with $G(j\omega) = $ constant additional distortions as 2, 3, 4
6	curve consists of pure e-functions with time constant T_1 oblique edges straight top	upper limiting frequency too low; $\omega_{g0} = 1/T_1$		short peaks are not correctly reproduced; heavily damped systems with proportional behaviour (idealized)
7	for $D > 1$: blurred corners for $D < 1$: oblique edges oscillation	upper limiting frequency too low; additionally for $D < 1$: resonance behaviour		short peaks are not correctly reproduced; $D \gtrapprox 1$: systems with spring–mass damping behaviour, e.g., vibration meters $D > 1$: system for temperature measurements with protective tube

continued

Table 4.13 (continued)

— Output variable $x_a(t)$
----- Input variable $x_e(t)$

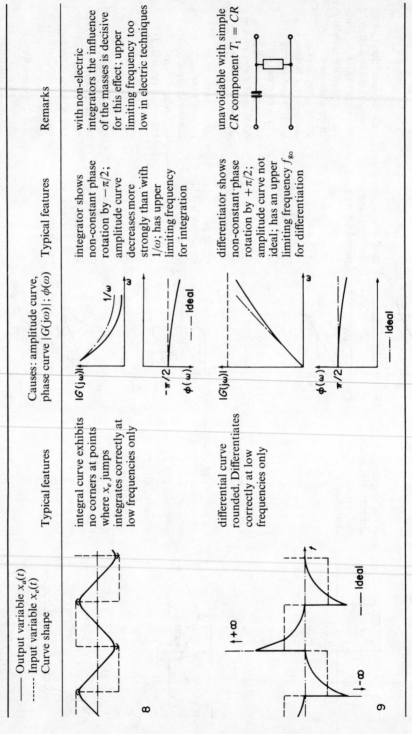

Curve shape	Typical features	Causes: amplitude curve, phase curve $|G(j\omega)|$; $\phi(\omega)$	Typical features	Remarks
8	integral curve exhibits no corners at points where x_e jumps; integrates correctly at low frequencies only		integrator shows non-constant phase rotation by $-\pi/2$; amplitude curve decreases more strongly than with $1/\omega$; has upper limiting frequency for integration	with non-electric integrators the influence of the masses is decisive for this effect; upper limiting frequency too low in electric techniques
9	differential curve rounded. Differentiates correctly at low frequencies only		differentiator shows non-constant phase rotation by $+\pi/2$; amplitude curve not ideal; has an upper limiting frequency f_{go} for differentiation	unavoidable with simple CR component $T_1 = CR$

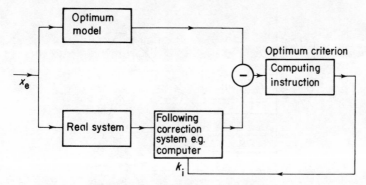

Figure 4.47   Principle of optimization

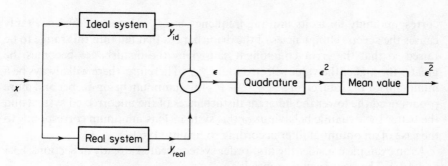

Figure 4.48   Mean square error generation

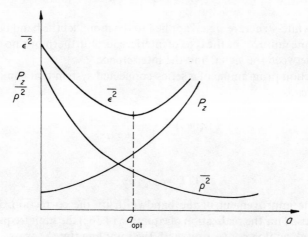

Figure 4.49   Behaviour of the components of the error as
a function of the degree of correction

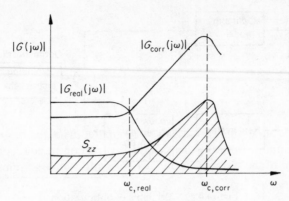

Figure 4.50   Physical explanation of the increase of the
noise in case of correction

correspondingly up to its limiting frequency $\omega_{c,\,corr}$. However, this inevitably causes the spectral frequencies of the disturbances that fall into this range to be raised so that the error component caused by the disturbance becomes the greater the higher the value selected for $\omega_{c,\,corr}$. Therefore, there will always be a minimum for the total error $\overline{\varepsilon^2} = \overline{\rho^2} + P_z$, this minimum being deeper and more pronounced the lower the inherent disturbances of the uncorrected system and the better the dynamic behaviour of this system. This minimum corresponds to the case of an optimum filter according to Schlitt (1960).

As an example consider the first-order system treated already in Sections 4.3.3 and 4.3.4. The frequency response is

$$G(j\omega) = \frac{1}{1 + j\omega T} \qquad\qquad (4.56b)$$

Let S_{x_0} be a white-structure signal applied to the input, let the band be limited to ω_x and let white noise S_{z_0} be the type of interference. Furthermore, no correlation is assumed between the signal and the interference.

The correction programme of a series-connected system that can be realized, in this case reads

$$G_k(j\omega) = \frac{1 + j\omega T}{1 + j\omega T_k} \qquad\qquad (4.56c)$$

the factor

$$T/T_k = \omega_{c,\,k}/\omega_c \qquad\qquad (4.56d)$$

indicating the improvement of the bandwidth by the correction. If a microcomputer is used for the realization of equation (4.56c) the limit frequency of the analog–digital–analog conversion, with the sampling time t_s, is

$$\omega_{c,\,s} = \pi|t_s$$

Thus, the mean square error is

$$
\overline{\varepsilon^2} = S_{x_0}\left(\int_0^{\omega_{c,s}} \frac{\omega^2 T_k^2}{1 + \omega^2 T_k^2}\, d\omega + \omega_x - \omega_{c,s}\right)
$$
$$
+ S_{z_0} \int_0^{\omega_{c,s}} \frac{1 + \omega^2 T^2}{1 + \omega^2 T_k^2}\, d\omega \qquad (4.56e)
$$

In order to obtain values that can be compared with the original system and enable an estimate of the efficiency of the correction to be generated, we calculate the error of the system without any correction while assuming a bandwidth limited to $\omega_c = 1/T$:

$$
\overline{\varepsilon_0^2} = S_{x_0}\left(\int_0^{\omega_c} \left|1 - \frac{1}{1 + j\omega T}\right|^2 d\omega + \int_{\omega_c}^{\omega_x} d\omega\right) + S_{z_0} \int_0^{\omega_c} d\omega
$$
$$
= S_{x_0}\omega_c\left(\frac{\omega_x}{\omega_c} - \frac{\pi}{4} - \frac{S_{z_0}}{S_{x_0}}\right) \qquad (4.56f)
$$

To find the more favourable solution in each case, and thus obtain suggestions for synthesis, consider the following two cases: (a) the limiting frequency $\omega_{c,s}$, i.e. the sampling frequency, be adapted to the limiting frequency of the corrected system $\omega_{c,k}$; and (b) the limiting frequency $\omega_{c,s}$ be adapted to the limiting frequency of the signal ω_x:

(a) $\omega_{c,s} = \omega_{c,k}$. From equation (4.56e), for $\omega_{c,s} = \omega_{c,k} < \omega_x$, one obtains

$$
\overline{\varepsilon^2} = S_{x_0}\omega_{c,k}\left[\left(1 + \frac{S_{z_0}}{S_{x_0}}\frac{\omega_{c,k}^2}{\omega_c^2}\right)\left(1 - \frac{\pi}{4}\right) + \frac{\omega_x}{\omega_{c,k}} - 1 + \frac{S_{z_0}}{S_{x_0}}\frac{\pi}{4}\right] \qquad (4.56g)
$$

and for $\omega_{c,s} = \omega_{c,k} > \omega_x$, respectively,

$$
\overline{\varepsilon^2} = S_{x_0}\omega_{c,k}\left[\frac{\omega_x}{\omega_{c,k}} - \arctan\left(\frac{\omega_x}{\omega_{c,k}}\right) + \frac{S_{z_0}}{S_{x_0}}\frac{\omega_{c,k}^2}{\omega_c^2} + \left(1 - \frac{\pi}{4}\right) + \frac{S_{z_0}}{S_{x_0}}\frac{\pi}{4}\right]
$$
$$
(4.56h)
$$

(b) $\omega_{c,s} = \omega_x$

$$
\overline{\varepsilon^2} = S_{x_0}\omega_{c,k}\left\{\left(1 + \frac{S_{z_0}}{S_{x_0}}\frac{\omega_{c,k}^2}{\omega_c^2}\right)\left[\frac{\omega_x}{\omega_{c,k}} - \arctan\left(\frac{\omega_x}{\omega_{c,k}}\right)\right]\right.
$$
$$
\left. + \frac{S_{z_0}}{S_{x_0}}\arctan\left(\frac{\omega_x}{\omega_{c,k}}\right)\right\} \qquad (4.56i)
$$

Any existing amplification or attenuation in the original system (static transmission factor) may be considered in the usual manner in the signal to noise ratio S_{x_0}/S_{z_0}. In Figure 4.51a,b,c, the results obtained from equations (4.56g,h)

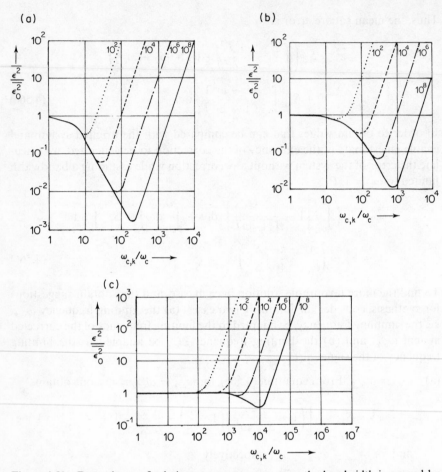

Figure 4.51 Dependence of relative mean square error on the bandwidth improved by correction: $\cdots\cdot\ S_{x_0}/S_{z_0} = 10^2$; $---\ S_{x_0}/S_{z_0} = 10^4$; $-\cdot-\ S_{x_0}/S_{z_0} = 10^6$; $\underline{\qquad}\ S_{x_0}/S_{z_0} = 10^8$; (a) $\omega_x/\omega_c = 10$; (b) $\omega_x/\omega_c = 10^2$; (c) $\omega_x/\omega_c = 10^4$

are shown for different values of the signal to noise ratio of 20, 40, 60, and 80 dB, i.e. for $S_{x_0}/S_{z_0} = 10^2$, 10^4, 10^6, and 10^8, in relation to the bandwidth increase by correction of $\omega_{c,k}/\omega_c$. The parameters of ω_x/ω_c were selected such that dynamically good systems (Figure 4.51a) as well as dynamically poor systems (Figure 4.51c) are involved. The values for the mean square error ε^2 are related to the error of the uncorrected system $\overline{\varepsilon_0^2}$ according to equation (4.56f) so as to indicate directly the reduction of the deviation.

All the diagrams for the adaptation of the bandwidth $\omega_{c,k}$ to the bandwidth of the programme for system correction $\omega_{c,s}$ (case (a)) reveal a minimum corresponding to the case of an optimum filter: while the dynamic portion of error

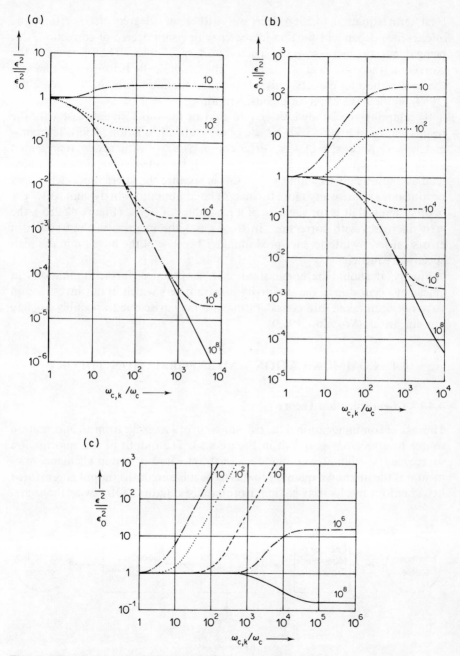

Figure 4.52 Continuation of Figure 4.51: $-\cdot-$ $S_{x_0}/S_{z_0} = 10$; $\cdots$ $S_{x_0}/S_{z_0} = 10^2$; $---$ $S_{x_0}/S_{z_0} = 10^4$; $-\cdot\cdot-$ $S_{x_0}/S_{z_0} = 10^6$; $\underline{\hspace{1cm}}$ $S_{x_0}/S_{z_0} = 10^8$; (a) $\omega_x/\omega_c = 10$; (b) $\omega_x/\omega_c = 10^2$; (c) $\omega_x/\omega_c = 10^4$

(first term equation (4.56e)) decreases with rising degree of correction, the interference-dependent portion increases with rising degree of correction, this being due to the increase in the spectral portions $\omega > \omega_c$. The efficiency of the correction, therefore, will be the higher, the dynamically better the system and the smaller the interferences. This substantiates the finding obtained by Woschini (1969) on the basis of physical considerations.

By adapting the bandwidth $\omega_{c,k}$ to that of the input signal (case (b)), the results shown in Figure 5.52a,b,c are obtained from equation (4.56i). The parameters were selected such that direct comparison with the results represented in Figure 4.51 is possible. In contrast to case (a), dependencies are obtained which tend asymptotically to a limit value; this is because the interference-dependent portion does not rise any more because of band limitation. In dynamically, very poor systems, that is for the case of large values of ω_x/ω_c (Figure 4.52b,c), the error increases with correction. In these cases the increase in the fraction of errors caused by interference predominates because of the increase in the high spectral frequencies.

Finally, it should be emphasized that, in practice, further limitations in efficiency occur due to the sensitivity of parameters which in this investigation have not been taken into consideration and may arise due to possibly existing non-linearities (Woschni, 1967).

4.4 COMMUNICATION AND INFORMATION THEORY

4.4.1 Communication Theory

The task of communication is the transmission of a message from an information source to a receiver as shown in Figure 4.53. The output of this information source may be a digital or continuous signal, as treated in Section 4.2. In measurement it is the unknown quantity that is to be measured: the output is generated by a random mechanism having a probabilistic nature. Otherwise the signal

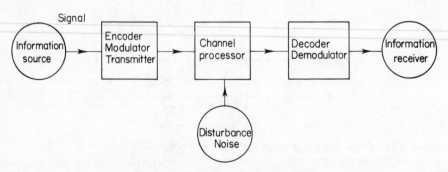

Figure 4.53   Communication system

would be completely known and there would be no need to obtain the output by means of measurement. The signal coming from the information source is the input signal of the encoder, modulator or transmitter. If the signal is digital the term encoder is used while for analogue signals modulator is used. This subsystem is treated in Sections 4.4.4 and 4.4.5. The function is to transform the signal, if there are several information sources to multiplex them onto the same transmission channel, or to make the signal immune to disturbances.

The modulated or coded signal is transmitted by the channel or processed by the processor. This channel may be a microwave or u.h.f. relay link, a wire or cable transmission as is today common in measurement, or a waveguide transmission for broad-band signals. In the future lightguide optical transmission will gain importance for transmission of measured data because of its high immunity to electromagnetic disturbances. The noise is considered to be additative. In measurement the signal is often processed by a microprocessor containing a memory. As in the case of an analog system the behaviour of this system may be described by a transfer function provided the computer program is a linear (Woschni, 1981).

The next link in the serial chain is the decoder or demodulator, the task of which is to construct an estimate of the original signal as correctly as possible. In the sense of describing the signal as a vector in the signal space (Section 4.2) this means that the distance between the output and input signals should be a minimum. Many similarities link this problem to the problems of character recognition (Finkelstein, 1976).

4.4.2 Information Theory

If the number of the different possible symbols that the information source is able to deliver is m, and if there is no *a priori* information about the probabilities of the different symbols at the receiver before receiving the signals, all possible symbols at the receiver have the same probability:

$$p_1 = p_2 = \cdots = p_i = 1/m \qquad (4.57a)$$

with the normalization

$$\sum_{i=1}^{m} p_i = 1 \qquad (4.57b)$$

The amount of information in message i is defined as

$$I = \log_{10}(1/p_i) \qquad (4.58a)$$

The base of the logarithm is often chosen as 2 (binary logarithm) because in many practical systems two stable states are used. The measure of the amount of information is therefore given by the information necessary to decide between two possible states and is called a *bit* (from 'binary digit').

The average information content over all source symbols is given by the source *entropy* H where

$$H = - \sum_{i=1}^{m} p_i \log_{10} p_i \qquad (4.58b)$$

H is a maximum when all symbols are equally likely. Then the maximum possible value H_0 is given by

$$H_0 = \log_2 m = \text{lb } m \text{ bit/symbol} \qquad (4.58c)$$

(as $\ln = \log_e$, then $\text{lb} = \log_2$, and $\text{ld} = \log_{10}$). For the two-symbol (binary) source, H_0 is equal to 1 (maximum uncertainty or average information per symbol). This average information content per symbol, the entropy H, gives the number of binary decisions which are necessary on the average to distinguish one state out of the ensemble of all possible states.

The above definitions are given for discrete sources. For analog signals a differential entropy of a continuous distribution is defined (Goldman, 1953):

$$H(x) = - \int_{-\infty}^{+\infty} w(x) \, \text{lb}[w(x)] \, dx \qquad (4.58d)$$

Entropy of a continuous distribution is a maximum (optimal coding: Shannon, 1948) for systems with amplitude limitation if $p(x) = $ constant and for systems with power limitation if $p(x)$ is a Gaussian distribution (Shannon, 1948; Feinstein, 1958; Woschni, 1973). As a measure of the difference between the maximum value H_0 and the real value H the redundancy ΔH, given by

$$\Delta H = H_0 - H \text{ bit/value} \qquad (4.59a)$$

or the relative redundancy Δh is used, where

$$\Delta h = \frac{\Delta H}{H_0} = \frac{H_0 - H}{H_0} = 1 - \frac{H}{H_0} \qquad (4.59b)$$

A binary source, for instance, has entropy $H(p)$ shown in Figure 4.54.

Generalizing the entropy for more than one variable, $x_1, x_2, \ldots, x_r, \ldots, x_n$, results in entropies of higher order. This plays an important role in analysing digital signals where correlation exists between several bits in the signal sequence. It is also of value in relating signals at the input and output of systems (problems of signal transmission). Let the joint probability be

$$p(x_1, x_2, \ldots, x_r, \ldots, x_n) \qquad (4.60a)$$

and the conditional probability be

$$p(x_1 | x_2, x_3, \ldots, x_r, \ldots, x_n). \qquad (4.60b)$$

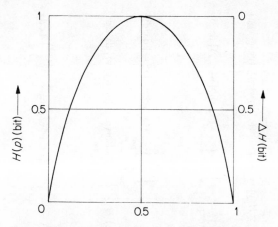

Figure 4.54   Entropy and redundancy of a binary
source

For the entropies it follows that the joint entropy is given by

$$H(x_1, x_2, \ldots, x_r, \ldots, x_n) = -\sum_{x_1} \sum_{x_2} \cdots \sum_{x_n} p(x_1, x_2, \ldots, x_n)$$
$$\times \ \mathrm{lb}[p(x_1, x_2, \ldots, x_n)] \qquad (4.60c)$$

and the conditional entropy is

$$H(x_1 | x_2, x_3, \ldots, x_n) = -\sum_{x_1} \sum_{x_2} \cdots \sum_{x_n} p(x_1, x_2, \ldots, x_n)$$
$$\times \ \mathrm{lb}[p(x_1 | x_2, \ldots, x_n)] \qquad (4.60d)$$

In Table 4.14 the equations of the entropy are placed together for the important measurement case of two fields of variables. In particular the problems of communication shown in Figure 4.53, lead to the concepts embodied in Figure 4.55 when expressed in terms of the entropies contained in Table 4.14. If a channel with Gaussian signals with power P_s and noise P_n is given the transinformation follows (Shannon, 1948; Goldman, 1953; Fano, 1961; Woschni, 1973).

$$H(x; y) = \tfrac{1}{2}\mathrm{lb}\left(1 + \frac{P_s}{P_n}\right) \text{ bit/value} \qquad (4.60e)$$

Examples in the field of measurement are treated in Section 4.4.3.

The transinformation $H(x; y)$ states the information content referred to one transmitted value (measure: bit/value). If the transient time of the channel is given as $t_{tr} = 1/2f_c$ the channel is transmitting

$$I = \frac{H(x; y)}{t_{tr}} = 2f_c H(x; y) \qquad (4.61a)$$

Table 4.14   Entropies for two events x and y

Definition	Entropy H — discrete signals	Entropy H — analog signals
entropy of event x	$H(x) = -\sum_i p(x_i)\,\text{lb}[p(x_i)]$	$H(x) = -\int_{-\infty}^{+\infty} w(x)\,\text{lb}[w(x)]\,\mathrm{d}x$
entropy of event y	$H(y) = -\sum_j p(y_j)\,\text{lb}[p(y_j)]$	$H(y) = -\int_{-\infty}^{+\infty} w(y)\,\text{lb}[w(y)]\,\mathrm{d}y$
joint entropy of events x and y	$H(x,y) = -\sum_i\sum_j p(x_i,y_j)\,\text{lb}[p(x_i,y_j)]$	$H(x,y) = -\int_{-\infty}^{+\infty}\int_{-\infty}^{+\infty} w(x,y)\,\text{lb}[w(x,y)]\,\mathrm{d}x\,\mathrm{d}y$
conditional entropy	$H(x\mid y) = -\sum_j p(y_j)\sum_i p(x_i\mid y_j)\,\text{lb}[p(x_i\mid y_j)]$ $= -\sum_i\sum_j p(x_i,y_j)\,\text{lb}[p(x_i\mid y_j)]$ $H(y\mid x) = -\sum_i p(x_i)\sum_j p(y_j\mid x_i)\,\text{lb}[p(y_j\mid x_i)]$ $= -\sum_i\sum_j p(x_i,y_j)\,\text{lb}[p(y_j\mid x_i)]$	$H(y\mid x) = -\int_{-\infty}^{+\infty}\int_{-\infty}^{+\infty} w(x,y)\,\text{lb}[w(x\mid y)]\,\mathrm{d}x\,\mathrm{d}y$ $H(y\mid x) = -\int_{-\infty}^{+\infty}\int_{-\infty}^{+\infty} w(x,y)\,\text{lb}[w(y\mid x)]\,\mathrm{d}x\,\mathrm{d}y$
transinformation	$H(x;y) = -\sum_i\sum_j p(x_i,y_j)\,\text{lb}\!\left(\dfrac{p(x_i)p(y_j)}{p(x_i,y_j)}\right)$	$H(x;y) = -\int_{-\infty}^{+\infty}\int_{-\infty}^{+\infty} w(x,y)\,\text{lb}\!\left(\dfrac{w(x)w(y)}{w(x,y)}\right)\mathrm{d}x\,\mathrm{d}y$
relations between the entropies	$H(x,y) = H(x) + H(y\mid x) = H(y) + H(x\mid y)$ $= H(x\mid y) + H(y\mid x) + H(x;y)$ $= H(x) + H(y) - H(x;y)$ $H(x\mid y) = H(x,y) - H(y) = H(x) - H(x;y)$ $H(y\mid x) = H(x,y) - H(x) = H(y) - H(x;y)$ $H(x;y) = H(x) + H(y) - H(x,y)$	

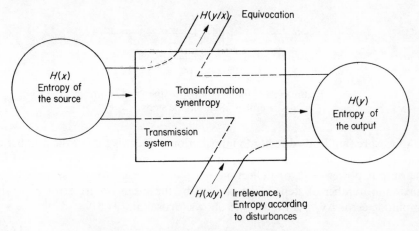

Figure 4.55   Entropies in communication systems

bits per second, known as information flow (Woschni, 1973). The optimum value of this information flow for optimal coding is the 'channel capacity' C_t (Shannon, 1948), given by

$$C_t = \frac{H(x; y)_{max}}{t_{tr}} = 2f_c H(x; y)_{max} \tag{4.61b}$$

From equation (4.60e) the channel capacity of a channel with white noise is

$$C_t = \frac{1}{2t_{tr}} \, \mathrm{lb}\left(1 + \frac{P_s}{P_n}\right) = f_c \, \mathrm{lb}\left(1 + \frac{P_s}{P_n}\right) \tag{4.61c}$$

This very important 'classical' relation was first published by Shannon (1948). The relation shows that it is possible to exchange signal-to-noise ratio P_s/P_n bandwidth f_c and vice versa.

Section 6.4 explores information theory applied to errors.

4.4.3   Applications to Measurement

In measurement the same general problem exists as occurs in a communication link (refer to Figure 4.53). The input is the value to be measured while the output is the measured value. Measurement systems often consist of subsystems connected in series (see Section 4.3.2). At the interconnection between two systems the problem of interface arises. At each of these points of interconnection the condition must be met that the information flow of the second system has to be the same as that of the first system; otherwise information would be lost. That means the channel capacities of the systems have to be equal. The interface problem itself is solved by means of coding (see Section 4.4.4).

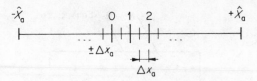

Figure 4.56 Explanation of the maximum
number of amplitude steps

A measure that leads directly to information content is that of the number of distinguishable amplitude steps m_a or power steps m_p. With a given limit for the output power $\pm P_s$ or deflection $\pm \hat{X}_a$ of the measuring instrument, the maximum number of steps is determined by the mean square error $\overline{\varepsilon^2}$ or the amplitude error Δx_a. As Figure 4.56 shows (Woschni, 1972b)

$$m_a = 1 + X_a/\Delta x_a; \quad m_p = 1 + P_s/\overline{\varepsilon^2} \tag{4.62a}$$

the fact being taken into account for addend 1 that '0' is also a possible measured value.

Using the results of information theory some very significant implications may be derived concerning storage of measured data in practical applications. For storing m_a equally probable values we require

$$s = \log_2 m_a = \mathrm{lb}\ m_a \tag{4.62b}$$

binary storage locations (bit), because with s binary storage locations a total of 2^s combinations can be represented. This number of storage locations is the decision content of information theory, $s = H_0$. Since with a given number m_p of power steps, m_a can be approximately computed as

$$m_a \simeq \sqrt{m_p} = \left(1 + \frac{P_s}{\overline{\varepsilon^2}}\right)^{1/2} \tag{4.62c}$$

The number of storage locations required for storing a measured value may be assessed by means of equation (4.62b). Thus, for instance, a measuring instrument with an amplitude error of $\Delta x/\hat{X} = 1\%$ necessitates $H_0 = s = \log_2 101 = 3.32193 \log_{10} 101 \simeq 6.64$ bits, that is, seven binary storage locations for storing one measured value.

On the other hand, a setting accuracy within, for instance, 10^{-6} error can be achieved with a punch tape having $3.32193 \log_{10} 10^6 \simeq 20$ punching possibilities per value. These intuitive considerations have a direct relation to information theory. According to equation (4.61a) one obtains, as the information flow provided by the measuring instrument in the most favourable conditions, an approximation with equation (4.62c):

$$I = f_c\ \mathrm{lb}\left(1 + \frac{P_s}{\overline{\varepsilon^2}}\right) \tag{4.62d}$$

If the error $\overline{\varepsilon^2}$ consists of the noise P_n only, equation (4.62d) becomes the relationship of Shannons' channel capacity C_t, introduced as equation (4.61c):

$$c_t = f_c \, \mathrm{lb}\left(1 + \frac{P_s}{P_n}\right) \tag{4.62e}$$

Thus, a measuring instrument with a signal-to-noise ratio of 60 dB ($\triangleq m_p = 10^6 \triangleq m_a = 10^3$) and a limiting frequency of 10 kHz yields a channel capacity of 200 kbit/s. A human being can consciously process only about 20 bit/s. Therefore, the measured values would have to be processed by a computer or stored on a magnetic tape ($C_t = 200$ kbit/s to 10 Mbit/s). In practice, the values for the information flow are mostly lower by 1 to 3 orders of magnitude, since the degree to which the measuring instruments can be adapted to the signals is far from the optimum; signals contain a large amount of redundant information.

4.4.4 Coding Theory

As treated in Section 4.4.1 the first subsystem of a communication system is the encoder, converting the input signal into a series of code words. The task of this coding is the adaptation (interfacing) of the information source to the channel or processor.

In communication systems a redundancy-diminishing (optimal) *source* coding, having the purpose of economizing the time for communication, often plays an important role. In measurements, security of the message against disturbances is the general criterion. Here, therefore, error-detecting or error-correcting codes are applied (Peterson, 1962).

For the representation of codes geometrical descriptions codegraphs, in the n-dimensional space, are used (see Section 4.2.6). Coding and decoding theorems exist. The decoding theorem deals with the problem of identifying a codeword by the receiver. For this purpose the decoder compares the incoming code words with the words of the code alphabet deciding which code word the transmitter has sent. In the case where the end of a code word is not marked by a special symbol, only the end-points of the codegraph may be filled with a code word, otherwise a part of a code word would be another code word. The equation that guarantees this is

$$\sum_{k=1}^{K} M^{-l_k} \leqslant 1 \tag{4.63a}$$

where M is the number of symbols and l_k is the length of the kth code word. In equation (4.63a) the equality sign represents the most advantageous case without code redundancy. Otherwise the factor c in the equation

$$c \sum_{k=1}^{K} M^{-l_k} = 1 \tag{4.63b}$$

is a measure of the code redundancy.

This decoding theorem is of great importance in measurement because codes with redundancy are often used. The theorem of optimal coding plays a great role in redundancy-diminishing coding used in communication, giving an equation for the optimum length of the source code words l_k:

$$-\frac{\text{lb } p_k}{\text{lb } M} \leqslant l_k < -\frac{\text{lb } p_k}{\text{lb } M} + 1 \tag{4.64}$$

where p_k is the probability of the appearance of the kth code word. Some important codes, including those applied in measurement, are now considered.

The most simple code is the counting code, mostly seen in decimal counting. Today this very easily learnable code is displaced in machines by the binary-coded decimal notation because of the smaller number of bits. As an example, in Figure 4.57, the 1-out-of-10 code is presented. For manual data coding (data input) a particular form of the binary code

$$Z_{\text{bin}} = A_n 2^n + A_{n-1} 2^{n-1} + \cdots + A_1 2^1 + A_0 2^0 = A_n A_{n-1} \cdots A_1 A_0 \tag{4.65}$$

the binary-coded decimal system (BCD-code), is used. Here digit-by-digit the decimal number is converted into the binary code. For each digit four bits, a so called tetrad, are necessary (Table 4.15).

In binary notation the complement, necessary for subtraction in a computer, sometimes leads to a non-existent code word. The compliment of $3(=0011)$, for instance, is 1100 which does not exist in the BCD code (Table 4.15). This disadvantage is avoided by use of the notation of the BCD code, also presented in Table 4.15. To each decimal number 3 is added. This code is usually used for data input.

Another group of codes are the reflected codes, arising by counting at first forward, that is from 0 to 9, and then backward from 19 to 10. One of this group is the Gray code. It is obtained from the binary code as shown in Table 4.16. The advantage of this code is that any two code words following each other always have unit distance between them, that is the two code words differ in one

```
    9 8 7 6 5 4 3 2 1 0
0   0 0 0 0 0 0 0 0 0 1
1   0 0 0 0 0 0 0 0 1 0
2   0 0 0 0 0 0 0 1 0 0
3   0 0 0 0 0 0 1 0 0 0
4   0 0 0 0 0 1 0 0 0 0
5   0 0 0 0 1 0 0 0 0 0
6   0 0 0 1 0 0 0 0 0 0
7   0 0 1 0 0 0 0 0 0 0
8   0 1 0 0 0 0 0 0 0 0
9   1 0 0 0 0 0 0 0 0 0
```

Figure 4.57   1-out-of-10
code

Table 4.15   BCD code in binary and excess-three notation

Decimal code	Binary-coded decimal code		Excess-three notation	
	First tetrad	Second tetrad $x_3 x_2 x_1 x_0$	First tetrad	Second tetrad $y_3 y_2 y_1 y_0$
0	0000	0000	0000	0011
1	0000	0001	0000	0100
2	0000	0010	0000	0101
3	0000	0011	0000	0110
4	0000	0100	0000	0111
5	0000	0101	0000	1000
6	0000	0110	0000	1001
7	0000	0111	0000	1010
8	0000	1000	0000	1011
9	0000	1001	0000	1100
10	0001	0000	0100	0011
11	0001	0001	0100	0100
20	0010	0000	0101	0011
50	0101	0000	1000	0011
51	0101	0001	1000	0100
76	0111	0110	1010	1001
99	1001	1001	1100	1100

digit only. For this reason this code is often used in measurement for encoding disks or linear encoding scales (Figure 4.58). The disadvantage of a distance greater than 1 between 9 and 10 occurring is avoided by using the improved Glixon code (Table 4.17). For special purposes other codes are used, for example, the teletype CCITT code No 3 or, for data transmission, the ISO-CCITT code No. 5 (Steinbuch and Weber, 1974).

During data input, transmission or processing procedures errors may arise because of such defects as a wrong perforation of a punched card. Error-detecting and error-correcting codes having additional code redundancy have been designed. The Hamming distance d_{min}, i.e. the minimum distance between

Table 4.16   Formation of the Gray code

Decimal number	0	1	2	3	4	5	6	7
Binary number	000	001	010	011	100	101	110	111
Shifted binary number	000	001	010	011	100	101	110	111
$\sum$	0000	00011	0110	0101	1100	1111	1010	1001
Gray code	000	001	011	010	110	111	101	100

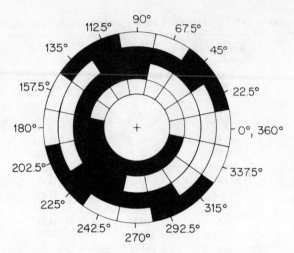

Figure 4.58   Coding disk (Gray code): code is read
radially from outside inward

two code words of an alphabet, has to be (Peterson, 1962), for error-detecting
codes with the degree f_d of errors to be detected,

$$d_{min} = f_d + 1 \qquad (4.66a)$$

and for error-correcting codes, if f_c is the degree of errors to be corrected

$$d_{min} = 2f_c + 1 \qquad (4.66b)$$

If correction to the degree $f_c^* < f_c$ only is used, it is possible to detect additional
errors up to degree f_d:

$$f_d^* = d_{min} - 2f_c^* - 1 \qquad (4.66c)$$

Table 4.17   Glixon code

Decimal number	Glixon code
0	0000
1	0001
2	0011
3	0010
4	0110
5	0111
6	0101
7	0100
8	1100
9	1000

A very simple but often used method for data inputting, is the addition of parity bits (parity check). This additional bit is chosen in such a way that the sum of all digits (the so called weight of the code) is either an even or an odd number. Table 4.18 presents the BCD code with parity check, detecting all errors with odd weight $(1, 3, \ldots)$. This code is often used, with excess-three basis, for data input as punched cards or punched tapes

Other and more complicated error-detecting or error-correcting codes are the selector code (w-out-of-n code), the Hamming codes produced by feedback shift registers (Hamming, 1950), the recurrent or cyclic codes (Peterson, 1962), and codes with block protection (Wozencraft and Reiffen, 1961).

As an example the conversion of the excess-three code to the binary code is now considered. With the relationships between x and y shown, Table 4.15 yields in Boolean algebra form

$$x_0 = \bar{y}_0$$
$$x_1 = (y_0 \cdot y_1) + (\bar{y}_0 \cdot y_1)$$
$$x_2 = (\bar{y}_1 \cdot \bar{y}_2) + (\bar{y}_0 \cdot \bar{y}_2) + (y_0 \cdot y_1 \cdot y_2)$$
$$x_3 = (y_2 \cdot y_3) + (y_0 \cdot y_1 \cdot y_3)$$

where

$\cdot$ is the logical AND operation
$+$ is the logical OR operation

The realization of these logic equations leads to the circuit of Figure 4.59.

Table 4.18 BCD code with parity check

First tetrad	Second tetrad	Parity check	Decimal number	$\sum 1$
0000	0000	1	0	1
0000	0001	0	1	1
0000	0010	0	2	1
0000	0011	1	3	3
0000	0100	0	4	1
0000	0101	1	5	3
0000	1001	1	9	3
0001	0000	0	10	1
0001	0001	1	11	3
0010	0000	0	20	1
0101	0000	1	50	3
1001	1001	1	99	5

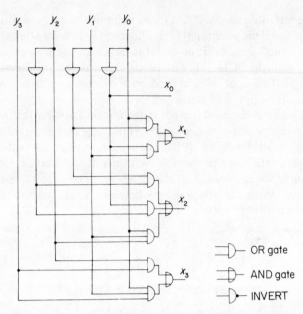

Figure 4.59   Circuit for the conversion of the excess-three
code to the binary code

4.4.5   Modulation Theory

A special form of adaptation of the source to the channel is that of modulation. Transmission of signals from several sources over a single channel can be accomplished using frequency-division or time-division multiplexing systems.

In measurement the problem of transmitting the output signals of many sensors or transmitters over one line is often solved by means of the time-division (time-sharing) method, shown in Figure 4.60 with the pulse interval t_p and the period t_o. Pulse modulation, with interleaving of the different signals, is applied there for parallel-to-serial conversion. In general a modulator may be interpreted as a controlled system with the carrier signal as one input and the modulation signal $x(t)$ as the control input (Figure 4.61). In the following a survey is first given of the several kinds of modulation.

In the analog modulation methods one, or several, parameters of the sinusoidal oscillation $u(t)$, termed the carrier oscillation, with the carrier frequency Ω_0:

$$u(t) = \hat{U} \sin(\Omega_0 t + \phi) = \hat{U} \sin \theta(t) \qquad (4.67a)$$

are caused to vary by the modulation signal $x(t)$. If the amplitude $\hat{U}$ is altered by the input signal $x(t)$, amplitude modulation (AM) results:

$$\hat{U} = f(x(t)) \qquad (4.67b)$$

(a)

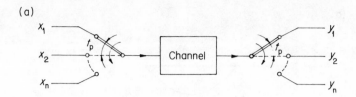

(b)

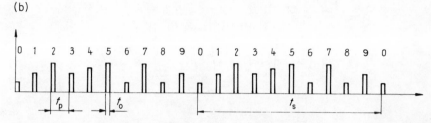

Figure 4.60   Time-division multiplexing: (a) system; (b) pulse frame (for ten transmitters)

Angle modulation is generated by using $x(t)$ to vary the argument of $u(t)$:

$$\theta(t) = f(x(t)) \tag{4.67c}$$

As $\theta = \Omega_0 t + \phi$, two kinds of angle modulation exist.

Frequency modulation (FM) occurs when Ω_0 is varied as

$$\Omega(t) = f(x(t)) \tag{4.67d}$$

Phase modulation (PM) occurs when ϕ is varied as

$$\phi(t) = f(x(t)) \tag{4.67e}$$

Figure 4.62 shows the modulation methods mentioned above for a sinusoidal
modulation signal with the modulation frequency ω:

$$x(t) = \hat{X} \sin(\omega t) \tag{4.67f}$$

Pulse modulation methods use the principle of sampling (see Chapters 5
and 12).

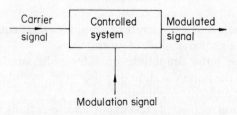

Figure 4.61   Generalized modulation system

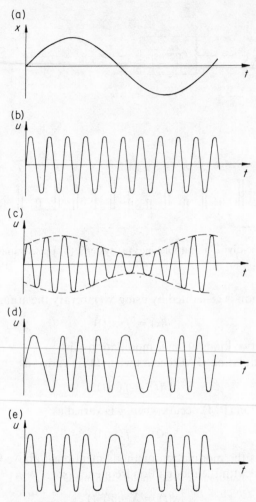

Figure 4.62 Analog modulation methods: (a)
modulation signal $x(t)$; (b) carrier oscillation; (c)
amplitude modulation; (d) frequency modulation;
(e) phase modulation

As shown in Figure 4.63 parameters of a regular pulse sequence are changed
by the modulation signal $x(t)$ in one of several ways.

Variation of the pulse amplitude provides pulse-amplitude modulation
(PAM), while with the pulse-duration or pulse-width modulation the lengths of
the pulses correspond to the modulation signal.

A pulse modulation method of great importance, especially in communication
because of its good signal-to-noise ratio, is pulse-code modulation (Figure 4.64).

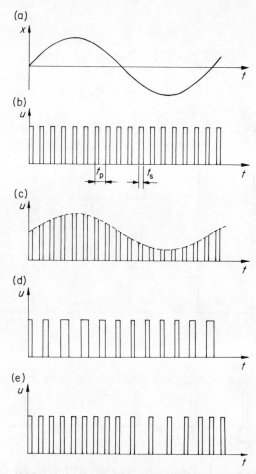

Figure 4.63   Pulse modulation methods (t_s = sampling time): (a) modulation signal $x(t)$; (b) carrier pulse sequence; (c) pulse-amplitude modulation; (d) pulse-duration or pulse-width modulation; (e) pulse-phase or pulse-position modulation

In the usual case a pulse-amplitude modulation signal is first generated, this then being converted to a coded pulse sequence using one of the codes of Section 4.4.4.

We now deal with some details such as the bandwidth necessary and applications to measurement.

A sinusoidal modulation signal

$$x(t) = \hat{X} \sin(\omega t + \varphi)$$

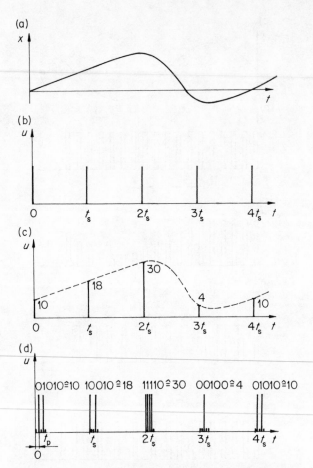

Figure 4.64 Pulse-code modulation: (a) modulation
signal; (b) carrier pulse sequence; (c) pulse-amplitude
modulation; (d) pulse-code modulation (binary code)

operating on a carrier $\hat{U}_0 \sin(\Omega_0 t)$ yields the amplitude modulated oscillation

$$u(t) = \hat{U}_0[1 + m \sin(\omega t + \varphi)]\sin(\Omega_0 t) \qquad (4.68a)$$

where the modulation depth $m = k\hat{X}$. The representation of equation (4.68a)
in the frequency domain is (Woschni, 1973; Wozencraft and Jacobs, 1965)

$$u(t) = \hat{U}_0\{\sin(\Omega_0 t) \pm \tfrac{1}{2}m \cos[(\Omega_0 \mp \omega)t \mp \varphi]\} \qquad (4.68b)$$

showing that the resultant is a signal with the carrier frequency and two side
frequencies (Figure 4.65a). The composition of the several spectral frequencies
provides the time function, as shown in Figure 4.65b.

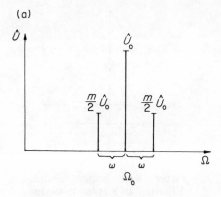

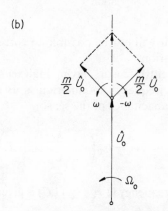

Figure 4.65 (a) Frequency spectrum of sinusoidal amplitude modulation; (b) indicator representation

In the general case of an input signal with bandwidth $\omega_1 - \omega_c$ to be modulated, the bandwidth needed by the transmission link is

$$b = 2\omega_c \triangleq \Omega_0 \pm \omega_c \tag{4.68c}$$

around the carrier frequency Ω_0. An amplifier has to have at least this bandwidth otherwise distortion of the original form of the modulated signal will arise in the later recovered signal (Woschni, 1981).

In measurement, amplitude modulation results at the output of a bridge operating with inductive sensors, as presented in Figure 4.66. To obtain satisfactory dynamic behaviour the condition between the limiting frequency of the measured input ω_c and the carrier frequency Ω_0 needs to be

$$\Omega_0 \geqslant 5\omega_c \tag{4.68d}$$

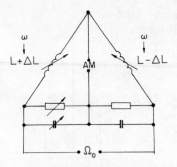

Figure 4.66 Bridge circuit, delivering an amplitude modulation

If this condition is not met it will not be possible to correctly demodulate the amplitude-modulated oscillation.

For the operation of capacitive sensors having high sensitivity and in certain cases for inductive sensors, frequency modulation is used (Figure 4.67). For sinusoidal variation of the capacitance of the sensor

$$C = C_0 + \Delta C = C_0\left(1 + \frac{\Delta C}{C_0}\sin(\omega t)\right) \qquad (4.69a)$$

the variation of the natural frequency $\Omega = 1/\sqrt{(LC)}$ is given by (Woschni, 1962)

$$\Omega = \Omega_0 + \Delta\Omega\sin(\omega t) = \frac{1}{\sqrt{(LC_0)}}\left[1 - \frac{\Delta C}{2C_0}\sin(\omega t) + \frac{3}{8}\left(\frac{\Delta C}{C_0}\right)^2\sin^2(\omega t) - \cdots\right]$$

$$(4.69b)$$

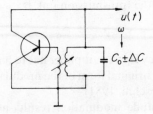

Figure 4.67 Circuit for the operation of a capacitive sensor to provide frequency-modulated output

Because of the non-linear characteristic differential capacitances sensing methods are used. The time function yields

$$u(t) = \hat{U}_0 \sin\left(\int [\Omega_0 + \Delta\Omega \sin(\omega t)] \, dt\right)$$

$$= \hat{U}_0 \sin\left(\Omega_0 t - \frac{\Delta\Omega}{\omega} \cos(\omega t)\right) \tag{4.69c}$$

The corresponding function for sinusoidal phase modulation is given by

$$u(t) = \hat{U}_0 \sin[\Omega_0 t + \Delta\phi \sin(\omega t)] \tag{4.69d}$$

A comparison between both equations shows that

$$\Delta\Omega/\omega = \Delta\phi \tag{4.69e}$$

represents the equivalent phase deviation (modulation index). Therefore the relationships contained in Table 4.19, between frequency and phase modulation,

Table 4.19 Relations between frequency and phase modulation

	Frequency modulation	Phase modulation
Frequency deviation	$\Delta\Omega$	$\Delta\Omega = \Delta\Phi\omega$
Phase deviation	$\Delta\Phi = \dfrac{\Delta\Omega}{\omega}$	$\Delta\Phi$

are valid (Woschni, 1962). The spectrum of a frequency- or phase-modulated signal is derived by means of a series expansion of Bessel functions (Woschni, 1962), leading to a bandwidth necessary for distortion-free transmission given by

$$b \simeq 2[\Delta\Omega + k\omega] \tag{4.69f}$$

where $1 < k < 2$.

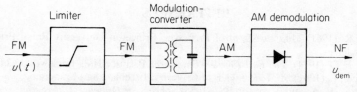

Figure 4.68 Principle of demodulation of frequency modulation

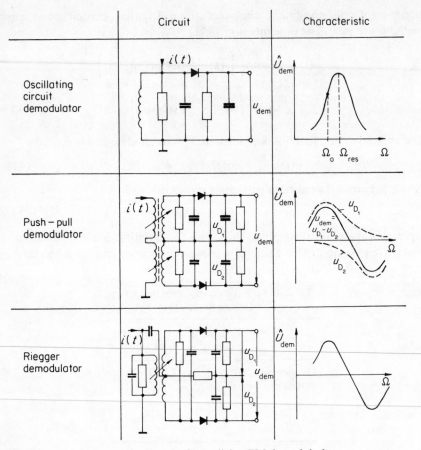

Figure 4.69   Circuits for realizing FM demodulation

The demodulation of a frequency-modulated oscillation is realized as demonstrated in Figure 4.68. A limiting stage is followed by a modulation converter converting the frequency modulation into an amplitude modulation which rectified by means of a diode. Some example circuits realizing the demodulation are shown in Figure 4.69 (Woschni, 1962).

REFERENCES

Bellman, R. (1961). *Adaptive Control Processes*, Princeton University Press, Princeton, New Jersey.
Birgham, E. O. (1974). *The Fast Fourier-Transform*, Prentice-Hall, Englewood Cliffs.
Blumenthal, L. (1961). *A Modern View of Geometry*, Freeman, San Francisco.
Coddington, E. A. and Levinson, N. (1955). *Theory of Ordinary Differential Equations*, McGraw-Hill, New York.

Davies, W. D. T. (1970). *System Identification for Self-Adaptive Control*, Wiley, Chichester.
Fano, R. M. (1961). *Transmission of Information*, Wiley, Chichester.
Feinstein, A. (1958). *Foundations of Information Theory*, McGraw-Hill, New York.
Feldtkeller, R. (1962). Einführung in die Vierpoltheorie, Hirzel, Stuttgart.
Finkelstein, L. (1976). *Paper* CTH 301, *Preprints IMEKO Congr.*, Institute of Measurement and Control, London.
Gelfand, I. M. and Schilow, G. E. (1960). *Verallgemeinerte Funktionen (Distributionen)*, Bd. 1, Deutscher Verlag d. Wiss., Berlin.
Goldman, S. (1953). *Information Theory*, Prentice-Hall, New York.
Goodman, J. W. (1968). *Introduction to Fourier Optics*, McGraw-Hill, New York.
Hamming, R. (1950). 'Error detecting and error correcting codes', *Bell. Syst. Tech. J.*, **19**, 147.
Harmuth, H. F. (1970). *Transmission of Information by Orthogonal Functions*, Springer, Berlin.
Jahnke, E. F. (1960). *Tafeln höherer Funktionen*, Teubner, Leipzig.
Kaplan, W. (1962). *Operational Methods for Linear Systems*, Addison-Wesley, Reading, Mass.
Koenig, H. E. and Blackwell, W. A. (1961). *Electromechanical System Theory*, McGraw-Hill, New York.
Newton, G. C., Gould, L. A. and Kaiser, J. F. (1957). *Analytical Design of Linear Feedback Controls*, Wiley, New York.
Olson, H. F. (1943). *Dynamic Analogies*, Van Nostrand, London.
Peterson, W. W. (1962). *Error Correcting Codes*, MIT Press, Cambridge, Mass.
Reichardt, W. (1960). *Grundlagen der Elektroakustik*, Akademische Verlagsgesellschaft, Leipzig.
Schlitt, H. (1960). *Systemtheorie für Regellose Vorgänge*, Springer, Berlin.
Shannon, C. E. (1948). 'A mathematical theory of communication,' Bell Syst. Tech. J., **27**, 379–423.
Steinbuch, K. and Weber, W. (1974). *Taschenbuch der Informatik*, Bd. 1, Springer, Berlin.
Tou, J. T. (1964). *Modern Control Theory*, McGraw-Hill, New York.
Ventcelj, J. S. (1964). *Elementary Dimamiceskogo Programmirovanijy*, Nauka, Moscow.
Woschni, E.-G. (1962). *Frequenzmodulation*, Verlag Technik, Berlin.
Woschni, E.-G. (1967). 'Parameterempfindlichkeit in der Meßtechnik, dargestellt an einigen typischen Beispielen', *Z. Messen-Steuern*, **10**, No. 4, 124–130.
Woschni, E. G. (1968). 'Einige Meßverfahren zur Messung stochastischer Größen', *Z. Messen-Steuern*, **11**, 428–32.
Woschni, E.-G. (1969). 'Inwieweit spielt die Qualität eines Meßgrößenaufnehmers beim Einsatz von On-Line-Rechnern noch eine Rolle?' *Z. Messen-Steuern*, **12**, No. 10, 384–5.
Woschni, E.-G. (1972a). *Meßdynamik*, Hirzel, Leipzig.
Woschni, E.-G. (1972b). *Sber. Plenums Klassen Akad. Wiss. DDR* No. 33–22.
Woschni, E. G. (1981). *Informationstechnik. Signal, System, Information*, Verlag Technik, Berlin.
Woschi, E.-G. and Kraus, M. (1975). *Meßinformationssysteme*, Verlag Technik, Berlin.
Woschni, E.-G. and Kraus, M. (1976). *Informationstechnik. Arbeitsbuch*, Verlag Technik, Berlin.
Wozencraft, J. M. and Jacobs, I. M. (1965). *Principles of Communication Engineering*, Wiley, New York.
Wozencraft, J. M. and Reiffen, B. (1961). *Sequential Decoding*, MIT Press, Cambridge, Mass.
Zadeh, L. A. and Desoer, C. A. (1963). *Linear System Theory*, McGraw-Hill, New York.

Handbook of Measurement Science, Vol. 1
Edited by P. H. Sydenham
© 1982 John Wiley & Sons Ltd.

Chapter

5   M. J. MILLER

Discrete Signals and Frequency Spectra

Editorial introduction

Rapid advances in data processing methods—both at the conceptual, procedural level of theoretical understanding and at the hardware implementation stage—have given measurement system designers truly great power to implement, for reasonable cost, advanced processing procedures. These advances have occurred predominantly for the discrete form of electric signal.

This chapter outlines the fundamentals required in understanding and making appropriate use of the digital techniques now rapidly coming into routine use even in instruments at the bottom end of the price range. It will be seen that practical implementation requires application of certain methodologies and that severe errors in interpreting the processed data can occur if the methods are not used appropriately.

It extends some of the material of the previous chapter. As time passes the material will become increasingly more important as advanced signal processing finds yet more application and greater use of the digital signal format. An appreciation of this trend is to be found in Oppenheim (1978), a text presenting chapters devoted to digital signal processing in a range of diverse application areas. An extensive review of the mathematics required is available in Rader and McClennan (1979). General textbooks, that include many worked examples, are Oppenheim and Schafer (1975) and Rabiner and Gold (1975).

5.1   INTRODUCTION

The availability of the microcomputer and other relatively inexpensive digital electronic hardware has resulted in the emergence of new approaches to the signal processing problems that occur in many measurement systems. *Digital signal processing* is fast replacing conventional analog techniques in spectrum analysers and other signal processors used in a wide variety of applications and across many disciplines. The availability of instrumentation capable of carrying out a fast and efficient calculation of the Fourier transform has provided the means for readily displaying signals as either time functions or in dynamically updated frequency spectrum form. This equipment finds applications in a wide range of disciplines such as in analysis of vibrations for geological research, mechanical engineering, sonar technology or chemistry. The time and frequency domain properties of signals are also important to the radio-astronomer in characterizing signal sources, to the neuropsychologist for analysing electro-encephalograms or to the engineer concerned with processing and synthesis of speech.

This chapter examines the principles involved in using digital processing for performing operations such as the Fourier transform. Such computer-based processing implies that the signals under consideration must be in *discrete* form, that is in the form of finite sequences of discrete quantities. Throughout this chapter it is assumed that the signals are available in discrete form but these techniques can be readily applied to continuous or analog waveforms provided appropriate analog-to-digital (A/D) conversion is used as discussed in Chapter 12. Furthermore it is assumed that the sequences have discrete values in time but continuous real amplitude values, that is as though there were no quantiza-tion of the sample amplitudes. (Some authors use the term *discrete* to describe such signals and *digital* to describe sampled and quantized signals.) This infinite-bit-precision assumption simplifies the study of processing procedures and permits a larger general body of theory to be used.

Fourier transform techniques, particularly the *discrete Fourier transform* (DFT) will be the central theme in the description of the processing tools available for the discrete signal domain. The DFT has been used in practical measurement applications for many years but it usually required relatively expensive special-purpose computer facilities with appropriate software development. The emergence in the 1960's of the *fast Fourier transform* (FFT) algorithm for drastically improving the computational speed associated with the DFT operation has completely changed the situation. The FFT and the availability of the microcomputer brought about a very rapid expansion of applications of these digital processing techniques. FFT analysers are now readily available in instrumentation form at moderate price.

In this chapter the principles of these discrete Fourier transform techniques will be discussed, particularly as they are applied in the area of *spectrum analysis*. It has now become relatively common for measurement systems to include FFT analyser instruments or else an equivalent software package on a main-frame computer. The nature of the DFT process is such however, that unless special care is taken, the resultant frequency spectrum estimates may be considerably in error and, even more annoying, calculations using different sets of data samples from the same signal or process may yield vastly different results. This chapter gives particular attention to the interpretation of the steps involved in discrete Fourier transformations and to the practical implications for the user. Initially the DFT processing relationships are stated without proof and their key properties summarized. Then a descriptive/graphical interpretation is given of the process to lay a firm foundation for understanding the theoretical development that follows.

Throughout it is assumed that the signals to be encountered in practice, whether they be radar or sonar echoes, electrical responses from voice or biomedical systems or outputs from mechanical transducers, will usually be *randomly* time varying. Use will therefore be made of statistical procedures (based on material presented in Chapter 6) especially in dealing with the interpretation of results and describing techniques for reducing errors in the spectral estimation procedures.

Another of the most notable areas of development in discrete system theory and techniques has been in the field of *digital filters*. Whilst reference is made in this chapter to some elementary ideas about digital filters, it is left to Chapter 10 to deal more specifically with that topic.

A great deal of the theory associated with discrete-time signals and linear systems will be familiar to the person who is well versed in traditional analog theory of linear systems; as presented in Chapter 4. The description of signals in the time and frequency domains and the use of the important connections between multiplication and convolution operations will be assumed well understood in what follows. Background is available in Chapter 4.

5.2   DISCRETE TIME SEQUENCES

A *discrete* signal is a sequence of numbers or sample values spaced usually at uniform intervals of time as illustrated by the sample sequence $x(kT_s)$ at the A/D converter output in Figure 5.1, where $x(kT_s)$ is the sequence of sample values $x(0)$, $x(T_s)$, $x(2T_s)$, ... and T_s is the sampling time interval. In practice most A/D converter outputs are in the form of sequences of multiple-bit binary words, one word per signal sample and each word appearing as a set of 1's and 0's on parallel output lines. The sequence $x(kT_s)$ as shown in Figure 5.1 is intended, therefore, only to represent a set of discrete sequential signal values and not necessarily an actual waveform. Furthermore, as previously mentioned, the amplitudes of the sequence values are assumed continuous, whereas any practical converter will have a finite number of quantization levels. Discrete sequences of interest here are not necessarily restricted to sampled values of analog signals. A sequence $x(kT_s)$ may be a set of numbers ... $x(0)$, $x(T_s)$,

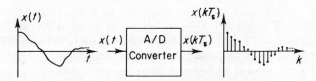

Figure 5.1   Continuous and discrete signals

$x(2T_s), \ldots$ corresponding, say, to vehicle speed data or sunspot observations being processed in a computer.

Discrete time signals are defined only at discrete values of the independent variable, time, that is at $t = kT_s$, where k is an integer and T_s is the interval between samples. For example T_s may be an interval of 24 hours if the sequence $x(kT_s)$ represents daily sunspot readings.

Several notational forms may be used to describe such a sequence including the following:

$x(kT_s)$ or $\{x(kT_s)\}$ which imply uniform spacing and x_k or $x(k)$ or $\{x(k)\}$ which may apply to uniform or non-uniform spacing.

Uniform spacing will be assumed in what follows and since the sampling interval (T_s) appears as a constant multiplier factor, it is often convenient to assume it unity for many calculation purposes.

5.3 THE DISCRETE FOURIER TRANSFORM SUMMARIZED

For a time sequence $x(kT_s)$ consisting of N samples uniformly spaced T_s seconds apart, the discrete-time to discrete-frequency Fourier transform pair most commonly used is given by

$$X\left(\frac{nf_s}{N}\right) = \sum_{k=0}^{N-1} x(kT_s)\,e^{-j2\pi nk/N} \quad \begin{array}{l} n = 0, 1, \ldots, N - 1 \\ f_s = 1/T_s \end{array} \tag{5.1a}$$

$$x(kT_s) = \frac{1}{N}\sum_{n=0}^{N-1} X\left(\frac{nf_s}{N}\right) e^{j2\pi nk/N} \quad k = 0, 1, \ldots, N - 1 \tag{5.1b}$$

Figure 5.2 illustrates an application of this transform in displaying the frequency components of a signal on a typical FFT analyser.

It is left to a later stage to show how the DFT equations (5.1) are developed for discrete signals. At this point led us summarize their important features:

(a) The DFT equations (5.1) are of similar form to the well known Fourier transforms for analog signals:

$$X(f) = \int_{-\infty}^{\infty} x(t)\,e^{-j2\pi ft}\,dt \tag{5.2a}$$

$$x(t) = \int_{-\infty}^{\infty} X(f)\,e^{j2\pi ft}\,df \tag{5.2b}$$

Note, however, that the exponential terms in (5.1) do not contain an explicit f_s or T_s term. As will be explained later, the n and k parameters have the connotations of frequency and time respectively.

(b) The DFT transforms an N-point *discrete*-time sequence

$$x(kT_s) = x(0), x(T_s), x(2T_s), \ldots, x((N - 1)T_s)$$

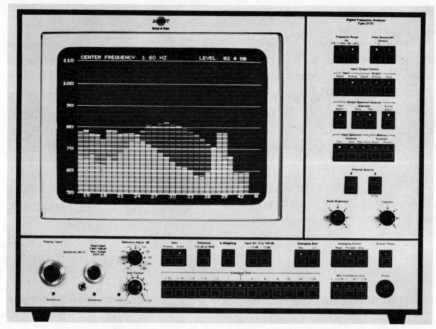

Figure 5.2   Typical FFT analyzer (by permission of Brüel and Kjaer)

into an N-point *discrete*-frequency domain sequence

$$X\left(\frac{nf_s}{N}\right) = X(0), X\left(\frac{f_s}{N}\right), X\left(\frac{2f_s}{N}\right), \ldots, X\left(\frac{(N-1)f_s}{N}\right)$$

(c)   An example of a DFT pair is illustrated in Figure 5.3. The DFT of a sequence of *real* time values results in a sequence of *complex* frequency values, commonly represented by separate plots of magnitude $|X(nf_s/N)|$ and phase $\theta_X(nf_s/N)$. (It is recommended as highly instructive for the uninitiated reader to check the values shown in Figure 5.3—a hand calculator would suffice.)

(d)   In practice, the signal sequence $x(kT_s)$ is only a segment of the much longer sequence which may possibly arise from the particular system generating $x(kT_s)$. The $x(kT_s)$ sample shown in Figure 5.3 can be thought of conviently in terms of the longer x signal sequence multiplied by a truncating *window function* $s_0(t)$ being $T_0 = NT_s$ seconds long and shown dotted in Figure 5.3 where

$$s_0(t) = \begin{cases} 1 & -\tfrac{1}{2}T_s < t < T_0 - \tfrac{1}{2}T_s \\ 0 & \text{otherwise} \end{cases}$$

The end points of the truncation window $s_0(t)$ are conceived to lie at the midpoint of adjacent samples.

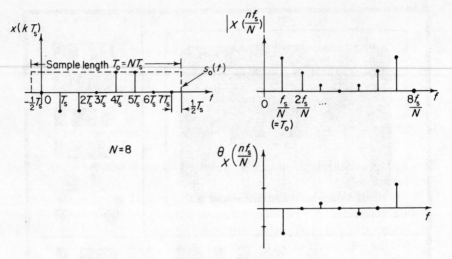

Figure 5.3   Typical discrete signal sample $x(kT_s)$ and transform $X(nf_s/N)$

(e) The spacing between the frequency sequence values is inversely proportional to the sample length (T_0) since

$$\text{Frequency spacing} = \frac{f_s}{N} = \frac{1}{NT_s} = \frac{1}{T_0}$$

Clearly, the longer the sequence length, the finer the frequency *resolution* in the transformed sequence.

(f) If the infinitely long sequence of which $x(kT_s)$ is a part, has random components throughout its entire history, then its frequency spectrum (that is, its Fourier transform) does not exist in any meaningful sense. However, as will be shown later, the sequence $X(nf_s/N)$ in the frequency domain can be used to estimate a quantity called the power spectral density function (to be defined), which does exist even for purely random sequences.

5.4   GRAPHICAL DEVELOPMENT OF THE DFT

5.4.1   General Comment

The discrete Fourier transforms in equation (5.1) are a convenient but by no means unique form of DFT pair that could be used. To see the significance of this let us examine the development of the discrete transform expressions and their user implications. At the expense of some small additional complication, let us begin by assuming that an analog time function $x(t)$ is the starting point. Recall that by no means all discrete sequences derive from sampled versions of

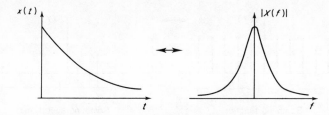

Figure 5.4   Illustrative Fourier transform pair

an analog signal but the approach here, based on Brigham (1974), provides important insights.

Consider the signal $x(t)$ and its Fourier transform in Figure 5.4. The symbol $\leftrightarrow$ is used to represent the Fourier operation. Actually the modulus $|X(f)|$ only is shown but this will illustrate the principles satisfactorily and in many practical cases is the only aspect of the frequency domain function that is of interest. Note also that this illustrative example assumes the form of $x(t)$ and $X(f)$ are known whereas, in practice, one or the other is usually unknown.

Consideration is now given to what happens when basic operations are performed.

5.4.2   Sampling: Aliasing Distortion

Sampling is required, as discussed in Chapters 4 and 12, to convert $x(t)$ to the (infinitely long) discrete signal $x_s(kT_s)$. This is shown in Figure 5.5 where k takes integer values from zero to infinity (for causal signals). Note that $x_s(kT_s)$ is taken to represent here a sampled function of time defined only for the discrete times $t = kT_s$. What now is the Fourier transform $X_s(f)$ of this sampled signal? As discussed in Chapter 4, we can write the sampled function $x_s(kT_s)$ as the product of the original signal $x(t)$ and a sampling function $s(t)$ the latter being a series of delta functions each of unit area. Hence

$$x_s(kT_s) = x(t)s(t) = x(t) \sum_{k=-\infty}^{\infty} \delta(t - kT_s) \tag{5.3}$$

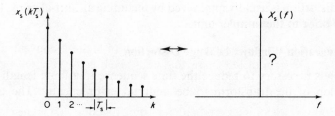

Figure 5.5   Discrete-time function

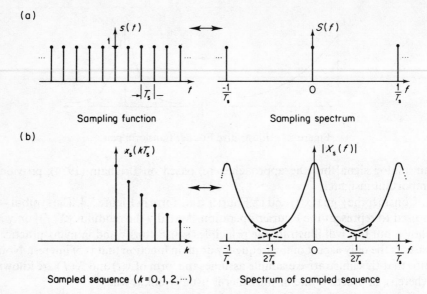

Figure 5.6   The effect of sampling: (a) Sampling function and transformation into frequency domain; (b) Sampled sequence and transformation into frequency domain

Furthermore, since multiplication in the time domain implies convolution in the frequency domain we can write the Fourier transform $X_s(f)$ of $x_s(kT_s)$ as

$$X_s(f) = X(f) * S(f) \tag{5.4}$$

where the symbol $*$ represents the convolution operation. Since $s(t)$ and $S(f)$ have the forms shown in Figure 5.6a it is a simple matter to carry out the necessary convolution of the original $X(f)$ function with each $S(f)$ delta function and then use superposition to obtain $X_s(f)$ in Figure 5.6b.

The following remarks can be made concerning the *sampling process*:

(a)   sampling $x(t)$ produced a sequence $x_s(kT_s)$ which has a *periodic* frequency spectrum such that knowledge of $X_s(f)$ over the interval 0 to $1/2T_s$ provides complete information about $X_s(f)$.

(b)   In the vicinity of $1/2T_s$ there is evidence of *aliasing distortion* which results from the fact that T_s could not satisfy the Nyquist criterion. In practice this distortion is usually minimized by including an anti-aliasing, low-pass, filter prior to the sampler unit.

5.4.3   Truncation Window: Leakage Distortion

Truncation is necessary to reduce the time sequence to a finite length to allow computation of the transform to be undertaken practically. The truncated function is

$$x(kT_s) \quad k = 0, 1, \ldots, N - 1$$

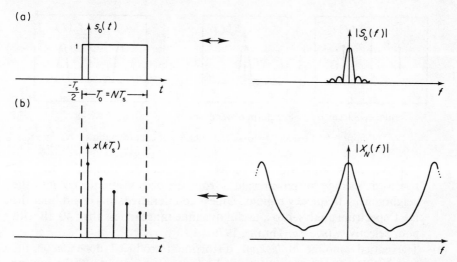

Figure 5.7   The effect of a truncation window: (a) truncation window and its transform;
(b) truncated sequence and its transform

as shown in Figure 5.7b. We can write $x(kT_s)$ as the product of the infinite sequence $x_s(kT_s)$ and some *truncation window function* $s_0(t)$ which is 1 over an interval from $-\frac{1}{2}T_s$ to $(N - \frac{1}{2})T_s$ and has spectrum $S_0(f)$. Hence

$$x(kT_s) = x_s(kT_s)s_0(t) = x(t)s(t)s_0(t) \tag{5.5}$$

with Fourier transform

$$X_N(f) = X(f) * S(f) * S_0(f) \tag{5.6}$$

where the modulus of $S_0(f)$ is the ubiquitous *sinc* function of f. Note that sinc $\alpha = \sin(\pi\alpha)/\pi\alpha$. The subscript N in the symbol $X_N(f)$ is intended to indicate that the Fourier transform is based on N samples only of $x(kT_s)$.

The following remarks can be made about the *truncation process*:

(a)   Truncation in time results in the second modification or distortion of the spectrum, called *leakage*. The effect of the rectangular time domain truncation window (Figure 5.7a) is equivalent in the frequency domain to convolving the sinc function $S_0(f)$ of the window with the spectrum $X_s(f)$. The effect is illustrated in Figure 5.7b by rippling distortion in the frequency domain.

(b)   If there were sharp transitions in the spectrum $X_s(f)$, the convolution with $S_0(f)$ will produce smoothing or blurring. For example, if the original time signal had contained a sinusoidal component (discrete-frequency component), convolution with the sinc function $S_0(f)$ would have resulted in each discrete spectral line (delta function) being replaced by a sinc function as illustrated in Figure 5.8.

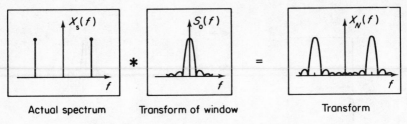

Figure 5.8   Leakage distortion due to a truncation window

This effect is called *leakage* because the truncation 'filtering' effect gives rise to a 'leakage' of power values from the original frequency into the neighbouring frequency regions. Unless counteracted, this could limit the DFT spectrum analyser to a useful dynamic range of less than 40 dB with poor selectivity (see e.g. Thrane, 1979).

(c) The actual amount of *leakage* distortion introduced depends on the length (T_0) of the truncation window in comparison to the sampling interval (T_s). In practice there will be trade-offs between the number of samples and the accuracy of the resultant transform. We will see in Section 5.6 how non-rectangular *smoothing windows* may be used to advantage to reduce undesirable effects of truncation.

5.4.4   Frequency Domain Sampling

We now have the Fourier transform $X_N(f)$ of $x(kT_s)$, the sampled truncated version of $x(t)$. However the continuous (periodic) frequency function $X_N(f)$ is still not in a suitable form to represent computer calculations which can only be represented by a finite number of samples. Hence we need to replace $X_N(f)$ by a discrete (sampled) version, say $X(nf_s/N)$. This can be considered as equivalent to the frequency domain sampling operation

$$X\left(\frac{nf_s}{N}\right) = X_N(f)S_f(f) \tag{5.7}$$

where, as illustrated in Figure 5.9a, $S_f(f)$ is a sampling (or discretization) function. Its time domain equivalent $s_f(t)$ is also shown. Reasons for choosing to sample the frequency function at integer multiples of $f_s/N(\text{Hz})$ will be discussed later where it will be shown that there are only N distinct frequency samples computable from the N time samples and that this frequency spacing will be adequate (in a Nyquist sense) to describe the spectrum.

The following remarks can be made about the process of discretization of the frequency function:

(a) The above operation is equivalent to the DFT analyser assuming that whatever sample segment of $x(t)$ it takes and processes, the rest of $x(t)$ is a

(a)

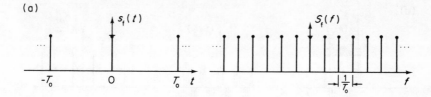

(b)

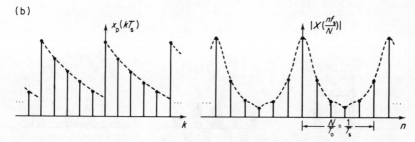

Figure 5.9 Effect of sampling in the frequency domain: (a) frequency sampling function $S_f(f)$ and its transform $S_f(t)$; (b) sampled spectrum $|X(nf_s/N)|$ and its transform $x_p(kT_s)$

periodic repetition of the sample function. To see this, note that the multiplication operation of equation (5.7) implies the convolution operation in the time domain which we can write as

$$x_p(kT_s) = x(kT_s) * s_f(t) \qquad (5.8)$$

If $s_f(t)$ is a series of relatively widely spaced delta functions (true since $T_0 \gg T_s$) it is easy to see that $x_p(kT_s)$, the result of this convolution, is a *periodic* sampled time function with period (T_0) of the form shown in Figure 5.9b, that is

$$x_p(kT_s) = x(kT_s + mT_0) \quad m \text{ an integer} \qquad (5.9)$$

Thus although the $x(t)$ values outside the truncation window interval 0 to NT_s(s) are unspecified, the DFT treats the original time function as though it were periodic with period $T_0 = NT_s$. Clearly there will, therefore, be differences between the spectrum of this fictitiously periodic signal and the actual signal being sampled.

(b) As a special case, if the original continuous signal $x(t)$ were *periodic*, this frequency sampling effect could be quite significant depending on whether or not the truncated sample of $x(t)$ contains exactly an integral number of periods of $x(t)$. That is, if the truncation window is 'fitted' over the periodic function as illustrated in Figure 5.10a then the actual $x(t)$ values are the

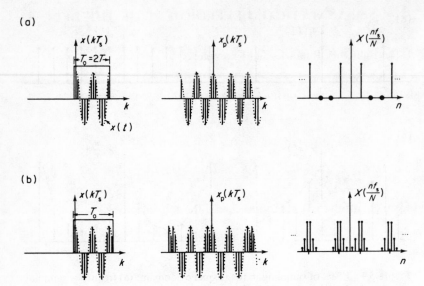

Figure 5.10 Effect of position of truncation window for a periodic signal:
(a) truncation window width equal to two periods of $x(t)$; (b) truncation window
not equal to nT

same as the fictitiously assumed values so no error is caused in the spectrum.
Consider, however, the situation as is depicted in Figure 5.10b where the
periodic input is truncated to an interval not equal to a period. The DFT
computes the transform of a periodic function $x_p(kT_s)$ with sharp discontinuities and as expected, these discontinuities give rise to additional
(invalid) components in the frequency domain. As will be discussed in
Section 5.6, these effects can be considerably reduced by employing *smoothing windows* which were referred to in the above discussion on leakage
distortion.

The important points of this section are summarized as follows.

Periodic sampling of the time function every $T_s(s)$ results in a periodic
frequency transform with period $1/T_s$. *Aliasing distortion* will occur unless
the time function is band-limited (or prefiltered) to frequencies less than
$1/2T_s$.

Truncation of the time function to a finite number (N) of sample values
results in *leakage* errors in the frequency domain. Sharp transitions in the
frequency spectrum will be smoothed out.

Discrete-sequence representation in the frequency domain implies that
the frequency domain values are the transforms of the original time
sequence treated as though it were repeated periodically with period
$NT_s(s)$. This effect, sometimes known as the 'picket fence' effect, may cause
errors in the measurement of periodic components in the spectrum.

5.5 ANALYTICAL DEVELOPMENT OF THE DFT

5.5.1 Introductory Remarks and Example

The above operations, presented graphically, can now be restated using analytical expressions. Consider the sampled infinite-length time function

$$x_s(kT_s) \quad k = \ldots, -1, 0, 1, \ldots$$

Its (continuous) Fourier transform $X_s(f)$ is given by

$$X_s(f) = \int_{-\infty}^{\infty} x_s(kT_s) e^{-j2\pi ft} dt$$

and from equation (5.3), reproduced here for convenience,

$$x_s(kT_s) = x(t) \sum_{k=-\infty}^{\infty} \delta(t - kT_s)$$

we obtain, on interchanging the order of summation and integration

$$X_s(f) = \sum_{k=-\infty}^{\infty} \int_{-\infty}^{\infty} x(t)\delta(t - kT_s) e^{-j2\pi ft} dt$$

Then using the sampling property of the delta function which can be expressed for any function $g(t)$ as

$$\int_{-\infty}^{\infty} g(t)\delta(t - kT_s) \, dt = g(kT_s)$$

we obtain the Fourier transform of the *infinite* sequence as

$$X_s(f) = \sum_{k=-\infty}^{\infty} x(kT_s) \exp(-j2\pi fkT_s) \qquad (5.10)$$

As was seen in Section 5.4, $X_s(f)$ is a complex continuous periodic function, with period $1/T_s$.

For *finite* duration sequences,

$$x(0), x(kT_s), x(2kT_s), \ldots, x((N - 1)kT_s)$$

the Fourier transform $X_N(f)$ based on N samples only, follows from the above as

$$X_N(f) = \sum_{k=0}^{N-1} x(kT_s) \exp(-j2\pi fkT_s) \qquad (5.11)$$

The *discrete Fourier transform* is the discrete frequency version of $X_N(f)$ taken at frequency points

$$0, f_s/N, 2f_s/N, \ldots$$

Since, as we have seen, $X_N(f)$ is periodic, N frequency points are sufficient so our N-point DFT becomes

$$X\left(\frac{nf_s}{N}\right) = X_N\left(f = \frac{nf_s}{N}\right) \quad n = 0, 1, \ldots, N - 1$$

$$= \sum_{k=0}^{N-1} x(kT_s)\exp\left[-j2\pi\left(\frac{nf_s}{N}\right)kT_s\right]$$

Therefore, since $f_s = 1/T_s$, we obtain the DFT

$$X\left(\frac{nf_s}{N}\right) = \sum_{k=0}^{N-1} x(kT_s)\,e^{-j2\pi nk/N} \qquad (5.12)$$

This is now applied, as an example, to the situation where it is desired to find $X_N(f)$ and $X(nf_s/N)$ for the sample waveform $x(kT_s)$ shown previously in Figure 5.3, values for which are tabulated for convenience as follows (also $T_s = 1$)

k	0	1	2	3	4	5	6	7
$x(kT_s)$	0	-1	-1	0	1	1	0	0

We have, from equation (5.11),

$$X_N(f) = \sum_{k=0}^{7} x(k)\,e^{-j2\pi fk}$$

$$= (-1)e^{-j2\pi f} + (-1)e^{-j4\pi f} + e^{-j8\pi f} + e^{-j10\pi f}$$

$$= e^{-j6\pi f}(-e^{j4\pi f} - e^{j2\pi f} + e^{-j2\pi f} + e^{-j4\pi f})$$

and using

$$\sin\theta = \frac{1}{2j}(e^{j\theta} - e^{-j\theta})$$

we obtain

$$X_N(f) = -2j\,e^{-j6\pi f}[\sin(2\pi f) + \sin(4\pi f)]$$

(Note that the variable f rather than $\omega(= 2\pi f)$ has been used throughout this work for consistency even though expressions such as the above become slightly more cumbersome.) The original time sequence and its Fourier transform are shown in Figure 5.11 where

$$|X_N(f)| = 2|[\sin(2\pi f) + \sin(4\pi f)]|$$

and

$$\arg[X_N(f)] = \begin{cases} -\frac{1}{2}\pi - 6\pi f & \text{when } \sin(2\pi f) + \sin(4\pi f) \geqslant 0 \\ \frac{1}{2}\pi - 6\pi f & \text{otherwise} \end{cases}$$

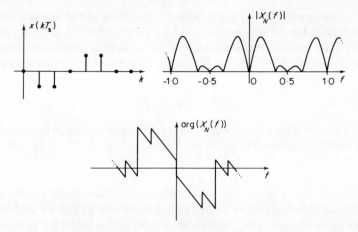

Figure 5.11 Example of Fourier transform $X_N(f)$ of a discrete sequence

Note that just as for Fourier transforms of continuous functions, since $x(kT_s)$ is a real-valued function, $|X_N(f)|$ has even symmetry about $f = 0$ and $\arg[X_N(f)]$ has odd. Reference also to Figure 5.3 in Section 5.3 shows the DFT $X(nf_s/N)$ of the same sequence as in this example for which it can be seen that

$$X\left(\frac{nf_s}{N}\right) = \begin{cases} X_N(f) & \text{for } f = nf_s/N \\ 0 & \text{otherwise} \end{cases}$$

Alternatively, equation (5.12) could have been used as will be discussed in more detail in a later section.

5.5.2 DFT Frequency Resolution

Some further comment is needed on the reasons for choosing the frequency spacing (f_s/N) in the DFT equation (5.12). The (continuous) Fourier transform $X_N(f)$ could, in principle, be evaluated in discrete form at any finite set of frequencies. Consider the evaluation of $X_N(f)$ at say L points spaced f_L(Hz) apart, that is at the frequencies

$$0, f_L, 2f_L, \ldots, (L - 1)f_L$$

using

$$X_N(rf_L) = \sum_{k=0}^{N-1} x(kT_s) \exp(-j2\pi rf_L kT_s) \qquad (5.13)$$

It is now important to ask what is the minimum value of the frequency spacing (f_L), that is, what is the best *frequency resolution* obtainable? The answer is that there is a likeness to the 'Nyquist theorem' operating in the frequency domain such that the values of the continuous Fourier transform $X_N(f)$ can be interpolated exactly at all f from a knowledge of values at only the discrete frequencies rf_L if the frequency spacing is small enough. In particular f_L must satisfy

$$f_L < \frac{1}{(N-1)T_s}$$

This can be appreciated by analogy with the Nyquist theorem in the time domain which provides a lower bound on the sampling rate $(1/T_s)$. As is well known, provided $x(t)$ is band-limited with maximum frequency $B(\text{Hz})$ then $x(t)$ can be reconstructed perfectly from sample values if the sample rate satisfies $1/T_s > 2B$. Such a finite length sequence of N samples will have total duration $T_0 = (N-1)T_s$, as shown in Figure 5.12, where $x(t)$ is shown defined for a symmetrical time interval to parallel the symmetrical positive and negative frequency spectra in the time-sampling theorem. The dual of this is that provided $X_N(f)$ is the transform of a finite duration function of length T_0, then $X_N(f)$ can be reconstructed perfectly from sample values if the frequency samples are spaced such that

$$1/f_L < 2T_0/2$$

or, since $T_0 = (N-1)T_s$, the upper bound on f_L is

$$f_L < \frac{1}{(N-1)T_s}$$

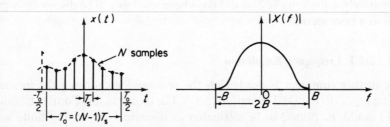

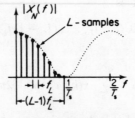

Figure 5.12   Illustration of Nyquist theorems in the time and frequency domains

It is usual in calculating DFT's to choose a frequency spacing

$$f_L = \frac{1}{NT_s} = \frac{f_s}{N} = \frac{1}{T_0} \tag{5.14}$$

where T_0 is the truncated sample length. This simplifies computation of the exponentials involved and represents an appropriate choice since in practice $N \gg 1$. Hence, the DFT denoted $X(nf_s/N)$ is calculated at frequencies (nf_s/N) for $n = 0, 1, \ldots$, that is, values at frequencies $0, f_s/N, 2f_s/N, \ldots$.

5.5.3 DFT Calculations

Consider the computations required for the DFT of the 8 point time sequence used in the example given in Section 5.5.1. The DFT was shown previously in Figure 5.3. It is instructive to consider the DFT calculations required using equation (5.12) which for convenience we write as

$$X\left(\frac{nf_s}{N}\right) = \sum_{k=0}^{N-1} x(kT_s)W^{nk} \quad \text{for } n = 0, 1, \ldots, N-1 \tag{5.15}$$

where the symbol W is used to replace the exponential 'weighting function'. That is

$$W = e^{-j2\pi/N}$$

In 'long-hand', the computations required for the DFT example are as follows: the term at $f = 0$ is $(n = 0)$

$$X(0) = x(0)W^0 + x(T_s)W^0 + x(2T_s)W^0 + \cdots + x(7T_s)W^0$$

the term at $f = f_s/8$ is $(n = 1)$

$$X(f_s/8) = x(0)W^0 + x(T_s)W^1 + x(2T_s)W^2 + \cdots + x(7T_s)W^7$$

the term at $f = 2f_s/8$ is $(n = 2)$

$$X(f_s/4) = x_0 W^0 + x(T_s)W^2 + x(2T_s)W^4 + \cdots + x(7T_s)W^{14}$$

finally, the term at $f = 7f_s/8$ is $(n = 7)$

$$X(7f_s/8) = x_0 W^0 + x(T_s)W^7 + x(2T_s)W^{14} + \cdots + x(7T_s)W^{49}$$

It is helpful to visualize the values of the exponentials (weights) plotted on a unit circle in the complex plane, namely

$$W^0 = 1, \ W^1 = e^{-j2\pi/8}, \ W^2 = e^{-j4\pi/8}, \ldots$$

These are shown in Figure 5.13. As can be seen $W^0 = W^8 = W^{16} = \cdots$ and $W^1 = W^9 = \ldots$, and in general

$$W^i = W^{n+i} \tag{5.16}$$

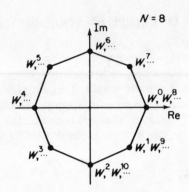

Figure 5.13 Exponential weights
in the complex plane

This fact can be used in programming a computer to carry out the above calculations. The FFT program algorithms first published by Cooley and Tukey in 1965 gave tremendous impetus to the DFT processing techniques because of the drastic reduction brought about in computer processing time required. As can be seen from the above, in the evaluation of each term for $n = 0, 1, \ldots,$ $(N - 1)$ certain exponential functions occur repeatedly. For example, in the above

$$W^4 = e^{-j\pi} \text{ occurs when } n = 1, k = 4$$
$$n = 2, k = 2$$
$$n = 4, k = 4$$

By calculating these repeated functions only once and working on all N summations at the same time, the FFT algorithms can give a time reduction factor of about $(\log_2 N)/N$ over an unsophisticated DFT calculation. As an example, for $N = 2^{14} = 16,384$ points, a time reduction factor of $14/16384$ results, (say, a reduction from an hour to a few seconds).

The FFT algorithms will be discussed further in a later section. Suffice to say at this point that the FFT makes it feasible to take many samples of time functions and to compute their Fourier transform almost as they occur. The advantage of this in applications such as real-time spectrum analysers will be examined in the next section.

Before doing so, it is important to clarify the frequency resolution question posed earlier as to the minimum frequency spacing between the frequency domain values. We have seen that the N-point DFT (or its FFT counterpart) yields an N-point frequency sequence with frequency spacing (f_N) equal to the inverse of the sample length. That is

$$f_N = \frac{1}{NT_s} = \frac{1}{T_0}$$

The *frequency resolution* can therefore be improved by the following techniques:

(a) For a given sampling rate, increase the number of sample points (N) used—at the expense, however, of computing time.

(b) For a finite length sequence, by artificially *adding zeros* to the end of the data string. For example, the above 8-point DFT considered previously could be written as a 16-point data set $x_a(kT_s)$ as follows

k	0	1	2	3	4	5	6	7	8	9	10	11	12	13	14	15
$x_a(kT_s)$	0	−1	−1	0	1	1	0	0	0	0	0	0	0	0	0	0

It is easy to see that both sequences $x(kT_s)$ and the augmented set $x_a(kT_s)$ have the same Fourier transform 'shape' since equation (5.11) becomes

$$X_N(f) = \sum_{k=0}^{7} x(kT_s)\exp(-j2\pi f kT_s)$$

$$= X_a(f) = \sum_{k=0}^{15} x_a(kT_s)\exp(-j2\pi f kT_s)$$

with

$$X_a(f) = X_N\left(f = \frac{nf_s}{16}\right) \quad n = 0, 1, 2, \ldots, 15$$

This result is illustrated in Figure 5.14

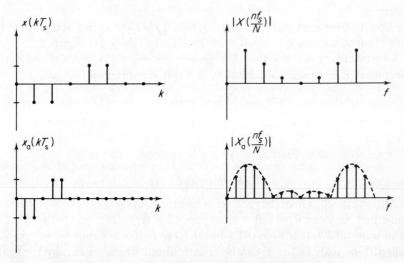

Figure 5.14   The effect of zero adding on frequency resolution

5.6 SPECTRAL ANALYSIS FROM SAMPLES OF SIGNALS

5.6.1 Introductory Remarks

The development of DFT techniques and particularly the application of the FFT algorithm in microprocessor-based hardware has led to a new generation of test equipment for measurement and displaying of frequency spectra of signals. Samples of an analog signal can be taken and the DFT calculations carried out so rapidly that snapshot displays of the magnitude of the frequency spectral distribution can be presented with no noticeable delay—hence the term *real-time spectrum analysers*. As further time sequence data are taken in with varying spectra (such as speech sounds), so the analyser frequency display can be sequentially updated. Such techniques find applications in most areas where signals are being processed such as in vibration analysis, pattern recognition, image processing, speech synthesis, medical electronics (analysis of EKG or EEG signals), seismic and other graphical measurements, radio astronomy, and in communication systems. This section discusses the principles of spectral analysis which should be seen as encompassing a variety of different measurements each centred, however, around the DFT.

Many signals of interest in real life are *random*. For deterministic signals, the frequency spectra can be determined by use of the Fourier series (for periodic signals) or the continuous Fourier transform (for analog signals which tend to zero for large time). The DFT provides a means for generating the voltage spectrum $X(f)$ of such signals. For random signals the *power spectral density* $S_x(f)$ is used and is approximated by appropriately averaging successive calculations of $|X(f)|^2$. Firstly, however definitions are needed in order to proceed.

A discrete-time random (or stochastic) signal $x(nT_s)$ or x_n can be described by a collection of samples or *snapshots* of the signal as shown in Figure 5.15.

At a particular time, say $T_s = 3$, a *random variable* X_3, say, is defined as the collection of sample value realizations. A sequence of random variables

$$X_0, X_1, \ldots, X_i, \ldots$$

can be visualized, each being defined in terms of the values at times $0, 1, \ldots, i, \ldots$.

The *autocorrelation function* $R_x(k)$ can be defined

$$R_x(k) = E[X_n X_{n-k}] \tag{5.17}$$

namely the expectation of the product of two random variables (RV's) separated in time by k samples. Real times are expressed in terms of n and the number of lags in terms of k rather than (nT_s) and (kT_s) to reduce the number of symbols required. The scale factor T_s can be re-introduced whenever required without any complication.

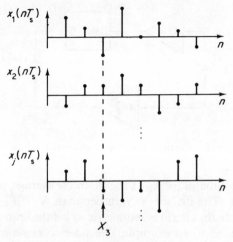

Figure 5.15 An ensemble for a discrete-time random process

The expectation of the product $X_n X_{n-k}$ is the average over the ensemble of all possible realizations of $X_n X_{n-k}$ at a particular epoch n, that is, the average for all possible time sequences which the system may be thought to be capable of producing. In practice, we only ever have available to us one of those realizations. Segments taken from further on in the signal sequence $x(kT_s)$ are, of course, parts of the same realization.

We, therefore, limit our attention to those signal sequences which are *stationary*, that is, those for which $E[X_n X_{n-k}]$ does not depend on the epoch n, but only on the lag k. (This has been implied above by writing $R_x(k)$ as a function of k alone.) Then we regard the expectation as a time average, that is, an average of the product $X_n X_{n-k}$ over all possible epochs n. It is not certain in general, even for stationary sequences, that this procedure is legitimate. We call those signals, for which the ensemble average is given by the time average, *ergodic*.

Provided the random signal is stationary, the power spectral density $S_x(f)$ can be defined as the Fourier transform of $R_x(k)$, which may be written

$$S_x(f) = \sum_{k=-\infty}^{\infty} R_x(k)\exp(-\mathrm{j}2\pi f k T_s) \qquad (5.18)$$

Since $R_x(k)$ has the dimensions of power, it can be seen, by analogy with the DFT of $x(kT_s)$ mentioned earlier, that values of $S_x(f)$ measure the contributions to signal power at each frequency f.

The time average for an infinite sequence is estimated from a finite segment of the sequence by

$$R_N(k) = \frac{1}{N} \sum_{n=1}^{N-|k|} X_n X_{n+|k|}$$

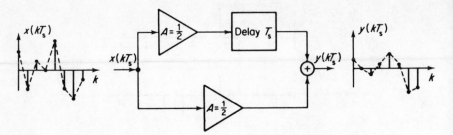

Figure 5.16   Averaging filter with discrete random input

and this is also the estimator for $R_x(k)$, the ensemble average, provided the signal sequence is ergodic. The divisor is N rather than $N - |k|$ for reasons to be discussed later. (Note that $R_x(k)$ is symmetric in k if the sequence is stationary.)

This is now applied to an example. Consider a zero-mean random signal $X(nT_s)$ or X_n being the input to a network which provides an output Y_n such that the RV at time (nT_s) is given by

$$Y_n = \tfrac{1}{2}(X_n + X_{n-1}) \tag{5.19}$$

(This is a simple averaging digital filter and is illustrated in Figure 5.16 where sample sequences are intended to show that the output can be expected to fluctuate less rapidly with time.) We wish to describe the autocorrelation functions and power spectral densities of the input and output signals.

Since the input signal X_n is purely random with zero mean, its autocorrelation function is zero for all lags except $k = 0$ since

$$R_x(k) = E[X_n X_{n+k}] = E[X_n]E[X_{n+k}] = 0 \quad k = \pm 1, \pm 2, \ldots$$

For the output

$$R_y(k) = E[\tfrac{1}{4}(X_n + X_{n-1})(X_{n+k} + X_{n+k-1})]$$

For lag $k = 0$

$$R_y(0) = E[\tfrac{1}{4}X_n^2] + E[\tfrac{1}{4}X_{n-1}^2] = \tfrac{1}{2}E[X_n^2]$$
$$= \sigma_y^2 = \sigma_x^2/2$$

For lag $k = 1$

$$R_y(1) = E[\tfrac{1}{4}X_n^2] = \sigma_x^2/4$$

The same result applies for $k = -1$ (as expected since autocorrelation functions are always even functions). For other values of lag k, the $R_y(k)$ is zero since the $Y_n Y_{n+k}$ product has no common sample RV's. Figure 5.17a illustrates these results.

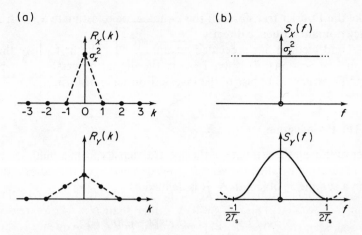

Figure 5.17   Filter input and output: (a) autocorrelation functions;
(b) spectral densities

To compute the spectral densities $S_x(f)$ and $S_y(f)$ we use equation (5.18)

$$S_x(f) = \sum_{k=-\infty}^{\infty} R_x(k)\exp(-j2\pi f k T_s) = R_x(0) = \sigma_x^2$$

since only the $k = 0$ term is non-zero. $S_x(f)$ is, therefore, a constant (white noise). Similarly

$$S_y(f) = \sum_{k=-1}^{1} R_y(k)\exp(-j2\pi f k T_s)$$

$$= \tfrac{1}{4}\sigma_x^2 \exp(j2\pi f T_s) + \tfrac{1}{2}\sigma_x^2 + \tfrac{1}{4}\sigma_x^2 \exp(-j2\pi f T_s)$$

$$= \tfrac{1}{2}\sigma_x^2[1 + \cos(\omega T_s)]$$

These results are plotted in Figure 5.17b. Clearly the averager has performed a filtering function (low-pass) on the incoming signal.

Spectral measurements of random signals based on finite length sample data sequences are complicated by questions such as:

(a)   Is the signal *stationary* or can any non-stationary trends be removed? Otherwise $S_x(f)$ cannot usually be defined. Signals may have slowly varying mean values, for example, which must first be removed.

(b)   Can we estimate $S_x(f)$ from the one realization available? We shall use statistical tests on our estimation procedures to see how well they perform.

We will examine the second problem first, assuming the signal is stationary with with zero mean. Two computational alternatives for estimating the spectral density $S_x(f)$ are to estimate $R_x(k)$ from a finite segment of one realization, and then take the Fourier transform of the estimate, or to estimate $S_x(f)$ from the frequency domain sequence directly.

The second method has a decided advantage over the first where the FFT algorithm (see Section 5.7) is available in the digital processor. (If necessary $R_x(k)$ can be estimated by use of the inverse Fourier transform.)

5.6.2 The Periodogram

However obtained, the estimate of the spectral density from a finite segment of a single realization is called the *periodogram*, denoted $S_N(f)$ to indicate that it is based on a sample x_n of length N. It is defined by

$$S_N(f) = \sum_{k=-N}^{N} R_N(k)\exp(-j2\pi nf\, T_s) \tag{5.20}$$

It can be shown (see Schwarz and Shaw, 1975, p. 160) that this is equivalent to

$$S_N(f) = \frac{1}{N}\left|\sum_{n=0}^{N-1} x_n \exp(-j2\pi nf\, T_s)\right|^2 \tag{5.21}$$

$S_N(f)$ may be plotted as a continuous function or for only discrete values of f. Note that we originally had (equation (5.9))

$$X_N(f) = \sum_{k=0}^{N-1} x(kT_s)\exp(-j2\pi f\, kT_s)$$

so

$$S_N(f) = \frac{1}{N}|X_N(f)|^2 \tag{5.22}$$

The periodogram estimate is found from the N-point DFT $X_N(f)$ which is used for deterministic or random signals.

Practical calculations based on equation (5.21) unexpectedly give very irregular results—the irregularities not diminishing as the length N of the sample is increased. It is important to understand and be able to overcome this problem. We have already seen how the sampling and truncation procedures give rise to modifications to the original true $X(f)$ function. The difficulty with the spectral estimation procedure springs from the *random* properties of the signals.

We begin by asking how accurately equation (5.21) estimates the true spectral density for random signals. One way of answering this question is by examining the *bias* and the *variance* of the estimate $S_N(f)$ to determine whether they approach zero as the number of samples N is increased. An estimator is said to be *unbiased* (good on the average) if the ensemble average of many N-point estimates approaches the true value of the parameter (random variable) being

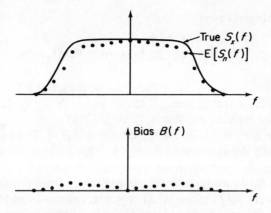

Figure 5.18   Frequency spectrum bias

estimated. The bias $B_N(f')$ at a particular frequency f', in the estimation of the true spectrum $S_x(f')$, is therefore given as

$$B_N(f') = S_x(f') - E[S_N(f')] \tag{5.23}$$

This is illustrated in Figure 5.18.

Now the estimator $R_N(k)$ mentioned earlier is a biased estimator of $R_x(k)$, since

$$E\left[\frac{1}{N} \sum_{n=1}^{N-|k|} X_n X_{n+|k|}\right] = \left(1 - \frac{|k|}{N}\right) R_x(k)$$

(and not $R_x(k)$ exactly). Since our estimate of $S_x(f)$ is obtained from this estimate, $S_N(f)$ is also a biased estimator. Specifically,

$$E[S_N(f)] = \sum_{k=-N}^{N} R_x(k)\left(1 - \frac{|k|}{N}\right)\exp(-j2\pi f k T_s) \tag{5.24}$$

An estimator is said to be *consistent* if its variance approaches zero as the number of samples N increases. In the case of the periodogram, we require

$$\lim_{N \to \infty} E[(S_N(f) - E[S_N(f)])^2]$$

to vanish. A consistent estimator is one which approaches the true value *smoothly* as N increases, that is, as new information about the sequence is received the estimate steadily improves.

Returning briefly to the autocorrelation estimator $R_N(k)$, it is true that its variance vanishes for large N, that is

$$\lim_{N \to \infty} E\left[\left\{R_N(k) - \left(1 - \frac{|k|}{N}\right)R_x(k)\right\}^2\right] = 0$$

Thus it is meaningful to say

$$\lim_{N \to \infty} \frac{1}{N} \sum_{n=1}^{N-|k|} x_n x_{n+|k|} = R_x(k)$$

for any realization $x_1, x_2, x_3, \ldots$ of the signal sequence. It is for this reason that the biased estimator was chosen, (If the divisor is $N - |k|$, the variance does not diminish (see Jenkins and Watts, 1968, p. 179).)

But the same does not hold for the estimator $S_N(f)$. The variance of $S_N(f)$ does not generally diminish even for large N. This has very serious implications for interpreting spectral diagrams produced by the DFT since successive calculations of $S_N(f)$ using different sample values from the same source can give quite different $S_n(f)$ plots. What is worse, for some random sources, the variance or uncertainty can be as large as the true value. For example, it can be shown (Jenkins and Watts, 1968, p. 233) that for white Gaussian noise the variance is *equal* to the expected value. Thus we have

$$S_N(f) = \sum_{k=-N}^{N} R_N(k) \exp(-j2\pi f k T_s)$$

and

$$S_x(f) = \sum_{k=-\infty}^{\infty} R_x(k) \exp(-j2\pi f k T_s)$$

but while it is true that

$$\lim_{N \to \infty} R_N(k) = R_x(k)$$

it is *not* true that

$$\lim_{N \to \infty} S_N(f) = S_x(f)$$

Such a result should come as no surprise when we recall that the $R_x(k)$ and $S_x(f)$ are each *ensemble* averages, taken over all the realizations that the system might possibly generate, and that we are attempting to estimate each by *time* averages from a *single realization* of *finite length*.

It is for this reason that DFT spectral diagrams may appear quite 'ragged' or irregular even though, say, the signal being transformed may have a smooth spectrum. For example, low-pass white Gaussian noise may appear to have a spectrum such as shown in Figure 5.19. (Users of conventional analog tuned radio frequency (TRF) types of spectrum analysers will be well familiar with the care needed for appropriate choice of analyser bandwidth and sweep speed to avoid similar variations when trying to measure the spectral properties of random signals.) With the DFT estimation of spectra there are two ways of reducing the errors (variance). The first, more traditional approach, is by the

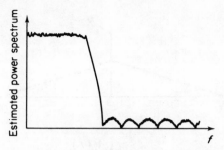

Figure 5.19   Estimated power spectrum for
low-pass noise

use of *smoothing windows* and the second more recently used method is by *averaging* over several periodograms.

Smoothing windows

Rewriting equation (5.24) as

$$E[S_N(f)] = \sum_{k=-\infty}^{\infty} R_x(k)v_k^N \exp(-\mathrm{j}2\pi f k T_s) \qquad (5.25)$$

where

$$v_k^N = \begin{cases} 1 - \dfrac{|k|}{N} & |k| \leqslant N \\ 0 & |k| > N \end{cases} \qquad (5.26)$$

The triangular function v_k^N plotted in Figure 5.20 is known as a *lag window*, and comes about because only a finite number (N) of samples is used. (Recall also frequency domain aliasing distortion due to finite sample length in calculating $X_N(f)$ for deterministic signals.)

The lag window factor v_k^N causes $S_N(f)$ to be a biased estimator of $S_x(f)$. Furthermore, equation (5.25) shows that the mean of the periodogram $E[S_n(f)]$ is the Fourier transform of two time functions multiplied together, namely $R_x(k)$ and v_k^N which we can write in short-hand form

$$E[S_n(f)] = F[R_x(k)v_k^N]$$

where $F[\ \ ]$ stands for the Fourier transform operation. Using the fact that multiplication in the time domain corresponds to convolution in the frequency domain we can write

$$E[S_n(f)] = F[R_x(k)] * F[v_k^N]$$

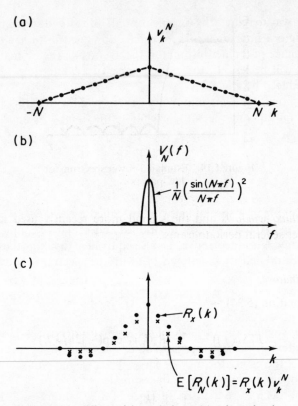

Figure 5.20 Effect of lag window: (a) triangular lag window function; (b) transform of lag window; (c) true and estimated autocorrelation functions of lag window

But $F[R_x(k)] = S_x(f)$ and letting $V_N(f) = F[v_k^N]$, we can write

$$E[S_N(f)] = S_x(f) * V_n(f) \tag{5.27}$$

The Fourier transform $V_n(f)$ of the triangular lag window is shown in Figure 5.20b. Figure 5.20c illustrates true and estimated autocorrelation functions, the difference between the two waveforms being the bias in the estimate of $R_k(k)$. However from equation (5.27) we conclude that the mean of the spectral estimate is the *convolution* of the true discrete spectrum and a frequency domain window function shown in Figure 5.20b. For large N, this window will have very narrow sidelobes and a very sharp peak (tending towards a delta function) so that for large N

$$E[S_N(f)] \simeq S_x(f)$$

and the estimator can be said to be *asymptotically unbiased*.

An obvious way to reduce the variance of $S_N(f)$ is to divide the signal sequence of length N into a number (say s) of equal subsequences of length $M = N/s$. Then the variance of the new estimator, the average of the estimators for each subsequence, will be less. In the extreme case of a white noise sequence, the variance is reduced by a factor s.

It can be shown (Jenkins and Watts, 1968, p. 241) that this procedure is equivalent to multiplying the autocorrelation function by the function

$$w(k) = \begin{cases} 1 - \dfrac{|k|}{M} & |k| \leqslant M \\[2mm] 0 & |k| > M \end{cases}$$

This is an example of a special lag window called a *smoothing window*, and is identical in form to the lag window mentioned above. (This example is known as Bartlett's window.) As noted above, it is also equivalent to taking the convolution of the true spectral density function $S_x(f)$ with the spectral window

$$W(f) = M\left(\frac{\sin(M\pi f)}{M\pi f}\right)^2$$

As M becomes smaller, the variance of the estimator decreases. However, there is a corresponding increase in the width of the pass band in the spectral window, and the bias of the estimator may be large. Thus we have to sacrifice bias for the sake of reducing the variance of the estimate.

Another approach to reduce the variance of $S_N(f)$ might be as follows. The variability may be due to the variability in $R_N(k)$ for $|k|$ close to N, since these are obtained by averaging over only $N - |k|$ points. So to consider only those values of $R_N(k)$ for small k (say $|k| < M$) must surely reduce the variance of $S_N(k)$.

Evidently this truncation procedure is equivalent to multiplying $R_N(k)$ by a *rectangular window*

$$w(k) = \begin{cases} 1 & |k| \leqslant M \\ 0 & |k| > M \end{cases}$$

The effect of this truncation is basically determined by the sinc function shape of the Fourier transform of this rectangular time window. The sharp discontinuities in the time rectangle give rise to undesirably high levels of sidelobes in its transform (the sinc function). Better spectrum analyser performance may result if a continuous and smooth time window were used to weight sample time functions (or autocorrelation functions). Many window functions have been suggested in the literature. The problem is essentially one of selecting a time-limited function with minimum energy outside some selected interval. Some well known window functions are listed in Table 5.1.

Table 5.1   Some well known window functions

Window function	Time domain weighting function $w(k)$	
rectangular	$w(k) = \begin{cases} 1 \\ 0 \end{cases}$	$-\frac{1}{2}(N-1) \leqslant k \leqslant \frac{1}{2}(N-1)$ otherwise
Hamming	$w(k) = \begin{cases} 0.54 + 0.46\cos(2\pi k/N) \\ 0 \end{cases}$	$-\frac{1}{2}(N-1) \leqslant k \leqslant \frac{1}{2}(N-1)$ otherwise
Hanning	$w(k) = \begin{cases} 0.5 + 0.5\cos(2\pi k/N) \\ 0 \end{cases}$	$-\frac{1}{2}(N-1) \leqslant k \leqslant \frac{1}{2}(N-1)$ otherwise

It has been noted that in respect of random signals, window function smoothing (multiplication in the time lag domain giving the effect of convolution or smoothing in the frequency domain) may in some cases reduce the variance at the expense of greater bias. In general, it is true to say that there is no single window function appropriate for all purposes. Thrane (1979), for example, says that only two are important in spectrum analyser applications, namely the rectangular (or flat) window and the Hanning window. The flat window is best for transient signals that can be completely contained within the truncation time window. (DFT analysers are very suitable for the analysis of transients since they are designed to work with finite block samples of time data.) On the other hand, Hanning weighting is recommended for continuous signals. A procedure for window smoothing involving three FFT operations is as follows:

(a)   calculate $S_N(f)$ from the data by FFT;
(b)   find $R_N(k)$ by inverse transform using the FFT;
(c)   multiply $R_N(k)$ values by an appropriate weighting window; and
(d)   use FFT to compute final estimate of $S_n(f)$.

More details are available in many of the references cited at the end of this chapter.

Averaging periodograms

An alternative approach to reducing the variance of $S_n(f)$ is to calculate several periodograms and to average them together. As we have seen, for random signals, successive spectra show significant fluctuations or differences due to the variance of the estimator. Averaging can improve the estimate and also enhance the detection of deterministic or periodic signals buried in noise. In the latter case one might expect the signal-to-noise ratio to be enhanced by the square root of the number of additions of complete spectra. Most DFT-type spectrum analysers have provision for selection of a specific number of spectra to be averaged or a running average with continuous updating of the spectral display. This assists in the detection of signals buried in noise or the separation of periodic and random components. An alternative to ensemble averaging of complete spectra would be to average partitions of a spectrum. As an illustration, $N = 16,384$ signal samples, say, could be sectioned into 16 segments. Then a 1024-point FFT could be performed on each, the results being averaged to obtain the final result. Also, overlapping the sequences (e.g. average 32 overlapping 1024-point transforms) can be used to further reduce the variance (see Rabiner and Gold, 1975 for details).

Other considerations

It was stated above that the use of spectral analysis instrumentation usually requires an intelligent operator approach to ensure that meaningful results are obtained. In this respect, the newer DFT analysers have much in common with their predecessors—the analog swept-tuned analysers for which an appropriate choice of such parameters as sweep speed, maximum dispersion and analyzer bandwidth are important in order to avoid gross errors. With digital processing analysers, it is likewise highly desirable to have some prior knowledge of the spectrum under test or at least to carry out several tests on typical records to avoid errors due to such factors as aliasing errors (sampling rate), poor resolution (sample window width), bias, and variance (smoothing windows or averaging). In addition, it was mentioned early in this section that the Fourier transform process assumed the signal under test was statistically *stationary*. For example, the signal sample shown in Figure 5.21 illustrates a systematic trend (time varying average) which should first be removed.

Such trends should be removed by least-squares fitting procedures to remove straight-line, parabolic or higher-order trends (see Schwartz and Shaw, 1975 for details).

In addition, if the signal contains discrete spectral components (sinusoidal or other periodic components), these would be expected to result in delta function responses in the frequency spectrum. This can be thought of in terms of the weighted sums in equation (5.1) adding up to excessively high values for some frequency terms. Preprocessing, for example by filtering, may be necessary to reduce bias errors once the peaks have become evident from preliminary

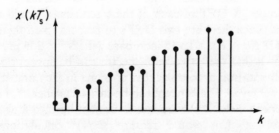

Figure 5.21   A non-stationary signal sample

spectral computations. More advanced texts, such as Jenkins and Watts (1968), provide details. The design of digital filters that could be employed are discussed in Chapter 10.

5.7 THE FAST FOURIER TRANSFORM ALGORITHM

Although the principles of discrete Fourier transforms have been known and used for many years, the range of applications has been restricted by the relatively large number of trigonometric calculations. The fast Fourier transform (FFT) algorithm developed by Cooley and Tukey in the 1960's for more efficiently carrying out the DFT calculation (equation (5.1)) completely altered the situation. Many manufacturers now produce digital processing spectrum analysers capable of measuring transients or of handling frequency spectra up to frequencies of the order of 100 kHz and to produce *real-time* updated displays of signal spectra as they vary with time. Figure 5.2 illustrates a typical analyser commonly used in the analysis of signals. Other applications in the medical field for analysis of electrocardiogram waveforms are becoming widespread.

The FFT algorithm is simply an efficient means of calculation of the discrete Fourier transform calculation

$$X(n/N) = \sum_{k=0}^{N-1} x(k)W^{nk} \tag{5.28}$$

where $W = e^{-j(2\pi/N)}$ and $T_s = 1$ for simplicity. We have seen that each of the N frequency terms $X(n)$ requires the sum of N terms $x(k)$ each multiplied by the complex exponential *weighting term* W^{nk}. To take a trivial case, a 16-point transform ($N = 16$) requires 16 complex multiplications for each frequency term and N complex additions, the former being more time-consuming to perform. The calculation of all 16 frequency terms, therefore, requires this process to be repeated $N = 16$ times. In total, $N^2 = 256$ multiplications are required for the whole spectral calculation—actually only $(N - 1)^2$ of these are *complex* multiplications because for $n = 0$ or $k = 0$ the W^0 term is unity.

Cooley and Tukey considered the decomposition of the transform into smaller groups. Consider initially the N-point $x(k)$ sequence broken up into two $N/2$ sequences. A DFT of each of these sequences would require $(N/2)^2$ multiplications. Since there are two DFT's to be performed there exists a total of $2(N/2)^2 = 128$ for N equal to 16 (compare with $N^2 = 256$ previously). Now, if it can be shown that these two semi-spectra can be appropriately combined by a simple computation procedure, then a saving in the number of multiplications has been achieved. The concept can be carried further. If N is a power of 2 then what happens if the $x(k)$ sequence is broken up into four equal groups? Each subtransform then would require $(N/4)^2$ calculations, a total of $4(N/4)^2 = 64$ multiplications for the whole spectrum. This idea of subdividing

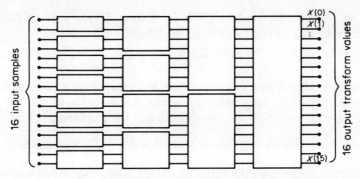

Figure 5.22   Stages involved in a 16-point DFT

can be continued down to sets of transforms of *pairs* of input data. The total number of times the N points can be split into halves is $\log_2 N$. Figure 5.22 illustrates the operations involved for $N = 16$.

This appears excellent in principle but the key question to ask is how can individual DFT's of separate sets of input data be combined to obtain the correct frequency transform values. After all, it must be recalled that each output frequency value is a weighted sum of *all* input values. For example the value $X(n = 1)$ is given by

$$X(1/N) = \sum_{k=0}^{N-1} x(k)W^k$$

and contains contributions from *all* N of the $x(k)$ terms. Consider the sequence $x(k)$ broken up into halves, one half containing even numbered samples and the other odd. The DFT then consists of two parts as follows

$$
\begin{aligned}
X(n) &= \sum_{k=0}^{N-1} x(k)W^{nk} \\[2mm]
&= \sum_{\substack{k=0,2,\ldots \\ (k\,\text{even})}}^{N-2} x(k)W^{nk} + \sum_{\substack{k=1,3,\ldots \\ (k\,\text{odd})}}^{N-1} x(k)W^{nk} \\[2mm]
&= \sum_{\substack{k=0 \\ \text{all}\,k}}^{N/2-1} x(2k)W^{2nk} + \sum_{\substack{k=0 \\ \text{all}\,k}}^{N/2-1} x(2k+1)W^{n(2k+1)} \\[2mm]
&= \underbrace{\sum_{k=0}^{N/2-1} x(2k)W^{2nk}}_{\substack{(N/2)\text{-point DFT of} \\ \text{even } x(k) \text{ points}}} + W^n \underbrace{\sum_{k=0}^{N/2-1} x(2k+1)W^{2nk}}_{\substack{(N/2)\text{-point DFT of odd} \\ x(k) \text{ points}}}
\end{aligned}
$$

Therefore, we have the combining rule for finding $X(n)$ in terms of two $(N/2)$-point transforms $X_1(n)$ and $X_2(n)$, say, as follows

$$X(n) = X_1(n) + W^n X_2(n) \quad n = 0, 1, \ldots, \tfrac{1}{2}N - 1 \qquad (5.29)$$

Note that since $X_1(n)$ and $X_2(n)$ have only $N/2$ values each, equation (5.29) can only provide $X(n)$ values for half the N-range, that is for values $0 \leqslant n \leqslant \tfrac{1}{2}N - 1$. For higher n-values $(\tfrac{1}{2}N \leqslant n \leqslant N - 1)$, we find the combining rule is

$$X(n) = X_1(n - \tfrac{1}{2}N) + W^n X_2(n - \tfrac{1}{2}N) \quad n = \tfrac{1}{2}N, \tfrac{1}{2}N + 1, \ldots, N - 1 \quad (5.30)$$

This follows from the periodic nature of the DFT; an N-point DFT is repetitive after N points and so also an $(N/2)$-point DFT is repetitive after $N/2$ points, i.e.

$$X_1(n + \tfrac{1}{2}N) = X_1(n) \quad n = 0, 1, 1, \ldots, \tfrac{1}{2}N - 1$$

$$X_1(n) = X_1(n - \tfrac{1}{2}N) \quad n = \tfrac{1}{2}N, \tfrac{1}{2}N + 1, \ldots, N - 1$$

and likewise for $X_2(n)$

Rewriting equation (5.29), replacing n by $n + \tfrac{1}{2}N$ yields

$$X(n + \tfrac{1}{2}N) = X_1(n + \tfrac{1}{2}N) + W^{n+N/2} X_2(n + \tfrac{1}{2}N) \quad n = 0, 1, \ldots, \tfrac{1}{2}N = 1$$

and since

$$W^{n+N/2} = e^{-j(2\pi/N)n} e^{-j\pi} = -W^n$$

we have the form

$$X(n + \tfrac{1}{2}N) = X_1(n) - W^n X_2(n) \quad n = 0, 1, \ldots, \tfrac{1}{2}N - 1$$

or, by change of variables (that is, make the range of n run from $N/2$ to $N - 1$), we obtain equation (5.30).

Summarizing the above derivations:

(a) the $x(k)$ sequence values are grouped into odd and even subsequences;
(b) these two subsequences are then transformed using $(N/2)$-point DFT's to give $X_1(n)$ and $X_2(n)$;
(c) the combining rules (equations (5.29) and (5.30)) are used to give $X(n)$.

For example, for $N = 8$

$$X(0) = X_1(0) + W^0 X_2(0) \quad \text{and} \quad X(4) = X_1(0) - W^0 X_2(0)$$

$$X(1) = X_1(1) + W^1 X_1(1) \quad \text{and} \quad X(5) = X_1(1) - W^1 X_2(1)$$

and so forth.

This is illustrated in Figure 5.23 where step (c) can be seen to contain only three different complex multiplications involving W^1, W^2, and W^3.

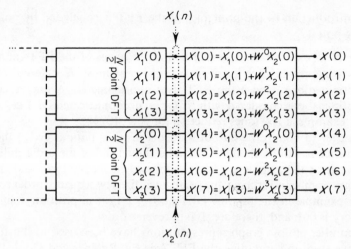

Figure 5.23 Final stage of an 8-point FFT

It is possible to improve the efficiency of step (b) by finding $X_1(n)$ and $X_2(n)$ each in terms of two half sequences. Note that the FFT process is best understood by logically working backwards from the final result $X(n)$ to the data. Let $X_1(n)$ be combined out of two $(N/4)$-point sequences $X_{11}(n)$ and $X_{12}(n)$, say, using

$$X_1(n) = \begin{cases} X_{11}(n) + W^n X_{12}(n) & n = 0, 1, \ldots, \tfrac{1}{4}N - 1 \\ X_{11}(n - \tfrac{1}{4}N) - W^n X_{12}(n - \tfrac{1}{4}N) & n = \tfrac{1}{4}N, \tfrac{1}{4}N + 1, \ldots, \tfrac{1}{2}N - 1 \end{cases}$$

and similar for $X_2(n)$.

This process is continued (back towards the data) until we are left with an array of 2-point transforms (of the data itself) to evaluate. For our 8-point transform example the complete FFT is summarized in Figure 5.24.

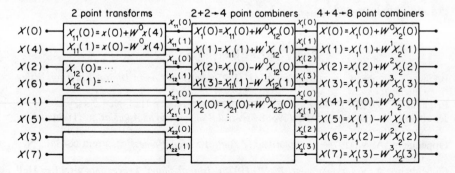

Figure 5.24 8-point FFT

This introduction to the principles of the FFT is concluded by noting the following points:

(a) For N a power of 2, there will be $\log_2 N$ stages of the FFT (number of subdivisions). In each stage there will be, at most, N different complex multiplications. Hence the FFT uses approximately $N \log_2 N$ complex multiplications compared with N^2 for an unsophisticated DFT calculation (for example, a 100 times reduction for $N = 1024$).

(b) In order to obtain the output sequence $X(n)$ in natural order, the input sequence has to be shuffled. It can be shown that if the order indexes (k) of each of the sample values is expressed in binary form and then the binary number is reversed, the result gives the new sequence order required. For example in the 8-point DFT above, $x(1)$ is in position 4 since 1 in binary is 001, and, therefore, 100 in reverse order.

(c) A number of flow graph representations have been used to illustrate the above steps in computing the FFT (see e.g. Rabiner and Gold, 1975 for details).

(d) Computer programs for efficient calculation of the FFT are now commonly available as software packages in many computer installations. They include the necessary data sequencing and efficient methods of computing the exponential (W^{nk}) factors.

Many references, such as Brigham (1974), Rabiner and Gold (1975), or Bogner and Constantinides (1975), provide further details on such matters as the use of flow graphs for representing FFT's and variations on the simple FFT scheme described above. Zoom-FFT, in which a specific part of a given spectrum can be given increased resolution, is covered in Thrane (1980).

ACKNOWLEDGEMENTS

The advice and suggestions of Professor N. T. Gaarder of the University of Hawaii and R. A. Frick of the South Australian Institute of Technology are gratefully acknowledged.

REFERENCES

Bogner, R. E., and Constantinides, A. G. (1975). *Introduction to Digital Filtering*, Wiley, Chichester.
Brigham, E. O. (1974). *The Fast Fourier Transform*, Prentice-Hall, New Jersey.
Jenkins, G. M., and Watts, D. G. (1968). *Spectral Analysis and its Applications*, Holden-Day, San Francisco.
Oppenheim, A. V. (1978). *Applications of Digital Signal Processing*, Prentice-Hall, New Jersey.
Oppenheim, A. V., and Schafer, R. W. (1975). *Digital Signal Processing*, Prentice-Hall, New Jersey.

Rabiner, L. R., and Gold, B. (1975). *Theory and Application of Digital Signal Processing*, Prentice-Hall, New Jersey.

Rader, C. M., and McClennan, J. H. (1979). *Number Theory in Digital Signal Processing*, Prentice-Hall, New Jersey.

Schwartz, M., and Shaw, L. (1975). *Signal Processing: Discrete Spectral Analysis, Detection and Estimation*, McGraw-Hill, New York.

Thrane, N. (1979). 'The discrete Fourier transform and FFT analyzers', *Brüel and Kjaer Tech. Rev.*, No. 1 (whole issue).

Thrane, N. (1980). 'Zoom FFT', *Brüel and Kjaer Tech. Rev.*, No. 2 (whole issue).

Handbook of Measurement Science, Vol. 1
Edited by P. H. Sydenham
© 1982 John Wiley & Sons Ltd.

Chapter

6

D. HOFMANN

Measurement Errors, Probability and Information Theory

Editorial introduction

The perfect measurement system does not exist nor does the perfect measurement circumstance. Furthermore the parameter being measured is, in the philosophical limit, an inexact entity. Thus all measurement situations are subject to disturbance from many error sources. When finally deciding just what the measured data represent it is absolutely essential to consider the various sources of error and how they can validly be combined into a simple statement.

Good measurement arises as much from the study of errors of measurement as it does from the choice of principle. All measurement situations should be considered as grossly in error until proved sufficiently accurate by theoretical and/or practical verification that they suit the need. Too often data are accepted without serious question that they may not be an adequate mapping for the physical variable into a representational equivalent. This chapter addresses the question of error estimation. It shows the sophistication that is available to apply when the demand of certainty of knowing how accurate a measurement really is, must be high.

6.1 INTRODUCTION: CLASSIFICATION OF MEASUREMENT ERRORS

6.1.1 General Remarks

Increasing automation in research, development, production, and consumption constantly creates new tasks for application of measurement engineering fundamentals. The number of sensors and measurement devices applied to a given task is constantly increasing. Their accuracy, application range, dynamics, reliability, and production life constantly need improving. The wide application of microcomputers and microelectronics has resulted in new possibilities for measurement engineering. The effective application of high-performance data-processing modules with microprocessors requires

(a)   mathematical formulation of measurement processes;
(b)   the generation of algorithms of the measurement strategy;
(c)   improvement of the precision of measurement-error analysis.

Models are preferred (Hofmann, 1979; Doebelin, 1975), for analyzing and synthesizing measurement processes.

Models, especially mathematical models, are simpler, cheaper, and more easily described and varied, compared to their originals (measurement signals, measurement systems, measurement processes). Moreover they are space or time transformable, can be optimized within certain limits, and they can easily be taught and learned.

The following considerations on measurement error, probability, and information theory deal with fixed (determined) and variable (stochastic) behaviour models of measurement objects (measurement signals, measurement systems, measurement processes). They are represented by mathematical algorithms. The objects and their models consist of elements and relations between them. Elements are functional groups or numbers. The organizational form of these elements of an object (model) is its structure. The element's behaviour is determined by the states of the elements of an object (model) being in interaction with other things and dependent on time.

In any case it is necessary to consider:

(a)   that different points of view might supply different models of one and the same object;
(b)   that different aims might result in an emphasis on or suppression of different structures and of the behaviour of one and the same object;
(c)   that several phenomena of the objects might be described more simply or, vice versa, in a more complicated manner, always being independent on their importance for the measurement-task solution;
(d)   that the model quality should always to be tested practically, with the help of the real object, not by an idealized approximation.

Every process type has got its own mathematical model theory (Peschel, 1978). This is caused by the nature of processes. Furthermore, model selection is influenced by knowledge attained, by certain opinions occurring in considering the model, and by calculational and experimental advantages.

The processes and models can be classified according to the following properties (Lange, 1978):

(a)   analog and discrete (amplitude-quantized) processes;
(b)   continuous and discontinuous (time-quantized) processes;
(c)   determined and undetermined (stochastic) processes;
(d)   decimal and binary (dual, BCD) processes;

(e)   periodic and non-periodic (aperiodic) processes;
(f)   linear and non-linear processes.

In the following, measurement error analysis is to be dealt with in detail, considering, in particular, those linear and quasilinear measurement systems having determined input quantities as well as undetermined disturbing signals that modify output quantities.

Published works on treatment of errors have tended to have been prepared to suit an area of application rather than be kept general. Texts, such as Barry (1964), Bendat and Piersol (1971), Halstead (1960), Helstrom (1968) and Topping (1972), may provide the specific treatment required.

6.1.2   Classification of Measurement Errors

Systematic measurement errors (bias) can be caused by use of imperfect measurement devices, measurement procedures, and standards, by environmental influences (influence quantities) existing during measurement, as well as by the influence of the measuring person. Systematic errors can be determined and eliminated according to a principle and are, in theory, predictable on a single measurement basis.

Random measurement errors (uncertainty) can occur in sensitive measurements conducted under repetitional conditions. Repetitional conditions exist when the measuring person is repeatedly measuring the same measuring quantity, using the same measuring device and the same measurement procedure. The scatter of the measured value is caused by temporally and locally non-constant error sources. Random measurement errors cannot be individually eliminated because it is not possible to predict the error magnitude at a given time.

Gross measurement errors are caused by mistakes, wrong or careless readings, a temporarily defective measuring device or strong disturbing influences from outside. The measuring person can avoid gross measurement errors during measuring and, therefore, these errors can be considered, and neglected, as degenerated ones.

Additive measurement errors are characterized by their property of being additively superimposed upon the measured value. They do not depend on the numerical value of the measured quantity. Additive measurement errors, therefore, occur as an undesirable zero-point displacement.

Multiplicative measurement errors are characterized by their property of being multiplicatively superimposed upon the measured value. These depend on the numerical value of the measured quantity. Multiplicative measurement errors, therefore, are based on sensitivity deviations of the measurement system from its desired value.

Absolute measurement errors are defined as the difference between the actual measured value and the error-free measured quantity sought.

Relative measurement errors are defined as absolute measurement errors divided by a reference quantity. Percent measurement errors are relative measurement errors multiplied by 100.

Reduced measurement errors, or error classes, are absolute measurement errors related to a measurement range multiplied by 100.

Static measurement errors are characterized as a time function with a constant value.

Dynamic measurement errors are characterized as a time function with changing values.

In measurement analysis two methodical error sources should be considered.

Frequently the measurement position in the surroundings of the desired quantity a (refer to Figure 6.1) is not accessible (for example, the inner temperature of a hot workpiece). The accessible quantity b (surface of a hot workpiece) is changed in its state by the measurement process (mounting of the surface thermometer). The acquired quantity c is equated to the desired measured quantity x_e. Between a and c, there exists a difference that can often not be ascertained.

Both systematic and random measurement errors can occur together, both arising from the same piece of physical apparatus. The physical–technical and mathematical investigations carried out separately on them are based on technical and arithmetic advantages. Random measurement errors are technically more easily comprehended. Moreover, there exist useful mathematical

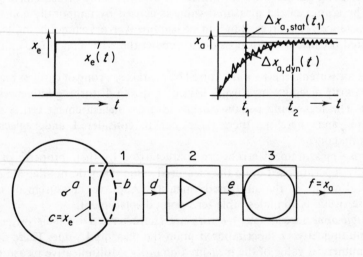

Figure 6.1   The influence of the measurement system on measurement-information acquisition and measurement error analysis. a, desired quantity; b, accessible quantity; c, attained quantity; d, picked-up value; e, comprehended value; f, measured value; 1, sensor; 2, processor; 3, display

algorithms for their arithmetic treatment. A practical possibility is that systematic measurement errors might be neglected without having been considered properly.

6.2 DETERMINISTIC ERROR MODELS

Deterministic error models are presumed to hold when there is a deterministic reason producing a deterministic result. Repeated measurements always result in the same measurement values. In the measurement process of this kind only systematic measurement errors can exist.

The model is considered to be justified under the condition that random measurement errors are negligible compared with systematic measurement errors. The practical, recommended value is represented by the ratio $q \leqslant \frac{1}{10}$, that is, when random errors are one tenth or less of the error budget.

Random measurement errors are not observed if measurements are accomplished with insensitive measurement devices. Adequate system discrimination is essential.

The absolute measurement error Δx is the difference between the incorrect actual value x_I and the correct nominal value x_S:

$$\Delta x = x_I - x_S \tag{6.1}$$

If the actual and the nominal values are time dependent, the systematic measurement error is

$$\Delta x(t) = x_I(t) - x_S(t) \tag{6.2}$$

The incorrect measurement value appears at the measurement-system output. This is designated by the index a:

$$\Delta x_a(t) = x_{aI}(t) - x_{aS}(t) \tag{6.3}$$

In system theory the transfer property of a system is usually designated by its transfer function $S(p)$ (often also written $G(p)$). It is defined as the quotient of the Laplace transform of the output quantity to the Laplace transform of the input quantity. The error-free transfer function of the measurement system is

$$S_S(p) = \tilde{x}_{aS}(p)/\tilde{x}_e(p) \tag{6.4}$$

It is defined by the measurement task. The actual transfer function of the measurement system is

$$S_I(p) = \tilde{x}_{aI}(p)/\tilde{x}_e(p) \tag{6.5}$$

The measurement-error transfer function

$$S_F(p) = \frac{S_I(p)}{S_S(p)} - 1 = \frac{\Delta \tilde{x}_a(p)}{\tilde{x}_{aS}(p)} \tag{6.6}$$

is typical for systematic errors in measurement processes. In the linear working range of the measurement system it does not depend on the input quantity magnitude.

In the complex frequency range (p-range) the systematic measurement error satisfies the equation

$$\Delta\tilde{x}_a(p) = \left(1 - \frac{S_S(p)}{S_I(p)}\right)\tilde{x}_{aI}(p) \tag{6.7}$$

Table 6.1   Laplace transforms of common transfer functions

No.	$F(p)$	$f(t)$
1	$\dfrac{1}{p}$	$1(t)$
2	$\dfrac{1}{p - a}$	e^{at}
3	$\dfrac{1}{p(p - a)}$	$\dfrac{1}{a}(e^{at} - 1)$
4	$\dfrac{1}{p(p + a)}$	$\dfrac{1}{a}(1 - e^{-at})$
5	$\dfrac{1}{1 + Tp}$	$\dfrac{1}{T}e^{-t/T}$
6	$\dfrac{1}{p(1 + Tp)}$	$1 - e^{-t/T}$
7	$\dfrac{1}{(1 + T_1 p)(1 + T_2 p)}$	$\dfrac{1}{T_1 - T_2}(e^{-t/T_1} - e^{-t/T_2})$
8	$\dfrac{1}{p(1 + T_1 p)(1 + T_2 p)}$	$1 - \dfrac{1}{T_1 - T_2}(T_1 e^{-t/T_1} - T_2 e^{-t/T_2})$
9	$\dfrac{1 + T_2 p}{1 + T_1 p}$	$\dfrac{T_1 - T_2}{T_1}e^{-t/T_1}$
10	$\dfrac{1 + T_2 p}{p(1 + T_1 p)}$	$1 - \dfrac{T_1 - T_2}{T_1}e^{-t/T_1}$
11	$\dfrac{1 + T_3 p}{(1 + T_1 p)(1 + T_2 p)}$	$\dfrac{1}{T_1 - T_2}\left(\dfrac{T_1 - T_3}{T_1}e^{-t/T_1} - \dfrac{T_2 - T_3}{T_2}e^{-t/T_2}\right)$
12	$\dfrac{1 + T_3 p}{p(1 + T_1 p)(1 + T_2 p)}$	$1 - \dfrac{1}{T_1 - T_2}[(T_1 - T_3)e^{-t/T_1} - (T_2 - T_3)e^{-t/T_2}]$

Transfer of the systematic measurement error to the time domain (t-domain) is accomplished by the Laplace transform. The transformation rules are

$$f(t) = \frac{1}{2\pi j} \int_{\sigma - j\omega}^{\sigma + j\omega} F(p)\, e^{pt}\, dp = L^{-1}\{F(p)\} \tag{6.8}$$

$$F(p) = \int_0^\infty f(t)\, e^{-pt}\, dt = L\{f(t)\} \tag{6.9}$$

Transformations are carried out practically with the use of correspondence tables (see Tables 6.1 and 4.8 and also Hofmann, 1979).

6.3 PROBABILISTIC ERROR MODELS

6.3.1 General Comment

Probabilistic error models are based on the practical experience that one cause may have several effects. Repeated measurements result in different measured values. In measurement processes random errors occur. Statistical regularities are the basis of all the measured values seemingly being scattering irregularly.

Statistical error models have to be used as a basis for consideration when complicated objects, having various influence quantities, are to be investigated without being able to realize determined properties of the object and of the influence quantities because of technical, physical, economic, and time reasons. The incomplete object is, therefore, described by an incomplete model. Practice decides that the models used are convenient to the aims.

Incomplete models are able to reflect the most important properties of an incomplete object with sufficient accuracy. The pertinent properties of random (stochastic) objects can be characterized by characteristic functions or by characteristic values derived from such functions. Typical characteristic functions are:

probabilities for discrete random variables

$$p_i = P(X = x_i) \tag{6.10}$$

probability densities for analog random variables

$$p(x_i) = P(x_i \leqslant X < x_i + \Delta x)/\Delta x \tag{6.11}$$

and the distribution function

$$F(x_i) = P(X < x_i) = \sum_{i,\, X < x_i} p_i = \int_{-\infty}^{x_i} p(x_i)\, dx \tag{6.12}$$

For estimating the performance of distribution laws the characteristic functions mentioned above are not always themselves used. Instead characteristic values (measures) derived from them are adequate. These are called *moments*. There are ordinary moments of kth order:

$$m_k = E[X^k] \tag{6.13}$$

and central moments of kth order:

$$\mu_k = E[(X - E[X])^k] \tag{6.14}$$

where E is called the *expected value* or *mathematical expectation*.

The ordinary first-order moment is also called the *arithmetic average value*:

$$m_1 = E[X] = \bar{x} = \frac{1}{n}\left(\sum_{i=1}^{n} x_i\right) \tag{6.15}$$

Random quantities are said to be *centred* if the arithmetic average has been subtracted from them. The central first-order moment is always

$$\mu_1 = E[(X - E[X])] = \overline{x_i - \bar{x}} = 0 \tag{6.16}$$

The central second-order moment (dispersion, variance, variation) is a measure of the scattering of measured values about the average value:

$$\sigma^2 = \mu_2 = D^2[X] = E[(X - E[X])^2] \tag{6.17}$$

The mean square deviation (standard deviation, variation) is

$$\sigma = +\sqrt{(D^2[X])} = +\sqrt{\sigma^2} \tag{6.18}$$

The term variation is not uniformly used. It is used for σ^2 in Smirnow and Dunin-Barkowski (1963), Storm (1967), and Müller (1975), but for σ in Lange (1978), Renyi (1971), and Gellert (1977).

The third-order central moment, divided by the third power of the standard deviation, characterizes the asymmetry of the distribution function and is called *skewness*. It is

$$\gamma_3 = \mu_3/\sigma^3 \tag{6.19}$$

with

$$\mu_3 = E[(X - m_1)^3] \quad \text{and} \quad \sigma = +\sqrt{(E[X - m_1)^2])}$$

The fourth-order central moment related to the fourth power of the standard deviation minus three characterizes the *gradation* of the distribution function.

The expression

$$\gamma_4 = (\mu_4/\sigma^4) - 3 = (\mu_4/\mu_2^2) - 3 \tag{6.20}$$

means *excess* of the random variable X, under the condition that the terms $\mu_4 = E[(X - m_1)^4]$ and $\sigma^2 = E[(X - m_1)^2]$ exist. For the normal distribution $\gamma_3 = \gamma_4 = 0$.

If p represents any real number $(0 < p < 1)$, a number Q_p with the properties $F(Q_p) = P(X < Q_p) \leqslant p$ and $P(X \geqslant Q_p) \leqslant 1 - p$ is called a *quartile* of order p (or p-quantile) of the random variable. The 0.5-quantile is the median.

6.3.2 Discrete Digital Random Variables

A discrete (digital) random variable X can represent many discrete enumerable values x_i in a finite interval. The random events of a one-dimensional random variable can be represented as dots (numbers) on a numerical straight line.

A discrete (digital) random variable X can be characterized in the following ways:

(a) By all possible, numerous values x_i, which can represent it in the interval $x_{max} - x_{min}$.

(b) By the probability

$$p_i = P(X = x_i) = P(x_i) = \lim_{n \to \infty} r_i/n \quad i = 1, 2, \ldots \tag{6.21}$$

with which the special value x_i occurs, where r_i is the number of realizations of x_i and n is the number of all realizations. These are

$$0 \leqslant p_i \leqslant 1; \quad \sum_{i = 1, 2, \ldots} p_i = 1 \tag{6.22}$$

(c) By the probability density

$$p(x) = \sum_{i = 1, 2, \ldots} p_i \delta(x - x_i)$$

where p_i is the probability that the random variable X will have the value x_i. The symbol $\delta(x)$ denotes the Dirac function (delta function) having the properties

$$\delta(x) = \begin{cases} \infty & \text{if } x = 0 \\ 0 & \text{if } x \neq 0 \end{cases}$$

and

$$\int_{-\infty}^{\infty} \delta(x) \, dx = 1; \quad \int_{-\infty}^{\infty} \varphi(x) \delta(y - x) \, dx = \varphi(y)$$

with $\varphi(x)$ being any function that is continuous at the point $x = y$. The function $\delta(x)$ is represented analytically by (Sweschnikow, 1970)

$$\delta(x) = \frac{1}{2\pi} \int_{-\infty}^{\infty} e^{j\omega x} \, d\omega$$

(d) By the distribution function

$$F(x_i) = P(X \leqslant x_i) = \sum_{i, X \leqslant x_i} p_i \tag{6.23}$$

which shows the probability of values occurring and being less than or equal to x_i. We have

$$P(x_1 < X \leqslant x_2) = \sum_{i, x_1 < X \leqslant x_2} p_i = F(x_2) - F(x_1)$$

$$F(x_1) \leqslant F(x_2) \quad \text{if } x_1 < x_2$$

$$\lim_{x_i \to -\infty} F(x_i) = F(-\infty) = 0 \quad \lim_{x_i \to \infty} F(x_i) = F(\infty) = 1$$

(e) By the arithmetic mean value (ordinary first-order moment)

$$m_1 = E[X] = \bar{x} = \frac{1}{n} \left(\sum_{i=1, 2, \ldots} x_i r_i \right) = \sum_{i=1, 2, \ldots} x_i P(x_i) = \sum_{i=1, 2, \ldots} x_i p_i \tag{6.24}$$

if absolute convergence exists, that is, if

$$\sum_{i=1}^{\infty} |x_i| p_i < \infty$$

(f) By the variance or dispersion (second-order central moment)

$$\sigma^2 = \mu_2 = D^2[X] = \sum_{i=1}^{\infty} (x_i - \bar{x})^2 p_i = \overline{(x_i - \bar{x})^2} \tag{6.25}$$

(g) By the mean square deviation (standard deviation)

$$\sigma = +\sqrt{\sigma^2} \tag{6.26}$$

6.3.3 Continuous Analog Random Variables

A continuous (analog) random variable can randomly realize innumerable values (dots, numbers) of the number straight line within a finite interval. An analog random process $X(x_i, t)$ provides n different realizations for n similar procedures (elementary events).

An analog stochastic process $X(x_i, t)$ becomes an analog random variable $X(x_i)$, such as $X(x_{a1}(1), x_{a2}(1), \ldots, x_{an}(1))$ at a fixed time of $t = 1$, or a random time function, the so called realization $x_i(t)$, such as $x_{a1}(t)$, at a fixed elementary event.

In ergodic processes the time means equal the statistically expected values (Beyer et al., 1976; Lange, 1973; Dietrich, 1973).

An analog random variable, or a degenerated analog stochastic process, can be characterized by several methods:

(a) By innumerably many values x_i that can occur in the interval $x_{max} - x_{min}$.

(b) By the probability density

$$p(x_i) = \lim_{x \to 0} [P(x_i \leqslant X < x_i + \Delta x)]/\Delta x = \lim_{\Delta x \to 0} r_i/n\Delta x \quad (6.27)$$

(c) By the probability

$$P(x_i \leqslant X \leqslant x_{i+1}) = \int_{x_i}^{x_{i+1}} p(x) \, dx \quad (6.28)$$

(d) By the distribution function

$$F(x_i) = \int_{-\infty}^{x_i} p(x) \, dx \quad (6.29)$$

We have

$$F(-\infty) = 0 \quad F(\infty) = 1$$

and for $P(x_1 \leqslant X \leqslant x_2)$

$$F(x) = \int_{x_1}^{x_2} p(x) \, dx$$

(e) By the linear average value (ordinary first-order moment)

$$\bar{x} = m_1 = \int_{-\infty}^{+\infty} xp(x) \, dx \bigg|_{t=t_i} = \lim_{T \to \infty} \frac{1}{2T} \int_{-T}^{+T} x(t) \, dt \bigg|_{x(t) = x_i(t)} \quad (6.30)$$

(f) By the mean square value (ordinary second-order moment)

$$\bar{x}^2 = m_2 = \int_{-\infty}^{+\infty} x^2 p(x) \, dx = \lim_{T \to \infty} \frac{1}{2T} \int_{-T}^{+T} x^2(t) \, dt \quad (6.31)$$

(g) By the variance of dispersion (second-order central moment)

$$\sigma^2 = \mu_2 = \int_{-\infty}^{+\infty} (x - \bar{x})^2 p(x) \, dx = \lim_{T \to \infty} \frac{1}{2T} \int_{-T}^{+T} [x(t) - \bar{x}]^2 \, dt \quad (6.32)$$

We have:

$$\sigma^2 = x^2 - \bar{x}^2$$

(h) By the standard deviation

$$\sigma = +\sqrt{\sigma^2} \quad (6.33)$$

The properties of stochastic, stationary, ergodic random processes can further be described as follows. The autocorrelation function is

$$R_{xx}(\tau) = \lim_{T \to \infty} \frac{1}{2T} \int_{-T}^{+T} x(t)x(t - \tau) \, dt = \overline{x(t)x(t - \tau)} \tag{6.34}$$

We have

$$R_{xx}(0) = \overline{x^2} \quad R_{xx}(\infty) = \bar{x}^2$$

$$\sigma^2 = R_{xx}(0) - R_{xx}(\infty) \quad \text{if } \bar{x} \neq 0$$

The Fourier transform is also applicable to the autocorrelation function, and as another characteristic function of random processes it results in the spectral power density given by

$$P_{xx}(j\omega) = \int_{-\infty}^{+\infty} R_{xx}(\tau) \, e^{-j\omega\tau} \, d\tau \tag{6.35}$$

The autocorrelation function and the spectral power density function are Fourier transforms according to the Wiener–Chinchine formula:

$$P_{xx}(j\omega) = \int_{-\infty}^{+\infty} R_{xx}(\tau) \, e^{-j\omega\tau} \, d\tau = F\{R_{xx}(\tau)\} \tag{6.36}$$

$$R_{xx}(\tau) = \frac{1}{2\pi} \int_{-\infty}^{+\infty} P_{xx}(j\omega) \, e^{j\omega\tau} \, d\omega = F^{-1}\{P_{xx}(j\omega)\} \tag{6.37}$$

The graphical representation of the spectral density function is called the *power density spectrum*.

Randomly scattered measured values usually occur between (at the best case) the normal distribution function, having the probability density (Beyer *et al.*, 1976) given by,

$$p(x) = \frac{1}{\sigma\sqrt{(2\pi)}} \exp\left(-\frac{(x - \bar{x})^2}{2\sigma^2}\right) \quad \text{with } -\infty < x_i < \infty \tag{6.38}$$

and the distribution function

$$F(x) = \frac{1}{\sigma\sqrt{(2\pi)}} \int_{-\infty}^{x_i} \exp\left(-\frac{(x - \bar{x})^2}{2\sigma^2}\right) dx \tag{6.39}$$

and (the most unfavourable case) the uniform distribution or rectangular distribution with the probability density

$$p(x) = \begin{cases} 1/(x_{max} - x_{min}) & \text{if } x_{min} \leqslant x \leqslant x_{max} \\ 0 & \text{if } x < x_{min} \quad \text{and} \quad x > x_{max} \end{cases} \tag{6.40}$$

and the distribution function

$$F(x) = \begin{cases} 0 & \text{if } x \leqslant x_{\min} \\ (x_i - x_{\min})/(x_{\max} - x_{\min}) & \text{if } x_{\min} < x_i \leqslant x_{\max} \\ 1 & \text{if } x > x_{\max} \end{cases} \qquad (6.41)$$

The true distributions cannot be exactly known; they are only estimated from the real, practical, measurement values. Statistical characteristics calculated by the formulae given are only estimations. The normal distribution function is usually assumed to apply ,unless circumstances arise which justify a closer study of the particular situation.

6.3.4 Characteristics of Random Measurement Errors

The following considerations are concerned with a situation in which the measured value x_{aI}, which is only influenced by systematic measurement errors, is made uncertain by additional stochastic disturbing influences. The deviation of each single measurement (of every event, of every value) x_{aIi} from the 'true' measured value x_{aI} is given by

$$\varepsilon_{ai} = x_{aIi} - x_{aI} \qquad (6.42)$$

cannot be calculated. The measured value x_{aI} is not comprehensible from measurement when random disturbances are influencing the measurement system. But, using the Gaussian method (method of the smallest error square sum) the arithmetic mean value (average value), see equation (6.15),

$$\bar{x}_{aI} = \frac{1}{n} \left(\sum_{i=1}^{n} x_{aIi} \right) \qquad (6.43)$$

is, from probability theory, the best value within the measured values x_{aIi} of a measurement series $x_{aI1}, \ldots, x_{aIi}, \ldots, x_{aIn}$. The mean value $\bar{x}_{aI}$ approaches the true value x_{aI}.

Instead of the mean value, other approximate values can be used which can be ascertained more easily. If the measurement series has n measured values arranged after their quantity, the median or central value x_{aI} is, for odd n, the measured value in the middle of the arranged series or, for even n, the arithmetic mean of both measured values of the arranged measurement series which are in the middle.

In most cases the difference between the median and the arithmetic mean value (calculated by equation (6.15)) is negligible. Simplification procedures however, are valuable.

The span-width mean value

$$R_M = (x_{\max aI} + x_{\min aI})/2 \qquad (6.44)$$

generally shows a greater deviation from the arithmetic mean value. For longer measurement series it can better be ascertained than the median, because only the greatest measured value $x_{\text{max al}}$ and the smallest measured value $x_{\text{min al}}$ have to be established from the ungrouped frequency table which need not be rearranged.

The mode D (modal value, density mean value) is the value of the measurement series that occurs most frequently at stated measured values.

The mean values of a measurement series, mentioned above (arithmetic mean value, median, span-width mean value, modal value), characterize a measurement series of n values by just one value. But they cannot be used for describing the empiric distribution of the measured values: deviation of the single measured values from the medium value also has to be given.

The deviation $v_{\text{al}i}$ of each single measurement $x_{\text{al}i}$ from the mean value $\bar{x}_{\text{al}}$,

$$v_{\text{al}i} = x_{\text{al}i} - \bar{x}_{\text{al}} \tag{6.45}$$

can be calculated. It is not, however, to be considered to be a representative random measurement error as it has a different value for each measurement and becomes zero in the sum.

On the other hand the mean square deviation of the single measurement $x_{\text{al}i}$ from the mean value $\bar{x}_{\text{al}}$, that is, the standard deviation,

$$s_{\text{al}} = \left[\left(\sum_{i=1}^{n} v_{\text{al}i}^2 \right) (n-1)^{-1} \right]^{1/2} \tag{6.46}$$

is more suitable for error information. The standard deviation can show limits including the confidence intervals. In those intervals near the single measured value $x_{\text{al}i}$ or mean value $\bar{x}_{\text{al}}$ the true value x_{al} is to be found with a known statistical certainty. For normally-distributed scattered measured values the confidence intervals of the single measurement are

$$v_{\text{al}E} = \pm a k_1 s_{\text{al}} \tag{6.47}$$

and of the mean value

$$v_{\text{al}M} = \pm t s_{\text{al}} / \sqrt{n} \tag{6.48}$$

The factor $k_1 = f_1(P, n)$ in equation (6.47) considers the uncertainty of ascertaining the standard deviation; the factor $t/\sqrt{n} = f_2(P, n)$ in equation (6.48) considers the uncertainty of ascertaining the mean value. The factors $a, k_1, t/\sqrt{n}$ can be obtained from tables (Hofmann, 1979; Hultzsch, 1971). Extracts are shown in Table 6.2.

The choice of the best value for P depends on the systematic measurement error as well as upon technical and economic aspects. For production inspection, in most cases, $P = 95\%$ and, therefore, $a = 1.96 \simeq 2$ are sufficient. For

Table 6.2 (a) Factors k_1 of the confidence limits of the standard deviation s_{aI} of the measurement series of n single measurements for statistical uncertainty P. (b) Factors $t/\sqrt{n}$ for half the confidence interval $ts_a/\sqrt{n}$ of the medium value $\bar{x}_{\text{aI}}$ of a measurement series with n single measurements and a known standard deviation s_{aI} and a chosen statistical certainty P. (c) Factor k of the outlier criterion for the elimination of gross errors

(a)		
$P = 95\%$	$P = 99\%$	$P = 99.73\%$

n	k_1	k_1	k_1
6	2.09	3.00	3.96
8	1.80	2.38	2.93
10	1.65	2.08	2.48
12	1.55	1.90	2.21
15	1.46	1.73	1.96
20	1.37	1.58	1.75
25	1.32	1.49	1.62
30	1.28	1.42	1.53
40	1.23	1.34	1.43
50	1.20	1.30	1.37

(b)		
$P = 95\%$	$P = 99\%$	$P = 99.73\%$

n	$t/\sqrt{n}$	$t/\sqrt{n}$	$t/\sqrt{n}$
6	1.05	1.65	2.25
8	0.838	1.24	1.60
10	0.715	1.03	1.29
12	0.635	0.898	1.11
15	0.555	0.700	0.938
20	0.468	0.640	0.772
25	0.412	0.560	0.668
30	0.374	0.504	0.599
40	0.320	0.429	0.506
50	0.284	0.379	0.447

(c)		
$P = 95\%$	$P = 99\%$	$P = 99.73\%$

n	k	k	k
9	4.42	7.10	11.49
10	4.31	6.99	10.26
12	4.16	6.38	8.80
15	4.03	5.88	7.66
20	3.90	5.41	6.73
25	3.84	5.14	6.25
30	3.80	5.00	5.95
40	3.75	4.82	5.56
50	3.73	4.70	5.34

precision measurements $P = 99\%$, and therefore $a = 2.58$. For $a = 3$ only 0.27% of all measurement errors are out of the tolerance band

$$v_{\text{aIE}} = \pm 3k_1 s_{\text{aI}}.$$

This confidence interval is generally called the practical maximum error of a single measurement.

If the normal distribution is assumed, the value $v_{\text{aIE}} = \sigma$ appears once in three observations, $v_{\text{aIE}} = 2\sigma$ once in twenty-two observations and $v_{\text{aIE}} = 3\sigma$ once in three hundred and seventy observations. This means that random measurement errors are, for the single measurement, given by

$$[x_{\text{aI}i} - x_{\text{aI}}]_{n, P} = \pm v_{\text{aIE}} = \pm ak_1 s_{\text{aI}} \tag{6.49}$$

and for the measurement series

$$[\bar{x}_{\text{aI}} - x_{\text{aI}}]_{n, P} = \pm v_{\text{aIM}} = \pm t s_{\text{aI}}/\sqrt{n} \tag{6.50}$$

The square brackets indicate that error information thus indicated are barriers (maximum errors) depending on the number n of the used measured values and the statistical certainty P.

The square of the standard deviation is called disperion or variance:

$$s_{\text{aI}}^2 = \left(\sum_{i=1}^{n} v_{\text{aI}i}^2 \right)(n - 1)^{-1} \tag{6.51}$$

In practical investigations, for instance, in control-chart technique (Hofmann, 1979), it is sufficient to mention the span width (scattering range, variation range)

$$R = x_{\text{max aI}} - x_{\text{min aI}} \tag{6.52}$$

This uses $x_{\text{max aI}}$, the maximum value of the measurement series, and $x_{\text{min aI}}$, the minimum value of the measurement series, instead of the standard deviation which cannot always be found as easily. The variation coefficient denotes the relative variation of the single values near the mean value. The variation coefficient

$$v_{\text{aI}} = (s_{\text{aI}}/\bar{x}_{\text{aI}}) \times 100\% \tag{6.53}$$

is a dimensionless number that is especially suitable for comparing the precision (scattering) of measurement series of different empirical distributions.

6.3.5 Gross Measurement Errors

In addition to random measurement errors there often occur gross measurement errors (outliers is the term adopted here) which cannot always be recognized immediately. Being considered as degenerated measurement errors, when identified, they are eliminated from the measurement series. Gross measurement-error recognition is simplified by representing the single measurement values x_{a1i} in a probability graph (Hofmann, 1979). If the measuring person is not able to decide whether the measurement error is a gross or a great random one, the *wild shot* or *outlier* criterion

$$|x_{ag} - \bar{x}_{a1}| > ks_{a1} \tag{6.54}$$

is the basis of the decision at a given number n of measured values and a given statistical certainty P. On this occasion, however, it is presumed that all the other measured values, with x_{ag} excepted, are scattered according to a normal distribution. This assumption has to be tested. The values $k = f(P, n)$ are to be seen in Table 6.2.

6.3.6 Imperfectly Known Systematic Measurement Errors

Insufficiently understood systematic measurement errors, as well as the uncertainties of calculated systematic measurement errors, must be estimated. They are summarized according to the Gaussian law of error propagation for random measurement errors. The resulting medium error $f_{a\Sigma}$ has to be assigned a double sign and be added to the measurement uncertainty (to the random measurement errors). Arriving at satisfactory values demands a highly qualified measuring person: estimated values are often disputed.

6.3.7 Formulation of Measurement Results

The complete measurement result can be in two forms. The result can be calculated from equations (6.1) and (6.49) for a single measured value x_{a1i}:

$$x_{aSE} = x_{a1i} - \Delta x_a \pm (ak_1 s_{a1} + f_{a\Sigma}) \tag{6.55}$$

and from equations (6.1), (6.43), and (6.50) for the mean value $\bar{x}_{a1}$ of the measurement series:

$$x_{aSM} = \bar{x}_{a1} - \Delta x_a \pm (ts_{a1}/\sqrt{n} + f_{a\Sigma}) \tag{6.56}$$

The difference between x_{aSE} and x_{aSM} is found in the different tolerance bandwidth. It is more precise to give x_{aSM}.

6.3.8 Error Propagation of Systematic Measurement Errors in Indirect Measurements

Linear error-propagation law

If the desired quantity y is a function of several measured values not depending on each other, $x_1, x_2, \ldots, x_n$ (indirect measurement of y), that is

$$y = f(x_1, x_2, \ldots, x_n) \tag{6.57}$$

the systematic measurement error of the desired quantity y must be calculated from the systematic measurement errors of the measured quantities $x_1, x_2, \ldots, x_n$ according to the *linear error-propagation* law

$$\Delta y = \sum_{i=1}^{n} \frac{\partial y}{\partial x_i} \Delta x_i \tag{6.58}$$

This results from development of the function f in a Taylor series and omission of all terms of second and higher power. For relative measurement errors

$$\frac{\Delta y}{y} = \sum_{i=1}^{n} \frac{\partial y}{\partial x_i} \frac{\Delta x_i}{y} \tag{6.59}$$

Procedures of logarithmic differentiation

If only the relative measurement error $\Delta y/y$ is to be calculated the *logarithmic differentiation* is a suitable procedure. The function

$$z = \log_a v \tag{6.60}$$

differentiated gives

$$dz = (\log_a e/v)\, dv \tag{6.61}$$

For natural logarithms equations (6.60) and (6.61) are

$$z = \ln v \tag{6.62}$$

and

$$dz = d(\ln v) = dv/v \tag{6.63}$$

For calculating the relative measurement error by applying logarithmic differentiation, equation (6.57) is converted into the logarithm format

$$\ln y = \ln f(x_1, x_2, \ldots, x_n) \tag{6.64}$$

Differentiating equation (6.64) yields

$$\frac{dy}{y} = \sum_{i=1}^{n} \frac{\partial(\ln y)}{\partial x_i}\, dx_i \tag{6.65}$$

If the differentials dx_i in equation (6.64) are formally substituted by the systematic measurement errors Δx_i of the measured values x_i, the desired relative measurement error is found:

$$\frac{\Delta y}{y} = \sum_{i=1}^{n} \frac{\partial(\ln y)}{\partial x_i} \Delta x_i \tag{6.66}$$

Using equation (6.63), equation (6.66) can be expressed in the form of equation (6.59). The partial derivatives $\partial(\ln y)/\partial x_i$ in equation (6.66) are relatively easily to determine if the x_i in equation (6.57) are mainly multiplicatively connected to each other.

Example. If the desired function is

$$y = ax_1^b x_2^{-c} x_3/(x_3 - x_4)$$

with the measured values x_1, x_2, x_3, x_4 and the constants a, b, c, then

$$\ln y = \ln a + b \ln x_1 - c \ln x_2 + \ln x_3 - \ln(x_3 - x_4)$$

and the relative measurement error of the function $y = f(x_1, x_2, x_3, x_4)$ is

$$\frac{\Delta y}{y} = b\frac{\Delta x_1}{x_1} - c\frac{\Delta x_2}{x_2} + \frac{\Delta x_3}{x_3} - \frac{\Delta x_3}{x_3 - x_4} + \frac{\Delta x_4}{x_3 - x_4}$$

Measurement errors of typical measuring circuits and measuring functions

Measurement-error analysis can be simplified if the absolute and relative measurement errors of typical measuring circuits and measuring functions are known. They have been grouped according to their characteristic measuring functions and measuring circuits in Tables 6.3 and 6.4 (Hofmann, 1979).

Table 6.3   Relative measurement errors of typical measurement circuits

No.	Circuit	Structure	Transfer function	Relative measurement error*
1	series		$S_1 S_2$	$\dfrac{\Delta x_1}{x_1} + \dfrac{\Delta x_2}{x_2}$
2	parallel		$S_1 \pm S_2$	$\dfrac{S_1}{S_1 \pm S_2}\dfrac{\Delta x_1}{x_1} \pm \dfrac{S_2}{S_1 \pm S_2}\dfrac{\Delta x_2}{x_2}$
3	feedback (positive feedback)		$\dfrac{S_1}{1 - S_1 S_2}$	$\dfrac{1}{1 - S_1 S_2}\dfrac{\Delta x_1}{x_1} + \dfrac{1}{(1/S_1 S_2) - 1}\dfrac{\Delta x_2}{x_2}$
4	feedback (reversed feedback)		$\dfrac{S_1}{1 + S_1 S_2}$	$\dfrac{1}{1 + S_1 S_2}\dfrac{\Delta x_1}{x_1} - \dfrac{1}{(1/S_1 S_2) + 1}\dfrac{\Delta x_2}{x_2}$

* The measurement errors $\Delta x_i/x_i$ are the relative measurement errors of the blocks i before their interconnection.

Table 6.4 Absolute and relative measurement errors of typical measurement functions

No.	Function $y = f(x_1, \ldots, x_n)$	Absolute measurement error Δy	Relative measurement error $\Delta y/y$
1	$x_1 + x_2 + x_3 + \cdots$	$\Delta x_1 + \Delta x_2 + \Delta x_3 + \cdots$	$\dfrac{\Delta x_1 + \Delta x_2 + \Delta x_3 + \cdots}{x_1 + x_2 + x_3 + \cdots}$
2	$x_1 - x_2 - x_3 - \cdots$	$\Delta x_1 + \Delta x_2 + \Delta x_3 + \cdots$	$\dfrac{\Delta x_1 + \Delta x_2 + \Delta x_3 + \cdots}{x_1 - x_2 - x_3 - \cdots}$
3	$x_1 \cdot x_2 \cdot x_3 \cdots$	$\Delta x_1 x_2 x_3 + x_1 \Delta x_2 x_3 + x_1 x_2 \Delta x_3 + \cdots$	$\dfrac{\Delta x_1}{x_1} + \dfrac{\Delta x_2}{x_2} + \dfrac{\Delta x_3}{x_3} + \cdots$
4	$\dfrac{x_1 x_2 x_3 \cdots}{x_a x_b x_c}$	$\dfrac{x_2 x_3 \cdots}{x_a x_b x_c \cdots}\Delta x_1 + \dfrac{x_1 x_3 \cdots}{x_a x_b x_c \cdots}\Delta x_2 + \cdots - \dfrac{x_1 x_2 x_3 \cdots}{x_a^2 x_b x_c \cdots}\Delta x_a + \dfrac{x_1 x_2 x_3 \cdots}{x_a x_b^2 x_c \cdots}\Delta x_b$	$\dfrac{\Delta x_1}{x_1} + \dfrac{\Delta x_2}{x_2} + \cdots - \dfrac{\Delta x_a}{x_a} - \cdots - \dfrac{\Delta x_b}{x_b} \cdots$
5	ax	$a\Delta x$	$\dfrac{\Delta x}{x}$
6	x^n	$nx^{n-1}\Delta x$	$\dfrac{n}{x}\Delta x$
7	$\sin x$	$\cos x \, \Delta x$	$\cot x \, \Delta x$
8	$\cos x$	$-\sin x \, \Delta x$	$-\tan x \, \Delta x$
9	$\tan x$	$\dfrac{1}{\cos^2 x}\Delta x$	$\dfrac{2}{\sin(2x)}\Delta x$
10	$\cot x$	$-\dfrac{1}{\sin^2 x}\Delta x$	$-\dfrac{2}{\sin(2x)}\Delta x$

Moreover, in connection with error calculation, it is frequently possible to obtain important simplifications to formulae by making approximations for small numerical values. The most important of these formulae are shown in Table 6.5.

6.3.9 Error Propagation of Random Measurement Errors in Indirect Measurements

Squared error-propagation law

If the desired quantity y is a function of several measured quantities $x_1, x_2, \ldots, x_n$ being independent from each other, that is

$$y = f(x_1, x_2, \ldots, x_n)$$

the characteristic values of random measurement errors of the quantity y result from the characteristic values of random measurement errors for the measured

Table 6.5 Approximation formulae for small number values
$$(\alpha, \beta, \gamma, \delta \ll 1)$$

No.	Equation	Approximation
1	$(1 \pm \alpha)(1 \pm \beta) \cdots$	$1 \pm \alpha \pm \beta \pm \cdots$
2	$(1 \pm \alpha)^n$	$1 \pm n\alpha$
3	$\dfrac{(1 \pm \alpha)(1 \pm \beta) \cdots}{(1 \pm \gamma)(1 \pm \delta) \cdots}$	$1 \pm \alpha \pm \beta \pm \cdots \mp \gamma \mp \delta \mp \cdots$
4	$\dfrac{1}{1 \pm \alpha}$	$1 \mp \alpha$
5	$\sqrt{(1 \pm \alpha)}$	$1 \pm \dfrac{\alpha}{2}$
6	$\sqrt[n]{(1 \pm n\alpha)}$	$1 \pm \alpha$
7	$\dfrac{1}{\sqrt{(1 \pm \alpha)}}$	$1 \pm \dfrac{\alpha}{2}$
8	e^{α}	$1 + \alpha$
9	a^{α}	$1 + \alpha \ln a$
10	$\sin \alpha; \tan \alpha$	α
11	$\cos \alpha$	1
12	$\sin(\varphi \pm \alpha)$	$\sin \varphi \pm \alpha \cos \varphi$
13	$\sqrt{(ab)} \quad a \simeq b$	$\frac{1}{2}(a + b)$

values x_i according to the squared error-propagation law (Gaussian error propagation)

$$\Delta y = \pm \left[\sum_{i=1}^{n} \left(\frac{\partial y}{\partial x_i} \Delta x_i \right)^2 \right]^{1/2} \tag{6.67}$$

If y is a function of only two measured values x_1 and x_2, then the linear law

$$|\partial y_{\text{max}}| = \left| \frac{\partial y}{\partial x_1} \Delta x_1 \right| + \left| \frac{\partial y}{\partial x_2} \Delta x_2 \right| \tag{6.68}$$

can be applied (Hultzsch, 1971).

Standard deviation of indirect measurements

The expression for the standard deviation s_{ya} of the desired quantity y_a follows equation (6.67):

$$s_{ya} = \pm \left[\sum_{i=1}^{n} \left(\frac{\partial y_a}{\partial x_{ia}} s_{ia} \right)^2 \right]^{1/2} \tag{6.69}$$

Confidence intervals of indirect measurements

The confidence intervals of the single measurement value (index E) as well as of the mean value (index M) of the measurement series are calculated according to

$$v_{ya\,E,\,M} = \pm \left[\sum_{i=1}^{n} \left(\frac{\partial y_a}{\partial x_{ia}} v_{ia\,E,\,M} \right)^2 \right]^{1/2} \tag{6.70}$$

6.4 INFORMATION-THEORETICAL ERROR MODELS

6.4.1 General Remarks

The informational content of an elementary event can be ascertained in two different ways. The first arises when one is only interested in knowing the existence or non-existence of an event (subject of the classic information theory of Shannon). A second situation arises when one additionally is interested in knowing the intensity, the amplitude, of the event (subject of modern measurement information theory).

Within discrete (digital) processes there are only discrete amplitudes to be found.

Continuous (analog) processes show seemingly continuous amplitude distributions. Because of the existence of the measurement error Δx_a only a Δx_a correct identification of the amplitude is possible. The amplitude, therefore, should be considered to be quantized. If the measurement error Δx_a is introduced with a double sign, the number m of the differentiable amplitude steps of analog measuring devices is

$$m = \frac{x_{\max a} - x_{\min a}}{2 \, | \Delta x_a |} + 1 \tag{6.71}$$

where $x_{\max a}$ is the maximum amplitude and $x_{\min a}$ the minimum amplitude.

The following equation can be applied for establishing the number m_{P_a} of the differentiable power steps:

$$m_{P_a} = \frac{P_{\max a} - P_{\min a}}{2\overline{\Delta x_a^2}} \tag{6.72}$$

where $P_{\max a}$ is the maximum power and $P_{\min a}$ the minimum power.

For digital measuring devices

$$m = f_Z T_E + 1 \tag{6.73}$$

where f_Z is the counting frequency and T_E the response time.

The additive 1 in equations (6.71) and (6.73) indicates that the lowest values $x_{\min a}$, $P_{\min a}$ can appear and can be recognized. They are equal to zero as a rule. Frequently $[(x_{\max a} - x_{\min a})/(2|\Delta x_a|)] \gg 1$ showing that the second term in equations (6.71) and (6.73) is also negligible.

As an example if a measurement system has a scale with a measurement range of $(x_{\max a} - x_{\min a}) = 100$ and a measurement error of $\Delta x_a = \pm 1\%$ then $m = (100/2) + 1 = 51$ amplitude steps can be differentiated.

6.4.2 Information Characteristics

Let m possible measured values be differentiable. Each value can be obtained by

$$q = \text{lb } m = \log_2 m = 3.32 \log_{10} m = 3.32 \lg m \tag{6.74}$$

in bit binary selection steps where these n possible values are coded by 0 or 1 signals of the redundancy-free dual code. On the other hand a measured value can be represented on an information storage basis with q binary storage locations with a structural module (error) of

$$m = 2^q \tag{6.75}$$

differentiable amplitude steps. For converting decimal numbers m into binary logarithms, lb m, the following chart can be used:

m	lb m	m	lb m	m	lb m	m	lb m
1	0.00000	4	2.00000	7	2.80734	10	3.32193
2	1.00000	5	2.32193	8	3.00000	10^2	6.64386
3	1.58496	6	2.58496	9	3.16993	10^3	9.96579

This means, say, for characterizing $m = 10$ values 4 bits are needed, that is, a four-figure binary number. For $m = 1000$ values 10 bits are needed—a ten-figure binary number. The number q of the selection steps needed for the representation of the decimal number m is called the *information quantity* or the *coding expense*. It describes the static properties of the informational system.

In any practical case the useful signal (information) at the measurement-system output is superimposed with a disturbance. For the receiver the information to be accepted is not determined in advance. The information is only given a greater or smaller probability. Information processes are stochastic ones. Their mathematical model is the probability theory.

The information content $I \to 0$ is very small if an event x_i, expected with a great probability $P_i \to 1$, occurs. The information content $I \to \infty$ is very great if an event x_i, expected with the very small probability $P_i \to 0$, occurs. This is modelled by the function $I = \log(1/P)$. If m values of x_i appear with the same probability in a stochastic process the probability for the appearance of a value x_i is given by

$$P_i(X = x_i) = 1/m \tag{6.76}$$

Putting equation (6.76) into the expression (6.74) leads to the information quantity (coding expense), in bit units,

$$q_i = \mathrm{lb}(1/P_i) = -\mathrm{lb}\, P_i \tag{6.77}$$

If m symbols s_i (numerals or numeral sequences, letters, words) appear with different probability P_i per symbol s_i they have different coding expenses. The medium's coding expense (medium information quantity) is called *information entropy* and is given by equations (6.24) and (6.77) for discrete amplitudes:

$$H = E[q] = \bar{q} = \sum_{i=1}^{n} P_i q_i = -\sum_{i=1}^{n} P_i\, \mathrm{lb}\, P_i \text{ bits/symbol} \tag{6.78}$$

The information entropy H is a measure having two meanings (Lange, 1978):

(a) for all symbols of the *alphabet*, it characterizes the medium coding expense per symbol in bits per symbol;
(b) for one symbol of the alphabet, it characterizes the medium's information content per symbol in bits per symbol.

Working with continuous (analog) signals we obtain, for discrete amplitude steps $\Delta x \to 0$, the probability density p_i for the appearance of an amplitude value in the range x_i to $x_i + \Delta x$. The probability is given by

$$P_i = p_i \Delta x \tag{6.79}$$

The medium's coding expense (medium information quantity or information entropy) is, for continuous signals, given by equations (6.78) and (6.79)

$$H = -\lim_{\Delta x \to 0} \sum_{i=-\infty}^{\infty} p_i \Delta x\, \mathrm{lb}(p_i \Delta x) \tag{6.80}$$

We say

$$\sum_{i=-\infty}^{\infty} p_i \Delta x = 1$$

For the transition to infinitesimally small quantization steps the information entropy is

$$H = -\int_{-\infty}^{\infty} p_i \, \text{lb} \, p_i \, dx + \lim_{\Delta x \to 0} \text{lb}(1/\Delta x) \tag{6.81}$$

The second term of equation (6.81) causes it to diverge and becomes infinitely great in the borderline case. This contradicts practical experience and the contents of equation (6.71). The second term is often not considered (Krauss and Woschni, 1975). When calculating information-entropy differences it disappears.

A continuous stochastic useful signal $x_e(t)$ carries a medium information quantity or input entropy according to

$$H(X_e) = -\int_{-\infty}^{\infty} p(x_e) \text{lb} \, p(x_e) \, dx + \lim_{\Delta x \to 0} \text{lb}(1/\Delta x) \tag{6.82}$$

at the input of an error-free measurement system. The information can be partly lost because of input disturbances. This part is called *equivocation*: $H(X_e/X_{al})$.

Output disturbances cause an output signal to contain misinformation. This information quantity of the message medium is called *irrelevance* or *dissipation*: $H(X_{al}|X_e)$.

The medium information quantity actually transferred is called *transinformation* $H(X_e; X_{al})$.

The medium information quantity at the measuring-system output, i.e. the output entropy $H(X_{al})$, is a combination of the irrelevance and the transinformation.

If both the useful signal $x_e(t)$ and the disturbing signal $r_e(t)$ follow the Gaussian probability density function then

$$p(x) = \frac{1}{\sigma \sqrt{(2\pi)}} \exp\left(-\frac{(x - \bar{x})^2}{2\sigma^2}\right) \tag{6.83}$$

From equations (6.83) and (6.82) it can be calculated that:

(a) input entropy

$$H(X_e) = \tfrac{1}{2} \text{lb}[2\pi e \, \overline{x_e^2(t)}] + \lim_{\Delta x \to 0} (1/\Delta x) \tag{6.84}$$

(b) irrelevance (dissipation)

$$H(H_{al}|X_e) = \tfrac{1}{2} \text{lb}(2\pi e \, \overline{r_e^2(t)} + \lim_{\Delta x \to 0} (1/\Delta x)) \tag{6.85}$$

(c) output entropy

$$H(X_{aI}) = \tfrac{1}{2}\{\text{lb } 2\pi e \ [\overline{x_e^2(t)} + \overline{r_e^2(t)}]\} + \lim_{\Delta x \to 0} (1/\Delta x) \qquad (6.86)$$

(d) transinformation

$$H(X_e; X_{aI}) = \tfrac{1}{2}\text{ lb}\{1 + [\overline{x_e^2(t)}/\overline{r_e^2(t)}]\} \qquad (6.87)$$

The transinformation is a property of the transferring channel, related to a certain source, but it is not only a typical property of the transferring channel.

For all calculations natural logarithms are preferred. It should also be considered that the numerical values of the information quantities depend on the units used. Calculations with binary logarithms (1b) have the unit *bit*, those calculated with natural logarithms (ln) the unit *nit* and calculations with decadic logarithms ($\log_{10}$) the unit *dit*. The following conversion formulae can be used (Novickij, 1978):

$$1 \text{ bit} = 0.69 \text{ nit} = 0.30 \text{ dit}$$
$$1 \text{ nit} = 1.45 \text{ bit} = 0.43 \text{ dit}$$
$$1 \text{ dit} = 2.30 \text{ nit} = 3.30 \text{ bit}$$

6.4.3 Information-theoretical Measurement Error Characteristics

Measures of errors occurring are the linear medium value of the measurement error:

$$\overline{\Delta x_a} = \int_{-\infty}^{\infty} \Delta x_a p(\Delta x_a) \, dx = \lim_{T \to \infty} \frac{1}{2T} \int_{-T}^{+T} [x_{aI}(t) - x_{aS}(t)] \, dt \qquad (6.88)$$

and the square medium value of the measurement error:

$$\overline{\Delta x_a^2} = \int_{-\infty}^{\infty} \Delta x_a^2 p(\Delta x_a) \, dx = \lim_{T \to \infty} \frac{1}{2T} \int_{-T}^{+T} [x_{aI}(t) - x_{aS}(t)]^2 \, dt \qquad (6.89)$$

where $x_{aI}(t)$ is the erroneous output signal of the measurement system and $x_{aS}(t)$ the error-free output signal of the measurement system.

Normally distributed desired and disturbing signals in a stationary, ergodic, stochastic process are the basis of the Bayes model. The square medium value of the measurement error in equation (6.89) is called the Bayes risk (Krauss and Woschni, 1975; Seidler, 1971).

Furthermore the medium square deviation s_{aI} of the single measurement value x_{aIi} from the medium value $\bar{x}_{aI}$, that is, the standard deviation according to equation (6.46), has been introduced as an error measure.

The standard deviation multiplied by a constant factor, $a \geqslant 1$, results in the confidence intervals that are indicators of the *maximum errors*. These maximum errors have two principal disadvantages:

(a) the selection of factor a can be accomplished arbitrarily;
(b) the longer the measurement series becomes, the greater the probability of a growing maximum error.

On the basis of the normal distribution the following distribution of the maximum error is to be expected:

$$\Delta \hat{x}_a = \begin{cases} s_{aI} & \text{on an average, once in 3 measurements} \\ 2s_{aI} & \text{on an average, once in 22 measurements} \\ 3s_{aI} & \text{on an average, once in 310 measurements} \\ 4s_{aI} & \text{on an average, once in 15000 measurements} \end{cases}$$

This results in different prerequisites to the estimates of measurement processes on the basis of short and long measurement series.

Moreover, having a limited number of measured values in many situations, it is often impossible to obtain the confidence interval (maximum error) itself. Only an approximation of the confidence interval, its estimated value, is calculable. When only a small number of measured values are available, it will not be possible to make any reasonable statement on the probably occurring maximum values. Therefore Novickij (1978) suggested use of the entropy value η of the measurement error.

For any distribution of the measurement error the information quantity can be described by equations (6.71) and (6.74). To simplify the calculation, with the help of logarithms,

$$q = \ln m = \ln[(x_{\max a} - x_{\min a})/d] = \ln(x_{\max a} - x_{\min a}) - \ln d \quad (6.90)$$

where d is the error interval. Furthermore

$$q = H(X_e) - H(X_e/X_{aI}) \quad (6.91)$$

The information entropy of the error is estimated under the simplifying supposition that equivocation and irrelevance are identical:

$$H(X_e|X_{aI}) = H(X_{aI}|X_e) = -\int_{-\infty}^{\infty} p(x)\ln p(x)\, dx \quad (6.92)$$

If the errors are normally distributed, see equation (6.83), then

$$\ln p(x) = -\ln[\sigma\sqrt{(2\pi)}] - x^2/(2\sigma^2)$$

and with

$$\int_{-\infty}^{\infty} p(x)\, dx = 1 \qquad \int_{-\infty}^{\infty} x^2 p(x)\, dx = \sigma^2$$

then

$$H(X_e/X_{al}) = \ln[\sigma\sqrt{(2\pi\, e)}]$$

To formally define the entropy error Δ_e the following relations are used:

$$H(X_e/X_{al}) = \ln d_e = \ln 2\Delta_e \qquad (6.93)$$

$$d_e = 2\Delta_e = \exp[H(X_e/X_{al})] \qquad (6.94)$$

$$\Delta_e = \tfrac{1}{2}\exp[H(X_e/X_{al})] \qquad (6.95)$$

From Novickij (1978):

(a)   for the normal distribution

$$\Delta_e = \sigma\sqrt{(\pi\, e/2)} = 2.066$$

(b)   for the equal distribution

$$\Delta_e = \sigma\sqrt{3} = 1.73$$

(c)   for the triangular distribution

$$\Delta_e = \sigma\sqrt{(3\, e/2)} = 2.02$$

The entropy coefficient is defined as

$$k = \Delta_e/\sigma \qquad (6.96)$$

From above

$$\Delta_e = k\sigma \qquad m = (x_{\max al} - x_{\min al})/2k\sigma \qquad q = \ln m$$

The commonly used confidence interval and the entropy error are similar. The measurement result is given using

$$x_{aS} = x_{ali} - \Delta x_a \pm \Delta_e \qquad (6.97)$$

6.4.4   Channel Capacity

The dynamic properties of measurement-information systems can be described by the information flow or the channel capacity per unit time:

$$C_T = \frac{1}{T_E}\,\text{lb}\, m \quad \text{bits/s} \qquad (6.98)$$

where T_E is the response time. The information flow (flow of binary numbers) is greater than the symbol flow if the symbols are each formed by several binary numbers (0, 1). Let

$$C_T = VI \quad \text{bits/s}$$

where V is the symbol flow in symbols/s and I the coding expense (decision contents) per symbol in bits/symbol. In compliance with the sampling theorem

$$T_E = 1/(2f_c) \qquad (6.99)$$

gives the relation between the sampling time T_A, or response time T_E, and the cut-off frequency f_c.

Equation (6.98) can also be stated as

$$C_T = 2f_c \text{ lb } m \quad \text{bits/s} \qquad (6.100)$$

For analog measurement systems the channel capacities have to be formed with equation (6.71). For digital measurement systems equation (6.73) is used. The channel capacity per unit time, C_T, is approximately equal to the maximum transinformation flow through a measurement system. It describes the static and dynamic behaviour of a measurement system. In doing so the channel capacity does not provide any other statements on the static and dynamic behaviour of a measurement system other than the differentiable amplitude step number m and the response time T_E.

6.5 EXAMPLE

6.5.1 Influence of Multiplicative Measurement Errors on the Measurement Result

A thermocouple, a thermostat, a moving-coil galvanometer, and a flow channel with a slide valve (Figure 6.2) are given. A step change of temperature is to be measured. The thermocouple has a static measured-value sensitivity of

$$S_{Th} = x_{aTh}/x_{eTh} = 0.05 \text{ mV/K} \qquad (6.101)$$

The moving-coil instrument has been calibrated with a sensitivity of $S_{NiCr-Ni} = x_{aTh}/x_{eTh} = 0.04$ mV/K for NiCr–Ni thermocouples and, therefore, it has the static sensitivity (Hofmann, 1977)

$$S_G = x_{aG}/x_{eG} = x_{eTh}/x_{aTh} = 25 \text{ K/mV} \qquad (6.102)$$

If temperature changes are to be indicated inertia-free and correctly by the moving-coil instrument, the following transfer function of the total system with thermocouple plus moving-coil instrument applies:

$$S_S(p) = 1 \text{ K/K} \qquad (6.103)$$

This is the transfer function for an inertia-free (zero-order) proportional system with the proportionality factor 1. Thermocouples, however, are slow-acting. If the dynamic properties of the total system (thermocouple plus moving coil instrument, see Figure 6.2) are determined only by the thermocouple—this

(a)

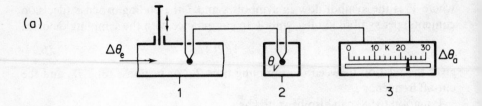

(b)

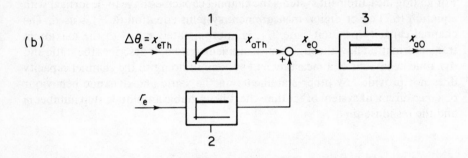

(c)

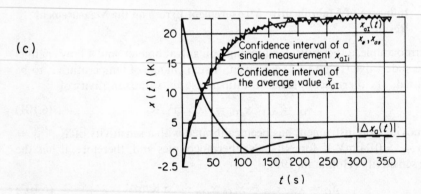

Figure 6.2   Progress of systematic and random measurement errors in some temperature measurements. (a) Components of the measurement circuit; (b) signal flow diagram of the measurement circuit; (c) progress of systematic and random measurement errors. The following table gives values of the measurement series 1 to 3

No.	x_{a11}	x_{a12}	x_{a13}	a_{a14}	x_{a15}	x_{a16}	x_{a17}	x_{a18}	x_{a19}	x_{a110}
1	22.56	22.65	22.35	22.53	22.47	22.59	22.40	22.57	22.45	22.48
2	22.45	22.53	22.48	23.49	22.51	22.47	22.65	22.35	22.40	22.57
3	20.01	20.15	19.85	20.03	19.97	20.09	19.90	20.07	19.95	19.98

is a simplification—the following dynamic measurement-system behaviour results from it.

The spherical measuring junction of the thermocouple is at a constant temperature θ_K, and at time $t = 0$ it is quickly exposed to a medium at a constant temperature $\theta_M > \theta_K$. The temperature change inside the sphere is a function of time and space given by a sum of exponential functions with different coefficients and negative real exponents (Hofmann, 1976). A small sphere radius, a large coefficient of thermal conductivity λ or a small coefficient of heat transfer α can imply a negligible development of all components of order greater than one. If this is the case the spherical sensor heats internally as a first-order inertial system and the thermocouple generates an output voltage (in mV) given by

$$x_{aTh}(t) = \hat{x}_{aTh}\left[1 - \exp\left(-\frac{3\alpha t}{c\rho R}\right)\right] \tag{6.104}$$

where

c = specific heat
ρ = density
R = radius of the thermocouple-sphere
$\hat{x}_{aTh}$ = output voltage in the stationary state.

The time constant of the thermoelement is

$$c\rho R/3\alpha = \tau \tag{6.105}$$

For $c = 5 \times 10^{-1}$ kJ/(kg K), $\rho = 1 \times 10^4$ kg/m³, $R = 3 \times 10^{-3}$ m, and $\alpha = 100$ W/(m²K) (stirred air) the time constant is $\tau = 50$ s.

Thus the actual measured-value transfer function of thermocouple is

$$S_{ThI}(p) = \frac{L\{x_{aTh}(t)\}}{L\{x_{eTh}(t)\}} = \frac{\hat{x}_{aTh}/\hat{x}_{eTh}}{1 + \tau p} \tag{6.106}$$

with (see Table 6.1, entry nos 1, 6)

$$\tilde{x}_{eTh}(p) = \frac{\hat{x}_{eTh}}{p} \quad \tilde{x}_{aTh}(p) = \frac{\hat{x}_{aTh}}{p(1 + \tau p)} \tag{6.107}$$

as well as

$$\hat{x}_{eTh} = \theta_M - \theta_K \quad \tau = 50 \text{ s}$$

and

$$\hat{x}_{aTh}/\hat{x}_{eTh} = 0.05 \text{ mV/K}$$

The actual transfer function of the total measurement system consisting of thermocouple and moving-coil instrument is described by equations (6.102) and (6.106), as well as Figure 6.2:

$$S_{xI}(p) = S_{ThI}(p)S_G(p) = 1.25/(1 + \tau p) \quad \text{K/K} \tag{6.108}$$

The internal error transfer function $S_{FE}(p)$ of the measurement system is, according to the equations (6.108) and (6.103),

$$S_{FE}(p) = \frac{S_{xI}(p)}{S_S(p)} - 1 = \left(\frac{1.25}{1 + \tau p} - 1\right) \text{K}/\text{K} \tag{6.109}$$

The absolute systematic internal error of the measurement system (the multiplicative measurement error) follows equation (6.109) in the p-range (complex frequency range):

$$\Delta \tilde{x}_{aE}(p) = S_{FE}(p)\tilde{x}_{aS}(p) = \left(\frac{1.25}{1 + \tau p} - 1\right) \frac{20}{p} \text{K} \tag{6.110}$$

with $\tilde{x}_{aS}(p) = 20 \text{ K}/p$ because $\tilde{x}_{aS}(p) = S_S(p)\tilde{x}_{eS}(p)$ and $S_S(p) = 1 \text{ K/K}$ as well as $\tilde{x}_{eS}(p) = 20 \text{ K}/p$ for a step change of temperature from $\theta_K = 50\,°\text{C}$ to $\theta_M = 70\,°\text{C}$ with $x_{eS} \equiv x_{eTh}$.

In the t-range the multiplicative measurement error (using equation (6.110) and Table 6.1, entry nos. 1, 6) has the behaviour

$$\Delta x_{aE}(t) = 5(1 - 5\,e^{-t/50\,s}) \text{ K} \tag{6.111}$$

6.5.2 Influence of Additive Measurement Errors on the Measurement Result

An additive measurement error appears when the reference junction is not at the prescribed temperature $\theta_v = 50\,°\text{C}$ but is at $\theta_v = 52\,°\text{C}$. The disturbance variable $\Delta x_{eF} = \theta_v - \theta_v^* = -2 \text{ K}$ produces a counter voltage by the reference junction which has the same temperature sensitivity $S_v = 0.05 \text{ mV/K}$ as the measuring junction; that means $\Delta x_{aF} = S_v \Delta x_{eF} = -0.1 \text{ mV}$. The inertia of the reference junction has not to be taken into consideration because the temperature of the reference junction has already reached its stationary value at the beginning of the measuring process and will remain constant during the measurement.

The absolute systematic external error is shown in Figure 6.2 for the measurement chain consisting of reference junction and moving-coil instrument with sensitivity S_G in equation (6.102):

$$\Delta x_{aF} = -0.1 \text{ mV} \times 25 \text{ K/mV} = -2.5 \text{ K} \tag{6.112}$$

6.5.3 Influence of Systematic Measurement Errors on the Measurement Result

The absolute systematic measurement error consists additively of the internal error (multiplicative measurement error) and the external error (additive measurement error), so that equations (6.111) and (6.112) will finally result in

$$\Delta x_{aI}(t) = \Delta x_{aE}(t) + \Delta x_{aF}(t) = 2.5(1 - 10\,e^{-t/50\,s}) \text{ K} \tag{6.113}$$

The step response from 50°C to 70°C at the thermocouple gives, instead of the *true* behaviour of

$$x_{aS}(t) = 20 \text{ K} \quad \text{for } t > 0 \tag{6.114}$$

the real behaviour of the sensor

$$x_{aI}(t) = x_{aS}(t) + \Delta x_{aI}(t) = [20 + 2.5(1 - 10 \, e^{-t/50 \, s})] \text{ K} \tag{6.115}$$

as simulated by the moving-coil instrument. For $t \gg \tau$ there exists a stationary value of 22.5 K (Figure 6.2)

6.5.4 Influence of Random Measurement Errors on the Measurement Result

Random measurement errors originate from such causes as parasitic irregular interference fields which can induce stochastic disturbing voltages in the unscreened wires of the thermocouple. Therefore, the behaviour of the output quantity oscillates randomly about the undisturbed behaviour (after equation (6.115)). The measured values (Figure 6.2) were ascertained in the stationary state ($t \gg \tau$) for repeated measurements.

On the basis of measurement series 1 the arithmetic mean value, using equation (6.43), is given by

$$\bar{x}_{aI} = \frac{1}{10} \left(\sum_{i=1}^{10} x_{aIi} \right) = 22.5 \text{ K} \tag{6.116}$$

The standard deviation is, from equation (6.46),

$$s_{aI} = \left(\sum_{i=1}^{10} (x_{aIi} - \bar{x}_{aI})^2 / 9 \right)^{1/2} = 0.09 \text{ K} \tag{6.117}$$

The confidence interval of a single measurement (see equation (6.47)) is, for the statistical certainty $P = 95\%$ and $n = 10$ measured values,

$$v_{aIE} = \pm a k_1 s_{aI} = 1.96 \times 1.65 \times 0.09 \text{ K} = 0.29 \text{ K} \tag{6.118}$$

The confidence interval of a mean value (see equation (6.48)) is, for the statistical certainty $P = 95\%$ and $n = 10$ measured values,

$$v_{aIM} = \pm s_{aI} t / \sqrt{n} = \pm 0.715 \times 0.09 \text{ K} = \pm 0.064 \text{ K} \tag{6.119}$$

The unknown systematic measurement errors $f_{a\Sigma}$ are to be zero for the measurement system in Figure 6.2.

6.5.5 Formation of the Complete Measurement Result

The complete measurement result consists of all components calculated up to now. The desired value of the output quantities is:

(a) for a single measured value with equations (6.55), (6.113) and (6.118),

$$x_{aE} = [22.56 - 2.5(1 - 10 \, e^{-t/50 \, s}) \pm 0.29] \text{ K} \tag{6.120}$$

(b) for the medium value with equations (6.56), (6.116) and (6.119),

$$x_{aM} = [22.5 - 2.5(1 - 10 e^{-t/50 s}) \pm 0.064] \text{ K} \qquad (6.121)$$

Concerning the measurement series in Figure 6.2 it should be said that the measured value x_{aI4} in measurement series 2 probably represents a gross measurement error (outlier). Application of the outlier criterion (equation (6.54)) says that if $(23.49 \text{ K} - 22.5 \text{ K}) > 4.42 \times 0.09 \text{ K}$, then x_{aI4} is to be excluded. Because $0.99 \text{ K} > 0.4 \text{ K}$ this assumption is confirmed. Before measurement series 3 in Figure 6.2 was taken a correction had been carried out. In the present case the comparison junction temperature was set at 50 °C, and the scale of the moving-coil galvanometer with a scaling of 25 K/mV was replaced by a scale with a scaling of 20 K/mV. This resulted in a 100% correction of the static measurement error. The correction of dynamic measurement errors has been discussed in detail in the literature (Hofmann, 1976; Hofmann, 1977).

REFERENCES

Barry, B. A. (1964). *Engineering Measurements*, Wiley, New York.

Bendat, J. S. and Piersol, A. G. (1971). *Random Data: Analysis and Measurement Procedures*, Wiley, New York.

Beyer, O., Hackel, H., Pieper, V., and Tiedge, J. (1976). *Wahrscheinlichkeitsrechnung und mathematische Statistik* (Engl. Transl. *Probabilistic Calculation and Mathematical Statistics*), Teubner Verlagsgesellschaft, Leipzig.

Dietrich, C. F. (1973). *Uncertainty, Calibration and Probability; the Statistics of Scientific and Industrial Measurement*, Adam Hilger, London.

Doebelin, E. O. (1975). *Measurement Systems: Application and Design*, revised edn., McGraw-Hill, New York.

Gellert, W. (1977). *Kleine Enzyklopädie Mathematik* (Engl. Transl. *Small Encyclopaedia of Mathematics*), VEB Bibliographisches Institut, Leipzig.

Halstead, H. J. (1960). *Introduction to Statistical Methods*, Macmillan, Melbourne.

Helstrom, C. W. (1968). *Statistical Theory of Signal Detection*, Pergamon, London.

Hofmann, D. (1976). *Dynamische Temperaturmessung* (Engl. Transl. *Dynamical Temperature Measurement*), VEB Verlag Technik, Berlin.

Hofmann, D. (1977). *Temperaturmessungen und Temperaturregelungen mit Berührungsthermometern* (Engl. Transl. *Temperature Measurement and Temperature Control with Contact Thermometers*), VEB Verlag Technik, Berlin.

Hofmann, D. (1979). *Handbuch Messtechnik und Qualitätssicherung* (Engl. Transl. *Handbook of Measurement Engineering and Quality Assurance*), VEB Verlag Technik, Berlin. (2nd ed. 1981).

Hultzsch, E. (1971). *Ausgleichsrechnung mit Anwendungen in der Physik* (Engl. Transl. *Averaging with Applications in Physics*), 2. Aufl., Akademische Verlagsgesellschaft Geest und Portig K. G., Leipzig.

Krauss, M. and Woschni, E. G. (1975). *Messinformationssysteme* (Engl. Transl. *Measurement-Information Systems*), 2. Aufl., VEB Verlag Technik, Berlin.

Lange, F. H. (1973). *Signale und Systeme*, Band 3, *Regellose Vorgänge* (Engl. Transl. *Signals and Systems*, Vol. 3, *Random Processes*), 2. Aufl., VEB Verlag Technik, Berlin.

Lange, F. H. (1978). *Methoden der Messtochastik* (Engl. Transl. *Methods of Stochastic Measurements*), Akademie-Verlag, Berlin.

Müller, P. H. (1975). *Lexikon der Stochastik* (Engl. Transl. *Dictionary of Stochastics*), Akademie-Verlag, Berlin.

Novickij, P. V. (1978). *Gütekriterien für Messeinrichtungen* (Engl. Transl. *Quality Criteria for Measurement Systems*), VEB Verlag Technik, Berlin.

Peschel, M. (1978). *Modellbildung für Signale und Systeme* (Engl. Transl. *Models for Signals and Systems*), VEB Verlag Technik, Berlin.

Renyi, A. (1971). *Wahrscheinlichkeitsrechnung* (Engl. Transl. *Probabilistic Calculation*), VEB Deutscher Verlag der Wissenschaften, Berlin.

Seidler, J. (1971). *Optimierung informationsübertragender Systeme* (Engl. Transl. *Optimization of Information-Transferring Systems*), VEB Verlag Technik, Berlin.

Smirnow, N. W. and Dunin-Barkowski, I. W. (1963). *Mathematische Statistik in der Technik* (Engl. Transl. *Mathematic Statistics in Engineering*), VEB Deutscher Verlag der Wissenschaften, Berlin.

Storm, R. (1967). *Wahrscheinlichkeitsrechnung, mathematische Statistik und statistische Qualitätskontrolle* (Engl. Transl. *Probabilistic Calculation, Mathematical Statistics and Statistical Quality Control*), 2. Aufl., VEB Fachbuchverlag, Leipzig.

Sweschnikow, A. A. (1970) *Wahrscheinlichkeitsrechnung und mathematische Statistik in Aufgaben* (Engl. Transl. *Probabilistic Calculation and Mathematical Statistics in Tasks*), Teubner, Leipzig.

Topping, J. (1972). *Errors of Observation and Their Treatment*, Chapman and Hall, London.

Handbook of Measurement Science, Vol. 1
Edited by P. H. Sydenham
© 1982 John Wiley & Sons Ltd.

Chapter

7

C.J.D.M. VERHAGEN, R.P.W. DUIN, F.A. GERRITSEN,
F.C.A. GROEN, J.C. JOOSTEN, AND P.W. VERBEEK

Pattern Recognition

Editorial introduction

The simplest form of measurement situation arises when variables that are to be mapped into measurement data are easily identified as singular quantities that are amenable to sensing by straightforward well-established procedures. Complex system situations often involve monitoring a great number of measurements in order to give ultimately a few output quantities.

Another approach is to attempt what the authors of this chapter so aptly term, a *many-to-one* transformation right at the measurement interface with the system. This procedure is being developed and reported in what has generally become known as pattern recognition. Patterns occur in many circumstances. Examples are found in ore-sorting, handwritten address recognition, defects in tin-plane stock, abnormalities of forest growth, structure of spoken language, artistic creations, social behaviour, economic trends, and so on. In cases such as these, the measurement need is often to decide into which of a few defined classes certain features of the pattern should be placed.

This chapter is concerned with the methodology and practice that has been developed. Although the subject has not, in general, matured into useful application to the extent that was envisaged a decade or so ago it has nevertheless provided workable, economic, solutions to many a real-world requirement. The material presented (compiled in 1979) will assist assessment of potential situations and provide a valuable springboard into this important aspect of many measurement systems.

7.1 INTRODUCTION

7.1.1 The Process of Pattern Recognition

Pattern recognition means here: *automatic* pattern recognition; pattern recognition in humans and animals is not taken into account. The field of pattern recognition is very broad notwithstanding this restriction, as both the terms *pattern* and *recognition* include very different entities or activities (Verhagen,

1975a). Pattern recognition systems, however, have some points in common: they are data or information processing systems, with fairly complex input data and fairly simple output data. The input data often are collected by means of physical measuring instruments from sources in the outside world, but other data (for instance, economic, sociographic) may also be used. The sources and the input data are presumed to contain 'patterns', mostly together with noise, disturbances, and background data. It is the task of the data processing system to assign the input data to a certain class in accordance with the source pattern. It is essential in pattern recognition that *different input data should be classified in the same class*. It is also true, however, that the same sources and the same input data may contain several different types of patterns and so—for different problems—have to be classified into different types of classes. A few examples may clarify these statements:

(a) All different ways of writing the character 'A' produce different inputs to, for example, a TV-scanner, but the output always has to be a class with indication 'A', one of the classes of the alphabet. In a different problem, however, a certain character 'A' has to be classified as belonging to a certain font, or written by a certain penman, and so to quite different types of classes: to the classes of the fonts or the classes of the individual penmen, respectively, instead of to the classes of the characters of the alphabet.

(b) Quite different scenes may all be classified as belonging to the class of scenes containing a certain vehicle, but in a different problem a certain scene with that vehicle may have to be classified as a sharp image, or as an image containing a vehicle with a certain number on its number plate, or driven by a certain driver.

The type of patterns and so the type of classes that are of interest at a certain moment to a certain observer is determined by his choice; this choice has to be made before a pattern recognition system can be developed. We assume from now on that this choice has been made.

A fundamental question is: what different representations of a certain pattern in a source are presumed to be indicated by a certain class? For instance, what different forms of the character 'A' have to be taken into account, and do ornamental characters have to be recognized too? This question is mostly very difficult to answer; it is related to the variability of the representations of the patterns of each class. Variability of the representations of the patterns has many causes (Section 7.1.2) and is mostly not very explicitly formulated. The pattern recognizer should analyse the variability and find attributes or features and decision methods to cope with the variability that is present.

As a pattern recognition system should map many different representations of a certain pattern, present in different sources, into the same class, one may describe this pattern recognition problem as a *many-to-one* mapping.

7.1.2 Variability

The variability of the representations of the patterns belonging to a certain class is an essential difficulty in pattern recognition. Several causes may exist for this variability.

Natural patterns from biological origin vary from specimen to specimen, and in humans and animals the shape or expression of one specimen can vary considerably as a function of time (consider face recognition).

Cultural patterns, such as printed or written characters, are roughly determined by human conventions, but conventions change, depend on personal interpretation, and are difficult to define. An artist may develop his own style of characters; as long as they are recognized by (some) humans they are acceptable.

In addition to these essential variabilities, the variability resulting from the *treatment and preparation* of the sources with the patterns may be important. The illumination of objects or the direction from which they are viewed greatly determine the input to the sensors of the pattern recognition system. The same is true for the way biological objects, like tissues or cells, are prepared (Figure 7.1).

Variability of the input data also originates from noise in the *communication channel* between the sources and the measuring instrument and from noise and imperfections of the *measuring system*. One specific source with a specific pattern may be led, by these causes, to many mutual different inputs to the pattern recognition system.

The natural and cultural variabilities have to be seen as essential to pattern recognition. The preparation and treatment of objects, the communication channel, and the measuring system, however, are preferably chosen in such a way that the variability added is—if possible—small in relation to the essential pattern variability. This is another way of stating a general rule of measurement, that is, that the measuring system should not change the situation more than corresponds with the accuracy pursued.

The situation in pattern recognition is quite complex because a great many types of distortions and noise may be present and the properties of the patterns may be quite different. If high spatial frequency details of an image are essential for the patterns under consideration the requirements are not comparable to the situation where the low-frequency shape of a curve has to be analysed. As a consequence, the preparation and treatment of the objects and the communication and measuring process require great attention for each individual pattern recognition problem. This sometimes means that standardization of preparation and treatment techniques and reduction of noise or distortion influences may be necessary. It also means that specialized measuring systems for pattern recognition may be appropriate. This is one of the reasons why it seemed useful to add this chapter on pattern recognition to this handbook.

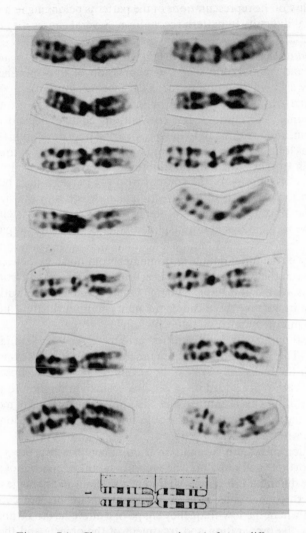

Figure 7.1 Chromosome number 1 from different individuals. Although variability can result from inherited variations in the size of certain bands, it is mainly caused by chromosome preparation and staining.
(Courtesy M. v. d. Ploeg, Leyden)

Later on, how to deal with variability will be discussed; two general approaches are:

(a) estimate statistical parameters to describe variability;
(b) use *a priori* physical knowledge of the sources, the patterns, the communication channel, and the measuring instrument.

7.1.3 Features

The input data are often very redundant. An image of 256×256 picture elements with 8 bits per element contains more than 5×10^5 bits, which have to be reduced to a few bits indicating the classes involved. In addition to this, the input data show much variability. For these reasons it is necessary to look for attributes or *features* in these data that may be more or less characteristic for the patterns to be recognized and that are far less redundant than the original input data. *Preprocessing* the input data has to produce a limited amount of feature values as a basis for the decision process, which results in classes at the output of the recognition system. The type of preprocessing needed and the type of features suitable cannot be found in a straightforward way. Physical and *a priori* knowledge about the sources of variability is essential as a guideline. Statistical methods may be helpful to select features (see Section 7.2.2) but in addition intuition, experience, and/or trial and error are sometimes necessary to find suitable features. This determines in great part the difficulty of the pattern recognition process.

7.1.4 Formal Description of Pattern Recognition

Pattern recognition has been characterized in Section 7.1.1 as a many-to-one mapping; 'many' indicates that pattern recognition deals with many different but equivalent representations of a certain pattern. In a similar way to that used in Chapter 1, pattern recognition may now formally be described by a mapping m from a set of sources S_i, containing all different but equivalent representations of the pattern i, into a class C_i, the name or code for this pattern; this has to hold for all patterns i under consideration. With '$\sim_{S_i}$' as indication for the *equivalence relation* on the set of sources S_i and '$=$' the identity relation we have $m: S_i \to C_i$ must be a *homomorphism* of $\langle S_i, \sim_{S_i} \rangle$ into $\langle C_i, = \rangle$, which means that if s_{ij} and s_{ik} are two representatives of S_i, so $s_{ij}, s_{ik} \in S_i$, and $s_{ij} \sim_{S_i} s_{ik}$ this implies that $m(s_{ij}) = m(s_{ik})$ (Finkelstein, 1975; Verhagen, 1975b).

If the set of sources S_i is mapped by m_1 (denoting sensing and preprocessing) into the set of features F_i, and further by m_2 (denoting decision, discrimination) into the set of classes C_i, the features have to provide the possibility of introducing an equivalence relation $\sim_{F_i}$ on the set of features, which allows mapping of the features in a homomorphic way into the classes. Figure 7.2 shows the situation in a global way.

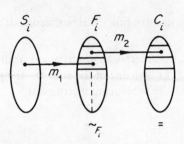

Figure 72. Formal description of pattern recognition by sensing and preprocessing m_1 and decision m_2 based upon an equivalence relation $\sim_{F_i}$ on the set of features F_i

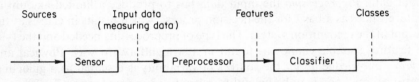

Figure 7.3   Block diagram of a pattern recognition system

A block diagram (Figure 7.3) represents, in a more technical language, the same idea as Figure 7.2.

7.1.5   Types of Patterns, Features, and Pattern Recognition Systems

It is difficult to describe the different types of patterns adequately, together with the related features and recognition systems, because there is a great variety of patterns. A very rough distinction may be made as follows:

(a)   *Patterns* that can be *described by analog or digital values of a number of features*. The variability of the patterns also causes variability in the feature values. This variability in the features can sometimes be handled by means of statistical methods. The corresponding mathematical tools for allocation to the classes are called *statistical discriminant methods* (see Section 7.2.2).

(b)   *Patterns* that can be *decomposed in subpatterns* and even smaller subpatterns (the smallest of these subpatterns are often called *primitives*) and that can be described by these subpatterns or primitives and their mutual relations. A line figure, for instance, can be composed of straight and curved lines placed in certain relation to each other. The patterns sometimes can be expressed in a language generated by a grammar. In that case *linguistic methods* (also called *structural* or *syntactic methods*) can be used to describe patterns. Recognition requires parsing after determination of the primitives and the mutual relations. In some special cases the grammar needed for analysing can be found based upon a finite set of characteristic sample patterns. This process is called *grammatical inference* (Section 7.2.3).

The patterns and the classes in which they are classified mostly concern *discrete classes*; a pattern belongs to a certain class or not. *Fuzzy patterns*, where each pattern can have a membership value between 0 and 1 of several classes at a time, currently receive much attention (Section 7.2.2). Sometimes an allocation to a *continuous* variable is possible by interpolation between classes.

A *learning set* of representations of the patterns under discussion is often available, together with labels indicating the class of each representation. This set should be used during the development and training of a pattern recognition system to learn about the variability of the representations of the patterns and to develop some (optimal) classification scheme. A large and representative learning set facilitates such a development but, unfortunately, such an ideal set is seldom available. If no representations with a label of their respective classes are given, an analysis of the input data may show whether so called clusters can be found, indicating the existence of regularities that might be called patterns (learning without a teacher).

Finally, one may distinguish between *pattern recognition in a strict sense*, where human recognition of visual, auditive or other patterns is automatized, and *pattern recognition in a very broad sense*, where all types of (even not yet known) regularities in data are taken into account and where pattern recognition techniques are used to analyse and classify the data. These data may have their source in such fields as the economy and scientific research.

7.1.6 Pattern Recognition and Nominal Measurement

The definition of nominal measurement is not unique. Originally (Stevens, 1946; Ellis, 1966) in a nominal measurement one must only be able to distinguish whether two representatives of a certain entity are equal in a certain aspect or not. No order in the entities whatever is under discussion. A standard collection of entities with names or codes is often available (standard colours for instance). This allows one to compare unknown entities with the items of the collection and to determine which one of the collection, if any, equals the unknown. The name or code of the item of the collection that equals the unknown is given to the latter. So one only has to be able to determine *identity* (sameness) and *difference*. These words are sometimes repeated in recent papers (Finkelstein, 1973, p. 18: 'Measures on a nominal scale merely describe whether two entities are identical or different'), but 'identical' (and 'equal') are broadly interpreted, as 'nearly equal', 'similar' and as 'equivalent' (in the sense of equivalence relation) (Finkelstein, 1973, p. 17). A relation with pattern recognition as described in the previous sections will be evident: *nominal measurement and pattern recognition share the basic idea of the equivalence relation.* Aspects like colour and hardness, are determined as equal or equivalent in nominal measurements; the aspect 'pattern' is regarded in the equivalence relation as used in pattern recognition.

The determination of equality, similarity or equivalence of items like colour and of visual or auditive patterns, as performed by human (and animal) perception seems to be quite easy as a product of evolution (and an immense learning set), but it is not well known yet how this works. Automatic and objective determination of equality, similarity or equivalence is difficult; computer evolution has not been directed toward perception and association.

No measuring scale like a ratio, interval or even ordinal scale was available for traditional nominal measurements. Much research was necessary to find, for instance, which physical quantities (with their scales) had to be used to compare colours, as experienced by man, objectively (luminance, hue, saturation). The situation to define objective equivalence measures for all types of patterns is still more difficult. In the terminology of the preceding sections it can be said that features have to be found which characterize the patterns involved in a suitable way; they have to discard the information which is considered irrelevant to the problem and to reveal what is relevant.

It may be remarked that an artificial pattern recognition system that successfully performs the classification tasks of a biological system also constitutes a model for it. Here lies a relation between perception and pattern recognition research.

7.1.7 Pattern Recognition and Measuring

The discussion in the preceding section already indicated a relation between pattern recognition and measuring. Another reason for this relation is that the input of pattern recognition systems often consists of physical measuring data and that, as treated in Section 7.1.2, the properties of the data acquisition system are important for the pattern recognition system.

There is a further reason to stress the relation between both fields. The ultimate purpose of a measurement is often not only to produce measured values of unknown quantities; the intention is mostly to interpret these values and to draw conclusions from them. They may be scientific conclusions (an hypothesis is justified or not by a measurement) or technical conclusions (a certain situation is all right or is not; in the latter case a readjustment of the situation by feedback might be necessary). Conclusions that can be drawn from the measured values in a simple way (for instance, the value of a temperature is above a threshold) are usually not described as a pattern recognition process. But in those cases where complex data and a complex analysis of the data are necessary in order to draw conclusions, and are such that pattern recognition methods are necessary, it will be clear that pattern recognition is essentially linked to the measuring process.

7.1.8 Pattern Recognition and Other Disciplines

Pattern recognition is not only connected with the measuring field, but also with many other fields, and this for two reasons: other disciplines apply automatic pattern recognition methods in order to solve some of their own problems, and pattern recognition borrows knowledge and methods from other fields (Verhagen, 1975a). In addition to some disciplines already mentioned before, like statistics, linguistics, and physics, (including measuring) pattern recognition uses knowledge from, for instance, mathematics, computer science, information theory, artificial intelligence, and perception science.

The following survey (current in 1979) tries to give some idea about fields where pattern recognition can be applied. A few examples per field are given and some short comments. In Verhagen *et al* (1980) some more details and literature references for the examples are given.

Medical field

Examples in the medical field are: diagnosis; analysis of cardiograms, and encephalograms; chromosome, blood cell, uterus cell or tissue classification; analysis of echocardiograms and other echograms; analysis of X-ray images to determine tumours, obstructions in blood vessels, shape of stomach wall. A great difficulty is to find features allowing a discrimination between normal and abnormal. Several systems are used in practice, but many more are still under development. It appears to be difficult to compete with the flexibility and reliability of humans, and often human labour is less expensive than automatic systems. Interactive systems are popular, allowing humans to take care of difficult and ill-defined situations, while the automatic system performs the routine and bookkeeping activities. Automatic systems have a lead if they can produce numerical values for interesting phenomena, for humans cannot do that.

Recognition of cultural patterns

Reading characters for administration and banking, for the blind and elderly people with decreasing vision; search and selection of drawings; information retrieval are further examples. Reading stylized and printed alphanumeric characters of several fonts is a problem well solved in practice. Handprinted characters from a limited number of penmen, written with certain restrictions, can be read with much success (95–99.9%, depending upon the restrictions and number of penmen). The success is substantially lower when no restrictions are demanded and very many people are involved. The problem of reading continuous handwriting is not yet solved.

Recognition of human environment

Fingerprints and footprints; scene analysis including identification of objects; speech and speaker recognition; remote sensing for geological data, status of the crop, pollution detection, production of maps, inventory of (sub)urban sites also present pattern recognition problems. The recognition of isolated words from a relatively small vocabulary (a few hundred words) and from a few speakers is possible with a high percentage of success (95–99%): the greater the size of the vocabulary and the number of speakers the lower the recognition rate. Continuous normal speech is only understandable automatically for special situations (programming a computer by a few people). Remote sensing applications are practical for specialized purposes.

Scientific research

Analysis of bubble-chamber images, flintstones, radioactivated mud, geological structures, meteorological data, images from materials science provide examples here. Relatively simple tracks and events in bubble-chamber images can be determined with a reasonable amount of success. Specialized systems are often used to determine quantitative data from thin material slices, etc.

Industrial field

Quality inspection of (fast) moving surfaces of steel, paper, banknotes; recognition of objects for sorting, quality control, and assembling; fidelity of loudspeakers; analysis of blurred X-ray images and radiograms; testing of materials; detection of abnormal, unsafe situations (such as of a nuclear reactor) lie in this class. The applications of pattern recognition in industry are increasing but are still in their infancy. Section 7.7 deals with some reasons for this situation (see also Verhagen, 1977).

Military field

Examples in this field are detecting military targets, planes, steamers, etc. Not many details have been published in the unclassified literature.

7.1.9   Contents of this Chapter

Most attention will be paid to sensors and (pre)processing for two-dimensional images within the framework of pattern recognition. Section 7.2 gives a survey of some important pattern recognition techniques; given the space available, only a general discussion without details is possible.

Section 7.3 is devoted to several types of sensors for two-dimensional images. The data produced by these sensors often have to be (pre)processed, if possible on-line. Some requirements and some processors especially suited for pattern recognition and image processing systems will be treated in Section 7.4.

The communication channel and the measuring system introduce distortions and noise in the input data of a pattern recognition system. The point-spread function of an optical system and the coupled electronic system (for example, as found in a TV-scanner) will blur the image and noise may be added. Section 7.5 discusses some restoration techniques like inverse filtering and noise reduction for two-dimensional data structures. In order to find features or to enhance them, special processing may be useful both for human observation and for automatic recognition; Section 7.6 is devoted to these activities.

Finally, some trends in the field of pattern recognition and the subjects treated in this chapter are given in Section 7.7.

7.2 SURVEY OF PATTERN RECOGNITION TECHNIQUES

7.2.1 Introduction

Many mathematical and heuristic techniques are used in pattern recognition, depending upon the problem. In accordance with Section 7.1.5 the rough distinction between statistical and linguistic methods will be continued. Only a short survey will be given, indicating some major methods used in pattern recognition. A slightly more extensive treatment is given in Verhagen *et al.* (1980) from which this survey borrowed heavily.

Many books on pattern recognition give detailed information, for example refer to Fukunaga (1972), Young and Calvert (1974), Bock (1974), and Steinhagen and Fuchs (1976) for statistical methods, Fu (1974), Fu (1977), and Gonzalez and Thomason (1978) for linguistic methods, and Duda and Hart (1973) and Fu (1976) for both; Batchelor (1978) discusses applications; Rosenfeld publishes a yearly survey on image processing in the journal *Computer Graphics and Image Processing*.

7.2.2 Statistical Methods

The starting point is a set of analog or digital values of k measurements from sources or preprocessed measurements giving k features that may be relevant for the patterns involved. A k-dimensional vector space can be constructed (*feature space*) where each dimension represents a feature. A point in this space indicates a representation of a pattern in a source. The variability of the representations

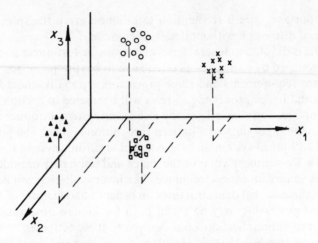

Figure 7.4   Example of a three-dimensional feature space
x_1, x_2, x_3 with four patterns in it; each pattern produces a
cluster of points which indicate equivalent representations
of a certain pattern

of the patterns (Section 7.1.2) means that equivalent representations of a certain
pattern will be indicated by different points in the feature space (Figure 7.4).

Intuitively one expects that sources with equivalent representations of a
certain pattern will be situated 'near' to each other in clusters (*compactness
hypothesis*), and that different patterns lie 'far' from each other. In order to use
notions like 'near' and 'far' one has to define a distance measure. Many distance
measures are in common use (Kanal, 1974), the simplest being the Euclidean
distance for a continuous space and the Hamming distance (number of different
bits) for a binary space.

Only in an ideal case will the points representing equivalent patterns lie close
to each other in small clustered regions as is shown in Figure 7.4, and the points
representing non-equivalent, different patterns, lie in remote regions; this also
indicates that really relevant features are chosen. In most cases the clusters
overlap, have irregular shapes, and are only given as a set of points for a learning
set (Section 7.1.5).

A still more difficult situation exists when no labelled points of a learning
set are given, and one has to analyse the positions of points in the feature space
to see whether some regularities may be found. This situation is similar to the
analysis of measuring data (now taken as features) originated by a physical
experiment or of economic data. Sometimes it is even not known how many
clusters, if any, are available. Now cluster analysis may be helpful. Many clustering methods have been developed in many sciences for many purposes (Jardine
and Sibson, 1971; Everitt, 1974; Bock, 1974; Hartigan, 1975). Subjective
judgments, especially when the number of clusters is not known beforehand,

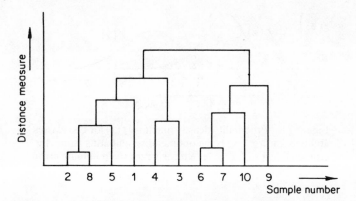

Figure 7.5 Dendrogram. Samples 2 and 8 have the smallest distance; a clustering in two groups gives the clusters (2, 8, 5, 1, 4, 3) and (6, 7, 10, 9); a clustering in three groups gives the clusters (2, 8, 5, 1), (4,3), (6, 7, 10, 9), and so on

are part of the procedure. A simple and often used method may illustrate this point. Firstly, a distance measure has to be chosen, guided by *a priori* knowledge, intuition, and/or trial and error. When n k-dimensional data vectors are available (so providing n points in a k-dimensional feature space), the simplest statement is to interpret them as n clusters. Next, the two points with the shortest distance are fused to a new cluster, giving $n - 1$ clusters. This procedure of combining the two clusters with the next shortest distance can be repeated until all the points form one cluster. To define the distance between clusters with more than one point one has to make a choice as to whether the distance between the means of the cluster points has to be taken, or the distance between the two nearest points in the clusters, or some other distance. The *dendrogram* of Figure 7.5 illustrates this method; the cluster fusions made successively are indicated. The height of the horizontal connection lines gives a measure of the distance between the clusters involved. The final degree of clustering has to be chosen again by subjective judgment.

From now on we will start with a learning set consisting of sources with the labels of the classes into which the patterns in the sources have to be classified. In the ideal case mentioned above, with small clusters per class and the clusters of different classes far away and not overlapping, a very simple recognition procedure is possible: classify an unknown source according to the cluster in which its point in feature space lies. Here it is taken for granted that the learning set is representative for all patterns, and that the clusters can be defined from them.

In practice, however, the clusters often overlap. When a statistical approach is used the values of the features are regarded as stochastic values because of the variations produced by all sources or variability (Section 7.1.2). The method

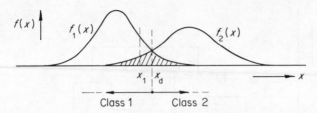

Figure 7.6   Probability density functions $f(x)$ for two classes
and one feature x; x_d is the optimal discriminating value; the
cross-hatched surface gives a measure for the minimum error

applied depends upon the knowledge available. If for each class and for each
feature the probability density function would be known, together with the
a priori probability of all classes and the 'loss' if a sample is classified in a wrong
class, statistics gives the methods to determine optimal *discriminant functions*
between classes, where optimum means a minimum expected loss. A very special
case for two classes and one feature x, with equal *a priori* probability and equal
loss, is indicated in Figure 7.6. In the region where the two density functions
$f_1(x)$ and $f_2(x)$ overlap, for instance at point x_1, the best decision one can make
according to the *Bayes classification rule* is to assign a sample to the class with
the highest value of the probability density, in this case class 1 (Fukunaga,
1972). The discrimination value x_d is that value of x in which the two densities
are equal. The cross-hatched surface is a measure for the minimum error. If the
a priori probabilities are p_1 and p_2 for the two classes, and if the loss when a
pattern of class 1 is classified in class 2 is l_{21}, and when a pattern of class 2 is
classified in class 1 is l_{12}, the best decision (minimum cost) is to assign a sample
x to class 1 if $p_1 l_{21} f_1(x) \geqslant p_2 l_{12} f_2(x)$; the discriminating value x_d now lies
where $p_1 l_{21} f_1(x) = p_2 l_{12} f_2(x)$. Points to the left of x_d are assigned to class 1
and points to the right to class 2.

For n-dimensional feature spaces similar procedures can be applied. For
normally distributed density functions, quadratic discriminant functions are
the result; linear discrimination functions are obtained in case the covariance
matrices of the classes are equal.

The probability density functions and the *a priori* probabilities are mostly
not known in practice, but have to be estimated by using statistical estimation
methods. If the type of the density function is known, the estimation concerns
the parameters of the function from the learning sample values. If the type is
unknown a guess may be made, or some non-parametric estimation technique
may be used. The *Parzen estimation*, by which each object in the feature space is
represented by a kernel and the class density estimation appears as the average
over the learning samples, is quite popular (Fukunaga, 1972). The kernels may
be normal density functions, uniform distributions, and others. In addition to

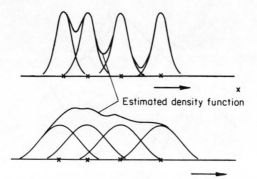

Estimated density function

Figure 7.7   Estimation of the density function
from learning samples, using Parzen estimation
with normal distribution and different width

the kerneltype, parameter values (for instance, width) have to be chosen—a
difficult job. When, for instance, for a one-dimensional example (Figure 7.7) the
normal densities are small, a very peaked estimate will result, broad-shaped
densities produce a very smooth distribution. Several optimizing techniques
using different criteria are possible (Duin, 1976).

The discriminant function between two classes again is determined by the
above formulation, giving non-linear functions; a linear approximation may be
useful (Specht, 1967).

A much more heuristic method does not estimate the density functions but
immediately refers to the intuitive idea that equivalent patterns lie near to each
other (of course with a distance measure defined). Now an unknown sample is
assigned to the class of the *nearest neighbour* among the learning samples, or to
the class of the majority of a certain number of neighbours.

Another heuristic method starts with an adopted type of discriminant function
(for instance, a linear function) and experimentally adapts its coefficients
sequentially in order to get the least wrongly classified learning samples as
possible. Systematic procedures producing convergent results have been developed (Nilsson, 1965; Minsky and Papert, 1969; Mendel and Fu, 1970).

In the preceding methods all features were taken into account at the same
time. Some other methods use the features successively. This can be represented
by a *decision tree*, where in each node a special feature is analysed. Each node
determines the choice of one of the branches which spring from that node. The
classification occurs at the bottom; the depth of the branches may be different
for several parts of the tree (Figure 7.8) (Fu, 1968; Kanal, 1974).

When a discriminant function has been determined by one method or another,
it has to be *tested* concerning the classification error it produces. To use, once
again, the points of the learning set and count the number of wrong classifications
is not advisable; a too optimistic result will be obtained because the discriminant

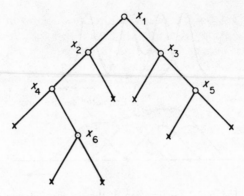

Figure 7.8 Decision tree; circles denote nodes with decision based on feature x_i: crosses denote the end point of classification

function is tailored for this set. It is better to use an independent test set but also one with labelled points. Now the result is too pessimistic, if afterwards a discriminant function is determined by using both the learning and the test set (and it is a pity not to use this set also for learning because the greater the learning set, the better the result). When all but one point of the learning set are used for determining the discrimination function and the one for testing, and this one point successively takes the place of all points of the learning set, an unbiased estimate for the error is obtained, though with a large variance. More methods exist (Toussaint, 1974).

It is not always necessary to classify all points into one of the classes. If a point lies far from one of the clusters or quite near a discriminating surface one may *reject* this point, so decreasing the possibility of making faults. A human decision may be made now or the sample may be left out.

Statistical methods may also be used to find *important features* among the ones originally used or to find interesting linear combinations of the features involved. A simple method determines for each feature the difference between the class means and compares this distance with the variances of the classes. A good feature has small variances in relation to this distance (refer to Figure 7.9). In

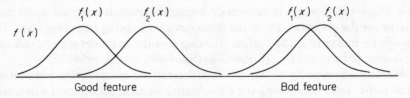

Figure 7.9   A good feature has small variance in relation to the distance between class means

this approach the relations between features are not taken into account. So, when such a relationship exists it is not always true that the m best individual features are the best m ones. More sophisticated methods have been developed for those cases (Kanal, 1974; Narendra and Fukunaga, 1977).

Finally, an interesting point has to be mentioned that is also important for measurement strategy. It concerns a relation between the number of features (measurements) and the size of the learning set. A small learning set does not allow one to estimate with a reasonable accuracy many properties of the classes; so only a few simple features are feasible. It can be proved that with a given size of the learning set an optimal number of features exist. Both a smaller and a larger number of features give worse results (Duin, 1978; Campenhout, 1978). This 'peaking effect' also indicates that it might not be useful to draw relevant conclusions from measuring more and more variables.

The *fuzzy set* approach can also be applied for pattern recognition purposes. Fuzziness can be introduced at many levels: the labels may be fuzzy, and/or the feature values, the classification results, the clusters, and the models may also be fuzzy. Many methods are being brought into existence (Backer, 1978; Gaines and Kohout, 1977). Indicating a membership value to the class labels for all sources of the learning set (a certain member is a rather typical representative of a certain pattern or has only a low membership grade) introduces more information in the learning set than when only indicating the class: it is to be expected that better classifications can result. The way the membership value is assigned, however, is often quite subjective, just as are (like in other methods) the distances defined in a feature space or the classification rules chosen. Fuzzy methods may be a worthwhile help; until now, however, they have not produced a final solution for pattern recognition.

7.2.3 Linguistic Methods

The features in linguistic pattern recognition are, according to Section 7.1.5, the primitives and their mutual relations. A pattern grammar describes the relations in an image, just as in linguistics a grammar describes the structure of a sentence. A few rules from a linguistic grammar are: a sentence consists of a noun phrase followed by a verb phrase; a verb phrase consists of a verb followed by a noun phrase; a noun phrase consists of an article followed by a noun phrase, or of a noun followed by an adjunct or of a noun, etc. Figure 7.10 gives an example of a *pattern grammar* (Shaw, 1969, 1970); Figure 7.10a shows the primitives, consisting of directed edges of a graph pointing from its tail node to its head node; the mutual relations between the primitives or conglomerations of primitives are defined in Figure 7.10b; Figure 7.10c represents some rules for generating a pattern with S as a start symbol and A and B as auxiliary symbols; with these rules the sentence of Figure 7.10d is derived as follows: $S \Rightarrow (S + S) \Rightarrow ((A * B) + (A * B)) \Rightarrow (((a + b) * (b + a)) + ((a + b) * (b + a)))$.

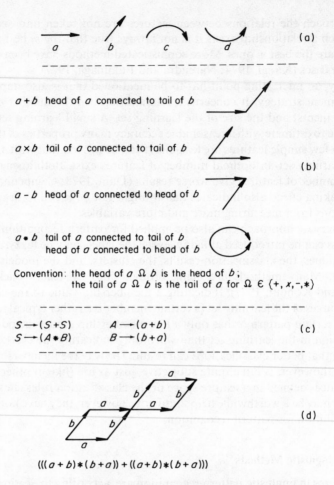

$a+b$ head of a connected to tail of b

$a \times b$ tail of a connected to tail of b

$a-b$ head of a connected to head of b

$a * b$ tail of a connected to tail of b
head of a connected to head of b

Convention: the head of $a \; \Omega \; b$ is the head of b;
the tail of $a \; \Omega \; b$ is the tail of a for $\Omega \in (+, x, -, *)$

$S \rightarrow (S+S)$ $A \rightarrow (a+b)$
$S \rightarrow (A*B)$ $B \rightarrow (b+a)$

$(((a+b)*(b+a)) + ((a+b)*(b+a)))$

Figure 7.10   Example of a pattern grammar: (a) primitives;
(b) definition of relations between primitives or conglomerations of primitives; (c) some rules for generating a pattern;
(d) example of a pattern generated by these rules

Many different types of grammars and primitives came into existence with,
for instance, primitives consisting of pixels instead of lines and with other types
of rules to handle pixels and lines than in the example. There is a great abundance
of theories about different types of grammars related to the types of rules used,
for instance, whether the rules are context-sensitive or context-free, whether
only special kinds of rules are allowed or whether no restrictions are present
(Salomaa, 1973).

A grammar may be derived from *a priori* knowledge of the images involved, for example, of well written characters, but sometimes an analysis of a number of representative samples of the patterns may provide a grammar (*grammatical inference*) (Bierman and Feldman, 1972; Gonzalez and Thomason, 1978). The easy incorporation of *a priori* knowledge in linguistic pattern recognition is a great advantage.

A priori knowledge and intuition are important for selecting primitives; simple and small primitives need many rules to compose an image, but allow the formation of many images; more complicated and extended primitives may facilitate the composition of certain images in an easy way but their structure is restricted. It can be difficult to extract primitives from scanned images. Hardware solutions to follow lines and segment them into primitives are reported (Wolff, 1977); software algorithms may be used to find primitives in digitized images.

Recognition first of all requires finding primitives and their relations. Then whether the pattern agrees with a certain grammar has to be analysed. The pattern has to be rejected if it does not result from the grammar involved. Two methods may be mentioned. In *bottom-up parsing* one starts with the description of the image by the primitives and the relations found, and by applying the rules of the grammar in a reversed direction one has to look for consistency with the rules of the grammar under discussion. A general procedure for this purpose is given by the theory of automata (Hopcroft and Ullman, 1969; Salomaa, 1973). Images accepted by automata agree with the different types of grammars indicated before.

Top-down parsing generates images in a directed way till the one to be recognized is composed. A special and simple, but often used procedure uses decision trees with a structure like that described in the previous section (Figure 7.8), although at each node there is not a statistical decision procedure, but a decision as to whether a certain rule is obeyed or not. At a certain node one has to decide, for instance, whether two primitives are connected head to tail or not; the next node may ask for a decision as to whether some other primitives have a specific relation or not, and so on.

For further literature on parsing see Aho and Ullman (1972, 1973), Fu (1974), and Gonzalez and Thomason (1978).

Until now, only well defined, deterministic images have been discussed. Stochastic grammars are appropriate for distorted images if well defined distortions are present. The removal of noise and distortions by image restoration (Section 7.5) is advisable before parsing. This removal will not be ideal, so in general, no exact match with the grammar will be found. This difficulty may be reduced by allowing some mismatch (Fu and Lu, 1977), but efficient procedures are hard to obtain. For corrections of errors in contex-free languages see also Thomason (1975).

7.3 SENSORS IN RELATION TO PATTERN RECOGNITION AND IMAGE PROCESSING

7.3.1 Introduction

A sensor in a pattern recognition system (refer to Section 7.1.4) connects the sources to the input of the recognition system, with its preprocessing and classification functions. The sensor produces measuring data, but often some preprocessing of these data already takes place in or near the sensor.

Quite often no special sensors are developed for a pattern recognition system; for instance: normal microphones are used for speech recognition or speaker identification, cardiographs are used for electrocardiogram or vectorcardiogram analysis, and normal X-ray, ultrasonic, microwave or nucleonic systems are used to produce images from technical or biological sources that are suited to image processing and recognition purposes. As most pattern recognition and processing systems use digital techniques, the addition or insertion of analog-to-digital converters, however, is necessary.

In some cases, a modification of existing sensor systems or the addition of some data processing in or near the sensor is very useful for pattern recognition purposes, especially when images are involved. Some examples are: TV cameras with added flexible digitization and storage facilities, flying-spot scanners with facilities for line following or scanning with different types of grids (rectangular or hexagonal) and one- or two-dimensional semiconductor photocell arrays with shift and memory facilities.

Finally, more specialized devices, such as laser-beam scanners or light-plane scanners, that scan objects in a predescribed way have been developed for pattern recognition purposes in order to get data in a suitable way, or such as distance measuring systems (using lasers or acoustics) in order to get data about the depth dimension of objects.

This section discusses some sensors for two-dimensional scanning of objects or images often used for pattern recognition purposes. Because no special attention is given to two-dimensional signals in the other chapters of this handbook, systems in this field not specialized for pattern recognition or without special modifications for this purpose are briefly treated as well. The same reason also justifies Sections 7.4, 7.5, and 7.6 of this chapter: because in the other chapters mostly one-dimensional, often time-dependent signals are treated, it is appropriate to pay some attention here to two-dimensional, space-dependent signals and their processing, where quite often similar, but adapted methods are used as with one-dimensional signals.

The systems to be discussed in this section all use some kind of radiation, for example, optical, ultrasonic or nucleonic radiation. Only contact by radiation takes place. This implies that, in practice, often a negligible or only a slight influence or distortion of the source takes place, and it also allows having a distance—sometimes a long distance—between the source and the sensor. This last property allows the collection of data from aircraft or satellites (remote sensing), from hot and—for man—dangerous places.

The radiation involved may originate in many different ways. The objects in the pattern source may produce the radiation itself (hot surfaces, infrared systems); or they may be illuminated and the reflected radiation used. One may choose between parallel measurement (like the eye) and sequential scanning. Scanning may be performed at the illumination side, for instance a point or light-plane scanning of the object (Section 7.3.6), at the detection side (like TV) with an illumination that is as homogeneous as possible, or at both.

Not all two-dimensional scanning systems could be treated; no attention has been given, for instance, to remote-sensing systems, side-looking radar and infrared scanners.

Important specifications for imaging systems are: spatial resolution (in both vertical and horizontal directions and depth) and number of pixels (*pic*ture *el*ements) in these directions; speed and method of scanning and the property of adapting speed, for example, slow scanning to reduce noise; sensitivity as a function of space (shading) and range; accuracy; dark current; linearity (both as to space and intensity); noise (both in time and space); flexibility and price.

7.3.2   Mechanical Scanners

A classical mechanical scanner is the *moving-stage microscope* (for example, the Zeiss SMP cytophotometer) (Ploeg *et al.*, 1974). The microscope stage with the preparation to be scanned, is driven by stepping-motors under computer control. Stages with steps of 0.25 μm are available. The transmitted light is measured by a photomultiplier and digitized into eight or more bits. The size of the measuring spot is determined by a diaphragm. A speed up to 1,000 points per second is possible.

A *drum scanner* uses a rotating drum; an image, for instance a photograph, tightened on the drum, is scanned sequentially by a photodiode, which moves, together with the illumination, in the direction of the axis of the drum. The aperture may be adjustable, say between 25 and 200 μm. The photocurrent is digitized into, for example, eight bits. High discrimination, stability, and reproducibility can be obtained, together with a good linearity, and without shading. The number of points to be scanned in 1 s is of the order of 10,000–20,000. The accuracy, and the price, depend on the quality of the mechanical construction and of the optical system.

7.3.3   TV-scanners

Television is the most experienced and widespread technique for conversion of images into analog electrical signals. TV-cameras are produced in large quantities and in many qualities, always at a relatively low price. This holds true both for the classical electron-beam scanners and for the photo-arrays that have more recently entered the television field. The former will be discussed shortly in this section, the latter in Section 7.3.5.

In a TV-scanner the source is projected onto a plain target where sequential scanning takes place. The scanning goes according to a sequence of equidistant lines distributed over two frames in such a way that the lines of the second frame lie between those of the first frame. The process of the two-frame interlaced scanning is repeated 25 or 30 times a second. In between the lines and the frames synchronization signals are inserted. The result of the scanning is a one-dimensional time signal with information concerning the light situation at the target, interrupted by synchronization signals. The time between two scans at a small region of the target—this region to be considered as a small capacitor—is $\frac{1}{25}$ or $\frac{1}{30}$ s; during this time the capacitor previously charged by the electron beam, is discharged with a current originated by the photoelectric effect and related to the light situation at that region. The resulting voltage of the capacitor depends on the integral of that current during the time mentioned (Polder, 1967). Fluctuations during this time and noise are reduced by the integration process.

In order to process the images produced by a TV-scanner, mostly the arising analog signal with information has to be digitized; often some analog high-frequency filtering is applied before to reduce noise; analog preprocessing is sometimes applied, such as gamma correction, crispening or pre-emphasis (Fink, 1957). The analog-to-digital conversion to digitize on-line must be quite fast if during one line-time many conversions have to take place. With 25 scans of two frames per second and 625 lines in the two frames the line-time is about 64 μs; 12 μs of it is suppressed for synchronization purposes. Sometimes only part of the line is used for the image in order to obtain a square image. A sampling frequency for 512 points per line of at least 10^7 Hz, and for 256 points per line of at least 5×10^6 Hz is necessary. Converters with 1 up to 8 bits are used, though the least significant bits are not always significant. If 512 lines are used and 512 points per line with 8 bits per point, about 1.3×10^6 bits would arrive per frame pair and about 32×10^6 bits per second. This speed is too high to communicate with most general purpose computers. Moving images require special purpose preprocessing or less points and bits per point.

Standing images require only one pair of frames, and with 256 lines only one frame. Many frames can also be used with digitization of only one or some columns, this allowing the use of a much slower converter and communication channel to the computer.

Instantaneous shots of moving images require a fast converter together with a fast memory able to accept the data on-line during scanning of one or two frames. This memory may also be used as working memory of a special fast image processor (Gerritsen et al., 1977) as a buffer to allow a computer to gather its contents according to its own speed, and at the same time it may be used for permanent display of the captured digitized image.

A TV-scanner produces several types of distortions and noise. The optical and electronic parts have a point-spread function. Shading is a space-dependent type of point degradation caused by the position-dependent sensitivity of the target. Time-dependent noise is produced by fluctuations of the intensity, target noise, and electronic noise (especially the first amplifier). A longer frame time, allowing slower scanning, can reduce the first two types of noise. Special low-noise amplifiers for the first amplification step reduce electronic noise. Averaging repeated shots reduces all types of time-dependent noise. Shading may also be caused by inhomogeneous illumination and by a non-ideal optical system. Shading correction is possible in hardware (both analog and digital) or in software; the signal obtained from a blank background with the same illumination is used to compensate for shading.

The scanning is deterministic as to the succession of scanned points, so no addresses of individual points are necessary. Two-dimensional filtering (Sections 7.4 and 7.5) requires points from several lines. In such cases it is advisable to have (random) access to these points after scanning via some convenient means, for instance, by temporarily storing some lines of points in a memory.

Many commercial digitized TV-scanners with divergent facilities and quality are offered for sale (Hougardy, 1976).

7.3.4   Flying-spot Scanners

A flying-spot scanner (Golab *et al.*, 1971; Eccles *et al.*, 1976a, b) uses sequential scanning of the object at the illumination side. The spot of a high-quality cathode ray tube, having a flat window, is focused by a lens onto a transparency or photographic negative to be analysed (Figure 7.11) or onto an opaque surface having varying reflectivity. A photomultiplier measures the light transmitted or reflected. As any position of the object may be selected without a fixed order of scanning one has random access. In the case of digitally coded positions, a fixed scan grid is mostly used. This may be square, rectangular or hexagonal, and its size and pitch can be chosen at will. The size and shape of the sample area can be influenced by electronic or optical means, for instance, by focusing or by introducing an astigmatism effect. Like TV systems the flying-spot scanner has a position-dependent sensitivity (shading) which originates from CRT screen inhomogeneity and geometrical and optical characteristics. Part of these error sources can be compensated by continuous monitoring of the spot brightness by a compensation photomultiplier and by using, for instance, this information as a reference of a *dual-slope A/D converter* which converts the photomultiplier current to time-interval length (Figure 7.12). If this length is measured by counting the number of periods of a high-frequency generator a digital value of the transmitted or reflected light is obtained. The current of the compensation

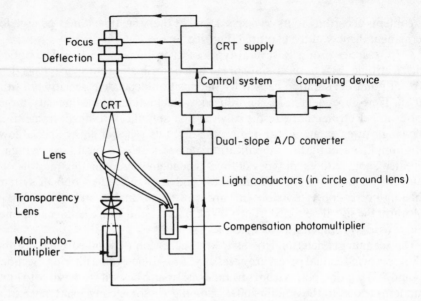

Figure 7.11   Flying-spot scanner

photomultiplier may be integrated for a certain time in the dual-slope device and brought back to zero by the current of the main photomultiplier. This has the additional advantage that approximately equal amounts of photons are used at different intensities of transmitted or reflected light such that the inherent Poisson noise is independent of intensity. Consequently, the scan speed now varies from point to point according to the density or reflectivity value to be measured and the accuracy desired. Typical scan times range from 10 μs to 1 ms per point for 1 % accuracy. Noise considerations are discussed in Billingsley (1975).

Spatial resolution depends on CRT spot size, optical magnification, and lens quality. Typical spot size is 40 μm on the screen and 1000×1500 sample points on a 24×36 mm^2 negative.

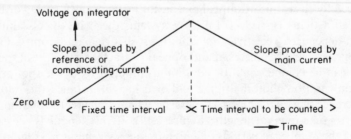

Figure 7.12   Dual-slope A/D converter

In the random-access mode the scan can be made to follow lines or edges as they appear in the object. This may bring about useful data reduction directly at the sensor stage (Wolff, 1977).

The flow of information may also be reversed when the object is replaced by photosensitive material such that the flying-spot scanner constitutes a *hard copy device*. Here the time of illumination depends upon the density desired.

7.3.5 Photo-arrays

Semiconductor technology has provided solid-state image sensors (Melen, 1973; Purll, 1976). Combining photodiodes in an array was a first step. Integration of the scanning function has followed since. The charge-coupled device (CCD) image sensor is a typical example (Howes and Morgan, 1978). Here the light-generated charges are transferred through carefully organized electrostatic peristaltic movements in semiconductor channels to end up in an high-impedance amplifier at the output. On the way the charge may become lost or mixed up with charges from other photodiodes. Special buried-channel techniques are applied for better insulation but these still fail when leakage is caused by intensity overflow for which *blooming* results. In order to prevent *smearing* the channels are often also shielded from light and thus prohibited from acting as photodiodes.

The simplest CCD consists of a linear channel lined with a row of photodiodes. These collect photons over a time period building up their charges until they are simultaneously dumped into the channel through which they are lumpwise transported to the output. Even such a simple configuration as this line scanner is useful in many industrial applications, especially when the product to be inspected passes the scanner on its way through the manufacturing process.

Typical diode size is 20 μm × 20 μm and resolution is of the same order. Array lengths vary from 256 to more than 1000 elements. Shading effects may have a range of less than 10 %. Calibration of a few single elements can give considerable improvement. Noise is typically a few hundreths of a per cent of maximum intensity allowed. Acquisition rates well in excess of TV-speeds are obtainable under favourable conditions.

Two-dimensional CCD scanners for television duties have been constructed. Array sizes range from 75 × 120 to 256 × 320 elements. Here the transfer-channels compete for area with the photosensitive elements. At least two read-out methods are in use. The first (interline transfer, ILT) employs a number of linear channels as described above. In order to create read-out time, interlacing is achieved by alternately dumping charge from one out of two sets of photo-diodes in a channel in between the two sets. The second method (frame transfer, FT) has linear channels that act as photodiodes at one half of the surface of the scanner, but are covered at the other half. Charge dumping now takes place by a quick shift of the charge train from the first half into the dark, to be further shifted to the output when read-out demands. Here it is separation of charge lumps that forbids simultaneous charge collection in the whole area.

Semiconductor and insulating materials are not primarily chosen for their optical properties. The silicon substrate would be preferable in that respect and attempts have been made to thin the chip for back phase imaging. Good spectral sensitivity between 300 nm and 700 nm can be obtained.

7.3.6 Special Optical Scanners for Three-dimensional Structures

The scanners treated above produce 2-D projections from 3-D structures. 3-D information can be found from one 2-D projection using distance cues (shadows, perspective, and other *a priori* knowledge), from more than one 2-D projection taken in different directions, and using stereopsis methods (stereology, a field intensively treated by the International Society for Stereology (Carpenter).* 3-D scene analysis attracts much attention (Shirai, 1978).

A number of scanners with a special illumination strategy and/or range-finding facilities have been developed especially for *robot systems* in order to produce data about the third dimension of more or less simple structures in an easy way. An example of such a scanner uses a light beam with known origin and changeable but known direction that scans 3-D structures, together with a TV system that determines where the beam is reflected at the structure. As point-by-point scanning takes a considerable time, a *plane of light* (also called 'sheet' or 'slit' of light) is generally used, that is, for instance, shifted in parallel by equal distances or is rotated around an axis with equally spaced angles thereby scanning the 3-D structure. A TV system is used again to determine the position of the reflected line positions (Shirai and Suwa, 1971; Röcker, 1974). A faster method uses the projection of *many simultaneous parallel or rotated light planes* or of a *square grid* on the structures (Will and Pennington, 1971). The evaluation of the television signals is then more complicated than with one light plane. Figure 7.13 explains the situation where a light plane is shifted in parallel across a simple object, a tetrahedron on a horizontal ground plane. The light plane is perpendicular to the ground plane. Intersections of the light planes and a plain tetrahedron surface are parallel, equally spaced, straight lines with directions and distances depending upon the angle between the light plane and the surface. A television camera looking at the tetrahedron receives data from the intersections about the position of the surfaces. Begin, end, and breakpoints of the straight lines are determined in a relatively easy way from the video signal. From this kind of information, together with data about the position of the light plane, the position of the surface can be computed. Curved

---

* A. M. Carpenter, Secretary/Treasurer, International Society for Stereology. IU–NWC Med. Ed., 3400 Broadway, Glen Park, IN 46408, USA.

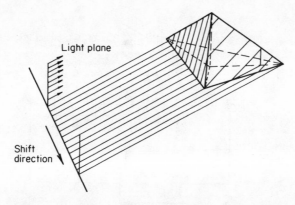

Figure 7.13 Tetrahedron scanned by parallel light
planes. Intersections of the light planes and two surfaces
of the tetrahedron are indicated

surfaces give curved, and not equally spaced intersections, allowing, by means
of more complicated computations, determination of information about these
surfaces.

The light planes may be generated in several ways. For instance, by light-plane
projectors rotated by motors, by standing light-plane projectors using rotating
mirrors, by laser beams shaped to light planes together with laser-beam de-
flecting devices. Instead of a television scanner a photocell array can also used
(Kieszling, 1976).

A different approach uses a type of radar *range-finding* technique (radar is
used for long distances) using a laser beam, and measures light flight distances
for the different light paths by means of time measurements during the scanning
of the structure surfaces. Pulsed laser methods require measurement of the time
between transmitting and receiving a pulse. Continuous lasers may also be used
in which case a light beam from a high-frequency, amplitude-modulated, laser
scans the 3-D structure, for example, by means of a scanning mirror (Figure
7.14). The reflected light in a direction approximately opposite to the incoming
light reflects from another mirror surface and is sensed with a photomultiplier,
thus producing a high-frequency signal. The amplitude depends upon the diffuse
surface reflectance of the points of the structure. Phase is related to the total
light path, so is dependent upon the distance of the mirror to the points of the
structure and provides range data (Duda and Nitzan, 1976).

A common point for all these systems is that no shadow effects are measured.
Preprocessing of the data to reduce noise influences is necessary but the
identification of edges and surfaces is simpler than with a homogeneous
illumination. Data about accuracy and other metrological parameters are
seldom given in the literature: this is not important if recognition of structures
composed out of (simple) surfaces is the main purpose.

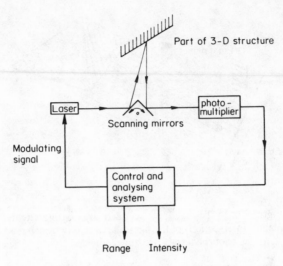

Figure 7.14   Block diagram of a range-finding system
using an amplitude-modulated laser

7.3.7   Devices for Determining Internal Structures

The optical devices discussed in the preceding sections mostly produce data
concerning the surfaces or objects, and consequently about the outside situation
of objects (optically transparent 3-D objects are an exception). To obtain data
concerning the inside situation of closed opaque objects a radiation source like
a nucleonic, X-ray or ultrasonic source may be used. By absorption or reflection of this radiation in the specimen an image may arise representing useful
information about the inside structure. This image has to be sensed in an
appropriate manner. Sometimes the sensors immediately produce data that
can be digitized and used as input data to an image processor and recognition
system (for instance, ultrasonic data); in other cases an intermediate step has
to be used (for instance, a photonegative produced by X-rays, that has to be
handled by one of the sensors described in the preceding sections). This section
briefly surveys some interesting systems used to produce images of the internal
structure of objects or specimens.

7.3.7.1   Radiography

A radiographic system consists of a nucleonic or X-ray source, an object with
specific absorption for the structures to be found and a recording medium.
When no specific absorption is present originally taking in of absorbing or
nucleonic material may be a solution as is often practised in medical applications. In the case of nucleonic take-in material the position of this material

inside the body and not the absorption properties of the structures is relevant and has to be determined. A general difficulty in this field is that the images obtained are often of poor quality, noisy, distorted and ill-resolved. Visual inspection is sometimes possible by trained observers, but here too the situation is not ideal. The automatic analysis of the images to obtain numerical data or to recognize the shapes of structures is mostly impossible without special measures. Some of the causes for the bad quality of the images are the finite source size, the finite duration of flash time of the X-ray source, no lenses are available for most cases, the point-spread function of the whole system is far from negligible and is space variant, movements inside the specimen or of the specimen as a whole may be present (for example, the dynamic action of the heart and veins, respiration). Additionally one has to deal with film response characteristics, noise, and background effects. Here the possibilities of image restoration and enhancement (Sections 7.5 and 7.6) may help greatly to produce better images, or indeed to make measurements possible. Interesting applications can be found in Hunt *et al.* (1973).

Total absorption depends on the absorption along the whole path of the rays. Projection in one direction only does not allow the reconstruction of the absorptivity as a function of position. Reconstruction is possible if many projections from different directions are available, though it needs considerable computer power. Many computing algorithms have been produced for this purpose—called tomography (Ter-Pogossian *et al.*, 1977). Depth information may also be obtained by a stepwise moving pseudo-random coded aperture between the object and measuring positions (Koral *et al.*, 1975).

7.3.7.2   *Ultrasonics*

Ultrasonics, as well as nuclear magnetic resonance (NMR) (see the following section), can immediately provide information about single small-volume elements, in principle this being without need for reconstruction from projections. In the medical field both are welcome low-hazard alternatives to nucleonic or X-ray sources. Ultrasound has also found numerous applications in material research (for instance, welding crack detection).

The use of sound for imaging dates from submarine detection (SONAR) in World War II. A sound pulse from a transmitter is partly reflected at any change of acoustical impedance of the media passed. At the transmitter, now used as a receiver, different reflections arrive at different times. Their echo time is a measure of the distance along the direction of transmission. The sound pulses are repeated periodically. This sonic analog of RADAR can also make use of the latter's display techniques by plotting echo intensity in brightness as a function of distance. The result is a one-dimensional section in the direction of transmission indicating changes in acoustical impedance.

A two-dimensional image can be generated via a slow, scanwise, mechanical variation of direction, this being applicable for objects that change or move at a still slower pace. Fast electronic beam steering through a phased-array method was developed in the wake of RADAR (Somer, 1968; Thurstone and Ramm, 1974). A multiple set of single transmitter–receivers, each working on its own for its own time interval, has been devised as an alternative fast 2-D technique (Bom *et al.*, 1973). The movement of heart structures (Figure 7.15), foetal breathing and the like can be studied in this way.

Discrimination is of the order of 1 mm at a frequency of 2.5 MHz given favourable conditions. The images obtained are often of too low a quality to provide automatic recognition of, for example, heart structures and to provide the determination of quantitative data about heart volume changes or the velocity of heart valves in an easy and reliable way. This is due to shielding of ultrasound by large acoustic impedance objects (such as the bone of the ribs), small differences in acoustical impedance between blood and tissue, the lack of reflection from surfaces not perpendicular to the direction of transmission, and noise. Image restoration and enhancement techniques, the use of time information from successive images, and interactive help from professional cardiologists, however, create more and more successful applications (Computers in Cardiology, commencing 1975, available from IEEE Computer Society).

A broad review on acoustic imaging is available in *Proc. IEEE*, April 1979.

7.3.7.3 *Zeugmatography*

Zeugmatography, spin imaging or spin mapping is a very new application of NMR (Lauterbur, 1973; Hinshaw, 1974; Andrew *et al.*, 1978). It detects the presence of hydrogen nuclei (protons) such as exist in water, fats, and carbohydrates, but may be tuned to any other type of nucleus that carries magnetic spin. When a sample is placed in a static magnetic field each of its protons may absorb or emit radiofrequency electromagnetic radiation of a well defined frequency determined by the local value of the field induction. The gross effect in normal circumstances is absorption. By special techniques (adding slant-shaped constant or varying fields) the induction can be given a unique value at a given point (and in a restricted volume around it). The absorption at the corresponding unique frequency then measures the proton density in the volume selected, so producing imaging.

Due to the different environments of protons in fats and water, respectively, their absorption characteristics, in particular the frequency selectivity, are different. Discrimination between malignant and normal tissue seems to be possible in some instances as their environments in relation to fat are different. Whole-body scanners require full-body embracing electromagnets.

Many different techniques have been developed, or are under development, in order to provide faster results (pulsed instead of continuous waves), higher

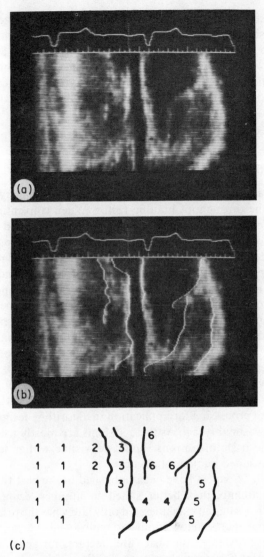

(c)

Figure 7.15   Echocardiogram produced by a multi-scan system. The top trace is an electrocardio-gram: (a) original image; (b) some heart structures, indicated by lines, found after processing; (c) interpretation of parts (a), (b). From left to right are shown in succession: 1, the anterior chest and heart walls; 2, the right ventricular cavity; 3, the inter-ventricular septum; 4, the left ventricular cavity; 5, the left ventricular posterior wall; and 6, the mitral valve leaflets

resolution (currently of the order of 0.5–10 mm depending on size and measuring circumstances), simpler hardware and software requirements and less distortion. Some distortion can be removed by restoration techniques. Static fields and radiofrequency waves of NMR are shielded by highly permeable materials and good conductors, respectively. In medical applications only conduction need be considered. At 10 MHz the attenuation is acceptable in a body-sized volume.

7.4 SPECIAL PROCESSORS

7.4.1 Introduction

In most image acquisition, storage, processing, and display systems, the processing part is the throughput limiting factor. When conventional computers are used, the large number of picture elements (pixels) involved causes the total number of memory references, computations and thus, total computation time, to be extremely large. Even in the case where a small number of relatively simple operations per point will suffice, processing by a conventional computer will be much slower than the acquisition rate. Assuming a TV-scanner is used as the input device, even a high-speed conventional computer will perform only one single instruction in the time the scanner needs to deliver several pixels.

Although throughput is not the sole factor for judging an image-processing system, in most cases achieving a reasonable throughput is one of the major problems. Sometimes the need is to process moving images, for example at 25 per second, which imposes very high demands on the throughput. In other cases very large numbers of images have to be processed with not too much delay, even image processing for application in algorithm design or in research often has to be reasonably fast. As these systems are mostly used interactively, the results of the basic operations should be visible within seconds, thereby allowing unconstrained human interaction.

A large number of papers have been published in the past twenty years that give designs of architectures better suited to number 'crunching' than the conventional Von Neumann machines, but only lately has the price–performance ratio of the components been low enough to realize such machines. Flynn (1972) gives an attractive review of the faster architectures for general-purpose computers. This chapter partly develops the same line of thought, but dwells more on the specific problems of two-dimensional information processing.

7.4.2 Alternatives for Obtaining Higher Processing Speeds

Software systems

Some systems use no special processing hardware. All processing is done by a general-purpose computer using software packages written in assembly and/or higher-level languages (Haralick and Currier, 1977; Johnson, 1970). Other systems use microcoded software packages (firmware), achieving a speed-up of 5–10 when compared to an all-assembler software package (Ito *et al.*, 1978). The primary advantage of these systems, their generality, is at the same time the most important reason for their lack of speed: most applications of general-purpose computers do not impose high demands on throughput, and special characteristics of two-dimensional processing have not been used in the design.

Faster technologies

The most obvious way of improving performance is, of course, to speed up the existing parts of hardware by using the fastest available electronic components. The physical constraint on this type of speed improvement (from mechanical switches to emitter-coupled logic gives a factor of 10^8) is, however, in sight.

Hardwiring

Other ways to obtain significant speed increases are hardwiring fundamental operations, for instance, adding a hardware multiplier; incorporating special floating-point logic; using faster and larger semiconductor memories to store image data, thereby minimizing swapping of data to and from disk storage during calculations. In many applications these are indeed found to be rather inexpensive ways of improving performance without affecting the (conventional) system architecture. However, by bringing more parallelism into the system architecture, by introducing a slave processor that meets the special demands of the most frequently used image processing operations, or by linking two or more computers (or microprocessors) for simultaneous operation on different parts of the same problem, it is possible to reduce the required computation times drastically.

Parallelism

Parallelism can be introduced at a number of different levels. Depending on the architecture, different jobs (tasks), independent parts of jobs, independent programs, independent subroutines, and even independent instructions or parts of instructions can be executed concurrently. An important problem is, however, the detection of such independent parts of programs or of independent instructions. Programs and programming languages have been tuned for execution on sequential machines. This is why some experts (see for instance Amdahl, 1967), believe that parallel machines have only a limited field of application. If that is still true, the field of pattern recognition and image processing is in any case very well suited to the application of parallel architectures. Most of the popular image-processing operations make extensive use of array operations. In the calculation of different pixels the operations may differ, but the operators will be the same all through the image. In other words most image-processing operations are parallel operations. The different image operations of an algorithm are, of course, still performed in series.

7.4.3 Parallel Architectures

The parallel computer architectures that have come of age are processor arrays, pipelined processors, associative processors, and multiprocessors.

Processor arrays

Processor arrays consist of a number of connected processing elements (PE's) with more or less limited possibilities of communication. An important characteristic of processor arrays is that control is mainly centralized in one single array control unit. The most popular interconnection patterns are given in Figure 7.16. The more complicated schemes (Figures 7.16b, c, and d), showing 4-connected, hexagonally connected and 8-connected interconnection patterns are the most suitable for image processing.

Ultimately one would like to use one PE per pixel, but because of size and economic restraints to date most realizations have been limited to a smaller number of PE's, whilst larger image arrays are processed by letting the smaller subarray scan the larger (for example 512 × 512) image. For the same reasons mostly simple, often only one bit wide, processors are used. Arithmetic computations are performed bit-serial, but image-parallel. Examples are 3 × 3 PPM (Kruse, 1973), 8 × 8 Illiac IV (Barnes *et al.*, 1968), 12 × 16 Clip 3 (Duff *et al.*, 1974), and the, now recently (1979) realized, 96 × 96 Clip 4 (Duff, 1976).

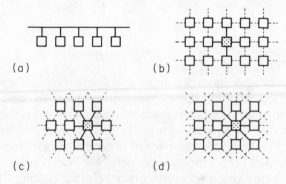

(a)

(b)

(c)

(d)

Figure 7.16 Interconnection patterns of processing elements: (a) no direct interconnection; (b) 4-connected interconnection pattern; (c) hexagonally connected interconnection pattern; (d) 8-connected interconnection pattern

Sometimes the term *array processor* is used for all machines that are merely better suited to perform array operations than conventional computers.

Pipelined processors

Pipelining techniques can be applied to use the different parts of hardware more optimally. In its simplest form, the fetch of the next instruction can be made to overlap the execution of the current instruction. More generally, when pipeline techniques are used (the name *assembly line techniques* would also have been appropriate), different parts of the computation are performed currently, but different datawords are handled serially. In evaluating, say, an inner product of two vectors $X(i)$ and $Y(i)$, as is necessary in the calculation of distances, convolutions, and correlations, a pipelined processor (in this example consisting of both a multiplier and an adder) concurrently multiplies two operands $x(k)$ and $y(k)$, adds the previous product $x(k-1) \cdot y(k-1)$ to the previous partial sum $\sum_{i=1}^{k-2} x(i) \cdot y(i)$ and at the same time fetches new operands $x(k+1)$, $y(k+1)$ from memory. Not suffering from size constraints, in pipelined processors one can afford to use sophisticated, very high-speed, logic. Floating-Point Systems Inc. AP-120B is an example of a pipelined processor (the manufacturer uses the name array processor that is meant to be used as a minicomputer's slave processor). Further examples can be found in Ledley *et al.* (1978), Haralick and Minden (1978), Lemkin *et al.* (1974), Asada *et al.* (1978), and Gerritsen *et al.* (1977).

Associative processors

Associative processors are a variation on processor arrays. The PE's are not directly addressed, but are activated if a program-specified match exists between a specified number and characteristic data inside the PE. Only the activated PE's perform the current instruction, all the others remain idle (see also Ramamoorthy *et al.*, 1978; Pao and Schulz, 1978).

Instruction streams and data streams

The three parallel architectures discussed have in common that the parallel units are controlled by one single control unit, but perform their operation on more than one data stream. This organization is generally referred to as single-instruction stream, multiple-data stream (SIMD). From this macroscopic point of view most conventional computers have the single-instruction stream, single-data stream organization (SISD), while the multiprocessor architecture to be discussed in the next paragraph will be classified as multiple-instruction stream, multiple-data stream (MIMD).

Multiprocessors

Multiple-instruction stream organizations can be split up into *real* multi-processors, that is configurations in which a number of complete and independent single-instruction processors on a certain level share memories to perform cooperatively a number of rather independent tasks and secondly the multiprocessor configurations in which so called *skeleton* processors share a larger number of system resources. In both cases each of the separate processors has its own program, but on a higher level the configuration is controlled and the processors are synchronized by an integral operating system. The shared memories function as mailboxes for intertask and interprocessor communication (Flynn, 1972; Boxer and Batchelor, 1978; Tojo and Uchida, 1978; Okada *et al.*, 1976).

Drawbacks

All architectures discussed suffer from unwanted effects that hamper the speed-up expected. In the cases of processor arrays and associative processors the length of the vector to be processed (logic size) has to be fitted to the size of the array of PE's (physical size). If the image to be processed is larger than the array of processors available, the scanning of the logic array by the physical array introduces considerable overhead. When the size of the vector to be processed is smaller than the size of the processor array (the worst case being the processing of scalar quantities) a number of processors will not be used. Similar vector fitting problems also arise when using pipelined processors. Branches and decisions based on calculations just performed also influence the performance of SIMD architectures. For instance, in the case of pipelined machines the current vector operation has to be completed (the pipe must be emptied) before the test can be performed and the next vector operations started.

In MIMD systems the most important problem is the overhead caused by intertask communication. If the separate processors each process a separate part of the data of a larger, common problem, then inevitably delays will arise when one processor is waiting for another's results, especially for variable execution times per task.

7.5   RESTORATION

7.5.1   Introduction

The image used in a pattern recognition system may be degraded. The process of correcting for this degradation is called *restoration*. If the degradation involves only the grey value of a point (pixel) it is called *point degradation*. If it also involves the neighbourhood of a point, it is called *spatial degradation*. Some books on image processing, including restoration, are Huang (1975), Rosenfeld and Kak (1976), Andrews and Hunt (1977), and Pratt (1978).

Distortions can be introduced by the sensor of the system (for example, by the optical transfer function), by movement of the source, or by atmospheric turbulence. Degradation is also produced by all types of noise. The simplest case occurs when this noise is additive. When the noise depends on the grey values in a known manner a (non-linear) rescaling of the grey values may compensate for this. Taking the logarithm of the grey value in the case of multiplicative noise results in additive noise.

Reduction of noise is sometimes as important as the correction for distortions.

7.5.2 Point-spread Function

When the degrading system is linear, the blurred image $g(x, y)$, can be given by a superposition integral

$$g(x, y) = \iint f(\xi, \eta) h(x, y, \xi, \eta) \, d\xi \, d\eta \tag{7.1}$$

where $f(\xi, \eta)$ is the original image and $h(x, y, \xi, \eta)$ is the so called *point-spread function* (PSF).

When the original image is a point source, $f(\xi, \eta)$ is a delta function and the resulting image $g(x, y)$ equals the point-spread function. When, at a shift of the source, the resulting image $g(x, y)$ is translated but otherwise unaltered, then

$$h(x, y, \xi, \eta) = h(x - \xi, y - \eta) \tag{7.2}$$

and the point-spread function is said to be *shift invariant* (SIPSF); otherwise the point-spread function is called *shift variant* (SVPSF).

The superposition integral results in a convolution integral for a shift invariant point-spread function

$$g(x, y) = \iint f(\xi, \eta) h(x - \xi, y - \eta) \, d\xi \, d\eta \tag{7.3}$$

Fourier transformation of both sides of equation (7.3) gives

$$G(u, v) = H(u, v) F(u, v) \tag{7.4}$$

Capitalized characters denote Fourier transformed functions.

Equations (7.3) and (7.4) provide the possibility of reconstructing the original image from the distorted one if the point-spread function or the transfer function are known (see next section). When the process of degradation is given, these functions can be calculated from theory, such as the point-spread function of a coherent or incoherent diffraction-limited ideal optical system (Goodman, 1968), the point-spread function of a moving source (Aboutalib et al., 1977).

When the process of degradation is not known, the point-spread function has to be estimated from experiments. Special sources can be used for this purpose such as point, line or step sources (Rosenfeld and Kak, 1976; Huang et al., 1971).

7.5.3 Inverse Filtering

When the degrading system is linear, the reconstruction can be obtained by inverse filtering. When the degraded image is the result of a spatial invariant point-spread function, $F(u, v)$ can be reconstructed according to equation (7.4) by multiplying $G(u, v)$ with $1/H(u, v)$. In other words, the inverse filter is then simply $1/H(u, v)$. Problems arise because $H(u, v)$ will be zero at some points (u, v) (limited passband of physical systems) and so, in the absence of noise, $G(u, v)$ is also zero and the ratio $G(u, v)/H(u, v)$ is not defined. When $H(u, v)$ would never be zero, the inverse filter reconstructs the original image exactly in the absence of noise.

When additive noise is present equation (7.4) changes to be

$$G(u, v) = H(u, v)F(u, v) + N(u, v) \qquad (7.5)$$

in which $N(u, v)$ is the noise term. So dividing $G(u, v)$ by $H(u, v)$ gives

$$\frac{G(u, v)}{H(u, v)} = F(u, v) + \frac{N(u, v)}{H(u, v)} \qquad (7.6)$$

For small values of $H(u, v)$ at points (u, v), $N(u, v)/H(u, v)$ can easily dominate $F(u, v)$. So a reconstruction filter $M(u, v)$ may only equal $1/H(u, v)$ for those values of (u, v) where the signal-to-noise ratio is sufficiently high, and for the other values a choice has to be made (for instance $M(u, v) = 1$) (Veen et al., 1978).

Another approach (which includes SVPSF) makes use of a one-dimensional representation $\mathbf{f}$ of the digital image. $\mathbf{f}$ is a vector of length n^2 when n is the size of the square image. The blurred image $\mathbf{g}$ is then given by

$$\mathbf{g} = [h]\mathbf{f} + \mathbf{n} \qquad (7.7)$$

in which $\mathbf{n}$ is the noise term. The reconstructed image is obtained by calculation of the pseudo-inverse of the matrix h, or by iterative filtering (Huang, 1975).

7.5.4 Linear Least-squares Filtering (Wiener Filtering)

An optimal reconstruction process can be obtained by minimizing the mean squared error between the original image and the reconstructed image. When it is assumed that the reconstruction filter is linear shift invariant the reconstruction $\hat{f}(x, y)$ is given by

$$\hat{f}(x, y) = \iint m(x - \xi, y - \eta)g(\xi, \eta) \, d\xi \, d\eta \qquad (7.8)$$

in which $m(x - \xi, y - \eta)$ is the reconstruction filter and $g(\xi, \eta)$ is the degraded image given by

$$g(\xi, \eta) = \iint h(\xi - r, \eta - s)f(r, s) \, dr \, ds + n(\xi, \eta) \qquad (7.9)$$

where $n(\xi, \eta)$ is the noise term.

Now $E = (f - \hat{f})^2$, is given by

$$E = \left(f(x, y) - \iint m(x - \xi, y - \eta)g(\xi, \eta) \, d\xi \, d\eta \right)^2 \qquad (7.10)$$

and has to be minimized for the whole image. This minimization results in a reconstruction filter $M(u, v)$ given by Rosenfeld and Kak (1976)

$$M(u, v) = S_{fg}(u, v)/S_{gg}(u, v) \qquad (7.11)$$

where $S_{gg}(u, v)$ is the spectral density of the degraded image and $S_{fg}(u, v)$ is the cross-spectral density of the degraded and the original images. When the noise and the images are uncorrelated and n has zero mean, equation (7.11) becomes

$$M(u, v) = \frac{1}{H(u, v)} \frac{|H(u, v)|^2}{|H(u, v)|^2 + [S_{nn}(u, v)/S_{ff}(u, v)]} \qquad (7.12)$$

where S_{nn} is the spectral density of the noise and S_{ff} is the spectral density of the image. The same result is obtained when f has a zero mean instead of n. As the grey value for physical systems is positive a subtraction of the mean grey value is then necessary. When no noise is present the result is again the inverse filter $M(u, v) = 1/H(u, v)$.

As appears from equations (7.11) and (7.12) it is necessary, for Wiener filtering, that the spectral densities of both the noise and the image are known or at least that good estimates are available. A further discussion for the discrete case is given by Pratt (1972), Pratt and Davarian (1977), and Ahmed and Rao (1975). The case where $h(x, y)$ is stochastic is described by Slepian (1967).

7.5.5 Discussion

Methods like inverse filtering and Wiener filtering can be performed by discrete Fourier transformations (DFT) of the degraded image, multiplication in the

Fourier domain and again a DFT to the spatial domain. Direct deconvolution in the spatial domain is also possible (with regard to computation time) when only a limited number of neighbours of a point is taken into account in the deconvolution (windowed filters). Special hardware to achieve convolution is mentioned in Section 7.4.

The digital Fourier transform will generally be computed by the fast Fourier transform (FFT) (Cochran *et al.*, 1967).

Besides the methods previously mentioned an abundant literature of reconstruction filters exists (Huang, 1975; Andrews and Hunt, 1977; Rosenfeld and Kak, 1976). An example is a recursive method in which the dependence of the image points is modelled by a Markov mesh (Kalman filtering) (Nahi and Assefi, 1972; Habibi, 1972; Jain, 1977; Rosenfeld and Kak, 1976).

A general problem of reconstruction filters is the assumption that the statistics (or spectral densities) of image and noise have to be constant over the whole image. This is not true in general. For instance, images used in pattern recognition may consist of objects and background with different statistics, separated by edges. These edges are important for recognition. The strategy adopted should, for instance, first subdivide the image into its constituent parts before filtering, this subdivision, however, depends upon the degraded image (Nahi and Habibi, 1975; Pratt, 1978).

Another problem lies in the least-square error criterion. Although this is a conveniently objective criterion, an image reconstructed by it may not be judged by a human observer as optimal. A human observer usually favours a noisy image with sharp edges more than a less noisy one with faint edges, as is produced by least-squared error filtering. It is important to take into account models of the human visual system to obtain measures which agree with the opinion of the human observer (Stockham, 1972).

7.6   ENHANCEMENT

7.6.1   Introduction

In the previous section restoration techniques were discussed that can be used to correct degradation present in the image. When pattern recognition is the purpose of image processing certain degradations can, in fact, be very helpful. For instance, the concept of different objects present in a background of an image is of primary importance in pattern recognition. So an operation which enhances the edges defining the objects is often necessary.

Grey-value rescaling is important not only because it may enhance the contrast in a particular grey-value range, but also because it gives the possibility of obtaining equally probable grey values. This is necessary in the computation of some texture parameters (Haralick *et al.*, 1973).

A third kind of operation starts with so called *segmented images*, images in which, for instance, objects are separated from the background. Several techniques exist, for example, to eliminate noise and to close gaps.

7.6.2   Grey-value Rescaling

Rescaling of the grey values can be applied to enhance the contrast in the image or to compensate for grey-scale characteristics. Besides stretching and compression the grey values can be requantized to obtain a flat histogram of requantized grey values. When, after requantization, K grey values must be equally probable, the cumulative histogram of the (old) grey values must be divided into K equal parts. However, because the cumulative histogram consists of discrete steps (the grey values are discrete), requantization results in not exactly equal quantization steps in the cumulative histogram. Several methods exist to reduce this requantization error (Troy *et al.*, 1973).

7.6.3   Spatial Grey-value Operations

Certain spatial frequencies may be important in the recognition of objects; these spatial frequencies can be enhanced by filtering. For instance, chromosomes can be recognized from their banding pattern. The spatial frequencies of the bands are known and can be used in a filter to enhance the banding pattern (Granum and Lundsteen, 1977). Low-pass filtering can be applied to reduce noise. Band-pass filtering and high-pass filtering can be applied to enhance edges. As with use of restoration filters, filtering can also be achieved by using Fourier transformation or by direct convolution with (windowed) filters in the spatial domain.

There is also a considerable interest in *non-linear filter methods*. These give the possibility of suppressing noise while still preserving sharp edges. Some examples are median filters (Huang *et al.*, 1978), where the median of the grey values of a window around a point is taken as the new grey value of that point (Figure 7.17) and the edge-preserving filter (Nagao and Matsuyama, 1978). In this latter filter, for each point the variance of the grey value is computed in differently oriented windows and the new grey value is the mean grey value of the direction with the lowest variance. Another variance filter is given by Kuwahara *et al.* (1976).

Special edge-locating filters are developed by Hueckel (1971, 1973) and Persoon (1976). Objects with a fixed shape or a fixed grey-value pattern can be located by matched filters (Rosenfeld and Kak, 1976). Straight lines in an image can be found by the Hough transform (Duda and Hart, 1972; Iannino and Shapiro, 1978). The Hough transform can be extended to certain other parametrized curves (Bazin and Benoit, 1965; Wechsler and Sklansky, 1977; Shapiro, 1978).

7.6.4 Segmented Image Operations

A segmented image is an image which is divided into its different constituent image parts. An image can, for instance, be divided into objects and background. In this case, the segmented image is a two-valued (binary) image: 1 for points belonging to an object and 0 for points belonging to the background. Segmentation of an image can be obtained by thresholding the grey value (Rosenfeld and Kak, 1976), by edge detection (Hueckel, 1971; Persoon, 1976) or by region growing (Brice and Fennema, 1970; Zucker, 1976).

Figure 7.17 Example of median filtering: (a) an image degraded by shot noise. Ten per cent of the pixels contain the maximum intensity value. (b) Image after median filtering (with a 3 × 3 neighbourhood)

Figure 7.17   (c) Image when it is filtered a second time. All
noise pixels have now disappeared but the edges are fainter
than after filtering once only

Objects smaller than a certain size can be eliminated by *erosion*. In this case
all object points which have neighbours belonging to the background are
assigned to the background. The size of the eliminated object is determined by
the number of times the erosion was applied before the object vanished. Small
bridges between objects will also disappear with this operation, see Figure 7.18a
and b.

The reverse operation is called *dilation* and is an erosion of the background.
All background points which have a neighbour belonging to an object are also
assigned to that object. Small gaps in objects are filled in with this operation, see
Figure 7.18c and d.

By applying dilation *n* times after *n* times of erosion (called *opening*) or *n*
times erosion after *n* times dilation (called *closing*), the same results are obtained
as mentioned above (elimination of small objects, filling in gaps), but now the
remaining objects again have about their original size. For a further discussion
see Hersant *et al.* (1976). Another important operation is *skeletonization*
(Hilditch, 1969; Stefanelli and Rosenfeld, 1971). This is an erosion, with the
restriction that the connectedness of the objects does not change. Skeletonization
results in one point thick lines, with the same connectivity as the original image,
Figure 7.19. Noisy contours of objects may lead to unwanted streaks to the
skeleton. Several methods exist to reduce this effect (Rosenfeld and Davis,
1976). Skeletons are important because, by their end points and vertices, they
easily reveal how an object is connected.

Erosion, dilation and skeletonization techniques can be extended to three
dimensions (Lobregt et al., 1979).

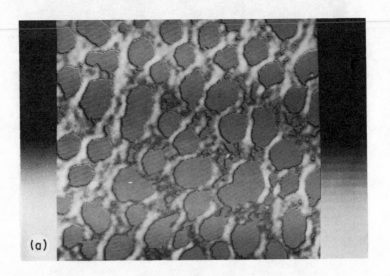

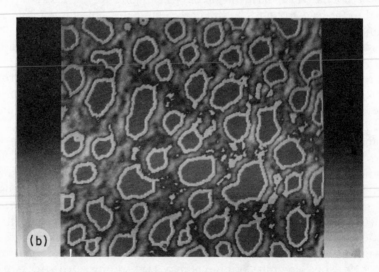

Figure 7.18 Example of erosion: (a) original CERMET image;
(b) eroded image of (a). All small particles and bridges are eroded
away (white pixels of (b))

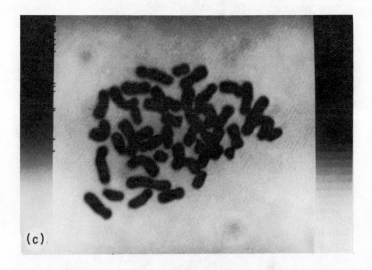

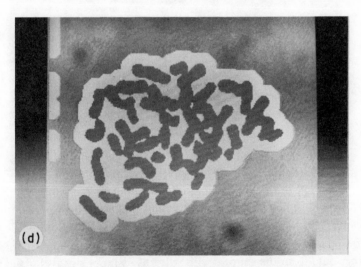

Figure 7.18 Example of dilation: (c) chromosome image; (d) complementary dilation applied. All gaps between the chromosomes are filled in (white pixels of (d))

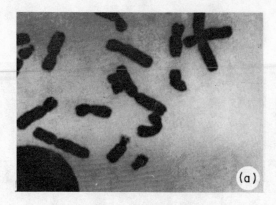

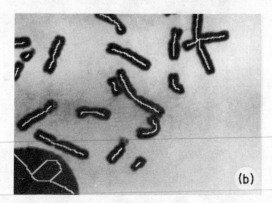

Figure 7.19   Example of skeletonization. Skeletons
of the chromosomes of (a) are shown in (b)

7.7   TRENDS

It is expected that the fields of pattern recognition and measurement will grow
closer together for the following reasons:

(a) Pattern recognition often requires physical measurements in order to
     gather data from the sources with patterns as input signals for the recog-
     nition system. In choosing special sensors one may sometimes avoid part of
     the preprocessing of the input data.
(b) The ultimate aim of measuring is often a classification or decision based
     upon the input data, for example, to distinguish between certain situations
     of the sources or to draw certain conclusions about them given the shape

of measured curves or a certain relation between input data. It is, therefore, quite natural that a pattern recognition procedure follows the measuring stage. Human creativity will mostly be necessary to interpret measuring results in the context of research. But in more standardized conditions, for instance during production, control of quality or of information content, an automatic classification will be appropriate.

(c) Advancing technology opens up possibilities for interactive design of processing methods for both measuring and recognition purposes. More complex measuring devices (such as echocardiographs) create a need for pattern recognition. Besides, miniaturized pattern recognition systems will become available to combine with measuring apparatus in order to provide data reduction, for example, for remote sensing and meteorology.

Real applications of pattern recognition systems are gradually gathering momentum, but progress is a little slower than was predicted one or two decades ago. A few reasons for this latter statement are:

(a) The choice of good, relevant features turned out to be very difficult in fields like character recognition, speech analysis, remote sensing, and biomedical engineering. Very many different choices are possible and are made in practice. Useful knowledge about human visual or auditive activities is only available in a very limited way and has not lead very often to the choice of appropriate features for automatic systems.

(b) After choosing features, much knowledge has then to be collected about their statistical or linguistical properties in order to be able to successfully apply one of the great variety of decision algorithms. Extensive and representative learning sets or data banks are necessary for this purpose (and also to compare results with other research workers) but they are difficult to obtain. So the quality and relevance of the features chosen cannot be tested easily; which results in a long-continued use of inappropriate features.

(c) Many practical problems involve images and the processing of images, and consequently the use of very many bits. Conventional computers are not well suited to this job and are too slow when on-line processing is desired. The same is true for speech recognition and industrial control.

(d) The reliability of pattern recognition systems, the flexibility to adapt to problems a little different from the original ones, and the level of maintenance ease were often too low for practical use.

(e) The cost of pattern recognition systems was often too high to compete with human beings, who are very experienced in recognizing patterns and who are at the same time very flexible, even with limited education.

(f) The introduction of pattern recognition systems, with all the consequences for the people involved, has not always been done in a sufficiently careful way.

The preceding enumeration indicates some difficulties but also suggests why it may be expected that automatic pattern recognition will gradually be applied to more tasks:

(a) Much research has been done in the last decade and is still being carried out. More knowledge about relevant features for certain fields of applications is becoming available gradually.

(b) Extensive data bases are coming into existence, for example, those concerning alphanumeric characters, speech, and chromosomes. Though the situation is unsatisfactory in many fields, this problem is recognized and progress is being made. It will become gradually less difficult to obtain learning sets in a number of fields.

(c) A great number of special data processing systems and components are becoming available that use microprocessors, pipeline architecture and some kind of parallel processing of data. These systems attain much higher speed and allow on-line processing.

(d) These hardware developments may also enhance the reliability of recognition systems and modular designs will enable maintenance in an easy way.

(e) As the price of digital LSI components is reduced, along with an increase in capabilities, potential application of new systems will improve. Increasing salary costs and the unwillingness of people to perform routine inspection tasks will also provide better prospects for automatic recognition systems.

(f) As it is well known now that the introduction of automatic systems needs very careful preparation and introduction; it may be expected that less failures will occur in the future.

A special point may be mentioned separately. It became clear in the past period what the difficult problems were and which were solvable. Knowing that reading continuous writing and unrestricted handprinted characters is difficult, one can pay attention to what type of restrictions in writing will be useful and acceptable to make automatic systems a success (Suen *et al.*, 1978). The idea is that people will be motivated to adapt to these restrictions if this facilitates the speed of banking operations or post handling. Speech understanding of connected words, an extensive vocabulary, and many-speaker situations is another difficult area. This may be avoided by streamlining a dialogue with a machine which only asks specialized questions with only a few possible answers and by repeating the question when the system did not understand the answer correctly. Such question–answering systems can be used for special tasks like seat reservations in trains (Shikano and Kohda, 1978).

A last optimistic aspect is the intensive cross-fertilization between methods and experiences from several subfields of pattern recognition and image processing. Combined use of statistical, fuzzy, and linguistical features and methods is becoming popular. Experience in one of the fields of biomedical engineering, character recognition, speech recognition, remote sensing or industrial applications may be useful also in other fields.

To summarize it can be expected that the field of pattern recognition in combination with the measuring field and image processing will gain momentum for practical applications in the future.

ACKNOWLEDGEMENTS

The authors acknowledge with pleasure the critical and detailed comments on this review by Mrs S. Massotty concerning the use of the English language.

REFERENCES

Aboutalib, A. O., Murphy, M. S., and Silverman, L. M. (1977). 'Digital restoration of images degraded by general motion blurs', *Trans. IEEE*, **AC-22**, 292–300.

Ahmed, N. and Rao, K. R. (1975). *Orthogonal Transforms for Digital Signal Processing*, Springer-Verlag, Berlin.

Aho, A. V. and Ullman, J. D. (1972, 1973). *The Theory of Parsing, Translation and Compiling*, Vols 1 and 2, Prentice Hall, Englewood Cliffs, New Jersey.

Amdahl, G. M. (1967). 'Validity of the single processor approach to achieving large scale computing capabilities', *Proc. Spring Joint Comput. Conf., AFIPS*, **30**, 483–490.

Andrew, E. R., Bottemley, P. A., Hinshaw, W. S., Holland, G. N., Moore, W. S., Simaroj, C., and Worthington, B. S. (1978). 'NMR imaging in medicine and biology', *Proc. XX Ampere Congr., Tallinn, USSR*.

Andrews, H. C. and Hunt, B. R. (1977). *Digital Image Restoration*, Prentice Hall, Englewood Cliffs, New Jersey.

Asada, H., Shinoda, H., Kidode, M., Watanabe, S., and Mori, K. (1978). 'Interactive image processing system with high-performance special processors', *Proc. 4th Int. Conf. Pattern Recognition, Kyoto*, pp. 1125–9.

Backer, E. (1978). Cluster analysis by optimal decomposition of induced fuzzy sets, *Ph.D. Thesis*, Delft University of Technology, Delftse Universitaire Pers, Delft.

Barnes, G. H., Brown, R. M., Kato, M., Kuck, D. J., Slotwick, D. K., and Stokes, R. A. (1968). 'The ILLIAC IV Computer', *Trans. IEEE*, **C-17**, 746–57.

Batchelor, B. G. (Ed.) (1978). *Pattern Recognition, Ideas in Practice*, Plenum, New York.

Bazin, M. J. and Benoit, J. W. (1965). 'Off-line global approach to pattern recognition for bubble chamber pictures', *Trans IEEE*, **NS-12**, 291–5.

Bierman, A. W. and Feldman, J. A. (1972). 'A survey of results in grammatical inference', in S. Watanabe (Ed.) *Frontiers of Pattern Recognition*, Academic Press, New York, pp. 31–54.

Billingsley, F. C. (1975). 'Noise considerations in digital image processing hardware', in T. S. Huang (Ed.) *Topics in Applied Physics*, Vol. 6, *Picture Processing and Digital Filtering*, Springer, Berlin, pp. 249–81.

Bock, H. H. (1974). *Automatische Klassifikation*, Vandenhoeck and Ruprecht, Göttingen.

Bom, N., Lancée, C. T., Zwieten, G van, Kloster, F. E., and Roelandt, J. (1973). 'Multiscan echocardiography, Part I: Technical description', *Circulation*, **48**, 1066–74.

Boxer, S. M. and Batchelor, B. G. (1978). 'Microprocessor arrays for pattern recognition', *Comput. Digital Tech.*, **1**, 60–6.

Brice, C. R. and Fennema, C. L. (1970). 'Scene analysis using regions', *Artificial Intelligence*, **1**, 205–26.

Campenhout, J. M. van (1978). 'On the peaking of the Hughes mean recognition accuracy: the resolution of an apparent paradox', *Trans. IEEE*, **SMC-8**, 390–5.

Cochran, W. T., Cooley, J. W., Favin, D. L., Helms, H. D., Kaenel, R. A., Lang, W. W., Maling, G. C., Nelson, D. E., Rader, C. M., and Welch, P. D. (1967). 'What is the fast Fourier transform?' *Trans. IEEE*, **AU-15**, 45–55.

Duda, R. O. and Hart, P. E. (1972). 'Use of the Hough transform to detect lines and curves in pictures', *Comm. ACM*, **15**, 11–15.

Duda, R. O. and Hart, P. E. (1973). *Pattern Classification and Scene Analysis*, Wiley, New York.

Duda, R. O. and Nitzan, D. (1976). 'Low-level processing of registered intensity and range data', *Proc. 3rd Int. Joint Conf. Pattern Recognition, Coronado*, pp. 598–601.

Duff, M. J. B., Watson, D. M., and Deutsch, E. S. (1974). 'A parallel computer for image processing', *Proc. IFIP Congress 1974, Information Processing, Stockholm*, pp. 94–7.

Duff, M. J. B. (1976). 'CLIP 4, a large scale integrated circuit array parallel processor', *Proc. 3rd Int. Joint Conf. Pattern Recognition, Coronado*, pp. 728–33.

Duin, R. P. W. (1976). 'On the choice of smoothing parameters for Parzen estimators of probability density functions', *Trans. IEEE*, **C-25**, 1175–9.

Duin, R. P. W. (1978). 'On the accuracy of statistical pattern recognizers', *Ph.D. Thesis*, Delft University of Technology, Dutch Efficiency Bureau, Pijnacker, Holland.

Eccles, N. J., McCarthy, B. D., Proffitt, D., and Rosen, D. (1976a). 'A programmable flying-spot microscope and picture preprocessor', *J. Microsc.*, **106**, 33–42.

Eccles, N. J., McCarthy, B. D., Proffitt, D. and Rosen, D. (1976b). 'The spatial and grey-level resolution of flying-spot microscopes', *J. Microsc.*, **106**, 43–8.

Ellis, B. (1966). *Basic Concepts of Measurement*, Cambridge University Press, Cambridge.

Everitt, B. (1974). *Cluster Analysis*, Heinemann, London.

Fink, D. G. (1957). *Television Engineering Handbook*, McGraw-Hill, New York.

Finkelstein, L. (1973). 'Fundamental concepts of measurement', *Acta Imeko*, North-Holland, Amsterdam, pp. 11–27.

Finkelstein, L. (1975). 'Fundamental concepts of measurement: definition and scales', *Measurement and Control*, **8**, 105–11.

Flynn, M. J. (1972). 'Some computer organisations and their effectiveness', *Trans. IEEE*, **C-21**, 948–60.

Fu, K. S. (1968). *Sequential Methods in Pattern Recognition and Machine Learning*, Academic Press, New York.

Fu, K. S. (1974). *Syntactic Methods in Pattern Recognition*, Academic Press, New York.

Fu, K. S. (Ed.) (1976). *Digital Pattern Recognition*, Springer-Verlag, Berlin.

Fu, K. S. (Ed.) (1977). *Syntactic Pattern Recognition: Applications*, Springer-Verlag, New York.

Fu, K. S. and Lu, S. Y. (1977). 'A clustering procedure for syntactic patterns', *Trans. IEEE*, **SMC-7**, 734–42.

Fukunaga, K. (1972). *Introduction to Statistical Pattern Recognition*, Academic Press, New York.

Gaines, B. R. and Kohout, L. J. (1977). 'The fuzzy decade: a bibliography of fuzzy systems and closely related topics', *Int. J. Man-Machine Studies*, **9**, 1-68.

Gerritsen, F. A., Spierenburg, J., and Sennema, H. (1977). 'A fast, multipurpose, micro-programmable image-processor', *Proc. 'SITEL-ULG' Seminar on Pattern Recognition, Liege*, SITEL, Ophain, Belgium, pp. 6.1.1.–8.

Golab, T., Ledley, R. S., and Rotolo, L. S. (1971). 'FIDAC-Film input to digital computer', *Pattern Recognition*, **3**, 123–56.

Gonzalez, R. C. and Thomason, M. G. (1978). *Syntactic Pattern Recognition, An Introduction*, Addison-Wesley, Reading, Mass.

Goodman, J. W. (1968). *Introduction to Fourier Optics*, McGraw-Hill, New York.

Granum, E. and Lundsteen, C. (1977). 'Progress on band pattern derivation from digitized chromosome images', *Proc. IV Nordic Meeting on Medical and Biological Engineering*, 63.1–63.3, Lyngby, Denmark.

Habibi, A. (1972). 'Two-dimensional Bayesian estimate of images', *Proc. IEEE*, **60**, 878–83.

Haralick, R. M., Shanmugan, K., and Dinstein, I. (1973). 'Textural Features for Image Classification', *Trans. IEEE*, **SMC-3**, 610–21.

Haralick, R. M. and Currier, P. (1977). 'Image discrimination enhancement combination systems (IDECS)', *Computer Graphics Image Processing*, **6**, 371–81.

Haralick, R. M. and Minden, G. (1978). 'Kandidats: an interactive processing system', *Computer Graphics Image Processing*, **8**, 1–15.

Hartigan, J. A. (1975). *Clustering Algorithms*, Wiley, New York.

Hersant, T., Jeulin, D., and Parniere, P. (1976). 'Basic notions of mathematical morphology used in quantitative metallography', in G. Ondracek (Ed.) *Newsletter '76 in Stereology* Gesellschaft für Kernforschung, Karlsruhe, pp. 17–78.

Hilditch, C. J. (1969). 'Linear skeletons from square cupboards', in B. Meltzer and D. Mitchie (Eds.), *Machine Intelligence*, Vol. 4, Edinburgh University Press, Edinburgh, pp. 403–20.

Hinshaw, W. S. (1974). 'Spin mapping: The application of moving gradients to NMR', *Phys. Lett.*, **48A**, 87–8.

Hopcroft, J. E. and Ullman, J. D. (1969). *Formal Languages and their Relation to Automata*, Addison-Wesley, Reading, Mass.

Hougardy, H. P. (1976), 'Recent progress in automatic image analysis, Instrumentation', *The Microscope*, **24**, 7–23.

Howes, M. J. and Morgan, D. V. (1978). *Charge-Coupled Devices and Systems*, Wiley, Chichester.

Huang, T. S., Schreiber, W. F., and Tretiak, O. J. (1971). 'Image processing', *Proc. IEEE*, **59**, 1586–609.

Huang, T. S. (Ed.) (1975). *Picture Processing and Digital Filtering*, Springer-Verlag, Berlin.

Huang, T. S., Yang, G. J., and Tang, G. Y. (1978). 'A fast two-dimensional median filtering algorithm', *Proc. IEEE Comput. Soc. Conf. Pattern Recognition Image Processing, Chicago, USA*, pp. 128–31.

Hueckel, M. H. (1971). 'An operator which locates edges in digital pictures', *J. ACM*, **18**, 113–25.

Hueckel, M. H. (1973). 'A local visual operator which recognizes edges and lines', *J. ACM*, **20**, 634–47.

Hunt, B. R., Janney, D. H., and Zeigler, R. K. (1973). 'Radiographic image enhancement by digital computers', *Mater. Eval. J. ASNT*, **31**, 1–5.

Iannino, A. and Shapiro, S. D. (1978). 'A Survey of the Hough transform and its extensions for curve detection', *Proc. IEEE Comput. Soc. Conf. on Pattern Recognition Image Processing, Chicago, USA*, pp. 32–8.

Ito, T., Akita, K., Fusaoka, A., Sato, K., Fukushima, M., Tsuji, H., Nakajima, H., Fukuda, Y., Shibayama, J., Hirayama, M., and Nakatsuka, K. (1978). 'A color picture processing system with firmware facility', *Proc. 4th Int. Joint Conf. on Pattern Recognition, Kyoto*, pp. 1130–4.

Jain, A. K. (1977). 'A semicausal model for recursive filtering of two-dimensional images', *Trans. IEEE*, **C-26**, 343–50.

Jardine, N. and Sibson, R. (1971). *Mathematical Taxonomy*, Wiley, New York.

Johnson, E. G. (1970). 'The PAX II picture processing system', in B. Lipkin and A. Rosenfeld (Eds.), *Picture Processing and Psychopictorics* Academic Press, New York, pp. 427–512.

Kanal, L. (1974). 'Patterns in pattern recognition: 1968–1974', *Trans. IEEE*, **IT-20**, 697–722.

Kieszling, A. (1976). 'A fast scanning method for three dimensions', *Proc. 3rd Int. Joint Conf. Pattern Recognition, Coronado*, pp. 586–9.

Koral, K. F., Rogers, W. L., and Knoll, G. F. (1975). 'Digital tomographic imaging with time-modulated pseudo-random coded aperture and Anger Camera', *J. Nuc. Med.*, **16**, 402–13.

Kruse, B. (1973). 'A parallel picture processing machine', *Trans. IEEE*, **C-22**, 1075–87.

Kuwahara, M., Hachimura, K., Eiho, S., and Kinoshita, M. (1976). 'Processing of RI-angiocardiographic images, in K. Preston and M. Onoe (Eds.), *Digital Processing of Biomedical Images*, Plenum, New York, pp. 187–202.

Lauterbur, P. C. (1973). 'Image formation by induced local interactions: examples employing nuclear magnetic resonance', *Nature*, **242**, 190–1.

Ledley, R. S., Park, C. M., Kulkarni, Y. G., Shiu, M. R., and Rotolo, L. S. (1978). 'Texac, a texture analysis computer', *Proc. 4th Int. Joint Conf. Pattern Recognition, Kyoto*, pp. 1119–23.

Lemkin, P., Carman, G., Lipkin, L., Shapiro, B., Schultz, M., and Kaiser, P. (1974). 'A real time picture processor for use in biologic cell identification', *J. Histochem. Cytochem.*, **22**, 725–31.

Lobregt, S., Verbeek, P. W., and Groen, F. C. A. (1979). '3-D skeletonization. Principle and algorithm', *Trans. IEEE*, **PAMI-2**, 75–7.

Melen, R. (1973). 'The tradeoffs in monolithic image sensors: MOS vs CCD', *Electronics*, **46**, 24 May, 106–10.

Mendel, J. M. and Fu, K. S. (1970). *Adaptive Learning and Pattern Recognition Systems: Theory and Applications*, Academic Press, New York.

Minsky, M. and Papert, S. (1969). *Perceptrons: an Introduction to Computational Geometry*, MIT Press, Cambridge, Mass.

Nagao, M. and Matsuyama, T. (1978). Edge preserving smoothing, *Proc. 4th Int. Joint Conf. on Pattern Recognition, Kyoto*, pp. 518–20.

Nahi, N. E. and Assefi, T. (1972). 'Bayesian recursive image estimation', *Trans. IEEE*, **C-21** 734–8.

Nahi, N. E. and Habibi, A. (1975). 'Decision-directed recursive image enhancement', *Trans. IEEE*, **CAS-22**, 286–93.

Narendra, P. and Fukunaga, K. (1977). 'A branch and bound algorithm for feature subset selection', *Trans. IEEE*, **C-26**, 917–22.

Nilsson, N. J. (1965). *Learning Machines*, McGraw-Hill, New York.

Okada, Y., Tajima, H., and Mori, R. (1976). 'A novel multiprocessor array', *Proc. Euromicro Symp. Venice*, pp. 83–90.

Pao, Y. and Schultz, W. H. (1978). 'An associative memory technique for the recognition of patterns', *Proc. 4th Int. Joint Conf. Pattern Recognition, Kyoto*, pp. 405–7.

Persoon, E. (1976). 'A new edge detection algorithm and its applications in picture processing', *Computer Graphics Information Processing*, **5**, 425–46.

Ploeg, M. van der, Duijn, P. van, and Ploem, J. S. (1974). 'High-resolution scanning densitometry of photographic negatives of human metaphase chromosomes, I, Instrumentation', *Histochemistry*, **42**, 9–29.

Polder, L. J. van der (1967). 'Target stabilization effects in television pick-up tubes', *Philips Res. Rep.*, **22**, 178–207.

Pratt, W. K. (1972). 'Generalized Wiener filtering computation techniques', *Trans. IEEE*, **C-21**, 636–41.

Pratt, W. K. and Davarian, F. (1977). 'Fast computational techniques for pseudo-inverse and Wiener image restoration', *Trans. IEEE*, **C-26**, 571–80.

Pratt, W. K. (1978). *Digital Image Processing*, Wiley, New York.

Purll, D. J. (1976). 'Automatic inspection and gauging using photodiode-arrays', *Chart. Mech. Eng.*, **23**, 79–82.

Ramamoorthy, C. V., Turner, J. L., and Wah, B. W. (1978). 'A design of a fast cellular associative memory for ordered retrieval', *Trans. IEEE*, **C-27**, 800–15.

Röcker, F. (1974). 'Localization and classification of three-dimensional objects', *Proc. 2nd Int. Joint Conf. on Pattern Recognition, Copenhagen*, pp. 527–8.

Rosenfeld, A. and Kak, A. C. (1976). *Digital Picture Processing*, Academic Press, New York.

Rosenfeld, A. and Davis, L. S. (1976). 'A note on thinning', *Trans. IEEE*, **SMC-6**, 226–8.

Salomaa, A. (1973). *Formal Languages*, Academic Press, New York.

Shapiro, S. D. (1978). 'Properties of transforms for the detection of curves in noisy pictures', *Computer Graphics, Information Processing*, **8**, 219–36.

Shaw, A. C. (1969). 'A formal picture description scheme as a basis for picture processing systems', *Information and Control*, **14**, 9–52.

Shaw, A. C. (1970). 'Parsing of graph-representable pictures', *J. ACM*, **17**, 453–81.

Shikano, K. and Kohda, M. (1978). 'A linguistic processor for train seat reservation in a conversational speech recognition system', *Proc. 4th Int. Joint Conf. on Pattern Recognition, Kyoto*, pp. 1039–41.

Shirai, Y. and Suwa, M. (1971). 'Recognition of polyhedrons with a range finder', *Proc. 2nd Int. Joint Conf. Artificial Intelligence*, British Computer Society, London, pp. 80–3.

Shirai, Y. (1978). 'Recent advances in 3-D scene analysis'. *Proc. 4th Int. Joint Conf. Pattern Recognition, Kyoto*, pp. 86–94.

Slepian, D. (1967). 'Linear least square filtering of distorted images', *J. Opt. Soc. Am.*, **57**, 918–22.

Somer, J. C. (1968). 'Electronic sector scanning for ultrasonic diagnosis', *Ultrasonics*, **6**, 153–9.

Specht, D. (1967). 'Generation of polynomial discriminant functions for pattern recognition', *Trans, IEEE*, **EC-16**, 308–19.

Stefanelli, R. and Rosenfeld, A. (1971). 'Some parallel thinning algorithms for digital pictures', *J. ACM*, **18**, 255–64.

Steinhagen, H. E. and Fuchs, S. (1976). *Einführung in die mathematischen Methoden der Zeichenerkennung*, VEB Verlag Technik, Berlin.

Stevens, S. S. (1946). 'On the theory of scales of measurement', *Science*, **103**, 677–80.

Stockham, T. G. (1972). 'Image processing in the context of a visual model', *Proc. IEEE*, **60**, 829–42.

Suen, C. Y., Berthod, M., and Mori, S. (1978). 'Advances in recognition of handprinted characters', *Proc. 4th Int. Joint Conf. Pattern Recognition, Kyoto*, pp. 30–44.

Ter-Pogossian, M. M., Phelps, M. E., Browell, G. L., Cox, J. R. jr., Davis, D. O., and Evens, R. G. (Eds.) (1977). *Reconstruction Tomography in Diagnostic Radiology and Nuclear Medicine*, University Park Press, Baltimore.

Thomason, M. G. (1975). 'Stochastic syntax-directed translation schemata for correction of errors in context-free languages', *Trans. IEEE*, **C-24**, 1211–6.

Thurstone, F. L. and Ramm, O. T. (1974). 'A new ultrasound technique employing two-dimensional, electronic beam steering', in *Acoustical Holography* Vol. 5, Plenum Press, New York, pp. 249–59.

Tojo, A. and Uchida, S. (1978). 'Multiprocessor system for picture processing with a multipurpose video processor', *Proc. 4th Int. Joint Conf. Pattern Recognition, Kyoto*, pp. 1116–8.

Toussaint, G. T. (1974). 'Bibliography on estimation of misclassification', *Trans. IEEE*, **IT-20**, 472–9.

Troy, E. B., Deutsch, E. S., and Rosenfeld, A. (1973). 'Gray-level manipulation experiments for texture analysis', *Trans. IEEE*, **SMC-3**, 91–8.

Veen, T. M. van, Groen, F. C. A., and Verbeek, P. W. (1978). 'Noise suppression in iterative restoration', *Proc. 4th Int. Joint Conf. Pattern Recognition, Kyoto*, pp. 505–8.

Verhagen, C. J. D. M. (1975a). 'Some general remarks about pattern recognition; its definition; its relation with other disciplines; a literature survey', *Pattern Recognition*, **7**, 109–16.

Verhagen, C. J. D. M. (1975b). 'Automatic pattern recognition', *Proc. Jurema*, part I, Zagreb, 113–19.

Verhagen, C. J. D. M. (1977). 'General survey of image processing applications, past and future', in *Automation and Inspection Applications of Image Processing Techniques, Proc. SIRA*, 8–17, London, Vol. 130, Society Photo Optical Instrumentation Engineers, Bellingham, WA, USA.

Verhagen, C. J. D. M., Duin, R. P. W., Groen, F. C. A., Joosten, J. C., and Verbeek, P. W. (1980). 'Progress report on pattern recognition', *Rep. Prog. Phys.*, **43**, 785–831.

Wechsler, H. and Sklansky, J. (1977) 'Finding the rib cage in chest radiographs', *Pattern Recognition*, **9**, 21–30.

Will, P. M. and Pennington, K. S. (1971). 'Grid coding, a preprocessing technique for robot and machine vision', *Proc. 2nd Int. Joint Conf. Artificial Intelligence*, British Computer Society, London, pp. 66–70.

Wolff, H. J. G. (1977). 'Recognition of handwritten capitals based on the use of a line follower', *Proc. 'SITEL-ULG' Seminar on Pattern Recognition, Liege*, SITEL, Ophain, Belgium, pp. 7.2.1–7.

Young, T. Y. and Calvert, T. W. (1974). *Classification, Estimation and Pattern Recognition*, American Elsevier, New York.

Zucker, S. W. (1976). 'Survey region growing: childhood and adolescence', *Computer Graphics for Image Processing*, **5**, 382–99.

Handbook of Measurement Science, Vol. 1
Edited by P. H. Sydenham
© 1982 John Wiley & Sons Ltd.

Chapter

8

A. VAN DEN BOS

Parameter Estimation

Editorial introduction

Given a measurement situation where it is known that a signal of a certain mathematical form exists buried in random noise, how can an estimate of the parameters, the coefficients, of the mathematical model be realized? This chapter provides a review of the mathematical philosophy and procedures with which such parameters can be estimated through application of methods of curve fitting to systematic data perturbed by stochastic processes. The mathematical expressions so realized do not necessarily describe the true internal physical behaviour of the process; that is, the estimated coefficients are not necessarily identifiable as real physical parameters. The output of the model so produced will, however, adequately describe the performance sought.

The subject matter of parameter estimation rests heavily on advanced mathematical material that has come into extended application over the past two decades. This came about because of the comparative ease with which the complex and lengthy mathematical operations needed can now be handled by digital machine computation. In order to provide a concise review, the author has needed to assume that the reader is familiar with determinant algebra, probability theory, discrete function mathematics, and spectral analysis in general. In addition to the references cited in the text readers will find relevant material in Griffiths *et al.*, (1973), Schwartz and Shaw (1975), Van Trees (1968, 1971a, 1971b).

8.1 INTRODUCTION

This chapter discusses application of statistical parameter estimation techniques to physical observations. A common characteristic of these techniques is that use is made of a more or less detailed parametric mathematical statistical model of the observations. The quantities to be measured are the parameters of this model.

The choice of model for the observations is closely connected to the objectives of the parameter estimation procedure. If the purpose is measurement of *physical* parameters the model of the observations arises from an, often detailed, physical analysis and is a careful description of the physical process that is supposed to generate the observations. The parameters then have a well defined physical meaning. An example is the model of the observations used in radioactive decay measurements. This is a weighted sum of exponentials. The weights and the decay constants are the parameters to be measured. The decay constants are specific for a particular radioactive component, while the weights are measures of component concentration. Models of observations used for measurement of physical parameters must be set up with utmost care. The reason is that systematic deviations of the observations from the model generally lead to systematic deviations of the measured parameters from their true values. An example of a systematic model error is using a bi-exponential radioactive decay model for observations which are, contrary to this assumption, tri-exponential. This example illustrates that models of observations used for estimation of physical parameters must be complete in the sense that they must include all major systematic contributions. As a result the establishment of a sufficiently complete parametric physical model often requires a substantial expert knowledge and is a rule more demanding than the selection and implementation of a suitable parameter estimation procedure. A much simpler class of parametric models may be used if the purpose is accurate quantitative description of the observations as a function of a, usually relatively precisely known, independent variable. These models are referred to as *curve fitting models*. They are often only loosely connected to the actual physical process generating the observations. Consequently it is not unusual if their parameters have no clear physical meaning. Well known examples are calibration curves obtained by fitting a linear or higher degree relationship to the observations. Further important examples are the modern dynamic parametric models for stochastic processes discussed in Section 8.4. Here the stochastic process is modelled as white noise that has passed a dynamic system described by a linear difference equation. The purpose of this model is not to explain the mechanism that generated the stochastic process. The purpose is a quantitative description of the spectral properties of the stochastic process in the form of a relatively small number of difference equation coefficients. This may be attractive for automatic control purposes or simply from data reduction point of view. Models used in practice may also be mixtures of curve fitting models and physical models. For example, in X-ray spectroscopy the measured spectrum is often modelled as a sum of gaussian peaks of unknown height, width and location. These parameters have a clear physical meaning. On the other hand, the undesirable background radiation in the measurements is usually modelled as a polynomial, which is a curve fitting model. Also, in experimental physics, physical models are used which contain a number of deliberate simplifications and approximations. For example, in dynamic modelling, models resulting from conservation laws may be complicated and nonlinear. They may, as a result, be very difficult to verify experimentally. Quantitative analysis and linearization may then result in a simplified model which is suitable for experimental validation.

It is important to realize that many parameter measurement methods used in experimental physics have been developed in an era when digital data acquisition and processing were non-existent or too expensive. As a result, in these conventional methods emphasis is on computational simplicity. Needless to say that this emphasis leaves little room for other considerations, for example, precision or numerical properties other than simplicity. A further striking characteristic of many of these conventional methods is that the observations are considered exact. The model of the observations does not include a model of systematic and/or non-systematic errors. Illustrative examples of conventional methods are found among graphical techniques, for instance, determination of time constants of linear dynamic systems using the asymptotes in the Bode diagram. Further examples of conventional methods are discussed in (Van den Bos, 1977). A disadvantage of conventional methods is that systematic and non-systematic errors in the resulting parameter measurements are difficult or virtually impossible to compute. This complicates a comparison of different methods both with respect to accuracy and precision. Moreover, these methods are often subjective. There is no clear-cut, objective procedure according to which the experimenter measures the parameter. Furthermore, the absence of errors in the model of the observations precludes the use of *a priori* knowledge about the errors for improving the precision with which the desired parameters are measured. Of course, these critical comments on conventional procedures are less relevant if only a rough estimate of a parameter is required. They should, however, be kept in mind whenever an efficient use of the available observations is to be made.

Modern statistical parameter estimation methods require a model of the observations that includes a model of both the systematic errors and the non-systematic errors. An example of a model of systematic errors is the background model in the above X-ray spectroscopy example. Non-systematic errors are modelled as zero-mean stochastic variables or, if their chronological ordering is relevant, as stochastic processes. Thus a particular set of observations is considered to be a set of observations made on stochastic variables or to be a realization of a stochastic process. Then the parameters to be estimated are parameters of probability density function defining those stochastic variables or stochastic processes. Thus the measurement problem is reformulated as estimation of parameters of probability density functions from observations of the corresponding stochastic variables. It has, therefore, taken the form of a statistical parameter estimation problem. In statistical parameter estimation the function of the observations that is used to compute the parameter is referred to as *estimator*. Thus an estimator is a stochastic variable. The value produced by an estimator for a particular set of observations is called an *estimate*. An estimate is a number.

The formulation of a measurement problem as a statistical parameter estimation problem implies that for its solution use can be made of the extensive collection of theories and methods available not only from pure mathematical statistics, but also from automatic control, econometrics, biology, and other fields. An outstanding text on stochastic variables and processes is Papoulis (1965). Authoritative in the field of mathematical statistics are Cramer (1961), Kendall and Stuart (1966, 1967, 1969), while for an excellent introductory text the reader is referred to Mood *et al.* (1974). Texts such as Eykhoff (1974) and Goodwin and Payne (1977) are mainly devoted to estimation of parameters of dynamical systems. An extensive collection of papers on the same subject is found in IEEE (1974), IFAC, (1967, 1970, 1973, 1978, 1979).

The statistical parameter estimation approach offers a number of advantages to the experimenter that are difficult to obtain if the conventional methods mentioned above are used. In the first place the expectation and the variance of an estimator can be computed. Then the bias, defined as the difference between this expectation and the true value of the parameter, represents the systematic error, that is, the accuracy of the estimator. Bias, although systematic itself, may be a result of non-systematic errors in the observations. Similarly, the variance, or rather its square root the standard deviation, is a measure of the non-systematic error, that is, the precision of the estimator. Bias and standard deviation are objective quantities suitable for comparison of different estimators of a parameter applied to the same observations. Simple examples of the computation of bias and standard deviation are discussed in Van den Bos (1977). Furthermore, statistically modelling the observations enables the experimenter to find relatively precise or even the most precise estimator for a particular problem. This is discussed in Sections 8.3 and 8.4; for static and dynamical models respectively. Also, if the *a priori* knowledge about the non-systematic errors in the observations is sufficiently detailed, a lower bound on the variance of any unbiased estimator of a particular parameter, for a given set of observations, can be computed. This lower bound, the minimum variance bound, is discussed in Section 8.2. It enables one to investigate the feasibility of a set of observations for the parameter estimation objectives concerned. Finally, a statistical model of the observations may be used for minimizing or reducing the variance of an estimator through experimental design, that is, through manipulating independent variables that can be freely chosen. This is, very briefly, discussed in Sections 8.3 and 8.4.

This introduction is concluded by the observation that parameter estimation methods are a useful addition to, and no substitute for, classical techniques for reducing systematic and non-systematic errors. For example, it is much better to avoid systematic errors in the observations than to include them in the model of the observations in the form of a parametric model. In the latter case the simultaneous estimation of the additional parameters increases the variances of the estimates of the remaining parameters, which were the primary objective.

Similarly, reduction of non-systematic errors in the observations is worthwhile, since the variance of most parameter estimators is proportional to the variance of these errors.

8.2 PRECISION

In this section the *minimum variance bound* (MVB), also called *Cramér–Rao lower bound*, is discussed. The MVB is a lower bound on the variance of any unbiased estimator. Since it is independent of any particular method of estimation, it provides a bound on the precision that can be achieved given the observations. The MVB is, therefore, a useful tool for investigating the feasibility of the observations for the parameter estimation objectives concerned, the more so as a class of estimators exists which, at least asymptotically, achieve this bound. These estimators are discussed in Section 8.3 of this chapter. In the following the underlined characters, e.g. $\underline{w}_1$, are stochastic variables. Vectors or matrices are shown in bold face type, e.g. $\mathbf{x}$. Before proceeding the notation that is used in what follows should be explained.

Let $\mathbf{x}$ and $\mathbf{y}$ be $K \times 1$ and $L \times 1$ vectors respectively and let $f(\mathbf{x})$ be a scalar function of the elements of $\mathbf{x}$. Then:

(a) the $1 \times K$ vector $\partial f(\mathbf{x})/\partial \mathbf{x}$ is defined by its m element $\partial f(\mathbf{x})/\partial x_m$;
(b) the $K \times K$ matrix $\partial^2 f(\mathbf{x})/\partial \mathbf{x}^2$ is defined by its m, n element $\partial^2 f(\mathbf{x})/\partial x_m \partial x_n$;
(c) the $L \times K$ matrix $\partial \mathbf{y}/\partial \mathbf{x}$ is defined by its m, n element $\partial y_m/\partial x_n$.

Let $\underline{\mathbf{x}} = (\underline{x}_1 \cdots \underline{x}_K)^{\mathrm{T}}$. Then the $K \times K$ covariance matrix $\mathrm{cov}(\underline{\mathbf{x}}, \underline{\mathbf{x}})$ is defined by its m, n element $\mathrm{cov}(\underline{x}_m, \underline{x}_n)$.

The MVB may be described as follows. Let $\underline{w}_1, \ldots, \underline{w}_N$ be the observations and define $\underline{\mathbf{w}}$ as the column vector of these observations. Furthermore let the probability density of $\underline{\mathbf{w}}$ be $f(\boldsymbol{\omega}; \boldsymbol{\theta})$ where the elements of $\boldsymbol{\omega}$ correspond with those of $\underline{\mathbf{w}}$ and the elements of the $K \times 1$ vector $\boldsymbol{\theta}$ are the unknown parameters. Suppose that the elements $r_1(\underline{\mathbf{w}}), \ldots, r_J(\underline{\mathbf{w}})$ of the $J \times 1$ vector $\mathbf{r}(\underline{\mathbf{w}})$ are unbiased estimators of the functions $\rho_1(\boldsymbol{\theta}), \ldots, \rho_J(\boldsymbol{\theta})$ and define $\boldsymbol{\rho}(\boldsymbol{\theta})$ as $(\rho_1(\boldsymbol{\theta}) \cdots \rho_J(\boldsymbol{\theta}))^{\mathrm{T}}$. In addition, define the *information matrix* $\mathbf{M}$ by

$$\mathbf{M} = E[(\partial \ln \underline{L}/\partial \boldsymbol{\theta})^{\mathrm{T}}(\partial \ln \underline{L}/\partial \boldsymbol{\theta})]$$

where $\underline{L} = f(\underline{\mathbf{w}}; \boldsymbol{\theta})$. Then under a number of, not too restrictive, conditions the *Cramér–Rao inequality* (Zacks, 1971) states that

$$\mathrm{cov}(\mathbf{r}(\underline{\mathbf{w}}), \mathbf{r}(\underline{\mathbf{w}})) \geqslant [\partial \boldsymbol{\rho}(\boldsymbol{\theta})/\partial \boldsymbol{\theta}] \mathbf{M}^{-1} [\partial \boldsymbol{\rho}(\boldsymbol{\theta})/\partial \boldsymbol{\theta}]^{\mathrm{T}}$$

expressing that the difference between the positive semi-definite left-hand and right-hand members is positive semi-definite. The right-hand member defines the *minimum variance bound* (MVB) or *Cramér–Rao lower bound* on the covariance of any unbiased estimator of $\rho(\theta)$. It can be shown that under certain regularity conditions $\mathbf{M}$ may alternatively be written

$$\mathbf{M} = -E(\partial^2 \ln \underline{L}/\partial\theta^2)$$

This form is often easier to compute. With respect to the Cramér–Rao inequality the following remarks can be made:

(a)   It can easily be shown that the variances of the elements of $\mathbf{r}$, which are the diagonal elements of $\text{cov}(\underline{\mathbf{r}}, \underline{\mathbf{r}})$, cannot be smaller than the corresponding diagonal elements of the MVB.
(b)   If $\rho(\theta) = (\theta_1 \cdots \theta_K)^T$ the MVB simplifies to $\mathbf{M}^{-1}$.
(c)   The lth diagonal element of the MVB, that is, the MVB on the variance of $\mathbf{r}_l$, equals

$$(\partial\rho_l/\partial\theta)\mathbf{M}^{-1}(\partial\rho_l/\partial\theta)^T$$

Since $\mathbf{M}^{-1}$ is the MVB for θ, it is concluded that the MVB for a function of θ follows a simple first-order error propagation law.
(d)   Suppose $\underline{t}$ is a biased estimator for θ. Let the bias be $\mathbf{b}(\theta)$. Then $\underline{t}$ is an unbiased estimator for $\theta + \mathbf{b}(\theta)$. Hence the MVB for $\underline{t}$ is

$$\{\mathbf{I} + [\partial\mathbf{b}(\theta)/\partial\theta]\}\mathbf{M}^{-1}\{\mathbf{I} + [\partial\mathbf{b}(\theta)/\partial\theta]\}^T$$

where $\mathbf{I}$ is the identity matrix.
(e)   Suppose that of the parameter vector $\theta = (\theta_1 \cdots \theta_{k'} \cdots \theta_K)^T$ only one element, $\theta_{k'}$, is unknown. Then the MVB on the variance of an unbiased estimator $t_{k'}$ of $\theta_{k'}$ equals $1/E[(\partial \ln \underline{L}/\partial\theta_{k'})^2]$. Next assume that all elements of θ have to be estimated and let the corresponding information matrix be $\mathbf{M}$. Then the k', k element of $\mathbf{M}$ is described by $m_{k'k'} = E[(\partial \ln \underline{L}/\partial\theta_{k'})^2]$. It can be shown that the product of the k'th diagonal element of a positive definite matrix and the corresponding diagonal element of its inverse is larger than or equal to one. Equality only occurs if the diagonal element is the only non-zero element in its row and column. It is concluded that the k', k' element of $\mathbf{M}^{-1}$ is larger than or equal to $1/m_{k'k'}$. So, as compared with the one-parameter case, the MVB generally increases.

This section is concluded by three simple examples which illustrate a number of aspects of the above theory.

Example 8.1   Estimation of the slope of a straight line.   Suppose that the parameter α is estimated from the observations $(\underline{w}_1, x_1), \ldots, (\underline{w}_N, x_N)$ described by the model

$$\underline{w}_n = \alpha x_n + \underline{v}_n$$

where the $\underline{v}_n$ are independent and identically distributed (iid) and are normally distributed with expectation $\mu = 0$ and standard deviation $\sigma(N(0, \sigma))$. Then the joint probability density of $\underline{v}_1, \ldots, \underline{v}_N$ is described by

$$f_{\underline{v}}(v) = (2\pi)^{-N/2} \sigma^{-N} \exp\left(-\tfrac{1}{2}\sigma^{-2} \sum_n v_n^2\right)$$

where $v = (v_1 \cdots v_N)^{\mathrm{T}}$. Hence the probability density of $\underline{w}_1, \ldots, \underline{w}_N$ is

$$f_{\underline{w}}(\omega; \alpha) = (2\pi)^{-N/2} \sigma^{-N} \det(\partial v/\partial \omega) \exp\left(-\tfrac{1}{2}\sigma^{-2} \sum_n (\omega_n - \alpha x_n)^2\right)$$

where $\omega = (\omega_1 \cdots \omega_N)^{\mathrm{T}}$ and $\partial v/\partial \omega$ is the Jacobian. Since $\omega_n = \alpha x_n + v_n$, $\partial v/\partial \omega = \mathbf{I}$ and hence $\det(\partial v/\partial \omega) = 1$. Therefore

$$\ln \underline{L} = -\tfrac{1}{2}N \ln(2\pi) - N \ln \sigma - \tfrac{1}{2}\sigma^{-2} \sum_n (\underline{w}_n - \alpha x_n)^2$$

Then

$$-E(\partial^2 \ln \underline{L}/\partial \alpha^2) = \sigma^{-2} \sum_n x_n^2$$

and therefore the MVB for α is $\sigma^2/\sum_n x_n^2$.

Example 8.2 Estimation of the parameters of an exponential model. Let $(\underline{w}_1, x_1), \ldots, (\underline{w}_N, x_N)$ be observations described by the model

$$\underline{w}_n = \sum_k \lambda_k \exp(-\mu_k x_n) + \underline{v}_n$$

where $\boldsymbol{\theta} = (\lambda_1 \cdots \lambda_K \mu_1 \cdots \mu_K)^{\mathrm{T}}$ are the unknown parameters, the x_n are exact and the $\underline{v}_n$ are iid and $N(0, \sigma)$. Then using the same argument as in Example 8.1 one obtains

$$\ln \underline{L} = -\tfrac{1}{2}N \ln(2\pi) - N \ln \sigma - \tfrac{1}{2}\sigma^{-2} \sum_n \underline{d}_n^2(\boldsymbol{\theta})$$

where

$$\underline{d}_n(\boldsymbol{\theta}) = \underline{w}_n - \sum_k \lambda_k \exp(-\mu_k x_n)$$

Then partial differentiating, taking expectations, and taking into account that $E[\underline{d}_n(\boldsymbol{\theta})] = 0$ yields

$$-E(\partial^2 \ln \underline{L}/\partial \mu_p \partial \mu_q) = \sigma^{-2} \sum_n \lambda_p \lambda_q x_n^2 \exp[-(\mu_p + \mu_q)x_n]$$

$$-E(\partial^2 \ln \underline{L}/\partial \lambda_p \partial \lambda_q) = \sigma^{-2} \sum_n \exp[-(\mu_p + \mu_q)x_n]$$

$$-E(\partial^2 \ln \underline{L}/\partial \lambda_p \partial \mu_q) = -\sigma^{-2} \sum_n \lambda_p x_n \exp[-(\mu_p + \mu_q)x_n] \qquad p, q = 1, \ldots, K$$

These expressions define the information matrix $\mathbf{M}$ and therefore the MVB, $\mathbf{M}^{-1}$.

Example 8.3   Influence of the estimation of an additional parameter on the MVB.   Suppose that the parameter β has to be estimated from the observations $(\underline{w}_1, x_1), \ldots, (\underline{w}_N, x_N)$ and that the model of the observations is described by

$$\underline{w}_n = \exp(-\beta x_n) + \gamma + \underline{v}_n$$

where the x_n are exact, γ is an unknown offset and the $\underline{v}_n$ are idd and $N(0, \sigma)$. Note that, since γ is unknown, both β and γ have to be estimated although γ is not of particular interest.

Using the expressions of Example 8.2 one obtains for the elements of the information matrix of $\boldsymbol{\theta} = (\beta \ \gamma)^{\mathrm{T}}$

$$-E(\partial^2 \ln \underline{L}/\partial \beta^2) = \sigma^{-2} \sum_n x_n^2 \exp(-2\beta x_n)$$

$$-E(\partial^2 \ln \underline{L}/\partial \beta \, \partial \gamma) = -\sigma^{-2} \sum_n x_n \exp(-2\beta x_n)$$

$$-E(\partial^2 \ln \underline{L}/\partial \gamma^2) = N/\sigma^2$$

If for example $\beta = 1$, $x_n = 0.1n$ and $N = 75$, these expressions yield

$$\mathbf{M} = \sigma^{-2} \begin{pmatrix} 2.500 & -9.942 \\ -9.942 & 75 \end{pmatrix} \quad \text{and} \quad \mathbf{M}^{-1} = \sigma^2 \begin{pmatrix} 0.846 & 0.112 \\ 0.112 & 0.028 \end{pmatrix}$$

The MVB for estimation of β is therefore $0.846\sigma^2$.

Next assume that, as a result of a different experimental set-up, no offset in the observations occurs. Then the model of the observations is described by

$$\underline{w}_n = \exp(-\beta x_n) + \underline{v}_n$$

So only β need be estimated. Then the information matrix is scalar and is equal to the element m_{11} of the above $\mathbf{M}$. Hence in this case the MVB for estimation of β is $1/m_{11} = 0.4\sigma^2$.

Discussion of the examples.   From its definition it follows that the MVB can only be computed if the probability density of the observations is known. For instance, in Example 8.1 use is made of the observations being independent and normally distributed with known variance. The resulting MVB is not a function of the unknown parameters and can, therefore, be computed beforehand for any set of values of the independent variables $x_1, \ldots, x_N$. Furthermore, if, within physical restrictions, the independent variables may be freely chosen, they may be chosen so as to minimize the MVB. For example, if the restriction is $|x_n| \leqslant 1$, the MVB is minimized if one chooses $|x_n| = 1$ for all n. In the literature a particular choice of independent variables is referred to as *experimental design*. For a general discussion of experimental design see Fedorov (1972).

In Example 8.2 the computation of the MVB is based on the same a priori knowledge about the observations as in Example 8.1. Unfortunately, as opposed to the MVB in Example 8.1, the MVB of Example 8.2 is a function of the unknown parameters and cannot, therefore, be computed beforehand. The absence of the unknown parameters in the MVB of Example 8.1 is a consequence of the fact that in this example the logarithm of the probability density is quadratic in the parameters. Example 8.1 is, therefore, an exceptional case. Nevertheless, even if the MVB is a function of the unknown parameters, it remains an extremely useful tool. For nominal values of the unknown parameters it enables one to quantify variances that might be achieved, to detect possibly strong covariances between parameter estimates and to select suitable experimental designs. For instance, for the exponential model of Example 8.2 large variances and covariances may occur when the difference between the decay constants is small. In such cases the conclusion may be that a hypothetical estimator attaining the MVB would still be not sufficiently precise for the estimation objectives concerned. A solution may be to decrease the MVB by a different experimental design, that is, to select different values of the x_n and N. If this is not possible the conclusion may be that the available observations are not suitable for the parameter estimation objectives concerned. With respect to the numerical computation of the MVB it should be noted that the information matrix may be ill-conditioned. Its inversion, needed to compute the corresponding MVB, must therefore be carried out with care.

Example 8.3 shows the increase of the MVB of the relevant parameter if, in addition, a further, possibly highly irrelevant, parameter is estimated. Therefore it is much more preferable to eliminate, if possible, the offset in the observations by a different physical experimental set-up than to include the offset in the model of the observations and estimate it along with the relevant parameters. This illustrates that parameter estimation is a complement to and no substitute for the conventional error reducing or eliminating techniques. For further, highly instructive, case studies of reducing the number of parameters in a practical physical experiment the interested reader is referred to Kaufmann and Akselsson (1975) and Van Espen et al. (1977).

8.3 PRECISE ESTIMATORS

8.3.1 Maximum Likelihood Estimators

Suppose that in a particular experiment the observations, considered as stochastic variables, are $\underline{\mathbf{w}} = (\underline{w}_1 \cdots \underline{w}_N)^{\mathrm{T}}$ and define $f_{\underline{\mathbf{w}}}(\boldsymbol{\omega}; \boldsymbol{\theta})$ as their probability density, where the elements of $\boldsymbol{\omega} = (\omega_1 \cdots \omega_N)^{\mathrm{T}}$ correspond with those of $\underline{\mathbf{w}}$ and $\boldsymbol{\theta} = (\theta_1 \cdots \theta_K)^{\mathrm{T}}$ is the vector of unknown parameters to be estimated from $\underline{\mathbf{w}}$. Now let $\mathbf{w} = (w_1 \cdots w_N)^{\mathrm{T}}$ be one particular realization of $\underline{\mathbf{w}}$, that is, the elements of $\mathbf{w}$ are numbers, not variables. Then for that particular realization the function $L = f_{\underline{\mathbf{w}}}(\mathbf{w}; \mathbf{t})$ with $\mathbf{t} = (t_1 \cdots t_K)^{\mathrm{T}}$ is defined as the *likelihood function* of the parameters. Thus L is a function of $\mathbf{t}$. Then the maximum likelihood estimate of the parameters $\mathbf{t}$ from $\mathbf{w}$ is defined as that value $\tilde{\mathbf{t}}$ of $\mathbf{t}$ that maximizes L. For a discussion of the extensive and, unfortunately, relatively complicated theory of maximum likelihood estimation see Norden (1972, 1973) and Zacks (1971). Here only a summary is given of a number of useful properties of maximum likelihood estimators. The first three properties listed below are asymptotic, that is, they are defined in terms of limits approached when the number of observations used goes to infinity. In the order in which they are listed these asymptotic properties require an increasing number of conditions to be met. A discussion of these conditions can be found in the literature cited above.

(a)   Under very general conditions maximum likelihood estimators are consistent. An estimator $\underline{t}_N$ is defined as consistent for θ if for any $\varepsilon > 0$ $P(|\underline{t}_N - \theta| < \varepsilon) = 1$ for $N \to \infty$ where N is the number of observations and $P(A)$ is the probability of the event A. Note that consistency is a property of the area under the probability density on a certain interval; it is not a property of the moments of the probability density. A consistent estimator is, therefore, not always asymptotically unbiased, nor has it always a finite asymptotic variance. Furthermore, even if a consistent estimator is asymptotically unbiased, this is no guarantee for unbiasedness for small numbers of observations. Many consistent maximum likelihood estimators are seriously biased for small numbers of observations.

(b)   The asymptotic probability density function of a broad class of maximum likelihood estimators is normal with expectation θ and covariance $\mathbf{M}^{-1}$ where θ are the true values of the parameters and $\mathbf{M}^{-1}$ is the MVB. Note that this is a property of the asymptotic probability density of $\underline{t}_N$, not of its moments.

(c)   Under certain additional conditions the asymptotic covariance matrix of a maximum likelihood estimator equals the MVB.

(d)   If $\tilde{\mathbf{t}}$ is a maximum likelihood estimator for the $K \times 1$ parameter vector θ, then $\mathbf{g}(\tilde{\mathbf{t}})$ is a maximum likelihood estimator of the $L \times 1$ vector $\mathbf{g}(\theta)$ of one-to-one functions of θ where $L \leqslant K$. This property is usually referred to as *invariance property*.

Example 8.4   Estimation of the slope of a linear relationship.   Consider the estimation of the parameter α of the model $\underline{w}_n = \alpha x_n + \underline{v}_n$ discussed in Example 8.1. Suppose that the observations $w_1, \ldots, w_N$ of $\underline{w}_n$ are available. Then the likelihood function of the parameter a is described by

$$L = (2\pi)^{-N/2} \sigma^{-N} \exp\left(-\tfrac{1}{2}\sigma^{-2} \sum_n (w_n - ax_n)^2\right)$$

Since L and $\ln L$ are monotonic, maximizing L and maximizing $\ln L$ with respect to a are equivalent. Hence the maximum likelihood estimate $\tilde{a}$ of α is that value of a that maximizes

$$\ln L = -\tfrac{1}{2}N \ln(2\pi) - N \ln \sigma - \tfrac{1}{2}\sigma^{-2} \sum_n (w_n - ax_n)^2$$

Equating the derivative with respect to a of this expression to zero yields

$$\tilde{a} = \frac{\sum_n w_n x_n}{\sum_n x_n^2}$$

The maximum likelihood estimator is therefore

$$\underline{\tilde{a}} = \frac{\sum_n \underline{w}_n x_n}{\sum_n x_n^2}$$

Since in this expression the x_n are exact, $\underline{\tilde{a}}$ is a weighted sum of independent stochastic variables $\underline{w}_n$. Hence var $\underline{\tilde{a}}$ is the quadratically weighted sum of the variances of the $\underline{w}_n$. These variances are all equal to σ^2. Hence

$$\text{var } \underline{\tilde{a}} = \frac{\sigma^2}{\sum_n x_n^2}$$

Moreover, $E(\underline{\tilde{a}}) = \alpha$ since $E(\underline{w}_n) = \alpha x_n$. So $\underline{\tilde{a}}$ is unbiased.

Example 8.5   Maximum likelihood estimation of the parameters of an exponential model.   For the estimation problem of Example 8.2 the logarithm of the likelihood function is

$$-\tfrac{1}{2}N \ln(2\pi) - N \ln \sigma - \tfrac{1}{2}\sigma^{-2} \sum_n d_n^2(\mathbf{t})$$

where

$$d_n(\mathbf{t}) = w_n - \sum_k l_k \exp(-m_k x_n)$$

where the elements of $\mathbf{t} = (l_1 \cdots l_K \, m_1 \cdots m_K)^{\mathrm{T}}$ correspond with those of $\boldsymbol{\theta}$. Then equating to zero the gradient with respect to $\mathbf{t}$ of $\ln L$ yields the following equations to be satisfied by the maximum likelihood estimate $\tilde{\mathbf{t}}$ of $\boldsymbol{\theta}$:

$$\sum_n d_n(\mathbf{t})\exp(-m_k x_n)|_{\mathbf{t}=\tilde{\mathbf{t}}} = 0 \quad k = 1,\ldots,K$$

$$\sum_k d_n(\mathbf{t})l_k x_n \exp(-m_k x_n)|_{\mathbf{t}=\tilde{\mathbf{t}}} = 0 \quad k = 1,\ldots,K$$

Discussion of the examples.   The maximum likelihood estimator of Example 8.4 has a number of favourable properties. In the first place, it is unbiased.

Furthermore, comparison of its variance with the MVB computed in Example 8.1 shows that it achieves the MVB for any number of observations. This is an illustration of a theorem stating that if there exists an unbiased estimator having the MVB as covariance, it is the maximum likelihood estimator. Moreover, the maximum likelihood estimator of Example 8.4 is closed form and corresponds to a unique maximum of the likelihood function. The estimator of Example 8.2, on the other hand, does not achieve the MVB for a finite number of observations, as can easily be shown. Furthermore, it may seriously be biased if the number of observations is small. For illustrative examples see Van den Bos (1979). Also the expressions for the estimator are implicit and non-linear and can, as a rule, only be solved numerically. Examples of relevant numerical techniques are described in Section 8.3.4. The solution is further complicated by the fact that a likelihood function, as that in Example 8.5, may have more than one maximum. From these the absolute maximum must be selected. The maximum likelihood estimators of Examples 8.4 and 8.5 have in common that they are both equivalent to least squares estimators. This important property will now be studied in a more general form.

Suppose that the observations $\mathbf{w} = (\underline{w}_1 \cdots \underline{w}_N)^{\mathrm{T}}$ are described by the general model

$$\underline{w}_n = g(\mathbf{x}_n; \boldsymbol{\theta}) + \underline{v}_n \quad n = 1, \ldots, N \tag{8.1}$$

where the $\mathbf{x}_n = (x_{1n} \cdots x_{Kn})^{\mathrm{T}}$ are known, $\boldsymbol{\theta} = (\theta_1 \cdots \theta_L)^{\mathrm{T}}$ is the vector of unknown parameters, $g(\mathbf{x}_n; \boldsymbol{\theta})$ belongs to a known parametric family of functions and $\underline{\mathbf{v}} = (\underline{v}_1 \cdots \underline{v}_N)^{\mathrm{T}}$ is normally distributed with $E(\underline{\mathbf{v}}) = \mathbf{0}$ and covariance $\mathbf{V} = \mathrm{cov}(\underline{\mathbf{v}}, \underline{\mathbf{v}})$, where $\mathbf{0}$ denotes the null matrix. Then the probability density of $\underline{\mathbf{v}}$ is described by

$$f_{\underline{\mathbf{v}}}(\mathbf{v}) = (2\pi)^{-N/2} (\det \mathbf{V})^{-1/2} \exp(-\tfrac{1}{2} \mathbf{v}^{\mathrm{T}} \mathbf{V}^{-1} \mathbf{v})$$

Hence, the log-likelihood function of $\mathbf{t}$, given the observations $\mathbf{w} = (w_1 \cdots w_N)^{\mathrm{T}}$, is

$$-\tfrac{1}{2} N \ln(2\pi) - \ln(\det \mathbf{V})^{1/2} - \tfrac{1}{2} (\mathbf{w} - \mathbf{g})^{\mathrm{T}} \mathbf{V}^{-1} (\mathbf{w} - \mathbf{g})$$

where $\mathbf{g} = (g(\mathbf{x}_1; \mathbf{t}) \cdots g(\mathbf{x}_N; \mathbf{t}))^{\mathrm{T}}$. Maximizing this function with respect to $\mathbf{t}$ is equivalent to minimizing $(\mathbf{w} - \mathbf{g})^{\mathrm{T}} \mathbf{V}^{-1} (\mathbf{w} - \mathbf{g})$. Hence, if in the model (equation (8.1)) the errors are normally distributed with known covariance matrix, maximum likelihood estimation of the parameters is equivalent to weighted least squares estimation with the elements of the inverse covariance matrix as weights. In practice least squares estimation techniques are also widely applied to estimation problems where the above conditions are not met. Therefore, Section 8.3.2 and 8.3.3 will be devoted exclusively to these important techniques.

8.3.2 Linear Least Squares

Suppose that in the model (equation (8.1)) the function $g(\mathbf{x}; \boldsymbol{\theta})$ has the particular form

$$g(\mathbf{x}; \boldsymbol{\beta}) = \beta_1 h_1(\mathbf{x}) + \beta_2 h_2(\mathbf{x}) + \cdots + \beta_L h_L(\mathbf{x})$$

where $\boldsymbol{\beta} = (\beta_1 \cdots \beta_L)^T$ is the vector of unknown parameters and the $h_l(\mathbf{x})$ are known functions which do not contain elements of $\boldsymbol{\beta}$. So the additively error corrupted observations $\underline{w}_n$ are linear in the parameters and can conveniently be summarized in the form

$$\underline{\mathbf{w}} = \mathbf{X}\boldsymbol{\beta} + \underline{\mathbf{v}} \tag{8.2}$$

where the n, l element of the $N \times L$ matrix $\mathbf{X}$ is defined as $h_l(\mathbf{x}_n)$. First three examples of this important model will be given.

Example 8.6   Linearity in the independent variables.   Suppose

$$\underline{w}_n = \beta_1 \xi_{n1} + \beta_2 \xi_{n2} + \cdots + \beta_L \xi_{nL} + \underline{v}_n \quad n = 1, \ldots, N$$

where $\beta_1, \ldots, \beta_L$ are the parameters and $\xi_1, \ldots, \xi_L$ are the independent variables. Then the observations are described by equation (8.2) with $\mathbf{X}$ defined as the $N \times L$ matrix with n, l element ξ_{nl}.

Example 8.7   Polynomial.   Let

$$\underline{w}_n = \beta_1 \xi_n + \beta_2 \xi_n^2 + \cdots + \beta_L \xi_n^L + \underline{v}_n \quad n = 1, \ldots, N$$

Then the observations are described by equation (8.2) with the n, l element of the $N \times L$ matrix $\mathbf{X}$ defined as ξ_n^l. Note that the observations are non-linear in the independent variables but linear in the parameters.

Example 8.8   Discrete-time impulse response.   Suppose $\underline{w}(n), n = 1, \ldots, N$, are error-corrupted observations of the response of a linear discrete-time system having impulse response $\beta_1, \ldots, \beta_L$ to an exactly known input $\xi(n)$ where n denotes discrete time. That is,

$$\underline{w}(n) = \beta_1 \xi(n) + \beta_2 \xi(n-1) + \cdots + \beta_L \xi(n-L+1) + \underline{v}(n) \quad n = 1, \ldots, N$$

where $\underline{v}(n)$ is a discrete-time stochastic process representing the errors. Then the observations may be summarized in the form of equation (8.2), where $\underline{\mathbf{w}} = (\underline{w}(1)\underline{w}(2)\cdots\underline{w}(N))^T$, $\mathbf{X}$ is the $N \times L$ matrix having n, l element $\xi(n + l + 1)$ and $\underline{\mathbf{v}} = (\underline{v}(1)\underline{v}(2)\cdots\underline{v}(N))^T$.

Consider again the model

$$\underline{\mathbf{w}} = \mathbf{X}\boldsymbol{\beta} + \underline{\mathbf{v}}$$

For the moment drop the assumption that $\underline{v}$ is normally distributed and that its covariance is known. Then least squares estimation of β from w can generally be described as minimizing the least squares criterion

$$(w - Xb)^T \Omega (w - Xb)$$

with respect to b where Ω is any positive definite symmetric weighting matrix. The solution $\hat{b}$ for b which minimizes the above criterion satisfies

$$\hat{b} = (X^T \Omega X)^{-1} X^T \Omega w$$

For a proof see Eykhoff (1974). This reference also discusses a number of useful properties of the estimator $\hat{\underline{b}}$ which may be summarized as follows:

(a) $\hat{\underline{b}}$ is closed-form and unique.
(b) $\hat{\underline{b}}$ is linear in the observations $\underline{w}$.
(c) $\hat{\underline{b}}$ is an unbiased estimator for β.
(d) $\text{cov}(\underline{b}, \underline{b}) = (X^T \Omega X)^{-1} X^T \Omega V \Omega X (X^T \Omega X)^{-1}$.
(e) If $\Omega = V^{-1}$ then $\text{cov}(\underline{b}, \underline{b}) = (X^T V^{-1} X)^{-1}$.
(f) If $\Omega = V^{-1}$, $\hat{\underline{b}}$ has smallest covariance among all unbiased estimators of β linear in $\underline{w}$. Therefore, in this particular case, $\hat{\underline{b}}$ is usually referred to as *best linear unbiased estimator* of β. Note that, loosely speaking, this particular choice of weighting matrix implies that in the least squares criterion the differences between the model Xb and the observations w are weighted according to the precision of the latter.
(g) If $\underline{v}$ is normally distributed with covariance V, the covariance $(X^T V^{-1} X)^{-1}$ of the estimator $\hat{\underline{b}} = (X^T V^{-1} X)^{-1} X^T V^{-1} \underline{w}$ is equal to the MVB for any number of observations.

If, in practice, the covariance of the errors is unknown, the identity matrix is often chosen as weighting matrix. This is usually referred to as *uniformly weighting*. Then the resulting least squares estimator is described by

$$\hat{\underline{b}} = (X^T X)^{-1} X^T \underline{w} \tag{8.3}$$

From the above properties it is concluded that generally this estimator has no optimal properties. It is, however, unbiased. If the covariance of the errors is known it is worthwhile to weight according to its inverse, since this yields the most precise estimator among all unbiased estimators linear in the observations. Note, however, that there may be more precise non-linear estimators or biased ones. Generally, the MVB is not achieved. On the other hand, in the special case that the errors are normally distributed, the best linear unbiased estimator coincides with the maximum likelihood estimator and, in addition, achieves the MVB for any number of observations. These considerations illustrate how *a priori* knowledge about the errors can be exploited to select an estimator or to improve the precision.

The numerical evaluation of the least squares estimator should be carried out with care, since the set of linear equations to be solved is often ill-conditioned, that is, the equations are nearly dependent. It may, therefore, be advisable to use algorithms that have especially been developed for linear least squares. For a discussion of these algorithms and corresponding ALGOL programs refer to Wilkinson and Reinsch (1971). Special programs for linear least squares are also included in scientific program libraries such as NAG (1978) and IMSL (1977).

Finally, it must be emphasized that the theory and the practical aspects of linear least squares discussed here are only a fraction of the material available. For most extensive discussions the reader is referred to Draper and Smith (1966) and Seber (1977). Fedorov (1972) is specialized to optimal design of linear least squares experiments.

Recursive linear least squares

The least squares estimate (equation (8.3)) computed from N observations may be written in the form

$$\hat{\mathbf{b}}_N = \mathbf{P}_N \mathbf{X}_N^T \mathbf{w}_N$$

where

$$\mathbf{P}_N = (\mathbf{X}_N^T \mathbf{X}_N)^{-1} \quad \text{and} \quad \mathbf{w}_N = (w_1 \cdots w_N)^T$$

with

$$\mathbf{X}_N = (\mathbf{x}_1 \vdots \cdots \vdots \mathbf{x}_N)^T \quad \text{and} \quad \mathbf{x}_n^T = (x_{n1} \cdots x_{nL}).$$

Now suppose that one additional observation w_{N+1}, $\mathbf{x}_{N+1}$ is made. Then it can be shown that $\hat{\mathbf{b}}_{N+1}$ satisfies

$$\hat{\mathbf{b}}_{N+1} = \hat{\mathbf{b}}_N + \mathbf{g}_{N+1}(w_{N+1} - \mathbf{x}_{N+1}^T \hat{\mathbf{b}}_N) \tag{8.4}$$

where

$$\mathbf{g}_{N+1} = \mathbf{P}_N \mathbf{x}_{N+1}/(1 + \mathbf{x}_{N+1}^T \mathbf{P}_N \mathbf{x}_{N+1}) \tag{8.5}$$

while

$$\mathbf{P}_{N+1} = \mathbf{P}_N - \mathbf{g}_{N+1}\mathbf{x}_{N+1}^T \mathbf{P}_N \tag{8.6}$$

For a proof see Goodwin and Payne (1977). Using these expressions the computation of $\hat{\mathbf{b}}_{N+1}$ from $\hat{\mathbf{b}}_N$, w_{N+1}, $\mathbf{x}_{N+1}$, and $\mathbf{P}_N$ may be carried out as follows. First the $L \times 1$ vector $\mathbf{g}_{N+1}$ is computed from $\mathbf{P}_N$ and $\mathbf{x}_{N+1}$. Note that in the expression for $\mathbf{g}_{N+1}$ the quantity $\mathbf{x}_{N+1}^T \mathbf{P}_N \mathbf{x}_{N+1}$ is scalar. The computation of $\hat{\mathbf{b}}_{N+1}$ from $\hat{\mathbf{b}}_N, \mathbf{g}_{N+1}, w_{N+1}$ and $\mathbf{x}_{N+1}$ is then straightforward. Finally, $\mathbf{P}_{N+1}$, which is required in the next recursion, is computed from $\mathbf{x}_{N+1}, \mathbf{g}_{N+1}$, and $\mathbf{P}_N$. Initial conditions $\mathbf{P}_{N_0}$ and $\hat{\mathbf{b}}_{N_0}$ can be non-recursively computed from the first N_0 observations. With this particular choice of initial conditions the recursive estimate $\hat{\mathbf{b}}_N$ for any N is equal to the estimate that would have been obtained non-recursively from the same observations. Furthermore, it is observed that in the expression for $\hat{\mathbf{b}}_{N+1}$ the difference $w_{N+1} - \mathbf{x}_{N+1}^T \hat{\mathbf{b}}_N$ is the deviation of the observation w_{N+1} from a predicted value based on the current estimate $\hat{\mathbf{b}}_N$ and $\mathbf{x}_{N+1}^T$, while $\mathbf{g}_{N+1}$ is a varying vector of weights determining to what extent this deviation is taken into account to modify $\hat{\mathbf{b}}_N$.

Recursive least squares may be particularly useful when the number of available observations is relatively large. The estimates may then be computed recursively for an increasing number of observations until they have satisfactorily converged. Thus often only a fraction of total number of available observations need be taken into account to meet the objectives of the measurements. Consequently computation time is saved.

A further application of recursive least squares is *parameter tracking* that is, estimation of time varying parameters. For that purpose a number of modified versions of the above algorithm have been developed. For a comprehensive description of these algorithms see Goodwin and Payne (1977). A characteristic example of a tracking algorithm will now be discussed.

Example 8.9   Exponential forgetting. In this example it will be supposed that the Nth observation w_N, $\mathbf{x}_N$ is made at time N, that is, at an integral multiple of a fixed unit time interval. Furthermore, the following least squares criterion will be chosen

$$(\mathbf{w}_N - \mathbf{X}_N \mathbf{b})^T \mathbf{\Omega}_N (\mathbf{w}_N - \mathbf{X}_N \mathbf{b})$$

where $\mathbf{\Omega}_N = \mathrm{diag}(\eta^{N-1}, \eta^{N-2}, \ldots, \eta, 1)$ with $0 < \eta \leqslant 1$. Thus in this criterion the weights of the quadratic deviations of the model from past observations are exponentially decreasing with time. Then the least squares estimator of $\boldsymbol{\beta}$ is described by

$$\hat{\mathbf{b}}_N = \mathbf{P}_N^T \mathbf{X}_N \mathbf{\Omega}_N \mathbf{w}_N$$

where $\mathbf{P}_N = (\mathbf{X}_N^T \mathbf{\Omega}_N \mathbf{X}_N)^{-1}$. This estimate satisfies the following recursive equations:

$$\hat{\mathbf{b}}_{N+1} = \hat{\mathbf{b}}_N + \mathbf{g}_{N+1}(w_{N+1} - \mathbf{x}_{N+1}^T \mathbf{b}_N)$$

where

$$\mathbf{g}_{N+1} = \mathbf{P}_N \mathbf{x}_{N+1}/(\eta + \mathbf{x}_{N+1}^T \mathbf{P}_N \mathbf{x}_{N+1})$$

while

$$\mathbf{P}_{N+1} = (\mathbf{P}_N - \mathbf{g}_{N+1} \mathbf{x}_{N+1}^T \mathbf{P}_N)/\eta$$

Note that substitution of $\eta = 1$ in these equations yields the original recursive least squares scheme. Then all past observations fully contribute to the current estimate. On the other hand, if η is small, only a relatively small number of past observations effectively contribute to the estimate. So for small η the estimate responds quickly to parameter changes, but is relatively imprecise. Consequently, in practice, the choice of η depends on the objectives of the measurement and is always a compromise between response rate and precision.

8.3.3 Non-linear Least Squares

Suppose that the problem is the estimation of the parameters θ from observations $\underline{w} = (\underline{w}_1 \cdots \underline{w}_N)^T$ described by

$$\underline{w}_n = g(\mathbf{x}_n; \theta) + \underline{v}_n \quad n = 1, \ldots, N \tag{8.7}$$

where the $\mathbf{x}_n$ are exactly known independent variables, the $\underline{v}_n$ are stochastic variables representing non-systematic errors and $g(\mathbf{x}; \theta)$ is a known function which, as opposed to the functions considered in Section 8.3.2, is *non-linear* in one or more elements of the vector of unknown parameters θ. For this model the weighted least squares estimate of θ is defined as that particular value $\hat{\mathbf{t}}$ of $\mathbf{t}$ that minimizes

$$J(\mathbf{t}) = (\mathbf{w} - \mathbf{g})^T \Omega (\mathbf{w} - \mathbf{g})$$

where $\mathbf{g} = (g(\mathbf{x}_1; \mathbf{t}) \cdots g(\mathbf{x}_N; \mathbf{t}))^T$ and Ω is a symmetric, positive definite weighting matrix.

With respect to this non-linear least squares estimator the following remarks can be made:

(a) If $\underline{v} = (\underline{v}_1 \cdots \underline{v}_N)^T$ is normally distributed with expectation zero and known covariance $\mathbf{V}$ then the least squares estimate $\tilde{\mathbf{t}}$ which minimizes

$$J(\mathbf{t}) = (\mathbf{w} - \mathbf{g})^T \mathbf{V}^{-1} (\mathbf{w} - \mathbf{g})$$

is the maximum likelihood estimate of θ (Section 8.3.1). The estimator $\tilde{\mathbf{t}}$ is, therefore, generally consistent and converges in distribution to a normal distribution with expectation θ and covariance $\mathbf{M}^{-1}$ where $\mathbf{M}$ is the information matrix. Moreover, it is easily shown that

$$\mathbf{M} = (\partial \mathbf{g}/\partial \mathbf{t})^T \mathbf{V}^{-1} (\partial \mathbf{g}/\partial \mathbf{t})$$

Again it is emphasized that consistency and convergence in distribution are asymptotic properties.

Example 8.10 Measurement of radioactive decay and concentration. The mathematico-physical model of radioactive particle count $\underline{w}_n$ at time x_n may be described by

$$\underline{w}_n = \sum_k \lambda_k \exp(-\mu_k x_n) + \underline{v}_n$$

In these expressions $\theta = (\lambda_1 \cdots \lambda_K \, \mu_1 \cdots \mu_K)^T$ where λ_k is a measure of the concentration and μ_k is the decay constant of the kth component. Furthermore, the counting result $\underline{w}_n$, is supposed to be a Poisson distributed stochastic variable. This implies that $E(\underline{w}_n) = \text{var}(\underline{w}_n)$. Hence, if $E(\underline{w}_n) \gg [E(\underline{w}_n)]^{1/2}$, $E(\underline{w}_n) \approx w_n$ and $\text{var}(\underline{w}_n) \approx w_n$. Then, by the central limit theorem, $\underline{w}_n$ is asymptotically $N(E(\underline{w}_n), [E(\underline{w}_n)]^{1/2})$. Supposing that $\text{cov}(\underline{w}_{n_1}, \underline{w}_{n_2}) = 0$ for $n_1 \neq n_2$, one obtains $\mathbf{V} \approx \text{diag}(w_1, \ldots, w_N)$. Hence minimizing the non-linear least squares criterion with the inverse of this $\mathbf{V}$ as weighting matrix asymptotically approximates the maximum likelihood estimator. This again illustrates the use of *a priori* knowledge to enhance the precision of an estimator.

(b) In practice, if $\underline{v}$ is supposed to be normally distributed with unknown covariance, $\mathbf{I}$ is often taken as weighting matrix $\mathbf{\Omega}$. The resulting estimator is not a maximum likelihood estimator unless $\mathbf{V} = \sigma^2 \mathbf{I}$.

(c) If $\underline{v}$ has known covariance $\mathbf{V}$ but is not normally distributed, non-linear least squares with $\mathbf{\Omega} = \mathbf{V}^{-1}$ is often used. The idea is again to give smallest weight to the largest deviations of the model from the observations. The estimator so obtained is generally not a maximum likelihood estimator.

(d) If the distribution and covariance of $\underline{v}$ are unknown, it is common practice to use least squares with $\mathbf{\Omega} = \mathbf{I}$. Again this is generally not a maximum likelihood estimator.

(e) Suppose that in equation (8.7) the $\underline{v}_n$ are iid with covariance $\mathbf{V} = \sigma_v^2 \mathbf{I}$. Then the estimator $\hat{\underline{t}}$ of θ which minimizes

$$J(\underline{t}) = (\mathbf{w} - \mathbf{g})^T (\mathbf{w} - \mathbf{g})$$

is, under very general conditions, asymptotically normally distributed with expectation θ and covariance $\mathbf{U}$ defined by

$$\mathbf{U} = \sigma_v^2 [(\partial \mathbf{g}/\partial \theta)^T (\partial \mathbf{g}/\partial \theta)]^{-1}$$

This result is due to Jennrich (1969). Note that if $\underline{v}$ is idd and normal, $\mathbf{U}$ is the MVB, while this is generally not true if $\underline{v}$ is non-normal. This result is a useful tool to investigate the precision of the uniformly weighted least squares estimator for idd errors and a large number of observations. The properties of non-linear least squares estimators for a relatively small number of observations may essentially deviate from those in the asymptotic case. For illustrative examples of this phenomenon see Van den Bos (1979).

(f) It is easily shown that

$$\text{grad } J(\mathbf{t}) = -2(\partial \mathbf{g}/\partial \mathbf{t})^T \mathbf{\Omega}(\mathbf{w} - \mathbf{g})$$

Since the elements of $\mathbf{g}$ are non-linear in $\mathbf{t}$ so are those of the gradient. This implies that there is generally no closed form solution for the equations grad $J(\hat{t}) = 0$ which are necessary conditions for a minimum of $J(t)$ at $t = \hat{t}$. Hence these equations must be solved numerically. Therefore in Section 8.3.4 a short discussion will be given of a number of practical numerical minimization techniques.

8.3.4 Numerical Minimization

In what follows three different methods for numerical minimization are briefly discussed. For a more complete discussion of these methods and many others the reader is referred to Murray (1972), which also includes a chapter on numerical non-linear least squares.

The first method to be described here is the *steepest-descent method*. This is a general minimization method. It has not especially been designed for non-linear least squares. The steepest descent method changes the current value t_c of the parameter vector by an amount

$$\Delta t_{SD} = -\Delta \text{ grad}[J(t_c)]/\|\text{grad } J(t_c)\|$$

where, since the gradient is normalized, Δ is the step size. So Δt_{SD} is opposite to the gradient. It can be shown that infinitesimally in this direction $J(t)$ at $t = t_c$ decreases most. The steepest-descent procedure in its most primitive form may be summarized as follows:

(a) Compute Δt_{SD} in $t = t_c$.
(b) Compute $J(t_c + \Delta t_{SD})$. If $J(t_c + \Delta t_{SD}) \leqslant J(t_c)$ select $t_c + \Delta t_{SD}$ as new t_c and repeat from (a). If not, reduce Δ and repeat (b).

Important properties of the steepest descent method are:

(a) Under very general conditions convergence to a minimum is guaranteed.
(b) Δt_{SD} is perpendicular to a contour of $J(t)$.

For non-linear least squares problems the latter property has some unfavourable consequences. In most non-linear least squares problems the minima are located in elongated, curved valleys, in the literature frequently referred to as *banana-shaped valleys*. Consequently, as a result of property (b), in the neighbourhood of a minimum the method proceeds along a zigzag path and converges extremely slowly. Far from the minimum, however, the method progresses rapidly.

The slow convergence of the steepest descent method near the minimum can be avoided by using alternative methods of which the *Gauss–Newton method*, a special method for non-linear least squares, will now be discussed as an example. Recall that the problem is minimizing

$$J(t) = (w - g)^T(w - g)$$

where $\mathbf{w} = (w_1 \cdots w_N)^T$ and $\mathbf{g} = (g(\mathbf{x}_1; \mathbf{t}) \cdots g(\mathbf{x}_N; \mathbf{t}))^T$. In the Gauss–Newton method $g(\mathbf{x}_n; \mathbf{t})$ is, in the neighborhood of $\mathbf{t} = \mathbf{t}_c$, approximated by its first-order Taylor expansion. Then

$$\mathbf{g}|_{t = t_c + \Delta t} \approx \mathbf{g}|_{t = t_c} + \mathbf{X}\Delta t$$

where $\mathbf{X} = \partial \mathbf{g}/\partial \mathbf{t}$ at $\mathbf{t} = \mathbf{t}_c$ and $\Delta \mathbf{t}$ is a small increment of $\mathbf{t}$. This expression, which is linear in $\Delta \mathbf{t}$, is substituted for $\mathbf{g}$ in the above expression for $J(\mathbf{t})$. The resulting approximate criterion, $\bar{J}(\mathbf{t})$, is subsequently minimized with respect to $\Delta \mathbf{t}$. This is a standard linear least squares problem which has the closed form solution

$$\Delta \mathbf{t}_{GN} = -\tfrac{1}{2}(\mathbf{X}^T\mathbf{X})^{-1} \text{ grad } J(\mathbf{t})|_{t = t_c}$$

where use has been made of the expression for the gradient described in Section 8.3.3. For what follows it is important to note that the contours of $\bar{J}(\mathbf{t}_c + \Delta \mathbf{t})$ are ellipsoids. In its most elementary form the Gauss–Newton iteration scheme may now be described as follows:

(a)   Compute $\Delta \mathbf{t}_{GN}$ from $\partial \mathbf{g}/\partial \mathbf{t}$ and $(\mathbf{w} - \mathbf{g})$ at $\mathbf{t} = \mathbf{t}_c$.
(b)   Select $\mathbf{t}_c + \Delta \mathbf{t}_{GN}$ as new $\mathbf{t}_c$ and repeat from (a).

Important properties of the Gauss–Newton method are:

(a)   As a rule the method diverges far from the minimum.
(b)   Far from the minimum the direction of the Gauss–Newton step is usually very close to the direction of the contour.
(c)   Close to the minimum the method converges rapidly.

The last property is a consequence of the fact that as the minimum is approached the contours of the criterion become ellipsoids and can, therefore, increasingly accurately be approximated by the quadratic criterion $\bar{J}(\mathbf{t}_c + \Delta \mathbf{t})$. Comparing the properties of the steepest-descent method with those of the Gauss–Newton method suggests combining both methods into one single method, which gradually changes from the steepest-descent direction to the Gauss–Newton direction as the minimum is approached and which, in addition, controls its step size. This is what is done by the *Marquardt method* described in Marquardt (1963). This method may be summarized as follows. In $\mathbf{t} = \mathbf{t}_c$ the Marquardt step $\Delta \mathbf{t}_{MQ}$ is defined by

$$(\mathbf{X}^T\mathbf{X} + \lambda \mathbf{I})\Delta \mathbf{t}_{MQ} = -\tfrac{1}{2} \text{ grad } J(\mathbf{t}_c)$$

where, as before, $\mathbf{X} = \partial \mathbf{g}/\partial \mathbf{t}$ at $\mathbf{t} = \mathbf{t}_c$ and λ is a positive scalar to be specified later. If for a particular λ this set of equations is solved for $\Delta \mathbf{t}_{MQ}$ the resulting solution can be shown to minimize $\bar{J}(\mathbf{t}_c + \Delta \mathbf{t})$ on a sphere $\|\Delta \mathbf{t}\|^2 = \|\Delta \mathbf{t}_{MQ}\|^2$. From the definition of the Marquardt step it follows that $\Delta \mathbf{t}_{MQ}$ approaches $\Delta \mathbf{t}_{GN}$ as λ approaches zero. On the other hand if λ is increased, $\Delta \mathbf{t}_{MQ}$ approaches $\Delta \mathbf{t}_{SD}$ while the step size goes to zero. Furthermore, it can be shown that both

$\|\Delta\mathbf{t}_{MQ}\|$ and the angle between $\Delta\mathbf{t}_{SD}$ and $\Delta\mathbf{t}_{MQ}$ are decreasing functions of λ. Hence $\Delta\mathbf{t}_{MQ}$ rotates towards $\Delta\mathbf{t}_{SD}$ as λ is increased.

The Marquardt iteration scheme may be summarized as follows. Let $v > 1$ and define $\mathbf{t}_c$ and λ_c as the current $\mathbf{t}$ and λ respectively. Then an iteration consists of the following steps:

(a)   Compute $\Delta\mathbf{t}_{MQ}$ for $\lambda = \lambda_1 = \lambda_c$ and for $\lambda = \lambda_2 = \lambda_c/v$ respectively. Next compute the corresponding values of $J(\mathbf{t}_c + \Delta\mathbf{t}_{MQ})$.

(b)   If $J(\mathbf{t}_c + \Delta\mathbf{t}_{MQ})|_{\lambda_2} \leqslant J(\mathbf{t}_c)$, select λ_2 and $\mathbf{t}_c + \Delta\mathbf{t}_{MQ}$ as new λ_c and $\mathbf{t}_c$ and repeat from (a). Otherwise proceed with (c).

(c)   If $J(\mathbf{t}_c + \Delta\mathbf{t}_{MQ})|_{\lambda_2} > J(\mathbf{t}_c)$ and $J(\mathbf{t}_c + \Delta\mathbf{t}_{MQ})|_{\lambda_1} \leqslant J(\mathbf{t}_c)$, select λ_1 and the corresponding quantity $\mathbf{t}_c + \Delta\mathbf{t}_{MQ}$ as new λ_c and $\mathbf{t}_c$ and repeat from (a). Otherwise proceed with (d).

(d)   Take $v\lambda_c$ as new λ_c and repeat from (a).

Note that, within restrictions imposed by convergence requirements, λ is kept as small as possible so as to retain as much as possible of the favourable convergence rate of the Gauss–Newton method.

Important properties of the Marquardt method are:

(a)   Under general conditions convergence to minimum is guaranteed.
(b)   Rapid convergence in the neighbourhood of the minimum.

Nowadays the Marquardt method, which works very well in practice, is included in a number of scientific program libraries (see e.g. NAG, 1978; IMSL, 1977).

8.4   ESTIMATION OF PARAMETERS OF DYNAMIC DIFFERENCE OR DIFFERENTIAL EQUATION MODELS

8.4.1   Introduction

This section discusses estimation of parameters of models of dynamic systems or stochastic processes. The class of models is restricted to ordinary difference or differential equations since these frequently occur and naturally arise in practical problems. Within this class one is still free to make a choice between difference of differential equation descriptions and with respect to the order of the right-hand and left-hand members of the equation. Again the decision between the alternatives is closely connected to the purpose of the estimation procedure. For example, for automatic control purposes an accurate description of the response of the system to an arbitrary input may be sufficient. In this case the difference or differential equation model need not be an accurate description of the *physical* system involved. The same applies to difference equation models of stochastic processes to be discussed below. In this case the purpose is often a compact quantitative description of the second-order properties of the stochastic process in the form of a small number of parameters. On the other hand, if the purpose is estimation of *physical* parameters, the differential equation model should accurately describe the relevant part of physical reality and follows directly from physical considerations, for example, conservation laws.

8.4.2 Discrete-time Difference Equation Models

First a number of relevant properties of discrete-time stochastic processes are summarized. Let $\underline{x}(n\Delta)$ and $\underline{y}(n\Delta)$, $n = 0, \pm 1, \pm 2, \ldots$, be stationary, discrete-time stochastic processes where Δ is a constant time interval. To simplify the notation in what follows the time scale will be chosen so that $\Delta = 1$. Then the *cross-covariance function* of $\underline{x}(n)$ and $\underline{y}(n)$ is defined as:

$$C_{xy}(k) = E\{[\underline{x}(n) - \mu_x][\underline{y}(n + k) - \mu_y]\}$$

where $\mu_x = E[\underline{x}(n)]$ and $\mu_y = E[\underline{y}(n)]$, while the *autocovariance function* of $\underline{x}(n)$ is defined by:

$$C_{xx}(k) = E\{[\underline{x}(n) - \mu_x][\underline{x}(n + k) - \mu_x]\}$$

Furthermore, the *cross-power-density spectrum* $S_{xy}(\omega)$ of $\underline{x}(n)$ and $\underline{y}(n)$ is defined as the infinite discrete Fourier transform of $C_{xy}(k)$:

$$S_{xy}(\omega) = \sum_{k=-\infty}^{\infty} C_{xy}(k)\exp(-j\omega k)$$

Note that $S_{xy}(\omega)$ is periodic with 2π. The corresponding inverse Fourier transform is defined as the sequence of Fourier coefficients $C_{xy}(k)$, $k = 0, \pm 1, \pm 2, \ldots$, of $S_{xy}(\omega)$:

$$C_{xy}(k) = \frac{1}{2\pi} \int_{-\pi}^{\pi} S_{xy}(\omega)\exp(jk\omega)\, d\omega$$

$S_{xx}(\omega)$ is referred to as the power-density spectrum or simply power spectrum of $\underline{x}(n)$.

An important relation between the power spectra of input processes and corresponding response processes of linear discrete-time systems concludes this summary. Let the response of a linear discrete-time system to a unit impulse input applied at time $k = 0$ be $\ldots, 0, 0, \hbar(0), \hbar(1), \hbar(2), \ldots$. This sequence defines the *discrete-time impulse response* of the system. Then the *discrete-time frequency response function* $\mathcal{H}(\omega)$ of the system is defined as the infinite discrete Fourier transform of the impulse response, that is,

$$\mathcal{H}(\omega) = \sum_{k=-\infty}^{\infty} \hbar(k)\exp(-j\omega k)$$

Next suppose that the stationary stochastic process $\underline{u}(n)$, $n = 0, \pm 1, \pm 2, \ldots$, is the input of a linear discrete-time system with frequency response function $\mathscr{H}(\omega)$. Furthermore, let $\underline{y}(n)$ be the response to $\underline{u}(n)$, $n = 0, \pm 1, \pm 2, \ldots$. Then it is not difficult to show that the relation between the power spectrum $S_{uu}(\omega)$ of $\underline{u}(n)$ and the power spectrum $S_{yy}(\omega)$ of $\underline{y}(n)$ is described by

$$S_{yy}(\omega) = |\mathscr{H}(\omega)|^2 S_{uu}(\omega)$$

If the autocovariance function of $\underline{u}(n)$ has the particular form

$$C_{uu}(k) = \sigma_u^2 \delta_{0,k},$$

where the Kronecker delta $\delta_{m,n}$ is given by

$$\delta_{m,n} = \begin{cases} 1 & \text{if } m = n \\ 0 & \text{otherwise} \end{cases}$$

$\underline{u}(n)$ is usually referred to as an *uncorrelated* or white process. Then the corresponding power spectrum satisfies

$$S_{uu}(\omega) = \sigma_u^2.$$

Hence, if $\underline{u}(n)$ is a white input process the power spectrum of the response is described by

$$S_{yy}(\omega) = |\mathscr{H}(\omega)|^2 \sigma_u^2$$

So in this particular case the shape of $S_{yy}(\omega)$ is completely determined by the frequency response of the system. This result will be used extensively throughout this section.

A model frequently used in practice to describe linear dynamic discrete-time transformations of an arbitrary stochastic process $\underline{u}(n)$ into a process $\underline{y}(n)$ is the following:

$$\underline{y}(n) + \alpha_1 \underline{y}(n - 1) + \cdots + \alpha_K \underline{y}(n - K)$$
$$= \beta_0 \underline{u}(n) + \beta_1 \underline{u}(n - 1) + \cdots + \beta_L \underline{u}(n - L)$$

In the literature this linear difference equation model is usually referred to as *autoregressive-moving average* (ARMA) *model*. If $\beta_1 = \cdots = \beta_L = 0$ the model is called *autoregressive* (AR), *all-pole* or *linear prediction* model. If $\alpha_1 = \cdots = \alpha_K = 0$ the model is referred to as *moving average* (MA) *model*. It is easily shown that the frequency response of an ARMA model is described by

$$\mathscr{H}(\omega) = \frac{\beta_0 + \beta_1 \exp(-j\omega) + \cdots + \beta_L \exp(-j\omega L)}{1 + \alpha_1 \exp(-j\omega) + \cdots + \alpha_K \exp(-j\omega K)}$$

Hence, if $\underline{y}(n)$ and $\underline{u}(n)$ are stationary

$$S_{yy}(\omega) = \left| \frac{\beta_0 + \beta_1 \exp(-j\omega) + \cdots + \beta_L \exp(-j\omega L)}{1 + \alpha_1 \exp(-j\omega) + \cdots + \alpha_K \exp(-j\omega K)} \right|^2 S_{uu}(\omega)$$

Now suppose that $\underline{u}(n)$ is uncorrelated. So $S_{uu}(\omega) = \sigma_u^2$. Then the expression for $S_{yy}(\omega)$ shows that the power density spectrum of any discrete-time, stationary stochastic process $\underline{y}(n)$ can be described as, or approximated by, the power-density spectrum of the response process of an ARMA model with a sufficiently large number of suitably chosen coefficients to an uncorrelated discrete-time stationary stochastic process $\underline{u}(n)$. Thus estimation of ARMA parameters from observations $\underline{y}(1), \ldots, \underline{y}(N)$ is a parametric alternative to classical non-parametric Fourier techniques as, for example, described by Jenkins and Watts (1968). The question now arises how the coefficients of the ARMA model can be estimated from observations of $\underline{y}(n)$. In what follows this will only be discussed for AR models since these are much more frequently applied in practice than ARMA models. One of the reasons is that for the AR parameters relatively simple closed form estimators are available while estimators for ARMA parameters require an iterative solution.

An extensive survey on estimation of parameters of AR models and related subjects including many references is Makhoul (1975). Haykin (1979) discusses estimation of parameters of both AR and ARMA models. The maximum likelihood estimator of the AR parameters discussed in Section 8.4.3 is due to Mann and Wald (1943).

8.4.3 Estimation of the Parameters of the Autoregressive Model

Suppose $\underline{y}(n)$ is a discrete time stationary stochastic process. For the moment also suppose that $\underline{y}(n)$ is normally distributed. Furthermore, assume that the following autoregressive model for the $\underline{y}(n)$ is chosen:

$$\underline{y}(n) + \alpha_1 \underline{y}(n-1) + \cdots + \alpha_K \underline{y}(n-K) = \underline{e}(n)$$

where the $\underline{e}(n)$ are idd and $N(0, \sigma_e^2)$. Then the maximum likelihood estimate $\tilde{\mathbf{t}} = (\tilde{a}_1 \cdots \tilde{a}_K \tilde{s}_e^2)^T$ of the parameters $\boldsymbol{\theta} = (\alpha_1 \cdots \alpha_K \sigma_e^2)^T$ from the $N + K$ observations $y(1-K), y(2-K), \ldots, y(N)$ may be computed as follows. In the first place it can be shown that $\tilde{a}_1, \ldots, \tilde{a}_K$ satisfy

$$c(0, k) + \tilde{a}_1 c(1, k) + \cdots + \tilde{a}_K c(K, k) = 0 \quad k = 1, \ldots, K \qquad (8.8)$$

and

$$\tilde{s}_e^2 = c(0, 0) + \tilde{a}_1 c(1, 0) + \cdots + \tilde{a}_K c(K, 0)$$

where

$$c(k_1, k_2) = \frac{1}{N} \sum_{n=1}^{N} y(n - k_1) y(n - k_2)$$

So the numerical procedure could consist of computing the $c(k_1, k_2)$ for k_1, $k_2 = 0, \ldots, K$, solving equation (8.8) for $\tilde{a}_1, \ldots, \tilde{a}_K$, and using this solution to compute s_e^2. However, it is observed that under very general conditions all $c(k_1, k_2)$ with $|k_1 - k_2| = k$ are consistent estimators of $C_{yy}(k)$. They are, therefore, asymptotically equivalent and are in practice often replaced by $c(k, 0)$. Then the above estimator may be approximated by

$$
\begin{pmatrix}
c(0) & c(1) & \cdot & \cdot & c(K-1) \\
c(1) & c(0) & \cdot & & \vdots \\
\vdots & & \vdots & c(0) & c(1) \\
c(K-1) & \cdot & \cdot & c(1) & c(0)
\end{pmatrix}
\begin{pmatrix}
\tilde{a}_1 \\
\vdots \\
\tilde{a}_K
\end{pmatrix}
= -
\begin{pmatrix}
c(1) \\
\vdots \\
c(K)
\end{pmatrix}
$$

where $c(k) = c(k, 0)$. Note that all elements on a particular north-west, south-east diagonal of the coefficient matrix are the same. It will be seen later on that this special structure can be exploited to reduce the number of operations required for the solution of the set.

Finally, it is easily shown that the MVB for unbiased estimation of $\theta = (\alpha_1 \cdots \alpha_K \sigma_e^2)^T$ is described by

$$
\frac{\sigma_e^2}{N}
\begin{pmatrix}
C_{yy}(0) & C_{yy}(1) & \cdot & \cdot & C_{yy}(K-1) & 0 \\
C_{yy}(1) & C_{yy}(0) & \cdot & & \vdots & 0 \\
\vdots & & \cdot & C_{yy}(0) & C_{yy}(1) & 0 \\
C_{yy}(K-1) & & \cdot & C_{yy}(1) & C_{yy}(0) & 0 \\
0 & 0 & \cdot & 0 & 0 & 2
\end{pmatrix}^{-1}
$$

For large N the covariance of $\tilde{t}$ can be estimated by replacing the $C_{yy}(k)$ and and σ_e^2 in this expression by their estimates.

The above results apply to normally distributed $\varrho(n)$. In some respects, however, they are more generally applicable. In the first place for idd, but not necessarily normal, $\varrho(n)$ the following relations are valid:

$$C_{yy}(k) + \alpha_1 C_{yy}(k-1) + \cdots + \alpha_K C_{yy}(k-K) = 0 \quad \text{where } k \geq 1$$

In the literature these relations are usually called *Yule–Walker equations*. Now suppose that α is estimated by replacing the $C_{yy}(k)$ in the first K Yule–Walker equations by their estimates $c(k)$ and solving these equations, Then it can be shown that the resulting estimator, the expression for which coincides with that for the approximate maximum likelihood estimator in the normal case, is asymptotically normal with an asymptotic covariance described by the same expression as the MVB in the normal case. Note, however, that in the non-normal case this asymptotic covariance matrix is generally not the MVB. For a discussion of a general class of estimators, called *prediction error estimators*, having these properties the reader is referred to Chapter 5 of Goodwin and Payne (1977).

Durbin's method for the computation of autoregressive coefficients

The first K estimated Yule–Walker equations and the equation for the estimated variance of the $\underline{e}(n)$ for an autoregressive model of order K are respectively

$$c(k) + \tilde{a}_1^{[K]}c(k - 1) + \cdots + \tilde{a}_K^{[K]}c(k - K) = 0 \quad k = 1, \ldots, K$$

and

$$\tilde{s}_e^{2[K]} = c(0) + \tilde{a}_1^{[K]}c(1) + \cdots + \tilde{a}_K^{[K]}c(K)$$

Using a method due to Durbin (1960) one can exploit the special structure of these equations to reduce the number of numerical operations. In Durbin's method $\tilde{a}_1^{[K]}, \ldots, \tilde{a}_K^{[K]}, \tilde{s}_e^{2[K]}$ and $c(K + 1)$ are used to compute $\tilde{a}_1^{[K + 1]}, \ldots, \tilde{a}_{K+1}^{[K + 1]}$, and $\tilde{s}_e^{2[K + 1]}$. So the current coefficient and variance estimates are used to compute the corresponding quantities for the next higher order model. This is continued until in the sense of some stopping criterion no further improvement is achieved by increasing the model order. The equations used in Durbin's method are as follows

$$\tilde{a}_{K+1}^{[K + 1]} = -[c(K + 1) + \tilde{a}_1^{[K]}c(K) + \cdots + \tilde{a}_K^{[K]}c(1)]/\tilde{s}_e^{2[K]}$$

$$\tilde{a}_k^{[K + 1]} = \tilde{a}_k^{[K]} + \tilde{a}_{K+1}^{[K + 1]}\tilde{a}_{K+1-k}^{[K]} \quad k = 1, \ldots, K$$

$$\tilde{s}_e^{2[K + 1]} = [1 - (\tilde{a}_{K+1}^{[K + 1]})^2]\tilde{s}_e^{2[K]}$$

where $K = 0, 1, \ldots$, while the initial conditions are described by $a_0^{[0]} = 0$ and $\tilde{s}_e^{2[0]} = c(0)$. From these equations it follows that the number of multiplications in one step of Durbin's method is roughly $2K$. Hence, computation of the coefficient estimates for all models of order up to and including K requires approximately K^2 multiplications. This is substantially less than with conventional methods. For example, Gauss elimination requires $K^3/3 + O(K^2)$ multiplications for a model of order K alone. Choleski decomposition, specialized to solution of symmetric, positive definite sets of equations uses $K^3/6 + O(K^2)$ multiplications for the same purpose. So with these methods computation of the coefficients of all models of order up to and including K requires a number of multiplications of order K^4. This is, of course, only a fair comparison if the estimates of the coefficients of all lower-order models are actually needed. Furthermore, whether or not a substantial reduction in the *total* computation time is achieved by Durbin's method, also depends on the number of observations taken into consideration since this determines the number of multiplications to be carried out for the computation of the required $c(k)$. For a large number of observations the latter number of multiplications may considerably exceed the number of multiplications required for solution of the Durbin equations. On the other hand, if the observations have been quantized by an analog-to-digital converter, as is often the case in practice, the $c(k)$ may be computed in fixed point. Fixed point operations are usually much faster than floating point operations, as are required for the solution of the Durbin equations. In some applications, for example, in geophysics, estimation problems are found giving rise to sets of equations similar to the Yule–Walker equations but with an arbitrary right-hand member. That is, the right-hand member does not necessarily consist of $-c(1), \ldots, -c(K)$. Then use can still be made of the particular structure of the left-hand member by using a method due to Levinson. This method requires twice as many multiplications as Durbin's method and is described by Robinson (1967).

Recursive estimation of the autoregressive coefficients

A recursive form for the estimator

$$\tilde{a}_1 c(1, k) + \cdots + \tilde{a}_K c(K, k) = -c(0, k) \quad k = 1, \ldots, K$$

discussed earlier in this section, can be obtained as follows. Let $\tilde{a}_N$ be the vector $(\tilde{a}_1 \cdots \tilde{a}_K)^T$ computed from $y(1 - K), \ y(2 - K), \ldots, y(0), \ y(1), \ldots, y(N)$. Furthermore, define

$$
\mathbf{w}_N = \begin{pmatrix} y(1) \\ y(2) \\ \vdots \\ y(N) \end{pmatrix} \quad
\mathbf{X}_N = - \begin{pmatrix} y(0) & y(-1) & \cdots & y(-K+1) \\ y(1) & y(0) & \cdots & y(-K+2) \\ \vdots & \vdots & & \vdots \\ y(N-1) & y(N-2) & \cdots & y(N-K) \end{pmatrix} = \begin{pmatrix} \mathbf{x}_1^T \\ \mathbf{x}_2^T \\ \vdots \\ \mathbf{x}_N^T \end{pmatrix}
$$

$$\mathbf{P}_N = (\mathbf{X}_N^T \mathbf{X}_N)^{-1} \quad \text{and} \quad w_N = y(N)$$

Then $\tilde{a}_N$ may equivalently be written

$$\tilde{a}_N^T = \mathbf{P}_N \mathbf{X}_N^T \mathbf{w}_N$$

Hence, using the recursive least squares formulae of Section 8.3.2 one obtains

$$\tilde{a}_{N+1} = \tilde{a}_N + \mathbf{g}_{N+1}(w_{N+1} - \mathbf{x}_{N+1}^T \tilde{a}_N)$$

where $w_{N+1} = y(N + 1)$ and $\mathbf{x}_{N+1}^T = -(y(N) \cdots y(N - K + 1))$. The vector $\mathbf{g}_{N+1}$ in this expression is computed from $\mathbf{P}_N$ and $\mathbf{x}_{N+1}$ while $\mathbf{P}_N$ is recursively computed from $\mathbf{P}_{N+1}$ and $\mathbf{x}_N$, as described in Section 8.3.2.

8.4.4 Estimation of Parameters of Dynamic Systems from Input–Output Observations

The difference equation models discussed up to now were models of stochastic processes. These stochastic processes were considered to be exactly measured responses of a system described by the difference equation to an unmeasured white input process. The problem to be discussed now is estimation of the parameters of a linear difference equation model from observations of both the input

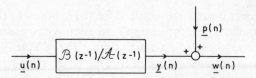

Figure 8.1 System model having stochastic input $\underline{u}(n)$ to the transfer function and non-systematic errors input, $\underline{p}(n)$

and the corresponding response. In addition, it will be supposed that the observations of the input are exact while the observations at the output consist of the sum of the true response and a stochastic process representing non-systematic errors. In measurement practice this model is frequently found to adequately describe observations of input and response of dynamic systems. The case that both the observations of the input and those of the response contain additive errors will be discussed later on.

The model to be studied first is shown in Figure 8.1. In this figure $\underline{u}(n)$ is the input which, for the moment, will be assumed to be a stationary, zero mean, stochastic process. The response to $\underline{u}(n)$ is $\underline{y}(n)$ which is also assumed to be a stationary stochastic process. This implies that the system is stable and that any transient responses to $\underline{u}(n)$ have disappeared. The process $\underline{p}(n)$ represents the non-systematic errors in the response observations $\underline{w}(n)$. Finally, $\mathscr{B}(z^{-1})/\mathscr{A}(z^{-1})$ represents the system's discrete-time transfer function to be specified below.

Now suppose that the system is described by the linear difference equation

$$\underline{y}(n) + \alpha_1 \underline{y}(n-1) + \cdots + \alpha_K \underline{y}(n-K)$$
$$= \beta_0 \underline{u}(n) + \beta_1 \underline{u}(n-1) + \cdots + \beta_M \underline{u}(n-M)$$

with $K \geqslant M$. Then the problem studied here will be estimating

$$\boldsymbol{\theta} = (\alpha_1 \cdots \alpha_K \, \beta_0 \cdots \beta_M)^{\mathrm{T}}$$

from input–output observations $w(1-K)$, $u(1-K)$, $w(2-K)$, $u(2-K)$, ..., $w(N)$, $u(N)$.

For convenience first the above difference equation is rewritten in the form

$$\mathscr{A}(z^{-1})\underline{y}(n) = \mathscr{B}(z^{-1})\underline{u}(n)$$

where

$$\mathscr{A}(z^{-1}) = 1 + \alpha_1 z^{-1} + \cdots + \alpha_K z^{-K}$$
$$\mathscr{B}(z^{-1}) = \beta_0 + \beta_1 z^{-1} + \cdots + \beta_M z^{-M}$$

while z is the forward shift operator defined by $zy(n) = y(n+1)$ or $z^{-1}y(n+1) = y(n)$. Then it follows from Figure 8.1 that

$$\mathscr{A}(z^{-1})\underline{w}(n) = \mathscr{B}(z^{-1})\underline{u}(n) + \underline{q}(n)$$

where

$$q(n) = \mathcal{A}(z^{-1})p(n) \tag{8.9}$$

This model suggests selecting as an estimator for θ that value $\hat{t}$ of

$$\mathbf{t} = (a_1 \cdots a_K \, b_0 \cdots b_M)^{\mathrm{T}}$$

that minimizes the least squares criterion

$$J(t) = \sum_{n=1}^{N} [A(z^{-1})w(n) - B(z^{-1})u(n)]^2$$

where $A(z^{-1}) = 1 + a_1 z^{-1} + \cdots + a_K z^{-K}$ and $B(z^{-1}) = b_0 + b_1 z^{-1} + \cdots + b_M z^{-M}$. The motivation for this choice is, of course, that $\hat{t}$ is a simple, closed form, linear least squares estimator. Unfortunately, it can be shown that for $N \to \infty$ the difference $\hat{t} - \theta$ only converges to zero if the covariance function $C_{pp}(k)$ of $p(n)$ satisfies

$$\mathcal{A}(z^{-1})C_{pp}(k) = 0 \quad k = 1, \ldots, K$$

For a proof for first-order $\mathcal{A}(z^{-1})$ see Goodwin and Payne (1977). This proof can easily be generalized to the above result. Recalling that $\mathcal{A}(z^{-1})p(n) = q(n)$, one observes that the above conditions for convergence are equivalent to the Yule–Walker equations if $q(n)$ is white. Conversely, it can also be shown that under the above conditions $q(n)$ is necessarily white. Hence a necessary and sufficient condition for the above least squares estimator to converge is that $q(n)$ is white. Note that this is *not* equivalent to the statement that $\hat{t}$ converges to θ if the errors in the observations of the response are white. It is clear that in most cases $q(n)$ will not be white and $\hat{t}$ will consequently be a biased estimator of θ. Therefore the use of $\hat{t}$ is generally not advisable, the more so as asymptotically unbiased alternatives for $\hat{t}$ are available which will now be discussed.

The generalized least squares method

Suppose that in the above model of the observations

$$\mathcal{A}(z^{-1})w(n) = \mathcal{B}(z^{-1})u(n) + q(n)$$

the process $q(n)$ is modelled as an autoregressive process

$$q(n) = \mathcal{G}^{-1}(z^{-1})e(n) \tag{8.10}$$

where $\mathcal{G}(z^{-1}) = 1 + \gamma_1 z^{-1} + \cdots + \gamma_L z^{-L}$ and $e(n)$ is white. Then

$$\mathcal{A}(z^{-1})w'(n) - \mathcal{B}(z^{-1})u'(n) = e(n)$$

where $\underline{w}'(n) = \mathscr{G}(z^{-1})\underline{w}(n)$ and $\underline{u}'(n) = \mathscr{G}(z^{-1})\underline{u}(n)$. Then, since $\underline{e}(n)$ is white, the value $\bar{\mathbf{t}}$ of $\mathbf{t}$ that minimizes

$$J(\mathbf{t}) = \sum_{n=1}^{N} [A(z^{-1})w'(n) - B(z^{-1})u'(n)]^2$$

is an asymptotically unbiased estimate of $\boldsymbol{\theta}$. Unfortunately, $\mathscr{G}(z^{-1})$ is usually unknown. To see this, suppose that the errors $\underline{p}(n)$ satisfy $\underline{p}(n) = \mathscr{D}^{-1}(z^{-1})\underline{e}(n)$. Then $\underline{q}(n) = \mathscr{D}^{-1}(z^{-1})\mathscr{A}(z^{-1})\underline{e}(n) = \mathscr{G}^{-1}(z^{-1})\underline{e}(n)$. Hence,

$$\mathscr{G}(z^{-1}) = \mathscr{D}(z^{-1})\mathscr{A}^{-1}(z^{-1}).$$

In this expression neither $\mathscr{D}(z^{-1})$, which represents the dynamic properties of the errors, nor $\mathscr{A}(z^{-1})$, which must be estimated, is known. The solution chosen in the generalized least squares method, due to Clarke (1967), is to estimate the parameters $\boldsymbol{\gamma} = (\gamma_1 \cdots \gamma_L)^{\mathrm{T}}$ of $\mathscr{G}(z^{-1})$ along with $\boldsymbol{\theta}$. For this estimation problem the generalized least squares method uses an iterative scheme which may be described as follows.

Suppose that the estimate of $\boldsymbol{\gamma} = (\gamma_1 \cdots \gamma_L)^{\mathrm{T}}$ obtained in the Ith iteration is $\hat{\mathbf{g}}^{[I]} = (\hat{g}_1^{[I]} \cdots \hat{g}_L^{[I]})^{\mathrm{T}}$ and that the polynomial $\mathscr{G}(z^{-1})$ having these coefficients is $\hat{G}^{[I]}(z^{-1})$. Then the $(I + 1)$th iteration consists of the following steps:

(a) Compute $w'(n) = \hat{G}^{[I]}(z^{-1})w(n)$ and $u'(n) = \hat{G}^{[I]}(z^{-1})u(n)$ for $n = 1, \ldots, N$.

(b) Minimize

$$J(\mathbf{t}) = \sum_{n=K+1}^{N} [A(z^{-1})w'(n) - B(z^{-1})u'(n)]^2$$

with respect to $\mathbf{t}$. Let the result be $\hat{\mathbf{t}} = (\hat{a}_1 \cdots \hat{a}_K \, \hat{b}_0 \cdots \hat{b}_M)^{\mathrm{T}}$.

(c) Compute $\hat{q}(n) = \hat{A}(z^{-1})w(n) - \hat{B}(z^{-1})u(n)$ for $n = 1, \ldots, N$, where $\hat{A}(z^{-1})$ and $\hat{B}(z^{-1})$ have been obtained by substituting $\hat{\mathbf{t}}$ in $A(z^{-1})$ and $B(z^{-1})$.

(d) Minimize

$$\sum_{n=L+1}^{N} [G(z^{-1})\hat{q}(n)]^2$$

with respect to $\mathbf{g} = (g_1 \cdots g_L)^{\mathrm{T}}$. Take the result as

$$\hat{\mathbf{g}}^{[I+1]} = (\hat{g}_1^{[I+1]} \cdots \hat{g}_K^{[I+1]})^{\mathrm{T}}$$

(e) Stop if convergence has been achieved. If not, repeat from (a). As initial condition $\hat{G}^{[0]}(z^{-1}) = 1$ may be used.

With respect to the generalized least squares method the following remarks can be made. In the first place this method is known to work satisfactorily in practice. Furthermore, as is shown by equations (8.9) and (8.10), the parameters characterizing the dynamic properties of the errors are estimated along with the parameters of the system. This may, for example, be of importance for automatic control applications. Finally, it is observed for normally distributed $\underline{e}(n)$ the solution is the maximum likelihood estimate. For other approaches to maximum likelihood estimation of the parameters of linear discrete-time dynamic systems, including the multi-input multi-output case, the reader is referred to Goodwin and Payne (1977). This reference also describes a recursive version of the generalized least squares method.

A further simple, straightforward method for estimation of parameters of linear discrete-time dynamic systems is the following.

The instrumental variable method

Suppose that the system is described by

$$y(n) + \alpha_1 y(n - 1) + \cdots + \alpha_K y(n - K)$$
$$= \beta_0 u(n) + \beta_1 u(n - 1) + \cdots + \beta_M u(n - M)$$

and that the problem is to estimate $\theta = (\alpha_1 \cdots \alpha_K \beta_0 \cdots \beta_M)^T$ from observations $v(1 - K), w(1 - K), \ldots, v(1), w(1), \ldots, v(N), w(N)$ where

$$w(n) = y(n) + p(n) \quad \text{and} \quad v(n) = u(n) + q(n)$$

where $q(n)$ and $p(n)$ are stochastic processes representing possibly correlated non-systematic errors in the observations of the input and the corresponding response respectively. Furthermore, suppose that in addition to these observations $K + M + 1$ further sequences of observations are available described by

$$f_1(1), \ldots, f_1(N)$$
$$f_2(1), \ldots, f_2(N)$$
$$\vdots \qquad \vdots$$
$$f_{K+M+1}(1), \ldots, f_{K+M+1}(N)$$

where the $f_l(n)$ are stochastic processes convariant with $y(n)$ and $u(n)$ but *not* with $p(n)$ and $q(n)$. The $f_l(n)$, $n = 1, \ldots, N$ will from now on be referred to as *instrumental sequences*. Multiplying the left-hand and the right-hand members of the system difference equation by the instrumental sequences and taking expectations yields

$$C_{yf_l}(0) + \alpha_1 C_{yf_l}(1) + \cdots + \alpha_K C_{yf_l}(K)$$
$$= \beta_0 C_{uf_l}(0) + \cdots + \beta_M C_{uf_l}(M) \quad l = 1, \ldots, K + M + 1$$

where the $C_{yf_l}(k)$ and the $C_{uf_l}(m)$ are the cross-covariance functions of $y(n)$ and $f_l(n)$ and of $u(n)$ and $f_l(n)$ respectively. So, if these exact cross-covariance functions were available the unknown system parameters could be computed by solving the set of linear equations

$$C_{yf_l}(0) + a_1 C_{yf_l}(1) + \cdots + a_K C_{yf_l}(K)$$
$$= b_0 C_{uf_l}(0) + \cdots + b_M C_{uf_l}(M) \quad l = 1, \ldots, K + M + 1 \quad (8.11)$$

for $a_1, \ldots, a_K, b_0, \ldots, b_M$. However, the exact cross-covariance functions are not known. Since $\underline{p}(n)$ and $\underline{q}(n)$ are not covariant with the $f_l(n)$, estimates of the system parameters can then be computed by replacing the covariances in equation (8.11) by the corresponding mean lagged product estimates

$$\hat{c}_{yf_l}(k) = \frac{1}{N} \sum_{n=1}^{N} w(n - k) f_l(n)$$

and

$$\hat{c}_{uf_l}(m) = \frac{1}{N} \sum_{n=1}^{N} v(n - m) f_l(n)$$

and subsequently solving the resulting set of linear equations

$$\hat{c}_{yf_l}(0) + \hat{a}_1 \hat{c}_{yf_l}(1) + \cdots + \hat{a}_K \hat{c}_{yf_l}(K)$$
$$= \hat{b}_0 \hat{c}_{uf_l}(0) + \cdots + \hat{b}_M \hat{c}_{uf_l}(M) \quad l = 1, \ldots, K + M + 1$$

for $\hat{\mathbf{t}} = (\hat{a}_1 \cdots \hat{a}_K \hat{b}_0 \cdots \hat{b}_M)^{\mathrm{T}}$. The estimator $\hat{\mathbf{t}}$ is usually referred to as *instrumental variable estimator*. As is shown in Jenkins and Watts (1968) mean lagged product estimators are under very general conditions consistent. Furthermore, a continuous function of a consistent estimator of a parameter is a consistent estimator of the function of that parameter (see Wilks, 1962). It is concluded that the instrumental variable estimator is consistent under very general conditions as well.

The above considerations show that computationally the instrumental variable estimator is very attractive. It is closed form, no iterative procedures are required and consequently convergence problems are avoided. On the other hand, generally the instrumental variable estimator has no optimal statistical properties, and does not estimate the parameters of the errors. Furthermore, the applicability of the estimator crucially depends on the availability of the required instrumental sequences. However, if one instrumental sequence has been found, the remaining instrumental sequences required may be chosen as time shifted versions of the first one. For example, if $f_1(n)$ is available the remaining instrumental sequences may be taken as $f_k(n) = f_1(n - k + 1)$, $k = 2, 3, \ldots$. This is illustrated in Examples 8.11 and 8.12.

Example 8.11   Open loop estimation using instrumental variables.   In the estimation problem of Figure 8.1, time shifted versions of the input $\underline{u}(n)$ may be taken as instrumental sequences if $\underline{u}(n)$ and $\underline{p}(n)$ are not covariant. Also time-shifted linearly filtered versions of the input are chosen. On the basis of the available knowledge about the dynamic properties of the system under investigation the filter is chosen so as to approximate these properties. The purpose is to ensure a relatively strong covariance between the observed response and the instrumental sequences. A recursive instrumental variable algorithm implementing this idea is described in Chapter 7 of Goodwin and Payne (1977).

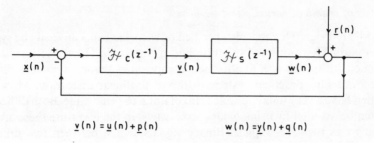

$$\underline{v}(n) = \underline{u}(n) + \underline{p}(n) \qquad \underline{w}(n) = \underline{y}(n) + \underline{q}(n)$$

Figure 8.2 Model used for estimation of a closed-loop control system

Example 8.12 Closed loop estimation using instrumental variables. Suppose that the problem is the estimation of the parameters of the transfer function $\mathscr{H}_s(z^{-1})$ in Figure 8.2. In this figure $\mathscr{H}_c(z^{-1})$ is the transfer function of a controller, $\underline{x}(n)$ is an input process at the set point, while $\underline{r}(n)$ is an additive stochastic process at the system output equivalent to all disturbances introduced anywhere in the loop. Furthermore, $\underline{u}(n)$ is the component of $\underline{v}(n)$ generated by $\underline{x}(n)$ while $\underline{y}(n)$ is the response of the system to $\underline{u}(n)$. The processes $\underline{p}(n)$ and $\underline{q}(n)$ result from $\underline{r}(n)$. It is assumed that $\underline{x}(n)$ and $\underline{r}(n)$ are independent. Now suppose that observations of $\underline{v}(n)$, $\underline{w}(n)$ and $\underline{x}(n)$ are available. Then shifted versions of the observations of $\underline{v}(n)$ may not be taken as instrumental sequences since $\underline{v}(n)$ is covariant with $\underline{r}(n)$, which is the part of $\underline{w}(n)$ not causally related to $\underline{v}(n)$ and which, therefore, represents the errors in the measured response. Shifted versions of $\underline{x}(n)$, however, are feasible instrumental sequences since these are independent from both $\underline{p}(n)$ and $\underline{q}(n)$.

8.4.5 Test Signals

In physical practice, the input $\underline{u}(n)$ is either a sampled version of a continuous time normal operating process or it is a sampled version of a continuous time test signal, that is, an externally generated process that is deliberately introduced into the system for estimation purposes. Test signals have a number of advantages over normal operating inputs. In the first place test signals need not be measured, they are accurately known and repeatable. Furthermore, it is usually fully justified to assume that they are independent of all other processes present in the system. Moreover, within physical limits their power spectrum and shape as a function of time can be freely selected. The synthesis of test signals will now briefly be discussed. This discussion will be restricted to periodic test signals since the most important practical test signals used in system parameter estimation belong to this class.

Synthesis of practical periodic test signals

A periodic test signal having any specified power spectrum can easily be synthesized by adding harmonics of suitably chosen amplitude. However, the resulting harmonic sum signals are, due to their complicated shape, often very difficult to introduce into practical systems without nonlinear distortion. Moreover, they often have a very unfavourable ratio of r.m.s. to peak value. Both difficulties can often be avoided by using *binary test signals*. In the literature these are also referred to as *two-level signals*. Binary signals switch between two different, fixed amplitude levels only. In addition, to facilitate generation using a general-purpose or special-purpose digital computer, practical test signals are preferably discrete interval signals, that is, they only change amplitude at integral multiples of a fixed time interval. Periodic, discrete-interval, binary signals combine the above desirable properties and will, therefore, now briefly be discussed.

The best known example of a periodic, discrete-interval, binary test signal is the *maximum length binary sequence*. These signals are extremely simple to generate and have, in addition, spectral properties which make them attractive for a variety of applications. The generation of maximum length binary sequences may be described as follows. The modulo-2 sum of the content of the last stage and that of one or more of the other stages of a binary shift register are fed back to the input of the first stage. For certain suitably chosen feedback combinations the sequence formed by the successive content of a particular stage repeats itself after $N = 2^K - 1$ shift pulses where K is the number of stages. It can be shown that this is the maximum achievable sequence length with modulo-2 sum feedback; hence the name. The periodic discrete-interval version of the maximum length binary sequence is obtained by applying the shift pulses equidistantly and taking the content of a particular stage, which is constant during the shift pulse interval, as output of the generator. Appropriate feedback combinations are tabulated in Peterson (1961). For a further discussion see Hoffmann de Visme (1971) and Davies (1970). The spectral properties of maximum length binary sequences may be summarized as follows. Define the complex Fourier coefficient γ_{ku} of the kth harmonic of a signal $u(t)$ periodic with T as

$$\gamma_{ku} = \frac{1}{T} \int_0^T u(t)\exp(-j2\pi kt/T) \, dt$$

Then it can be shown that for any maximum length binary sequence having amplitude levels $+1$ and -1

$$|\gamma_{ku}|^2 = \left(1 + \frac{1}{N}\right)\left(\frac{\sin(\pi k/N)}{\pi k/N}\right)^2 \quad k \neq 0$$

and

$$|\gamma_{0u}|^2 = 1/N^2$$

(see Hoffmann de Visme, 1971). So for any maximum length binary sequence the envelope of the power spectrum is similar. It can also be shown that the relatively flat part of this spectrum around the origin contains the major part of the total power. The latter property makes maximum length binary sequences very suitable to, more or less uniformly, cover the bandwidth of the system under test.

As will be discussed later in this section, the power spectrum of the test signal influences the precision of an estimator of the parameters of a dynamic system. So to improve the precision, the use of a discrete-interval binary test signal having a power spectrum specified by the user and generally different from that of a maximum length binary sequence may be preferred. The same applies if certain estimation methods, specialized to periodic test signals and discussed below, are used. It may then be advantageous to employ periodic test signals that have the major part of their power concentrated in a specified way in a small number of relatively widely spaced harmonics. Periodic binary signals having this property are usually referred to as *binary multifrequency signals*. A simple method for synthesis of periodic, discrete-interval, binary signals, multifrequency signals in particular, having approximately a given power spectrum is described in Van den Bos and Krol (1979). This method minimizes in the least squares sense the difference between the Fourier amplitude spectrum of the signal and a specified amplitude spectrum. Finally, it is worthwhile to mention that the computation of the Fourier coefficients of a periodic discrete-interval signal from the Fourier coefficients of discrete time, that is, sampled versions and the converse is particularly simple and is described in Van den Bos and Krol (1979).

Estimation using periodic inputs

The discrete-time estimation methods to be briefly discussed now are only applicable if the test signal is periodic. The complex Fourier coefficients γ_{ix} of a discrete-time signal $x(n)$ periodic with I are defined by

$$\gamma_{ix} = \frac{1}{I} \sum_{n=0}^{I-1} x(n) \exp(-j2\pi in/I) \quad i = 0, \ldots, I - 1$$

For an extensive discussion of the theory of discrete Fourier analysis, which in many respects is analogous to its continuous time counterpart, the reader is referred to Oppenheim and Schafer (1975). Chapter 5 provides a review.

Now consider again the estimation problem of Example 8.12 and suppose that $\underline{x}(n)$ is periodic with I. Then $\underline{u}(n)$ and $\underline{y}(n)$ are also periodic with I. In addition, assume, as is suggested in Example 8.12 that the required instrumental sequences are chosen as $\underline{x}(n - l + 1)$, $l = 1, \ldots, K + M + 1$. Then using elementary properties of discrete Fourier coefficients one can easily show that the equations (8.11) are equivalent to

$$\sum_{i=0}^{I-1} [A(s_i^{-1})\gamma_{iy}^* - B(s_i^{-1})\gamma_{iu}^*]s_i^{-l}\gamma_{ix} = 0 \tag{8.12}$$

where $s_i = \exp(j2\pi i/N)$ and $l = 1, \ldots, K + M + 1$. These are linear equations in the unknown parameters. They could be solved if the exact Fourier coefficients γ_{iu} and γ_{iy}, $i = 0, \ldots, I - 1$, were available, which is not the case. However, by replacing the γ_{iu} and γ_{iy} in equation (8.12) by corresponding estimates and subsequently solving, estimates of the system parameters are obtained. Convenient estimators of γ_{iy} and γ_{iu} are

$$\hat{\gamma}_{iy} = \frac{1}{N}\sum_{n=0}^{N-1} \underline{w}(n)\exp(-j2\pi in/I)$$

and $\hat{\gamma}_{iu}$, defined correspondingly. In this expression N is supposed to be an integral multiple of the period I. Then under very general conditions with respect to $\underline{p}(n)$ and $\underline{q}(n)$, $\hat{\gamma}_{iy}$ and $\hat{\gamma}_{iu}$ are consistent (See Levin 1959). Since the proposed estimator of the system coefficients is a continuous function of the $\hat{\gamma}_{iu}$ and $\hat{\gamma}_{iy}$, it is consistent if the $\hat{\gamma}_{iu}$ and $\hat{\gamma}_{iy}$ are.

A related method is the following. Consider the least squares criterion

$$J(\mathbf{t}) = \sum_{i=0}^{I-1} |A(s_i^{-1})\gamma_{iy} - B(s_i^{-1})\gamma_{iu}|^2$$

where $\mathbf{t} = (a_1 \cdots a_K b_0 \cdots b_M)^{\mathrm{T}}$. Then equating the gradient of $J(\mathbf{t})$ to zero and subsequently using some elementary properties of discrete Fourier coefficients yields the following necessary conditions for a minimum:

$$\sum_{i=0}^{I-1} [A(s_i)\gamma_{iy}^* - B(s_i)\gamma_{iu}^*]s_i^{-k}\gamma_{iy} = 0 \quad k = 1, \ldots, K \tag{8.13}$$

and

$$\sum_{i=0}^{I-1} [A(s_i)\gamma_{iy}^* - B(s_i)\gamma_{iu}^*]s_i^{-m}\gamma_{iu} = 0 \quad m = 0, 1, \ldots, M \tag{8.14}$$

Since the exact Fourier coefficients γ_{iu} and γ_{iy} satisfy

$$\mathscr{A}(s_i^{-1})\gamma_{iy} - \mathscr{B}(s_i^{-1})\gamma_{iu} = 0$$

it follows that at $\mathbf{t} = \mathbf{\theta} J(\mathbf{t})$ is zero. Moreover, since $J(\mathbf{t})$ is quadratic in $\mathbf{t}$, $\mathbf{t} = \mathbf{\theta}$ is a unique minimum. Now, if, as above, the γ_{iu} and γ_{iy} are replaced by corresponding consistent estimators, equations (8.13) and (8.14) become computationally simple, closed form, consistent estimators of the system parameters. Moreover, comparing equation (8.12) with equations (8.13) and (8.14) one observes that the latter estimator employs shifted versions of the estimated periodic component of input and response as instrumental sequences in equations (8.13) and (8.14) respectively.

Both methods described have the advantage that the Bode diagram can directly be estimated from the estimated Fourier coefficients. This is useful as a check of the feasibility of the observations. Also, if observations are missing or erroneous, the total loss of observations may be restricted to the loss of one period only. Furthermore, for suitably chosen I the computation of $\hat{y}_{iu}$ and $\hat{y}_{iy}$ from $v(n)$ and $w(n)$ can very efficiently be carried out using a fast Fourier transform (FFT) algorithm. FFT algorithms are described in Oppenheim and Schafer (1975). In this volume the FFT is discussed in Chapter 5. On the other hand, if periodic test signals are used having all or the major part of their power concentrated in a small number of harmonics, both the amount of data and the computation time may considerably be reduced if only the Fourier coefficients corresponding to these dominant harmonics are computed and included in the estimators, equations (8.12) or (8.13) and (8.14). Finally, as opposed to virtually all other discrete-time estimation methods, these periodic test signal methods have continuous time counterparts, that is, methods for estimation of parameters of *differential* equations. An example of such a continuous time method will briefly be discussed in Section 8.4.7.

Precision of dynamic system parameter estimation

This section discusses the MVB for estimation of parameters of linear discrete-time systems. In particular the dependence of the MVB on the input will be studied. This study will be restricted to the estimation problem of Figure 8.3. In this figure the stationary stochastic process $\underline{y}(n)$ is the response of the system to the stationary input process $\underline{u}(n)$. Furthermore, it is supposed that the response observations $\underline{w}(n)$ are the sum of $\underline{y}(n)$ and a stationary stochastic process $\underline{v}(n)$ which represents non-systematic errors. The process $\underline{v}(n)$ is modelled as the stationary response of a system having a linear transfer function to an iid, $N(0, \sigma_e)$ process $\underline{e}(n)$. Now assume that in Figure 8.3

$$\mathscr{H}_s(z^{-1}) = \mathscr{B}(z^{-1})/\mathscr{A}(z^{-1}) \quad \text{and} \quad \mathscr{H}_D(z^{-1}) = \mathscr{G}(z^{-1})/\mathscr{D}(z^{-1})$$

where

$$\mathscr{A}(z^{-1}) = 1 + \alpha_1 z^{-1} + \cdots + \alpha_K z^{-K}$$

$$\mathscr{B}(z^{-1}) = \beta_0 + \beta_1 z^{-1} + \cdots + \beta_M z^{-M}$$

$$\mathscr{D}(z^{-1}) = 1 + \delta_1 z^{-1} + \cdots + \delta_{K'} z^{-K'}$$

$$\mathscr{G}(z^{-1}) = 1 + \gamma_1 z^{-1} + \cdots + \gamma_{M'} z^{-M'}$$

Now suppose that $\boldsymbol{\theta}_S = (\alpha_1 \cdots \alpha_K \, \beta_0 \cdots \beta_M)^T$ is estimated from observations $\underline{u}(1), \underline{w}(1), \ldots, \underline{u}(N), \underline{w}(N)$ and that, in addition, the error parameters $\boldsymbol{\theta}_D = (\delta_1 \cdots \delta_{K'} \, \gamma_1 \cdots \gamma_{M'} \, \sigma_e^2)^T$ are unknown. Then the subject of this section is the MVB for unbiased estimation of $\boldsymbol{\theta}_S$.

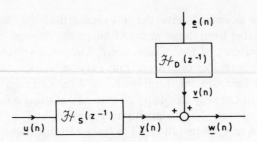

Figure 8.3   Model used for study of dependence
of MVB on the input

The information matrix $\mathbf{I}$ for the problem considered is defined by

$$\mathbf{I} = -E(\partial^2 \ln \underline{L}/\partial \boldsymbol{\theta}^2)$$

where $\boldsymbol{\theta} = (\boldsymbol{\theta}_S^T \mathop{\vdots} \boldsymbol{\theta}_D^T)^T$ and $\underline{L}$ is the likelihood function of the observations $\underline{w}(1), \ldots, \underline{w}(N)$. $\underline{L}$ can be computed from the probability density function of $\underline{e}(1), \ldots, \underline{e}(N)$ defined by

$$f_{\underline{e}}(\boldsymbol{\varepsilon}) = (2\pi)^{-N/2} \sigma_e^{-N} \exp\left(-\tfrac{1}{2}\sigma_e^{-2} \sum_{n=1}^N \varepsilon_n^2\right)$$

and the relation

$$\underline{e}(n) = \mathcal{H}_D^{-1}(z^{-1})[\underline{w}(n) - \mathcal{H}_S(z^{-1})\underline{u}(n)]$$

which follows from Figure 8.3. Then, as is shown in Goodwin and Payne (1977), the information matrix computed from this likelihood function assumes the following form

$$\mathbf{I} = \begin{pmatrix} \mathbf{I}_S & \vdots & \mathbf{0} \\ \cdots & + & \cdots \\ \mathbf{0} & \vdots & \mathbf{I}_D \end{pmatrix}$$

where

$$\mathbf{I}_S = -E(\partial^2 \ln \underline{L}/\partial \boldsymbol{\theta}_S^2) \quad \text{and} \quad \mathbf{I}_D = -E(\partial^2 \ln \underline{L}/\partial \boldsymbol{\theta}_D^2)$$

Note that $\mathbf{I}_S$ is the information matrix for estimation of $\boldsymbol{\theta}_S$ if $\boldsymbol{\theta}_D$ is known. Then the MVB for unbiased estimation of $\boldsymbol{\theta}$ becomes

$$\mathbf{I}^{-1} = \begin{pmatrix} \mathbf{I}_S^{-1} & \vdots & \mathbf{0} \\ \cdots & + & \cdots \\ \mathbf{0} & \vdots & \mathbf{I}_D^{-1} \end{pmatrix}$$

Hence, for the model of system and errors concerned, the MVB for the system parameters $\boldsymbol{\theta}_S$ does not depend on whether or not the error parameters $\boldsymbol{\theta}_D$

are known. Therefore, in what follows only I_S will be studied. Goodwin and Payne (1977) show that I_S may be put in the form

$$I_S = N \int_{-\pi}^{\pi} J(\omega) S_{uu}(\omega) \, d\omega \qquad (8.15)$$

where the elements of the matrix $J(\omega)$ are functions of the frequency and are parametric in θ_S and θ_D while $S_{uu}(\omega)$ is the power density spectrum of the input $\underline{u}(n)$. Then the feasibility of an experiment for the pertinent parameter estimation objectives can be investigated by computing the MVB from equation (8.15) for given $S_{uu}(\omega)$ and nominal values of θ_S and θ_D. Also, if $\underline{u}(n)$ is a test signal, the achievable precision with different test signals under the same circumstances may be computed and subsequently compared. The question may then arise how to design an optimal test signal, that is, a test signal the power spectrum $S_{uu}(\omega)$ of which optimizes a chosen criterion defined on the elements of the MVB. This problem is extensively discussed in Goodwin and Payne (1977). Here only some important aspects of optimal test signal design will be reviewed.

Design of optimal test signals

Equation (8.15) shows that the elements of the MVB can be made arbitrarily small by increasing $S_{uu}(\omega)$. In practice, however, the amplitude or the power of the test signal is always restricted. Here it will be assumed that the input power σ_u^2 satisfies the constraint

$$\sigma_u^2 = \frac{1}{2\pi} \int_{-\pi}^{\pi} S_{uu}(\omega) \, d\omega = 1$$

Now consider the class of $L \times L$ information matrices described by the general form of equation (8.15). Furthermore, define two $S_{uu}(\omega)$ satisfying the above constraint as equivalent if the corresponding information matrices are equal. Then it can be shown that for any power spectrum $S_{uu}(\omega)$ there exists an equivalent line power spectrum $S'_{uu}(\omega)$ consisting of at most $\frac{1}{2}L(L + 1) + 1$ lines. Then $S'_{uu}(\omega)$ is of the form

$$S'_{uu}(\omega) = \sum_{i=-L'}^{L'} s_i \delta(\omega - \omega_i) \quad i \neq 0$$

where $s_i \delta(\omega - \omega_i)$ is a Dirac delta function of area s_i at $\omega = \omega_i$, $L' = \frac{1}{2}L(L + 1) + 1$ and $\omega_{-i} = -\omega_i$. So for any input one can find an equivalent input consisting of a finite number of sinusoids. It is emphasized that the above-mentioned number of lines is the maximum required number. It can often be substantially reduced by making use of special properties of the information matrix for the estimation problem at hand (see Chapter 6 of Goodwin and Payne, 1977). The importance of the above considerations for optimal test signal design is that one need only optimize the chosen criterion with respect to a relatively small number of variables ω_i and s_i. Frequently used criteria are the trace or the determinant of the MVB.

Often, in practice, *a priori* knowledge about the system and errors is available, being obtained from previous experiments or mathematico-physical analysis. Optimal test signal design techniques provide a means to exploit this knowledge to improve the precision. Further applications of these techniques may be establishing theoretically justifiable rules of thumb for test signal selection and comparing the performance of test signals frequently used in practice to what optimally can be achieved under the same circumstances.

8.4.6   Connections Between Differential and Difference Equation Models

This subsection discusses a number of aspects of estimating the parameters $\boldsymbol{\theta} = (\alpha_0 \cdots \alpha_{K-1} \, \beta_0 \cdots \beta_K)^{\mathrm{T}}$ of the differential equation model

$$\frac{d^K y(t)}{dt^K} + \alpha_{K-1} \frac{d^{K-1} y(t)}{dt^{K-1}} + \cdots + \alpha_0 y(t) = \beta_K \frac{d^K u(t)}{dt^K} + \cdots + \beta_0 u(t) \quad (8.16)$$

from observations of the input $u(t)$ and the response $y(t)$. This type of model directly results from continuous-time physical laws, for example, conservation laws. Consequently, the coefficients usually have a clear physical meaning. In applied and experimental physics it is often these coefficients or functions of them that are to be measured, as opposed to in automatic control applications where any model, perhaps discrete-time or of an order lower than that of the physical model, is feasible as long as it accurately describes the response to any input that can be expected.

Analysis of the discrete-time estimation methods described earlier in this chapter reveals that the continuous time analogs of these would all require numerical differentiation of the observations with the exception of the frequency domain methods discussed in Section 8.4.5. The continuous-time analog of one of the latter methods is described in Section 8.4.7. If the observations are in analog form, numerical differentiation without bandwidth limitation usually gives rise to unacceptable signal-to-noise ratios. On the other hand, bandwidth-limited differentiation is an approximative technique, the effect of which on the estimation results is difficult to trace. Also, differentiation schemes for discrete-time observations are essentially approximative. A further possibility, the use of hybrid computer methods for estimation of continuous-time system parameters, will not be discussed here. The interested reader is referred to a survey by Piceni and Eykhoff (1975). As compared with the number of applications of estimation methods using digital computers, nowadays the number of pertinent hybrid computer applications is relatively small.

From the above considerations it may be concluded that using discrete-time methods for estimation of parameters of continuous-time systems would be attractive. However, this requires a one-to-one relation between the discrete-time and continuous-time parameters. For discrete-interval inputs such a relation exists and will now briefly be discussed.

First the differential equation (8.16) is rewritten in a state-space form. The particular state-space representation chosen is

$$\mathbf{x}(t) = \mathbf{A}\mathbf{x}(t) + \mathbf{b}u(t)$$
$$y(t) = \mathbf{c}\mathbf{x}(t) + du(t)$$

(8.17)

where

$$\mathbf{A} = \begin{pmatrix} 0 & 0 & 0 & \cdots & 0 & -\alpha_0 \\ 1 & 0 & 0 & \cdots & 0 & -\alpha_1 \\ 0 & 1 & 0 & \cdots & 0 & -\alpha_2 \\ \vdots & \vdots & \vdots & & \vdots & \vdots \\ 0 & & & \cdots & 1 & -\alpha_{K-1} \end{pmatrix} \qquad \mathbf{b} = \begin{pmatrix} \beta_0 - \alpha_0\beta_K \\ \vdots \\ \beta_{K-1} - \alpha_{K-1}\beta_K \end{pmatrix}$$

$$\mathbf{c} = (0 \ \cdots \ 0 \ 1) \qquad\qquad d = \beta_K$$

and

$$\mathbf{x}(t) = (x_1(t) \cdots x_K(t))^{\mathrm{T}}$$

For a general discussion of the state-space description of linear systems see Kwakernaak and Sivan (1972). In this reference it is also shown that the solution of the above state-space equations is described by

$$\mathbf{x}(t) = \exp[\mathbf{A}(t - t_0)]\mathbf{x}(t_0) + \int_{t_0}^{t} \exp[\mathbf{A}(t - t_0)]\mathbf{b}u(t) \, dt$$

(8.18)

where $\mathbf{x}(t_0)$ is the vector of initial conditions and $\exp \mathbf{X}$ is defined as the Taylor expansion

$$\exp \mathbf{X} = \mathbf{I} + \mathbf{X}/1! + \mathbf{X}^2/2! + \cdots$$

Now suppose that $u(t)$ is discrete-interval with interval Δ, that is $u(t)$ is constant for $n\Delta < t < (n + 1)\Delta$ for all n, and define $u'(n)$ as the value of $u(t)$ on this interval. Furthermore, define $x'(n)$ as $x(n\Delta)$. It then follows from equation (8.18) that

$$\mathbf{x}'(n + 1) = \mathbf{A}'\mathbf{x}'(n) + \mathbf{b}'u'(n)$$
$$y'(n) = \mathbf{c}'\mathbf{x}'(n) + d'u'(n)$$

(8.19)

where

$$\mathbf{A}' = \exp(\mathbf{A}\Delta) \quad \mathbf{b}' = \mathbf{A}^{-1}[\mathbf{A}' - \mathbf{I}]\mathbf{b}$$
$$\mathbf{c}' = \mathbf{c} \qquad\qquad d' = d$$

(8.20)

For a proof see Kwakernaak and Sivan (1972). The discrete-time system (equation (18.19)) and the continuous-time system (equation (8.17)) are called *equivalent* systems since at $t = n\Delta$ their respective responses coincide.

The relations (equation (8.20)) show how the equivalent discrete-time state-space description can be computed from the continuous-time description. The solution of the converse problem will now be considered. For a more extensive discussion see Harris (1979). Suppose that the system under test is described by equation (8.16) and that the input $u(t)$ is discrete-interval. Furthermore, suppose that the parameters $\theta = (\alpha'_1 \cdots \alpha'_K \beta'_0 \cdots \beta'_K)^T$ of the equivalent difference equation

$$y'(n) + \alpha'_1 y'(n-1) + \cdots + \alpha'_K y'(n-K) = \beta'_0 u'(n) + \cdots + \beta'_K u'(n-K)$$

have been computed from samples $y'(n)$ and $u'(n)$ of $y(t)$ and $u(t)$ respectively. Then this difference equation is first rewritten in a state-space form. The particular form chosen here is

$$\mathbf{x}'(n+1) = \mathbf{A}'\mathbf{x}'(n) + \mathbf{b}'u'(n)$$

$$y(n) = \mathbf{c}'\mathbf{x}'(n) + d'u'(n)$$

where

$$\mathbf{A}' = \begin{pmatrix} 0 & 0 & 0 & \cdots & 0 & -\alpha'_K \\ 1 & 0 & 0 & \cdots & 0 & -\alpha'_{K-1} \\ 0 & 1 & 0 & \cdots & 0 & -\alpha'_{K-2} \\ \vdots & \vdots & \vdots & & \vdots & \vdots \\ 0 & & & \cdots & 1 & -\alpha'_1 \end{pmatrix} \quad \mathbf{b}' = \begin{pmatrix} \beta'_K - \alpha'_K \beta'_0 \\ \vdots \\ \beta'_1 - \alpha'_1 \beta'_0 \end{pmatrix}$$

$$\mathbf{c}' = (0 \quad \cdots \quad 0 \quad 1) \quad d' = \beta'_0$$

and

$$\mathbf{x}'(n) = (x'_1(n)x'_2(n) \cdots x'_K(n))^T$$

Now let the eigenvalues $\lambda'_1, \ldots, \lambda'_K$ of $\mathbf{A}'$ be distinct. For the case of multiple eigenvalues see Harris (1979). Furthermore, define s'_k as the eigenvector corresponding with λ'_k. Then it can be shown that the eigenvalues $\lambda_1, \ldots, \lambda_K$ of the $\mathbf{A}$ matrix of an equivalent continuous-time description satisfy

$$\lambda_k = \frac{1}{\Delta} \ln \lambda'_k \quad k = 1, \ldots, K \tag{8.21}$$

while

$$\mathbf{A} = \mathbf{S}'\mathbf{\Lambda}\mathbf{S}'^{-1} \tag{8.22}$$

where $\mathbf{\Lambda} = \text{diag}(\lambda_1, \ldots, \lambda_K)$ and $\mathbf{S}'$ is the matrix having the $\mathbf{s}'_k$ as columns. Moreover, it follows from equation (8.20) that

$$\mathbf{b} = (\mathbf{A}' - \mathbf{I})^{-1}\mathbf{A}\mathbf{b}' \qquad (8.23)$$

while

$$\mathbf{c} = (0 \quad \cdots \quad 0 \quad 1) \quad \text{and} \quad d = \beta'_0$$

So an equivalent continuous-time description can be obtained from $\mathbf{A}'$ and $\mathbf{B}'$ by first computing the eigenvalues and corresponding eigenvectors of $\mathbf{A}'$ and subsequently using equations (8.21), (8.22), and (8.23) for the computation of $\mathbf{A}$ and $\mathbf{B}$. The computation of the scalar differential equation (8.16) from these matrices is then straightforward.

Generally, the eigenvalues λ'_k are complex. So

$$\lambda_k \Delta = \ln \lambda'_k = \ln|\lambda'_k| + \mathrm{j}(\arg \lambda'_k + 2n\pi)$$

where $n = 0, \pm 1, \pm 2, \ldots$. Hence, if for the imaginary part of $\lambda_k \Delta$ the principal value $-\pi < \text{Im}(\ln \lambda'_k) < \pi$ is chosen, this is justified only if $-\pi < \Delta \, \text{Im} \, \lambda_k < \pi$, that is, if $|\text{Im} \, \lambda_k| < \pi/\Delta, k = 1, \ldots, K$. It is concluded that the sampling frequency in rad s^{-1} should exceed twice the largest imaginary part found among the eigenvalues $\lambda_k, k = 1, \ldots, K$.

In practice, the above relations between equivalent representations are also used if $u(t)$ is not strictly discrete-interval but does not change substantially between consecutive sampling instants. The results are then, of course, approximations. Finally, it is observed that the computation of the continuous-time description from its discrete-time equivalent is relatively complicated and time-consuming. In the next subsection it is shown that these computational difficulties can be avoided if $u(t)$ is a *periodic test signal*, since in that case the parameters of the continuous-time system can directly be estimated from the observations without estimating the parameters of the equivalent discrete-time system first.

8.4.7 Continuous-time Estimation Using Periodic Test Signals

Suppose that the problem is the estimation of the parameters

$$\mathbf{\theta} = (\alpha_0 \cdots \alpha_{K-1} \, \beta_0 \cdots \beta_K)^{\mathrm{T}}$$

of the model

$$\frac{\mathrm{d}^K y(t)}{\mathrm{d}t^K} + \alpha_{K-1} \frac{\mathrm{d}^{K-1} y(t)}{\mathrm{d}t^{K-1}} + \cdots + \alpha_0 y(t) = \beta_K \frac{\mathrm{d}^K u(t)}{\mathrm{d}t^K} + \cdots + \beta_0 u(t)$$

from observations $\underline{w}(t) = y(t) + \underline{p}(t)$ and $\underline{v}(t) = u(t) + \underline{q}(t)$ where $\underline{p}(t)$ and $\underline{q}(t)$ are stationary stochastic processes representing non-systematic errors and $u(t)$ is a test signal periodic with Θ. In addition, assume that the observations are available for $0 \leqslant t \leqslant T$ where T is an integral multiple J of Θ and that $y(t)$

is also periodic with Θ. The latter assumption implies that transient responses have vanished.

The exact Fourier coefficients γ_{my} and γ_{mu} of $u(t)$ and $y(t)$, defined in Section 8.4.5, satisfy

$$\mathscr{A}(r_m)\gamma_{my} - \mathscr{B}(r_m)\gamma_{mu} = 0$$

where

$$\mathscr{A}(r) = \alpha_0 + \alpha_1 r + \cdots + \alpha_{K-1} r^{K-1} + r^K,$$

$$\mathscr{B}(r) = \beta_0 + \beta_1 r + \cdots + \beta_K r^K$$

and $r_m = j2\pi m/\Theta$. Then, using similar arguments as in Section 8.4.5, one can easily show that necessary conditions for a minimum $\hat{\mathbf{t}} = (\hat{a}_0 \cdots \hat{a}_{K-1} \, \hat{b}_0 \cdots \hat{b}_K)^T$ of the least squares criterion

$$J(\mathbf{t}) = \sum_{\substack{l=-L \\ l \neq 0}}^{L} |A(r_{ml})\gamma_{mly} - B(r_{ml})\gamma_{mlu}|^2$$

are

$$\sum_l \{\hat{A}(-r_{ml})\gamma^*_{mly} - \hat{B}(-r_{ml})\gamma^*_{mlu}\} r^k_{ml}\gamma_{mly} = 0 \quad k = 0, \ldots, K-1 \quad (8.24)$$

and

$$\sum_l \{\hat{A}(-r_{ml})\gamma^*_{mly} - \hat{B}(-r_{ml})\gamma^*_{mlu}\} r^k_{ml}\gamma_{mlu} = 0 \quad k = 0, \ldots, K \quad (8.25)$$

where $l = \pm 1, \pm 2, \ldots, \pm L$, the m_l are the harmonic numbers of the harmonics taken into consideration and $m_l = -m_{-l}$. Moreover, this minimum is unique and $\hat{\mathbf{t}} = \boldsymbol{\theta}$. So the equations (8.24) and (8.25) are a set of $2K + 1$ linear equations in the $2K + 1$ unknown parameters $\boldsymbol{\theta}$. They can, therefore be explicitly solved for the exact parameters $\boldsymbol{\theta}$ if the γ_{mly} and γ_{mlu} are known which is, a result of the errors in the observations, not the case. However, if the γ_{m_lu} and γ_{m_ly} in equations (8.24) and (8.25) are replaced by corresponding consistent estimators, (8.24) and (8.25) become a consistent closed-form estimator of $\boldsymbol{\theta}$. Practical estimators for γ_{my} and γ_{mu} are

$$\hat{\gamma}_{my} = \frac{1}{J\Theta} \int_0^{J\Theta} \underline{w}(t)\exp(-j2\pi mt/T)\, dt$$

and $\hat{\gamma}_{mu}$ defined analogously. Levin (1959) has shown that under very general conditions with respect to $\underline{p}(t)$ and $\underline{q}(t)$ $\hat{\gamma}_{my}$ and $\hat{\gamma}_{mu}$ are unbiased and consistent.

With respect to the proposed estimator of $\boldsymbol{\theta}$ the following remarks can be made. In the first place, comparing equations (8.24) and (8.25) to equations

(8.13) and (8.14) shows that it may be considered to be an instrumental variable estimator using estimated differentiated versions of the relevant periodic component of the response and the input as instrumental sequence. Furthermore, this continuous-time estimator has the same advantages, computational simplicity in particular, as its discrete-time counterpart described in Section 8.4.5. For a more detailed discussion of the estimator, including numerical examples and an evaluation of its precision, the reader is referred to Van den Bos (1970, 1974).

REFERENCES

Clarke, D. W. (1967). 'Generalized-least-squares estimation of the parameters of a dynamic model', in *Preprints 1st IFAC Symp. on Identification in Automatic Control Systems*, Academia, Prague. Paper 3.17.

Cramér, H. (1961). *Mathematical Methods of Statistics*, Princeton University Press, Princeton.

Davies, W. D. T. (1970). *System Identification for Self-Adaptive Control*, Wiley, London.

Draper, N. R. and Smith, H. (1966). *Applied Regression Analysis*, Wiley, New York.

Durbin, J. (1960). 'The fitting of time-series models', *Revue Inst. Int. de Stat.*, **28**, 233–43.

Eykhoff, P. (1974). *System Identification*, Wiley, London.

Fedorov, V. V. (1972). *Theory of Optimal Experiments*, Academic Press, New York.

Goodwin, G. C. and Payne, R. L. (1977). *Dynamic System Identification: Experiment Design and Data Analysis*, Academic Press, New York.

Griffiths, J. W. R., Stocklin, P. L. and Schooneveld, C. Van. (1973). *Signal Processing*, Academic Press, London.

Harris, E. L. (1979). 'Using discrete models with continuous design packages', *Automatica*, **15**, 97–100.

Haykin, S. (1979). *Nonlinear Methods of Spectral Analysis*, Springer, Berlin.

Hoffmann de Visme, G. (1971). *Binary Sequences*, The English Universities Press, London.

IEEE (1974). *IEEE Trans.*, **AC-19**, No. 6., complete issue.

IFAC (1967). *Preprints 1st IFAC Symp. on Identification in Automatic Control Systems*, Academia, Prague.

IFAC (1970). *Preprints 2nd IFAC Symp. on Identification and Process Parameter Estimation*, Academia, Prague.

IFAC (1973). *Proc. 3rd IFAC Symp. on Identification and System Parameter Estimation*, North-Holland, Amsterdam.

IFAC (1978). *Proc. 4th IFAC Symp. on Identification and System Parameter Estimation*, North-Holland, Amsterdam.

IFAC (1979). *Proc. 5th IFAC Symp. on Identification and System Parameter Estimation*, Pergamon, Oxford.

IMSL (1977). *IMSL Library 1*, International Mathematical and Statistical Libraries, Houston.

Jenkins, G. M. and Watts, D. G. (1968). *Spectral Analysis and Its Applications*, Holden-Day, San Francisco.

Jennrich, R. I. (1969). 'Asymptotic properties of non-linear least squares estimators', *Ann. Math. Stat.*, **40**, 633–43.

Kaufmann, H. C. and Akselsson, R. (1975). 'Non-linear least squares analysis of proton-induced X-ray emission spectra', *Adv. X-Ray Anal.*, **18**, 353–61.

Kendall, M. G. and Stuart, A. (1966). *The Advanced Theory of Statistics*, Vol. 3, 1st Ed., Griffin, London.

Kendall, M. G. and Stuart, A. (1967). *The Advanced Theory of Statistics*, Vol. 2, 2nd Ed., Griffin, London.

Kendall, M. G. and Stuart, A. (1969). *The Advanced Theory of Statistics*, Vol. 1, 3rd Ed., Griffin, London.

Kwakernaak, H. and Sivan, R. (1972). *Linear Optimal Control Systems*, Wiley, New York.

Levin, M. J. (1959). 'Estimation of the characteristics of linear systems in the presence of noise', *D.Sc. Thesis*, Electrical Engineering Dept., Columbia University.

Makhoul, J. (1975). 'Linear prediction: a tutorial review', *Proc. IEEE*, **63**, 561–80.

Mann, H. B. and Wald, A. (1943). 'On the statistical treatment of linear stochastic difference equations', *Econometrica*, **11**, 173–220.

Marquardt, D. W. (1963). 'An algorithm for least-squares estimation of nonlinear parameters', *J. Soc. Indust. Appl. Math.*, **11**, 431–41.

Mood, A. M., Graybill, F. A., and Boes, D. C. (1974). *Introduction to the Theory of Statistics*, McGraw-Hill, Tokyo.

Murray, W. (1972). *Numerical Methods for Unconstrained Optimization*, Academic Press, London.

NAG (1978). *NAG Fortran Library Manual, Mark 6*, Numerical Algorithms Group, Oxford.

Norden, R. H. (1972). 'A survey of maximum likelihood estimation', *Int. Stat. Rev.*, **40**, 329–54.

Norden, R. H. (1973). 'A survey of maximum likelihood estimation', Part 2, *Int. Stat. Rev.*, **41**, 39–58.

Oppenheim, A. V. and Schafer, R. W. (1975). *Digital Signal Processing*, Prentice-Hall, Englewood Cliffs.

Papoulis, A. (1965). *Probability, Random Variables, and Stochastic Processes*, McGraw-Hill, New York.

Peterson, W. W. (1961). *Error Correcting Codes*, Wiley, New York.

Piceni, H. A. L. and Eykhoff, P. (1975). 'The use of hybrid computers for system-parameter estimation', *Ann. de l'AICA*, **17**, no. 1, 9–22.

Robinson, E. A. (1967). *Statistical Communication and Detection*, Griffin, London.

Schwartz, M. and Shaw, L. (1975). *Signal Processing: Discrete Spectral Analysis, Detection and Estimation*, McGraw-Hill, London.

Seber, G. A. F. (1977). *Linear Regression Analysis*, Wiley, New York.

Van den Bos, A. (1970). 'Estimation of linear system coefficients from noisy responses to binary multifrequency signals', *Preprints 2nd IFAC Symp. on Identification and Process Parameter Estimation*, Academia, Prague. Paper 7.2.

Van den Bos, A. (1974). 'Estimation of parameters of linear systems using periodic test signals', *D. Tech. Sc. Thesis*, Delft University of Technology.

Van den Bos, A. (1977). 'Application of statistical parameter estimation methods to physical measurements', *J. Phys. E: Sci. Instrum.*, **10**, 753–60.

Van den Bos, A. (1979). 'Small sample properties of a class of nonlinear least squares problems', in *Proc. 5th IFAC Symp. on Identification and System Parameter Estimation*, Pergamon, Oxford. Paper M2.6.

Van den Bos, A. and Krol, R. G. (1979). 'Synthesis of discrete-interval binary signals with specified Fourier amplitude spectra', *Int. J. Control*, **30**, 871–84.

Van Espen, P., Nullens, H., and Adams, F. (1977). 'A computer analysis of X-ray fluorescence spectra', *Nucl. Instrum. Meth.*, **142**, 243–50.

Van Trees, H. L. (1968). *Detection, Estimation and Modulation Theory*, Part 1, *Detection, Estimation and Linear Modulation Theory*, Wiley, London.

Van Trees, H. L. (1971a). *Detection, Estimation and Modulation Theory*, Part 2, *Non-linear Modulation Theory*, Wiley, London.

Van Trees, H. L. (1971b). *Detection, Estimation and Modulation Theory*, Part 3, *Radar, Sonar Signal Processing and Gaussian Noise Signals in Noise*, Wiley, London.

Wilkinson, J. H. and Reinsch, C. (1971). *Handbook for Automatic Computation, Vol. II: Linear Algebra*, Springer, Berlin.

Wilks, S. S. (1962), *Mathematical Statistics*, Wiley, New York.

Zacks, S. (1971). *The Theory of Statistical Inference*, Wiley, New York.

Handbook of Measurement Science, Vol. 1
Edited by P. H. Sydenham
© 1982 John Wiley & Sons Ltd.

Chapter

9

W. J. KERWIN

Analog Signal Filtering and Processing

Editorial introduction

At some stage in the information flow chain of an instrumentation system the need will arise to alter the frequency content of the signals. This, and the following chapter, discuss how signals in the electric form can be processed: this chapter explains procedures based on the analog, linear, signal form; Chapter 10 explains those based on the digital alternative. Examples of this need are reduction of unwanted frequencies, such as mains-induced noise, bandwidth extension by selective processing of gains of the various frequencies, selective attenuation as needed in audio compensation systems, and bandwidth limitation for signal detection systems.

Filtering was, in the main, originally developed to fill telecommunications needs arising from the 1880's onward. By the 1930's the design of passive network filters had reached a level of maturity. The arrival of computers and the inexpensive solid-state semiconductor active amplifier element made new and more complex networks viable. Thus developed the active filter in which a passive network was combined with an active amplifier. Active methods enabled circuits to be devised which did not need the expensive, lossy, inductor component and in which the numerical value of capacitors could effectively be multiplied up. During the 1970's electronic filter design emerged from the realm of the circuit theorist to become an easy-to-use tool. Design books greatly helped this transition. This chapter reviews some dominant design methods in use and, as such, cannot replace the many design texts, a selection of which are mentioned.

9.1 PASSIVE SIGNAL PROCESSING

9.1.1 Introduction

With the possible exception of certain applications where extensive optical filtering is valuable and cases where very limited spectral processing can be used in a more appropriate energy domain (examples are the tuning of mechanical systems, the selective insulation of thermal systems) the electric form of signal will be preferred as the input to a data processing stage. It is easier to implement, the least expensive alternative, small in size, and can employ very powerful filtering strategies.

To begin a design study of a suitable filtering stage it will be necessary to first decide the specifying requirements. What changes to the amplitude and phase of the components of input signal are required over the band of frequencies involved? This information is generally specified as the transfer function of the filter stage. This can be realized by inspection of known responses, types, or by circuit theory derivation.

Additionally the stage may require uniform gain change (called *amplification* if unity or greater, *attenuation* if less than unity). It may be essential that phase shift is kept in control or that it is shifted uniformly.

As will be seen in this chapter, the totally general and economic filter design does not exist. Design involves choice of a selected design that best suits the above realized specification.

As a general rule compromise must be made between sharpness of the frequency roll-off away from the band-pass region and the time domain response. Fortunately today it is a relatively simple, straightforward, and inexpensive matter to implement filter designs at the prototype stage. Once tested for general suitability they can then be further refined to make them less sensitive to circuit component tolerance spreads and temperature dependence.

This much said the chapter begins with some definitions and an example.

Laplace notation

We are concerned here primarily with signal modification and amplification, both steady state and transient. We will discuss both passive and active networks. The Laplace transform notation, $s = \sigma + j\omega$, will be used throughout. To differentiate between a steady state problem and a transient problem, we will use $p = j\omega$ for steady state and $s = \sigma + j\omega$ for the transient case. Impedance is then Lp or $1/Cp$, and admittance $1/Lp$ or Cp for an inductor (L) or a capacitor (C).

Transfer functions

Discussion is restricted to lumped, linear, finite systems, both passive and active. Transfer functions will be given as output response over excitation, that is,

$$T(p) = \frac{\text{signal output}}{\text{signal input}}$$

The roots of the numerator of the transfer function (the zeros of the function) are therefore the zeros of transmission of the system. The poles of the transfer function (the roots of the denominator) are the points of infinite system response.

The steady state magnitude response of the transfer function is easily obtained by separating $T(p)$ into real and imaginary parts ($p = j\omega$), so as to obtain

$$T(p) = \frac{A_1 + jB_1}{A_2 + jB_2} \tag{9.1}$$

then the magnitude of $T(p)$ is

$$\left| T(p) \right|_{p=j\omega} = \left(\frac{A_1^2 + B_1^2}{A_2^2 + B_2^2} \right)^{1/2} \tag{9.2}$$

The phase is given by

$$\overline{T(p)}_{p=j\omega} = \tan^{-1}\left(\frac{B_1}{A_1}\right) - \tan^{-1}\left(\frac{B_2}{A_2}\right) \tag{9.3}$$

Network analysis

The usual mesh or nodal analysis could be used for all of the systems to be discussed here; however, the vest majority are ladder networks and much simpler analysis methods will therefore be used. The most appropriate method is commonly called linearity. It uses the fact that in a linear network the transfer function is independent of the signal level. Thus, a value of 1 volt or 1 ampere is usually assumed. In addition, analysis proceeds from output to input.

Example 9.1. Determine the steady state transfer voltage ratio v_o/v_i and the magnitude and phase of the network in Figure 9.1.

Let $v_o = 1$ volt, then

$$i_3 = \tfrac{3}{2}p(1) = \tfrac{3}{2}p \quad v = 1 + (\tfrac{4}{3}p)(\tfrac{3}{2}p) = 1 + 2p^2$$

$$i_2 = (\tfrac{1}{2}p)v = \tfrac{1}{2}p(1 + 2p^2) \quad i_1 = i_2 + i_3$$

$$v_i = v + i_1(1) = 1 + 2p^2 + 2p + p^3$$

$$T(p) = \frac{v_o}{v_i} = \frac{1}{v_i} = \frac{1}{p^3 + 2p^2 + 2p + 1}$$

$$|T(p)|_{p=j\omega} = \frac{1}{\sqrt{[(1 - 2\omega^2)^2 + (2\omega - \omega^3)^2]}} = \frac{1}{\sqrt{(\omega^6 + 1)}}$$

$$\overline{T(p)}_{p=j\omega} = 0° - \tan^{-1}\left(\frac{2\omega - \omega^3}{1 - 2\omega^2}\right)$$

The above example used normalized element values (near or at unity Ω, H, F) and has a cut-off frequency (-3 dB) which occurs at $\omega = 1$ rad s^{-1}, as can be seen from the magnitude expression $|T(p)|_{\omega=1} = 1/\sqrt{2}$. Note also that $|T(p)|_{\omega \to \infty} \to 0$ and has a magnitude of 1 at d.c. so that it is a low-pass filter.

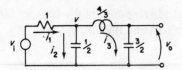

Figure 9.1 Third-order, low-pass, filter (units in Ω, H, F)

Since the numerator of $T(p)$ is a constant, there are no finite zeros and since the denominator is a cubic, there are three poles and the system is of third order. In a later section we will discuss scaling procedures to obtain any desired source impedance and cut-off frequency. In addition, we will transform low pass prototype filters to high-pass or band-pass filters.

9.1.2 Low-pass Filter Functions

Butterworth function

The Butterworth function (Butterworth, 1930) is one of a class of functions that have maximally flat magnitude (MFM). In addition to the MFM property, they are all pole functions (i.e. no finite zeros). The MFM function is one that has as many derivatives equal to zero as possible at $\omega = 0$. A much simpler way of defining and deriving the Butterworth polynomials is to realize that this definition is satisfied if the term by term *magnitude* coefficients of the numerator and denominator polynomials have the same ratio as exists at $\omega = 0$. For the Butterworth this means that all ω coefficients in the magnitude expression except the highest one are equal to zero. In addition, all Butterworth functions are normalized so that the cut-off frequency $\omega_{-3\,dB} = 1$ rad s^{-1}.

Example 9.2. Find the values of the coefficients for a third-order Butterworth function having unity gain at d.c.

Let

$$T(p) = \frac{1}{ap^3 + bp^2 + cp + 1}$$

thus

$$|T(p)|^2 = \frac{1}{a^2\omega^6 + (b^2 - 2ac)\omega^4 + (c^2 - 2b)\omega^2 + 1}$$

therefore

$$b^2 - 2ac = 0 \quad c^2 - 2b = 0$$

and since then

$$|T(p)|^2 = \frac{1}{a^2\omega^6 + 1}\bigg|_{\omega=1} = \frac{1}{2};$$

that is, at $\omega = 1$ the function is down 3 dB. Therefore $a = 1$ so that $b = 2$, $c = 2$ and

$$T(p) = \frac{1}{p^3 + 2p^2 + 2p + 1}$$

Depending on the design problem, we may be working directly from the polynomial coefficients or from the pole positions so that both forms are of value. The normalized pole positions are given by: (a) they are on a unit circle; (b) they are all $180°/n$ apart; (c) there is a single real pole at -1 for all odd order functions and complex pairs only for all even order functions; and (d) they are in the left half plane.

The magnitude response of all normalized Butterworth functions (unterminated) is given by

$$|T(j\omega)|^2 = \frac{1}{\omega^{2n} + 1} \tag{9.4}$$

thus

$$A(\text{dB}) = 20 \log\left[\left(\frac{1}{\omega^{2n} + 1}\right)^{1/2}\right] \tag{9.5}$$

or

$$\omega = \sqrt[2n]{(10^{-A/10} - 1)} \tag{9.6}$$

$$n = \frac{\log(10^{-A/10} - 1)}{2 \log \omega} \tag{9.7}$$

so that given A at a certain ω, the required n can be determined.

Example 9.3. A magnitude of -40 dB is required at $\omega = 3.5$ rad s^{-1} for a filter with a cut-off frequency of 1 rad s^{-1}. What order Butterworth filter is required?

In this case,

$$n = \frac{\log(10^{40/10} - 1)}{2 \log 3.5} = 3.676$$

Thus, a fourth-order filter will be required.

Table 9.1   Butterworth denominator polynomials

$$p + 1$$
$$p^2 + p\sqrt{2} + 1$$
$$p^3 + 2p^2 + 2p + 1$$
$$p^4 + 2.61313p^3 + 3.41421p^2 + 2.61313p + 1$$
$$p^5 + 3.23607p^4 + 5.23607p^3 + 5.23607p^2 + 3.23607p + 1$$

The Butterworth function is one that will normally be used when flat magnitude is most important (without ripple), and where linearity of phase or freedom from transient overshoot to a step input is of lesser importance. The slope approaches $6n$ dB/octave as $\omega \to \infty$. A few Butterworth polynomials are given in Table 9.1.

Thomson functions

The Thomson function (Thomson, 1959) is an all pole function whose time delay is maximally flat (MFD). The time delay is defined as $TD = -d\phi(\omega)/d\omega$ where $\phi(\omega)$ is the steady state phase. These functions can be derived in a similar manner to that used for the MFM functions after first finding the phase and then differentiating before applying the maximally flat procedure previously shown for the MFM functions. These functions are frequently called linear phase functions.

The usual applications are in those systems where linearity of phase or lack of overshoot or oscillation in response to a step input is of primary importance. Their magnitude response is a gradual decrease in amplitude to the -3 dB point and then an increasing slope until $6n$ dB/octave is obtained.

Storch (1954) has given a simple method of finding these polynomials (also called Bessel polynomials). Table 9.2 gives the first five MFD denominators, and their -3 dB frequencies (all are normalized to unity time delay).

Chebyshev functions

The Chebyshev function is similar to the Butterworth and Thomson functions in that it is an all pole function. It is defined in terms of the steady state magnitude such that the magnitude varies between tolerance limits throughout the pass band. These limits have equal maxima and minima and so the function is referred to as an equal ripple magnitude function. The specification is the peak-to-peak magnitude of the ripple in dB. These functions are very efficient in that they offer a faster transition from the pass band to the stop band than the previous functions, and, of course, maintain a specified tolerance in the pass band. The disadvantage is a highly non-linear phase change with frequency worsening as the ripple increases, thereby seriously worsening the transient

Table 9.2   Thomson denominator polynomials (MFD)
and $\omega_{-3\,dB}$ (in rad s^{-1})

$p + 1$	1.0000
$p^2 + 3p + 3$	1.3617
$p^3 + 6p^2 + 15p + 15$	1.7557
$p^4 + 10p^3 + 45p^2 + 105p + 105$	2.1139
$p^5 + 15p^4 + 105p^3 + 420p^2 + 945p + 945$	2.4274

response. These functions have a high degree of overshoot to a step input and are highly oscillatory as they return to final value.

Since the functions are dependent on the amount of ripple, it is necessary to tabulate them as a function of the specified ripple. It is customary to specify the pass band as the end of the equal ripple band and to normalize this to 1 rad s^{-1}. Since the functions change as the degree of ripple changes, it is usual to tabulate a few specific ripple magnitudes. Here we will specify the polynomials in terms of the ripple and thus one set of equations will cover all possible ripples.

Figure 9.2 shows an odd and an even Chebyshev function with a magnitude of one at $\omega = 0$, as it would be in an unterminated filter, and the definition of ε. Note that $20 \log[\sqrt{(1 + \varepsilon^2)}] = A$(dB) of ripple. A is positive in all cases. Table 9.3 lists a few Chebyshev polynomials as a function of v, normalized to $\omega = 1$ rad s^{-1} at the end of the ripple. v is defined in terms of the ripple magnitude A, for a given order n, as

$$v = \frac{1}{n} \sinh^{-1}\left(\frac{1}{\varepsilon}\right) \tag{9.8}$$

where

$$\varepsilon = \sqrt{(10^{A/10} - 1)} \tag{9.9}$$

To determine the attenuation at any particular stop band frequency, we have

$$|T(j\omega)|^2 = \frac{1}{1 + \varepsilon^2 \cosh^2(n \cosh^{-1} \omega)} \tag{9.10}$$

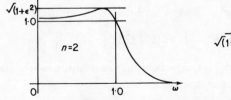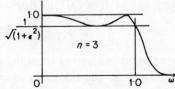

Figure 9.2   Chebyshev magnitude versus frequency

Table 9.3   Chebyshev polynomials

$$p + \sinh v$$
$$p^2 + (\sqrt{2} \sinh v)p + \sinh^2 v + \tfrac{1}{2}$$
$$(p + \sinh v)[p^2 + (\sinh v)p + \sinh^2 v + \tfrac{3}{4}]$$
$$[p^2 + (0.76537 \sinh v)p + \sinh^2 v + 0.85355][p^2 + (1.84776 \sinh v)p + \sinh^2 v + 0.14645]$$
$$(p + \sinh v)[p^2 + (0.61803 \sinh v)p + \sinh^2 v + 0.90451]$$
$$\times \,[p^2 + (1.61803 \sinh v)p + \sinh^2 v + 0.34549]$$

The -3 dB frequency is also of interest for the Chebyshev functions and is

$$\omega_{-3\,dB} = \cosh\!\left(\frac{\cosh^{-1}(1/\varepsilon)}{n}\right) \tag{9.11}$$

where ε is defined by equation (9.9).

Example 9.4.   What order Chebyshev function (normalized) is required if the ripple A is 0.1 dB and the attenuation is to be 60 dB at $\omega = 4$ rad s^{-1}? What is the -3 dB frequency?
Solving for ε (equation (9.9)):

$$\varepsilon = \sqrt{(10^{4/10} - 1)} = 0.15262$$

Therefore, from equation (9.10), ε as above, and $\omega = 4$ rad s^{-1}

$$-60 \text{ dB} = 20 \log\!\left[\left(\frac{1}{1 + \varepsilon^2 \cosh^2(n \cosh^{-1} \omega)}\right)^{1/2}\right]$$

Solving for n, we get $n = 4.595$. Thus, $n = 5$ is required. From equation (9.11), we have

$$\omega_{-3\,dB} = \cosh\!\left(\frac{\cosh^{-1}(1/\varepsilon)}{n}\right) = 1.135 \text{ rad s}^{-1}$$

Example 9.5.   Determine the denominator polynomial for a fourth-order Chebyshev ($n = 4$) of 0.03 dB ripple (i.e. $A = 0.03$ dB).

$$\varepsilon = \sqrt{(10^{4/10} - 1)} = 0.083257$$

From equation (9.8), $v = (1/n)\sinh^{-1}(1/\varepsilon) = 0.795175$ and $\sinh v = 0.881663$.
From Table 9.3 the Chebyshev polynomial is then

$$(p^2 + 0.6748p + 1.6309)(p^2 + 1.6291p + 0.9238)$$

Inverse Chebyshev function

The inverse Chebyshev function has an MFM pass band and an equal ripple stop band produced by finite complex conjugate $j\omega$ axis zeros. In this case, the MFM characteristic is obtained by making the ratio of the numerator coefficient to the denominator coefficient in the magnitude expression (term by term) equal to the value of $|T(p)|$ at $\omega = 0$. All inverse Chebyshev functions are normalized to $\omega_{-3\,dB} = 1$ rad s^{-1}. The finite $j\omega$ axis zeros allow a much steeper roll-off and therefore greater discrimination; however, a stop band return occurs following the zero or zeros.

A general form of a $2z$–$3p$ function having no d.c. loss is:

$$T(p) = \frac{dp^2 + 1}{ap^3 + bp^2 + cp + 1} \tag{9.12}$$

For this to be a normalized inverse Chebyshev function, we have three constraints: (a) MFM; (b) $\omega_{-3\,dB} = 1$ rad s^{-1}; and (c) specification of the finite zero position, $\omega_{-\infty}$.

Finding the square of the magnitude of $T(p)$ (equation (9.12)):

$$|T(p)|^2_{p=j\omega} = \frac{d^2\omega^4 - 2d\omega^2 + 1}{a^2\omega^6 + (b^2 - 2ac)\omega^4 + (c^2 - 2b)\omega^2 + 1} \tag{9.13}$$

Thus

$$\omega_{-\infty} = \sqrt{(1/d)} \quad \text{or} \quad d = 1/\omega_{-\infty}^2 \tag{9.14}$$

Equating coefficients (since $|T(p)|_{\omega=0} = 1$) to obtain MFM, we get

$$c^2 - 2b = -2d \tag{9.15}$$

$$b^2 - 2ac = d^2 \tag{9.16}$$

at $\omega = 1$ rad s^{-1} at which $|T(p)|^2 = \frac{1}{2}$ (-3 dB), we have

$$|T(p)|^2 = \frac{d^2 - 2d + 1}{a^2 + d^2 - 2d + 1} = \frac{1}{2} \qquad a = 1 - d \tag{9.17}$$

then substituting $a = 1 - d$ into equation (9.16) and solving equation (9.15) and equation (9.16) simultaneously, we find

$$b^2 - 2(1 - d)\sqrt{(2b - 2d)} = d^2 \tag{9.18}$$

Thus a specification of the desired d (from $\omega_{-\infty}$) allows b to be determined, as well as a from equation (9.17) and c from equation (9.15). A minimum practical d exists which depends on the network used to realize this function, but is not a limitation.

Example 9.6.   Determine the $2z$–$3p$ inverse Chebyshev function having zero output at 3.0 rad s^{-1} (i.e. $\omega_{-\infty} = 3$ rad s^{-1}).

We have

$$d = 1/\omega^2_{-\infty} = \tfrac{1}{9}$$

thus

$$a = 1 - d = \tfrac{8}{9}$$

and from equation (9.18)

$$b^2 - 2(\tfrac{8}{9})\sqrt{(2b - \tfrac{2}{9})} = (\tfrac{1}{9})^2$$

This is easily solved using a programmable hand calculator, since we want only positive real roots. Thus $b = 1.8150$; and from equation (9.15)

$$c = \sqrt{(2b - \tfrac{2}{9})} = 1.8460$$

and the function is

$$T_1(p) = \frac{\tfrac{1}{9}p^2 + 1}{\tfrac{8}{9}p^3 + 1.8150p^2 + 1.8460p + 1}$$

Similarly, other values of the zeros can be specified. We can also set up $2z - 4p, 2z - 5p, \ldots,$ inverse Chebyshev functions, as will be discussed shortly.

The more nearly the zero position approaches 1 rad s^{-1}, the higher is the peak return in the stop band as shown in Figure 9.3.

For the curve shown with a zero at $\omega = 2$ rad s^{-1}, the transfer function is

$$T_2(p) = \frac{\tfrac{1}{4}p^2 + 1}{\tfrac{3}{4}p^3 + 1.5856p^2 + 1.6344p + 1} \qquad (9.19)$$

The magnitude at the peak in the stop band (at ω_{max}) is given by equation (9.20) for all unterminated $2z$–$3p$ inverse Chebyshev functions:

$$|T(p)|_{\omega_{max}}(\text{dB}) = 20 \log\left[\left(\frac{1}{1 + (27/4)\omega^2_{-\infty}(\omega^2_{-\infty} - 1)^2}\right)^{1/2}\right] \qquad (9.20)$$

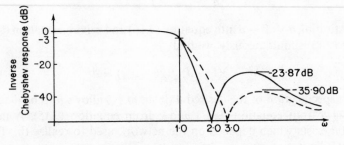

Figure 9.3   Inverse Chebyshev responses

The frequency of the peak stop band return is

$$\omega_{max} = \omega_{-\infty}\sqrt{3} \tag{9.21}$$

This function can be set up so as to remove a specific frequency or to give the best cut-off slope consistent with a specific stop band minimum attenuation as determined by equation (9.20).

A $2z$–$4p$ function can be set up as

$$T(p) = \frac{ep^2 + 1}{ap^4 + bp^3 + cp^2 + dp + 1} \tag{9.22}$$

having

$$\omega_{-\infty} = \sqrt{(1/e)} \quad \text{or} \quad e = 1/\omega_{-\infty}^2 \tag{9.23}$$

Now finding the magnitude of equation (9.22) and equating the corresponding numerator and denominator coefficients, as before, as well as setting $\omega_{-3\,dB} = 1$ rad s^{-1}, we obtain the following results:

$$a = 1 - e \tag{9.24}$$

$$d^4 + 4d^2e - 8d\sqrt{[2(1 - e)(d^2/2 + e)]} + 8(1 - e) = 0 \tag{9.25}$$

$$c = d^2/2 + e \tag{9.26}$$

$$b = \sqrt{[2(1 - e)c]} \tag{9.27}$$

Example 9.7. Determine a $2z$–$4p$ inverse Chebyshev having $\omega_{-\infty} = 2$ rad s^{-1}.

From equation (9.23)

$$e = 1/\omega_{-\infty}^2 = \tfrac{1}{4}$$

then, from equation (9.25)

$$d^4 + d^2 - 8d\sqrt{[\tfrac{3}{2}(d^2/2 + \tfrac{1}{4})]} + 6 = 0$$

This is easily programmed and iterated to find the real positive values of d which are, $d = 0.94048$ and $d = 2.25390$. The higher value of d is the correct one in all cases, since it gives the polynomial with left half plane poles. Then, from equation (9.26), $c = d^2/2 + \tfrac{1}{4} = 2.7900$, and from equation (9.27), $b = \sqrt{[2(1 - \tfrac{1}{4})c]} = 2.0457$, and from equation (9.24), $a = 1 - e = \tfrac{3}{4}$. Thus

$$T(p) = \frac{\tfrac{1}{4}p^2 + 1}{\tfrac{3}{4}p^4 + 2.0457p^3 + 2.7900p^2 + 2.2539p + 1}$$

This allows the determination of any $2z$–$4p$ inverse Chebyshev function based on a specification of $\omega_{-\infty}$ only. For the $2z$–$4p$ function the stop band maximum occurs at

$$\omega_{max} = \omega_{-\infty}\sqrt{2} \tag{9.28}$$

and the magnitude at that point is

$$|T(p)|_{\omega_{max}}(\text{dB}) = 20 \log\left[\left(\frac{1}{1 + 16\omega_{-\infty}^4(\omega_{-\infty}^2 - 1)^2}\right)^{1/2}\right] \qquad (9.29)$$

For the previous example for $\omega_{-\infty} = 2$ rad s^{-1}, $\omega_{max} = 2\sqrt{2} = 2.8284$ rad s^{-1}, and the magnitude at ω_{max} is -33.63 dB. For the $2z$–$3p$ function having $\omega_{-\infty} = 2$ rad s^{-1}, the peak return was -23.87 dB. In addition, the magnitude of the slope following the peak reaches 12 dB/octave as $\omega \to \infty$, whereas the $2z$–$3p$ has an ultimate slope as $\omega \to \infty$ of only 6 dB/octave.

The detailed derivation of the $2z$–$5p$ inverse Chebyshev function will not be presented here; however, a very useful result is the $\omega_{-\infty} = 2$ rad s^{-1} function. It is

$$T(p) = \frac{\frac{1}{4}p^2 + 1}{\frac{3}{4}p^5 + 2.5159p^4 + 4.2199p^3 + 4.3975p^2 + 2.8801p + 1} \qquad (9.30)$$

for which

$$\omega_{max} = \omega_{-\infty}\sqrt{(5/3)} \qquad (9.31)$$

and

$$|T(p)|_{\omega_{max}} = \left[1 + \frac{(9)(5)^5}{(4)(3)^5}\omega_{-\infty}^6(\omega_{-\infty}^2 - 1)^2\right]^{-1/2} \qquad (9.32)$$

For the $2z$–$5p$ function of equation (9.30), $\omega_{max} = 2.5820$ rad s^{-1} and $|T(p)|_{\omega_{max}} = -42.22$ dB. Of course, the ultimate slope as $\omega \to \infty$ is improved to 18 dB/octave in this case.

A comparison of the magnitude responses versus frequency of all of the functions presented above is given in Figure 9.4. Functions are scaled to $\omega_{-3} = 1$ rad s^{-1} for the third-order case.

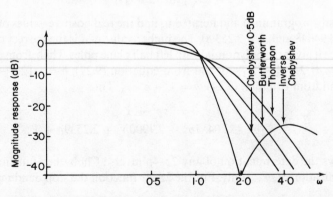

Figure 9.4 Magnitude response versus frequency for third-order filters

9.1.3 Low-pass Filter Design as the Basis

The element values given here will be restricted to the two most important cases: the unterminated filter and the filter with equal terminations ($R_S = R_L = 1\ \Omega$). We begin by study of design of normalized, low-pass procedures, extending these designs to whatever scale or frequency pass type is needed.

Butterworth Filters

For the terminated filter, the element values are given in Figure 9.5 for $n = 2, 3, 4,$ and 5. Values are in Ω (ohm), H (henry), F (farad). All cut-off frequencies are $\omega_{-3\,\mathrm{dB}} = 1\ \mathrm{rad\ s}^{-1}$.

In the unterminated case for $R_S = 1\ \Omega$, $R_L = \infty$, the element values are given in Figure 9.6. Values are in Ω, H, F; again $\omega_{-3\,\mathrm{dB}} = 1\ \mathrm{rad\ s}^{-1}$.

Thomson filters

For the terminated filter, values are given in Figure 9.7 for $n = 2, 3, 4,$ and 5. For the unterminated case, the values are shown in Figure 9.8. All filters are normalized to one second time delay. All values are in Ω, H, F; see Table 9.2 for the value of $\omega_{-3\,\mathrm{dB}}$.

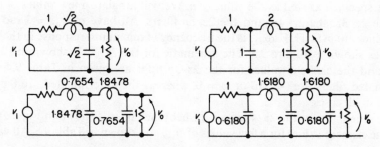

Figure 9.5   Equal termination Butterworth filters (units in Ω, H, F)

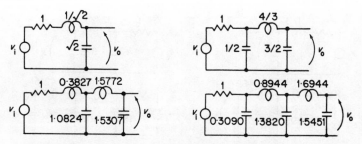

Figure 9.6   Unterminated Butterworth filters (units in Ω, H, F)

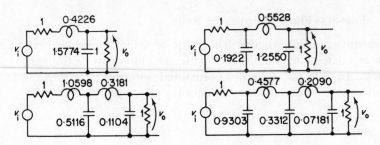

Figure 9.7   Equal termination Thomson filters (units Ω, H, F)

Chebyshev filters

Since the amount of ripple can be specified as desired, the number of possible Chebyshev filters is doubly infinite. A few Chebyshev filters are given in Figure 9.9. All are normalized to end of ripple at $\omega = 1$ rad s^{-1}; all values are in Ω, H, F.

Inverse Chebyshev filters

Again we have a doubly infinite set of possible filters since the stop band ripple can be specified as well as the value of n. We will tabulate a few values of $\omega_{-\infty}$ for the $2z-3p$, the $2z-4p$, and the $2z-5p$ filters. All have $\omega_{-3dB} = 1$ rad s^{-1}. All values are in Ω, H, F. ω_{max} is the frequency of maximum response in the stop band as shown in Figure 9.3. The schematic for the $2z-3p$ is shown in Figure 9.10, and the element values for the $2z-3p$ filter are given in Table 9.4. The terminated filters are -6 dB at $\omega = 0$, whereas the unterminated are 0 dB at $\omega = 0$ in all cases.

The schematic for the $2z-4p$ inverse Chebyshev filter is shown in Figure 9.11, and the element values for a few values of $\omega_{-\infty}$ are shown in Table 9.5. All values are in Ω, H, F.

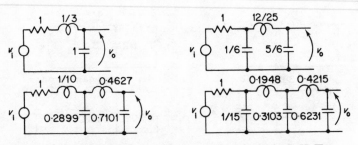

Figure 9.8   Unterminated Thomson filters (units in Ω, H, F)

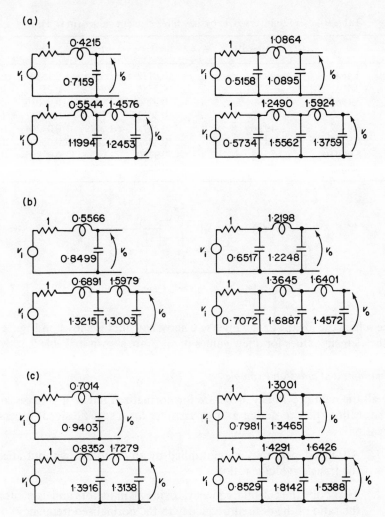

Figure 9.9   Chebyshev filter element values (units in Ω, H, F): (a) 0.10 dB
ripple; (b) 0.25 dB ripple; (c) 0.50 dB ripple

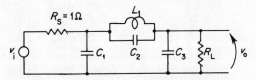

Figure 9.10 $2z$–$3p$ inverse Chebyshev filter

Table 9.4 $2z$–$3p$ inverse Chebyshev filter element values (in Ω, H, F)

$\omega_{-\infty}$	ω_{max}	$|T(p)|_{\omega_{max}}$ (dB)	R_L	C_1	L_1	C_2	C_3
2	3.4641	−29.87	1	0.8172	1.6344	0.1530	0.8172
2	3.4641	−23.87	∞	0.2556	0.9687	0.2581	1.3787
3	5.1962	−41.90	1	0.9230	1.8460	0.06019	0.9230
3	5.1962	−35.90	∞	0.4013	1.1794	0.09421	1.4447
4	6.9282	−49.86	1	0.9574	1.9149	0.03264	0.9574
4	6.9282	−43.86	∞	0.4461	1.2482	0.05007	1.4688
5	8.6602	−55.88	1	0.9730	1.9459	0.02056	0.9730
5	8.6602	−49.88	∞	0.4659	1.2793	0.03127	1.4800

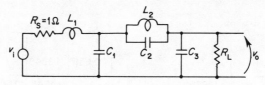

Figure 9.11 $2z$–$4p$ inverse Chebyshev filter

The schematic for the $2z$–$5p$ inverse Chebyshev filter is shown in Figure 9.12 and the element values for a few values of $\omega_{-\infty}$ are shown in Table 9.6.

Scaling laws and a design example

Since all the data previously given are for normalized filters, it is necessary to use the scaling rules to design a low-pass filter for a specific signal processing application.

Rule 1.   All impedances may be multiplied by any constant without affecting the transfer voltage ratio.

Rule 2.   To modify the cut-off frequency, divide all inductors and capacitors by the ratio of the desired frequency to the normalized frequency.

Table 9.5 $2z$–$4p$ inverse Chebyshev filter element values (in Ω, H, F)

$\omega_{-\infty}$	ω_{max}	$|T(p)|_{\omega_{max}}$ (dB)	R_L	L_1	C_1	L_2	C_2	C_3
2	2.8284	−39.63	1	0.7350	1.6723	1.5189	0.1646	0.5816
2	2.8284	−33.63	∞	0.3666	0.8649	1.2338	0.2026	1.3890
3	4.2426	−55.19	1	0.7500	1.7703	1.7121	0.06490	0.6918
3	4.2426	−49.19	∞	0.3761	0.9928	1.4327	0.07756	1.4693
4	5.6568	−65.65	1	0.7570	1.8050	1.7726	0.03526	0.7247
4	5.6568	−59.65	∞	0.3791	1.0332	1.4973	0.04174	1.4965
5	7.0711	−73.60	1	0.7600	1.8205	1.8001	0.02222	0.7396
5	7.0711	−67.60	∞	0.3804	1.0512	1.5265	0.02620	1.5089

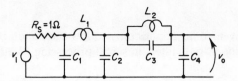

Figure 9.12 $2z-5p$ inverse Chebyshev filter

Table 9.6 $2z-5p$ inverse Chebyshev filter element values (in Ω, H, F)

$\omega_{-\infty}$	ω_{max}	$|T(p)|_{\omega_{max}}$(dB)	C_1	L_1	C_2	L_2	C_3	C_4
2	2.5820	−42.22	0.2981	0.8649	1.1824	1.3678	0.18278	1.3996
3	3.8730	−61.30	0.3045	0.8824	1.2991	1.5568	0.07137	1.4828
4	5.1640	−74.26	0.3066	0.8878	1.3364	1.6184	0.03862	1.5105
5	6.4550	−84.16	0.3075	0.8902	1.3530	1.6461	0.02430	1.5231

Example 9.8. Design a low-pass filter of MFM type (Butterworth) to operate from a 600 Ω source into a 600 Ω load, with a cut-off frequency of 500 Hz. The filter must be at least 36 dB below the d.c. level at 2 kHz, that is, -42 dB.

Since 2 kHz is four times 500 Hz, it corresponds to $\omega = 4$ rad s^{-1} in the normalized filter. Thus at $\omega = 4$, we have

$$-42 \text{ dB} = 20 \log\left[\frac{1}{2}\left(\frac{1}{\sqrt{(4^{2n} + 1)}}\right)\right]$$

therefore, $n = 2.99$, so $n = 3$ must be chosen. Thus a third-order terminated Butterworth is required. From Figure 9.5 we have the normalized network shown in Figure 9.13a.

The impedance scaling factor is $600/1 = 600$ and the frequency scaling factor is $2\pi500/1 = 2\pi500$, that is, the ratio of the desired radian cut-off frequency to the normalized cut-off frequency (1 rad s^{-1}). Note that the impedance scaling factor increases the size of the resistors and inductors, but reduces the size of the capacitors. The result is shown in Figure 9.13b.

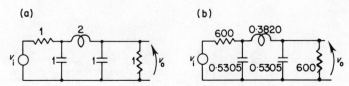

Figure 9.13 Third-order Butterworth, low-pass, filter: (a) normalized, (units in Ω, H, F); (b) scaled, (units in Ω, H, μF)

Transformation rules

All information given so far applies only to low-pass filters; yet we frequently need high-pass or band-pass filters in signal processing (Weinberg, 1962).

(*a*) *Low-pass to high-pass.* To transform a low-pass filter to high-pass, we first scale it to a cut-off frequency of 1 rad s^{-1}, if it is not already 1 rad s^{-1}. This allows a simple frequency rotation about 1 rad s^{-1} of $p \to 1/p$. All L's become C's, all C's become L's, and all values reciprocate. The cut-off frequency does not change.

Example 9.9. Design a third-order high-pass Butterworth filter to operate from a 600 Ω source to a 600 Ω load and having a cut-off frequency of 500 Hz.

Starting with the normalized filter of Figure 9.5 for which $\omega_{-3} = 1$ rad s^{-1}, we reciprocate all elements and all values to obtain the filter shown in Figure 9.14a, for which $\omega_{-3} = 1$ rad s^{-1}. Now we apply the scaling rules to raise all impedances to 600 Ω and the radian cut-off frequency to $2\pi 500$ rad s^{-1} as shown in Figure 9.14b.

(*b*) *Low-pass to band-pass.* To transform a low-pass filter to a band-pass filter we will first scale the low-pass so that the cut-off frequency is equal to the bandwidth of the normalized band-pass filter for which $\omega_0 = 1$ rad s^{-1} (ω_0 is the centre frequency of the band-pass filter). Then we apply the transformation $p \to p + 1/p$. For an inductor

$$Z = Lp \quad \text{transforms to} \quad Z = L(p + 1/p)$$

For a capacitor

$$Y = Cp \quad \text{transforms to} \quad Y = C(p + 1/p)$$

The first step is then to determine the Q of the band-pass filter where

$$Q = \frac{f_0}{B} = \frac{\omega_0}{B_r}$$

(f_0 is the centre frequency and B is the 3 dB bandwidth in Hz), scale the low-pass filter to a cut-off frequency of $1/Q$ rad s^{-1}, then series tune every inductor

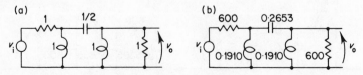

Figure 9.14   Third-order Butterworth, high-pass, filter: (a) normalized
(units in Ω, H, F); (b) scaled (units in Ω, h, μF)

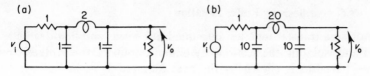

Figure 9.15   Low-pass filter (units in Ω, H, F)

L with a capacitor of value $1/L$ and parallel tune every capacitor C with an inductor of value $1/C$.

Example 9.10.   Design a band-pass filter centred at 100 kHz having a 3 dB bandwidth of 10 kHz starting with a third-order Butterworth low-pass filter. The source and load resistors are each to be 600 Ω.
   The Q required is

$$Q = \frac{100 \text{ kHz}}{10 \text{ kHz}} = 10 \quad \text{or} \quad \frac{1}{Q} = 0.1$$

Scaling the normalized low-pass filter of Figure 9.15a to $\omega_{-3\text{dB}} = 1/Q = 0.1$ rad s^{-1}, we obtain the filter of Figure 9.15b. Now, converting to band-pass with $\omega_0 = 1$ rad s^{-1}, we obtain the normalized filter of Figure 9.16a. Next, scaling to an impedance of 600 Ω and to a centre frequency of $f_0 = 100$ kHz ($\omega_0 = 2\pi 100\text{k}$ rad s^{-1}), we obtain the filter of Figure 9.16b.

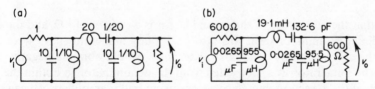

Figure 9.16   Band-pass filter ($Q = 10$): (a) normalized, $\omega_0 = 1$ rad s^{-1}
(units in Ω, H, F); (b) scaled, $f_0 = 100$ kHz

9.2   ACTIVE SIGNAL PROCESSING

9.2.1   Introduction

The addition of active elements can be done in two ways: (a) as a buffer or gain element only; or (b) to provide feedback so as to eliminate the need for inductance and still be able to realize the functions we previously discussed (passive *RC* circuits cannot realize any of those functions beyond first order). This second category of networks is called *active RC*.

9.2.2 *RLC* Synthesis by Buffer Isolation

By factoring a transfer function into quadratic terms and first-degree terms as needed and isolating the sections by buffer amplifiers, the synthesis problem becomes trivial as is shown by the next example. It is, in fact, synthesis by inspection. This is useful when we wish to realize functions not covered in the preceding sections.

Example 9.11.   Design a network realizing

$$T(p) = \left(\frac{p+3}{p^2 + p + 4}\right)\left(\frac{\frac{1}{2}p}{p^2 + \frac{1}{2}p + 2}\right)$$

This can be split into two sections; first,

$$T_1(p) = \frac{p+3}{p^2 + p + 4} = \frac{1 + 3/p}{p + 1 + 4/p}$$

and, second

$$T_2(p) = \frac{\frac{1}{2}p}{p^2 + \frac{1}{2}p + 2} = \frac{\frac{1}{2}}{p + \frac{1}{2} + 2/p}$$

Note that all quadratics when divided by p have terms which are inductive, resistive, and capacitive in that order, and we can realize each of the above factors with a simple L section voltage divider by inspection as in Figure 9.17. By separating the two sections with a buffer amplifier, we prevent interaction between the sections which would otherwise change the transfer function.

Note that the numerator of the second factor is a resistor of $\frac{1}{2}$ Ω, and the corresponding denominator term was also $\frac{1}{2}$ Ω. If any numerator term were larger than the corresponding denominator term, we would obtain a negative result when the numerator was subtracted from the denominator to obtain the series branch of the L section, but since we are using buffer amplifiers, we can divide the numerator by any arbitrary number so that this does not occur and then use a corresponding buffer amplifier gain to exactly achieve the desired transfer

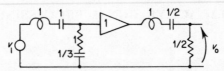

Figure 9.17   *RLC* synthesis using buffer amplifiers (units in Ω, H, F). The transfer function is given by

$$T(p) = \frac{v_o}{v_i} = \left(\frac{1 + 3/p}{p + 1 + 4/p}\right)\left(\frac{\frac{1}{2}}{p + \frac{1}{2} + 2/p}\right)$$

function. Another highly useful result using this method of synthesis is that different impedance scaling factors can be applied to the separate sections without affecting the result. Frequency scaling factors must, of course, be the same for all sections.

This method is not the most economical of elements, but this is of little consequence in the early stages of the design of a measurement system. After the design is complete and it has been determined that the particular filter is the one required, more sophisticated design methods can then be used to save elements if desired.

9.2.3 Active Feedback

Introduction

A very large number of active feedback synthesis methods have been developed in order to eliminate the need for inductance in filter networks. Each method has applicability to certain frequency ranges or to a particular pole Q. For a quadratic factor $p^2 + \beta p + \gamma$, the pole Q is $Q = (\sqrt{\gamma})/\beta$. For many networks the Q sensitivity to amplifier gain K defined as

$$S_K^Q = \frac{K}{Q} \frac{\partial Q}{\partial K} \tag{9.33}$$

is a strong function of Q, thereby limiting the application of those networks to low-Q systems. In addition, some networks have a passive element sensitivity problem which also limits their applicability to low-Q designs:

$$S_R^Q = \frac{R}{Q} \frac{\partial Q}{\partial R} \qquad S_C^Q = \frac{C}{Q} \frac{\partial Q}{\partial C} \tag{9.34}$$

To simplify this very large field (Kerwin *et al.*, 1972), we will consider only three active RC configurations: first, networks suited to low-Q applications (the Q of the poles of most low-pass filters is quite low) that use low-gain voltage amplifiers, so that operation is possible over a very wide frequency range, second, a network suited to high-Q applications, third, a single amplifier network that provides $j\omega$ axis zeros.

Low-gain second-order low-pass active RC filters

An early class of active RC filters (Sallen and Key, 1955) used individual quadratic RC sections with feedback from a voltage amplifier as shown in Figure 9.18. The transfer function is given by

$$T(p) = \frac{K/C}{p^2 + (1.1/C + 1 - K)p + 1/C} \tag{9.35}$$

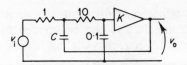

Figure 9.18   Second-order active
RC, low-pass, filter (units in Ω, F)

As can be seen, the amplifier gain reduces the p coefficient by direct subtraction and when this coefficient becomes less than 2 (for $C = 1$ F), we have complex poles and therefore can duplicate an RLC network of the same order. These structures have high sensitivity to amplifier gain change. From the transfer function of equation (9.35), we can determine that

$$S_K^Q = \frac{K}{1.1/C + 1 - K} \tag{9.36}$$

and thus for $C = 1$ F, $Q = 10$, $K = 2.0$ we have $S_K^Q = 20$, that is a 1% change in K produces a 20% change in Q! We must therefore restrict use of this network to low-Q systems.

Example 9.12.   Using the positive-gain active RC structure of Figure 9.18, design a fourth-order low-pass Chebyshev filter of 0.03 dB ripple, and determine the d.c. gain and the Q sensitivities to amplifier gain change.
We derived the fourth-order Chebyshev polynomial of 0.03 dB ripple in Example 9.5 and in factored form, it is

$$(p^2 + 0.6748p + 1.6309)(p^2 + 1.6291p + 0.9238).$$

Thus, we find for the first factor

$$\frac{1.1}{C_1} + 1 - K_1 = 0.6748 \quad \frac{1}{C_1} = 1.6309$$

$$C_1 = 0.6132 \text{ F} \quad K_1 = 2.1191$$

and for the second factor

$$\frac{1.1}{C_2} + 1 - K_2 = 1.6291 \quad \frac{1}{C_2} = 0.9238$$

$$C_2 = 1.0825 \text{ F} \quad K_2 = 0.3871$$

Cascading these two networks, we obtain the complete fourth-order Chebyshev shown in Figure 9.19.
This normalized fourth-order Chebyshev low-pass filter can now be scaled in impedance and frequency as desired. The normalized cut-off frequency is

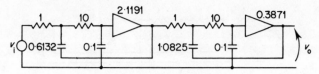

Figure 9.19 Fourth-order Chebyshev active filter having 0.03 dB ripple (units in Ω, F)

1 rad s^{-1}. The d.c. gain is $(2.1191)(0.3871) = 0.8203$, and the sensitivity to changes in gain of the first operational amplifier is

$$S_{K_1}^Q = \frac{K_1}{1.1/C_1 + 1 - K_1} = 3.14$$

and for the second is

$$S_{K_2}^Q = \frac{K_2}{1.1/C_2 + 1 - K_2} = 0.24$$

Higher-order single-amplifier low-pass filter

Higher-order transfer functions can be obtained with a single amplifier. The most useful are the third- and fourth-order functions as shown in Figures 9.20 and 9.21 (Aikens and Kerwin, 1972; Kerwin et al., 1972).

State variable second order

The state variable active RC synthesis method was developed in 1967 (Kerwin et al., 1967). It is a more complex structure requiring three amplifiers for a low-pass, band-pass, or high-pass second-order structure and four amplifiers for a

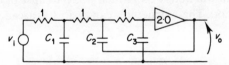

Figure 9.20 Third-order active RC, low-pass, filter (units in Ω, F)

	C_1	C_2	C_3
Butterworth	1.7058	0.8671	0.6761
Thomson	0.7064	0.3100	0.3046
0.5 dB Chebyshev	2.2932	0.9940	0.6130

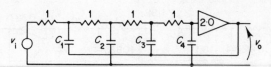

Figure 9.21 Fourth-order active *RC*, low-pass, filter (units in Ω, F)

	C_1	C_2	C_3	C_4
Butterworth	1.1746	2.3041	0.8519	0.4338
Thomson	0.3391	0.7413	0.2500	0.1516
0.5 dB Chebyshev	1.9860	3.0640	1.0367	0.4181

complete biquadratic function. The reason for considering such a complex structure is that the sensitivity to both active and passive elements is very low (all < 1 *and* independent of the Q). In addition, the minimum number of capacitors is used (2) even when a complete biquadratic is required. A normalized schematic is shown in Figure 9.22 for the three-amplifier structure.

Even though all three types of filters are obtained simultaneously, the primary use of this structure is for band-pass filters since that is when we need low-Q sensitivity. The band-pass output transfer function (assuming ideal operational amplifiers), the Q, and the centre frequency ω_0 are

$$T(p) = -\left(\frac{R_2(1 + R)}{1 + R_2}\right)\left(\frac{p}{p^2 + [(1 + R)/(1 + R_2)]p + R}\right) \qquad (9.37)$$

$$Q = \frac{(1 + R_2)\sqrt{R}}{1 + R} \qquad (9.38)$$

$$\omega_0 = \sqrt{R} \qquad (9.39)$$

As can be seen no subtractions exist, which was the cause of the high sensitivity in the previous method. These structures can easily be cascaded for higher-order transfer functions.

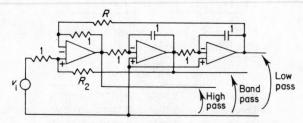

Figure 9.22 State variable second-order filter in normalized form (units in Ω, F)

Example 9.13. Design a second-order band-pass filter having a $Q = 50$ at a centre frequency of 10 kHz using the state variable network. Use a minimum resistor value of 10 kΩ. What is the centre frequency gain?

From equation (9.39), we normalize to $\omega_0 = 1$ rad s^{-1}:

$$\omega_0 = \sqrt{R} = 1 \quad R = 1\,\Omega$$

and from equation (9.38), for $R = 1\,\Omega$,

$$Q = \frac{1 + R_2}{2} = 50 \quad R_2 = 99\,\Omega$$

The centre frequency gain is

$$-\frac{R_2(1 + R)}{1 + R_2}\left(\frac{1 + R_2}{1 + R}\right) = -99$$

Scaling to 10 kΩ, we have a scaling factor of 10k, and from $\omega_0 = 1$ rad s^{-1} to $\omega_0 = 2\pi 10$k, we have a frequency scaling factor of $2\pi 10$k. The result is shown in Figure 9.23.

The amplifier gain must be at least twice the Q required at the operating frequency, and must be much greater than twice the Q if the value of R_2 calculated is to be correct. If we used an operational amplifier having a gain of 10^5 and a -3 dB frequency of 10 Hz, then at 10 kHz the gain would be 100 or just barely adequate for the above example even if we removed the 990k resistor. So we can see that this method is restricted to primarily the audio frequencies, unless we use high performance operational amplifiers.

jω Axis zeros

To design filters such as the inverse Chebyshev, we must have jω axis zero capability. The most commonly used RC method for jω axis zeros is the twin-T. By providing appropriate feedback and network loading a pair of jω axis zeros and a pair of *independent* complex poles can be obtained (Kerwin and Huelsman, 1966). The schematic is shown in Figure 9.24 for the case where the zeros are

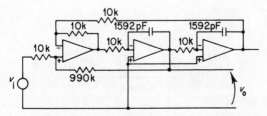

Figure 9.23   State variable second-order filter of Example 9.13

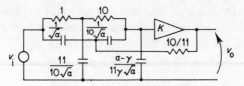

Figure 9.24 $2z-2p$ network (units in Ω, F):

$$K = 2 + \frac{10}{11}\left(\frac{\alpha}{\gamma} - 1 + \frac{\beta}{\gamma}\sqrt{\alpha}\right)$$

$$T(p) = \frac{v_o}{v_i} = \frac{K\gamma}{\alpha}\left(\frac{p^2 + \alpha}{p^2 + \beta p + \gamma}\right)$$

beyond the poles (greater radial distance from the origin). The transfer function is also shown in Figure 9.24.

Example 9.14.   Design an inverse Chebyshev $2z-4p$ active RC low-pass filter with a cut-off frequency of 5 kHz and zero output at 10 kHz. Scale the impedance level by 5k.

Using the $T(p)$ from Example 9.7, clearing of fractions, and factoring, we obtain:

$$T(p) = \frac{p^2 + 4}{(3p^2 + 6.1293p + 3.8375)(p^2 + 0.6845p + 1.0423)}$$

This function has a cut-off frequency of 1 rad s^{-1} and a zero at 2 rad s^{-1}. When scaled by $2\pi 5k$ in frequency, we will have a cut-off frequency of 5 kHz and a zero at 10 kHz as required.

We will realize the $2z-2p$ network first, and factoring out $\frac{1}{3}$, we have

$$T_1(p) = \frac{1}{3}\left(\frac{p^2 + 4}{p^2 + 2.0431p + 1.2792}\right) \triangleq \frac{1}{3}\left(\frac{p^2 + \alpha}{p^2 + \beta p + \gamma}\right)$$

With these values of α, β, γ and from Figure 9.24, we obtain the network shown in Figure 9.25. There is a multiplier of (γ/α) $K = 0.3293$, but this has no effect

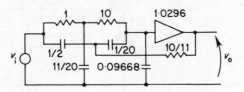

Figure 9.25 $2z-2p$ RC network (units in Ω, F)

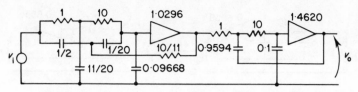

Figure 9.26   Complete $2z$–$4p$ inverse Chebyshev filter (units in Ω, F)

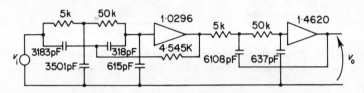

Figure 9.27   Scaled $2z$–$4p$ inverse Chebyshev filter

on the pole and zero positions. Now we must realize the second factor, $T_2(p)$ where

$$T_2(p) = \frac{1}{p^2 + 0.6845p + 1.0423}$$

and cascade it with $T_1(p)$ to realize $T(p)$. Since we are using building blocks in which the amplifier is at the output, it does not matter which network is placed first. Using the network of Figure 9.18, we find

$$1/C = 1.0423 \quad C = 0.9594$$

$$\frac{1.1}{C} + 1 - K = 0.6845 \quad K = 1.4620$$

Cascading these two networks as shown in Figure 9.26, we obtain the complete $T(p)$. Note that the d.c. gain achieved is $(1.0296)(1.4620) = 1.5053$.

Scaling the impedance upward by 5k and to a cut-off frequency of $2\pi 5k$ rad s^{-1} (5 kHz), we obtain the network shown in Figure 9.27.

Low-pass to high-pass transformation

We cannot use the low-pass to high-pass transformation since we would obtain an RL network, not an RC network. If, however, we multiply all impedances by p (the impedance scaling rule tells us that we can multiply by any constant, real or complex), we obtain an RL network which when transformed to high-pass gives us the RC high-pass we want. This is illustrated in Figure 9.28. We start with the normalized low-pass filter (Butterworth, $\omega_{-3\,\mathrm{dB}} = 1$ rad s^{-1}) and obtain the normalized high-pass Butterworth filter ($\omega_{-3\,\mathrm{dB}} = 1$ rad s^{-1}).

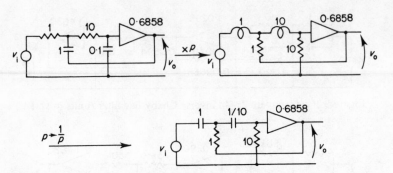

Figure 9.28   Low-pass to high-pass transformation of an active RC second-order Butterworth filter (units in Ω, H, F)

Low-pass to band-pass transformation

We have no general rule here; however, earlier in this section we designed a band-pass filter using the state variable method. This is a very practical band-pass synthesis method and does not require starting with a low-pass prototype. Higher-order band-pass filters only require cascading several second-order sections which are individually designed by using each of the quadratic factors of the higher-order function. These can be obtained from a low-pass function by making the substitution $p = Q\lambda$, where Q is the band-pass filter Q desired. Then substitute $\lambda = p + 1/p$ to convert the function to band-pass, factor into quadratic sections, realize each one with the state variable network, and connect them in cascade.

Example 9.15.   Transform the normalized second-order Butterworth low-pass function to a $Q = 10$ band-pass function ($\omega_0 = 1$ rad s^{-1}), and factor it into two band-pass quadratics.
   Given:

$$T(p) = \frac{1}{p^2 + p\sqrt{2} + 1}$$

substitute

$$p = Q\lambda = 10\lambda$$

to yield

$$T(\lambda) = \frac{1}{100\lambda^2 + 10\lambda\sqrt{2} + 1}$$

Let

$$\lambda = p + 1/p$$

then

$$T(p) = \frac{1}{100}\left(\frac{p^2}{p^4 + 0.1414p^3 + 2.01p^2 + 0.1414p + 1}\right)$$

factoring gives

$$T(p) = \frac{1}{100}\left(\frac{p}{p^2 + 0.0682p + 0.9317}\right)\left(\frac{p}{p^2 + 0.0732p + 1.0733}\right)$$

9.3 TIME DOMAIN CONSIDERATIONS

To this point we have discussed only the steady state aspects of signal processing. In considering the time domain response, the most common test signal is the step function. The response to a step input is illustrated in Figure 9.29 shown for a final value of unity.

One of the features mentioned earlier was the difference in time domain performance of the Thomson as compared to the other filter functions. It has very little overshoot or oscillation to a step input. For those applications where this is of importance, it is the function to choose. This does not come free, however; the Thomson has a much less selective filtering characteristic than the others. The transient performance of various third-order filters to a step input is compared in Table 9.7. Note particularly the excellent settling time of the Thomson. *All* of the functions have been scaled to a -3 dB frequency of 1 rad s^{-1}. This is essential if a fair comparison is to be made.

Scaling in the time domain is very similar to that in the frequency domain, but is inverted. If we were to increase $\omega_{-3\,\mathrm{dB}}$ from 1 to 10^6 rad s^{-1}, then all times in Table 9.7 would be divided by 10^6; that is, they would be in microseconds.

The networks of Table 9.7 are shown in Figure 9.30; all are scaled to $\omega_{-3\,\mathrm{dB}} = 1$ rad s^{-1}! All values are in Ω, H, F.

Many useful design texts have been published on active filter design. A selection is Budak (1974), Daryanani (1976), Hilburn and Johnson (1973)

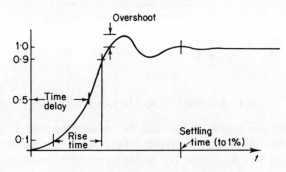

Figure 9.29   Step response definitions

Table 9.7   Third order system step response for $\omega_{-3\,\mathrm{dB}} = 1$ rad s^{-1} (all times in seconds)

	Rise time	Delay time	Overshoot (%)	Settling time (to within 1%)
Butterworth	2.29	2.14	8.15	9.42
Thomson	2.18	1.68	0.75	3.78
Chebyshev (0.1 dB)	2.39	2.38	10.19	12.50
Chebyshev (0.5 dB)	2.50	2.55	8.93	13.54
Inverse Chebyshev ($\omega_{-\infty} = 2$ rad s^{-1})	2.59	1.88	10.96	9.24

Johnson and Hilburn (1975), and Huelsman (1977). Commercially available integrated circuits, backed by application notes, have greatly simplified active filter application.

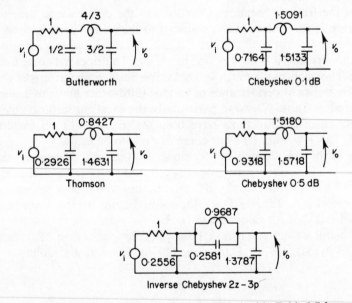

Figure 9.30   Third-order networks characterized in Table 9.7 for $\omega_{-3\,\mathrm{dB}} = 1$ rad s^{-1} (units in Ω, H, F)

9.4   COMPUTER-AIDED DESIGN

The term 'computer-aided design' can be misleading and it is sometimes replaced with 'computer-aided analysis'. In general, the available computer programs are analysis programs, but of course accurate and fast analysis does aid the designer.

One of the most useful and readily available programs is SPICE developed by Drs B. Cohen and D. O. Pederson of the University of California at Berkeley (Cohen and Pederson, 1976). This circuit simulation program works from a nodal description of element locations, and for the purpose of analysing the kinds of networks we have been concerned with includes voltage-controlled voltage sources (as well as others). The output can be a magnitude, or phase, or transient response to a pulse input, either as a print-out or plot. The program is limited only by the computer it is used with. Non-linear d.c. analysis is also included, as is noise analysis and distortion analysis.

The program is very useful for determining changes in performance of a circuit as a function of element value change either due to tolerance, or change with temperature. A temperature coefficient can be included with each element specified and the analysis performed at various specified temperatures.

The program also includes modelling capability for four types of semi-conductor devices: diodes, bipolar transistors, field effect transistors, and MOSFETs. Other versions are available: I-SPICE (for interactive SPICE) and T-SPICE (for thermal SPICE). T-SPICE allows a complete thermal analysis, given the chip thermal characteristics and the complete circuit.

Another useful program for signal processing applications is GOSPEL developed by Dr Lawrence P. Huelsman at The University of Arizona (Huelsman, 1968). This is an optimization program which can be used to solve sets of simultaneous non-linear equations such as are encountered in trying to determine circuit element values to realize a particular function. This program was used by Aikens and Kerwin (1972) to determine the element values for the single amplifier active *RC* structures given in Section 9.2.3.

REFERENCES

Aikens, R. S., and Kerwin, W. J. (1972). 'Single amplifier, minimal *RC*, Butterworth, Thomson and Chebyshev filters to sixth order', *Proc. Int. Filter Symp., Santa Monica, April.*

Budak, A. (1974). *Passive and Active Network Analysis and Synthesis*, Houghton Mifflin, Boston.

Butterworth, S. (1930). 'On the theory of filter amplifiers', *Wireless Engr*, 7, 536–41.

Cohen, B. and Pederson, D. O. (1976). *Users Guide for SPICE*, Department of Electrical Engineering and Computer Sciences, University of California, Berkeley.

Daryanani, G. (1976). *Principles of Active Network Synthesis and Design*, Wiley, New York.

Hilburn, J. L. and Johnson, D. E. (1973). *Manual of Active Filter Design*, McGraw-Hill, New York.

Huelsman, L. P. (1968). *GOSPEL, A General Optimization Software Package for Electrical Network Design*, University of Arizona Engineering Experiment Station.

Huelsman, L. P. (1977). *Active RC Filters: Theory and Application: Benchmark Papers in Electrical Engineering*, Vol. 15., Academic Press, New York.

Johnson, D. E. and Hilburn, J. L. (1975). *Rapid Practical Design of Active Filters*, Wiley, New York.

Kerwin, W. J., Aikens, R. S., and Gross, W. H. (1972). 'A sensitivity comparison of single and multiple amplifier equal capacitor active *RC* structures', *Proc. 6th Asilomar Conf. on Circuits and Systems, Asilomar, November.*

Kerwin, W. J. and Huelsman, L. P. (1966). 'Design of high performance band pass filters', *Proc. IEEE 1966 Int. Conv., New York,* IEEE, New York.

Kerwin, W. J., Huelsman, L. P., and Newcomb, R. W. (1967). 'State variable synthesis for insensitive integrated circuit transfer functions', *IEEE J. Solid-st. Circuits,* **SC-2**, 87–92.

Kerwin, W. J., *et al.* (1970). In L. P. Huelsman (Ed.), *Active Filters: Lumped, Distributed, Integrated, Digital, and Parametric,* McGraw-Hill, New York.

Sallen, R. P. and Key, E. L. (1955). 'A practical method of designing *RC* active filters', *IRE Trans. Circuit Theory,* **CT-2**, 74–85.

Storch, L. (1954). 'Synthesis of constant delay ladder networks using Bessel polynomials', *Proc. IRE,* **42**, 1666–75.

Thomson, W. E. (1959). 'Maximally flat delay networks', *IRE Trans.,* **CT-6**, 235.

Weinberg, L. (1962). *Network Analysis and Synthesis,* McGraw-Hill, New York.

Handbook of Measurement Science, Vol. 1
Edited by P. H. Sydenham
© 1982 John Wiley & Sons Ltd.

Chapter

10 A. G. BOLTON

Filtering and Processing of Digital Signals

Editorial introduction

Signals in the processing chain of a measurement system may require that their frequency component characteristics be modified. The previous chapter described how this can be done using analog electronic circuitry operating with analog signals. Despite the great internal circuitry and system complexity of digital data processing systems they are often capable of carrying out processing operations on a signal at lower cost than their analog counterpart. Analog signals are first converted into the digital signal domain; in many cases the signal to be processed already exists in digital form.

Digital filtering is not a recent concept but it did emerge after the analog methods arriving predominantly as a basic concept in the application of the early digital computing machines to such fields of use as geophysics (see e.g., Robinson and Treital, 1964). It finds increasing use as digital computing hardware costs fall and because many signals needing filtering now appear in the digital format. This form is also often chosen because many more systems designers today are more familiar with digital hardware than with the linear counterpart, analog system.

With analog filtering methods there has emerged a well-defined subset of knowledge directed very specifically toward what have become known as filters. The term *digital filtering* is largely synonomous with the description *digital signal processing*. For this reason digital filtering information is contained within general digital signal processing texts or, if presented under the specific title, will usually include a wide range of basic signal processing knowledge. Suitable introductory texts are Bogner and Constantinides (1975) and Hamming (1977). Here digital filtering is explained in terms of implementation of systems analogous to the analog filters discussed in Chapter 9.

10.1 IMPLEMENTING ANALOG TECHNIQUES IN DIGITAL FORMATS

10.1.1 Introduction

Digital techniques are very flexible because of programming and memory facilities. The digital components and equipment available provide a variety of new techniques for measurement systems. This chapter introduces some properties of digital systems by outlining how analog techniques can be implemented digitally. Also provided is a brief introduction to Z domain analysis. Chapter 5 presents an alternative approach to digital signal processing.

The performance of a digitally implemented filter can often resemble that of an analog equivalent so closely that for all practical purposes they are identical. This provides a useful basis for introducing digital filters. Special features can be introduced gradually.

10.1.2 Review of Analog Filters

The previous chapter discusses the range of analog filters available and their implementation in some detail. A summary is given here so the relevant aspects can be identified for application to digital filters.

A filter is usually designed in low-pass form and transformed to high-pass, band-pass or band-stop as required. The low-pass filter is selected to fulfil the frequency discrimination or step response requirements: often both aspects must be considered together. This is because a filter which removes unwanted high-frequency components can distort the desired signal considerably within the passband. Graphs can be used to select filters or approximate algebraic relationships applied to realize the component values.

Chapter 9 includes diagrams of overshoot in the step response and ripple in the frequency response. Figure 10.1 shows the overshoot and ripple properties for various second-order filters. Higher-order filters improve the frequency discrimination with a little increase in the overshoot. These are implemented in both active analog and digital filters by adding cascaded second-order stages. Stages with the highest damping factors should be earliest in the sequence, otherwise the signal levels within the filter can be quite large.

Table 10.1 gives values for the pole locations for several types of low-pass filters. The Bessel and Chebyshev filters are normalized so that the 3 dB point is at unity frequency. Various normalizing relationships are in use and should be checked before using any tables. Note that the ripple in the even-order Chebyshev filters is above the steady state gain whilst it is below for filters of odd order.

Using the filter design tables the analog filter can be implemented using pure integration. A useful arrangement is given in Figure 10.2. It has a pass-band

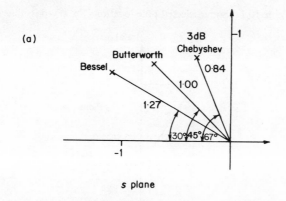

s plane

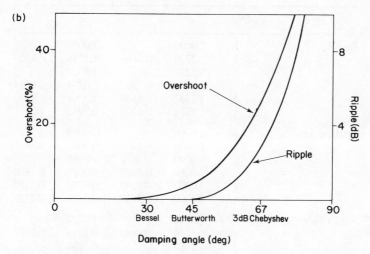

Figure 10.1   (a) Pole locations and (b) corresponding overshoot
and ripple

gain of unity and the signal levels at the outputs of the integrators are nearly equal. Also the design relationships are very convenient to implement.

Digital integration approximates analog integration for large sampling rates. In its simplest form it is implemented using the addition operation. This allows a useful range of digital filters to be implemented quite conveniently. Peculiarities of digital filters will be examined after basic design techniques have been described.

10.1.3   Basic Design Techniques

In Figure 10.3 the analog signal to be filtered is converted to digital form using an analog-to-digital (A/D) converter, processed using a computing element

Table 10.1   Some selected pole locations for low-pass filters

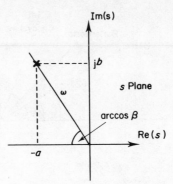

Type	Order	ω	β	a	b
Bessel	1	1.0000		1.0000	
	2	1.2720	0.8660	1.1016	0.6360
	3	1.3227		1.3227	
		1.4476	0.7235	1.0474	0.9993
	4	1.4302	0.9580	1.3701	0.4102
		1.6034	0.6207	0.9952	1.2571
Butterworth	1	1.0000		1.0000	
	2	1.0000	0.7071	0.7071	0.7071
	3	1.0000		1.0000	
		1.0000	0.5000	0.5000	0.8660
	4	1.0000	0.9239	0.9239	0.3827
		1.0000	0.3827	0.3827	0.9239
3 dB Chebyshev	1	1.0024		1.0024	
	2	0.8414	0.3882	0.3224	0.7772
	3	0.2986		0.2986	
		0.9161	0.1630	0.1493	0.9038
	4	0.4427	0.4645	0.2056	0.3920
		0.9503	0.0896	0.0852	0.9465

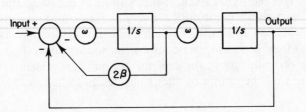

Figure 10.2   Analog filter implementation using integration:

$$T(s) = \frac{\omega^2}{s^2 + 2\beta\omega s + \omega^2}$$

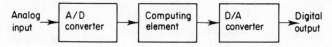

Figure 10.3 One basic digital filtering arrangement

such as a microprocessor, and again converted to analog form. Various other sources of digital signal data, including a transmission system, data storage, or a previous digital filter, could arise. Similarly the output could be used in a digital form in a variety of ways. The simple analog input and output system serves here as a useful introduction.

The analog quantity is represented within the digital system in *binary weighted* form. It is usually most convenient to represent negative quantities using two's complement notation (Hill and Peterson, 1974). This allows the same arithmetic program instructions to be used with both positive and negative quantities. Replication of the analog system by the digital system relies on various assumptions. These assumptions will be examined in detail after the basic design procedure has been outlined. They are, with respect to time, that

(a) the *sampling rate* is larger than any frequency component present at the analog input;
(b) the sampling interval is smaller than any time constant within the filter;

and with respect to magnitudes, that:

(c) the binary data represent the analog signal with an arbitrary degree of precision;
(d) the analog quantity is never too large to be represented within the digital system;

and with respect to coefficients that:

(e) the coefficient values can be implemented precisely.

Digital integration can be performed by adding successive input values. The program flowchart of Figure 10.4 will provide digital integration which nearly approximates analog integration given the previous assumptions. The sampling interval T determines the constant of integration. The output from the digital integrator will be $1/T$ times greater than that from an analog integrator. In a microprocessor the time delay T can be obtained using a fixed number of instructions with a known execution time or an external clock which periodically interrupts another program to perform the filter algorithm. The filter time constants are proportional to the sampling interval T. It is, therefore, important to fix its value. This gives an advantage to digital filters. The interval can be varied to tune filters of arbitrary complexity with precision. In the digital filter design it is convenient to normalize frequencies and time constants to the sampling interval.

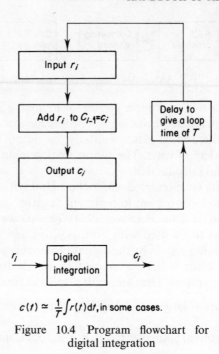

$$c(t) \simeq \frac{1}{T}\int r(t)dt, \text{in some cases.}$$

Figure 10.4 Program flowchart for digital integration

The useful first order filter given in Figure 10.5 uses the integration algorithm. It approximates an analog filter having the transfer function

$$T(s) = \frac{1}{\tau s + 1} \quad \text{where } \tau = 1/a_a$$

The subscript a stands for analog. In the corresponding digital filter the coefficient a must be adjusted to allow for the digital integration because it has an output $1/T$ times greater than that of the analog integrator. It is convenient to normalize the time constant and cut-off frequency of the filter to the sampling interval:

$$\tau_n = \tau/T \quad \text{and} \quad \omega_n = \omega T$$

The coefficient of the digital filter is then $a = 1/\tau_n$.

Example 10.1. A first-order, low-pass digital filter with a cut-off frequency of 100 Hz ($\omega_c = 628$ rad s^{-1}, $\tau = 1.6$ ms) is to be implemented.
The steps necessary to design such a filter are:

(a) Choose the sampling interval. Use about $\frac{1}{10}$ times the time constant

$$T = 0.1 \text{ ms}$$

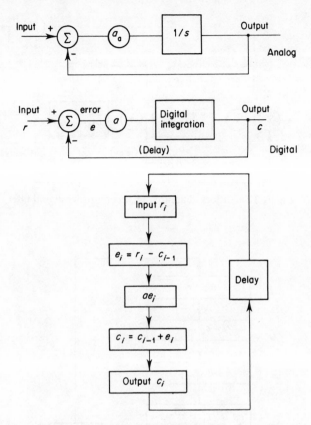

Figure 10.5 Block diagram and flowchart for a first-
order, low-pass digital filter

(b) Calculate the normalized time constant:

$$\tau_n = \tau/T = 16$$

(c) Evaluate the coefficient

$$a = 1/\tau_n = 2^{-4}$$

(d) Use the flowchart of Figure 10.5 with $a = 2^{-4}$ and $T = 0.1$ ms to implement a digital processor.

Extension of this design procedure to second order systems uses the block
diagram and corresponding program flowchart of Figures 10.6 and 10.7. In
this case it is necessary to use tables to find the pole locations. Again the cut-off
frequency can be normalized to the sampling interval as shown in the following
example.

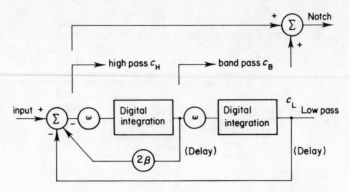

Figure 10.6   Block diagram of a second-order digital filter

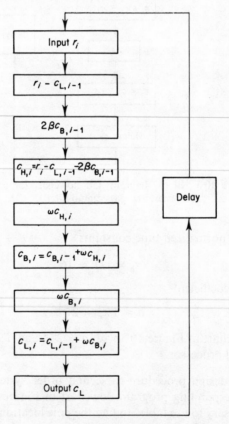

Figure 10.7   Program flowchart for a
second-order digital resonator

Example 10.2. A second-order, 3 dB, Chebyshev low-pass filter with a cut-off frequency of 40 Hz ($\omega_c = 251$ rad s^{-1}) is required.

(a) Choosing a sampling interval less than one tenth the time constant:

$$T = 4 \text{ ms}$$

(b) Calculate the cut-off frequency normalized to the sampling interval:

$$\omega_{cn} = \omega_c T = 0.1004$$

(c) Select the required pole location using the tables for analog filters:

$$\omega = 0.8414 \quad \beta = 0.3882$$

(d) Evaluate the filter coefficients:

$$\omega_f = \omega_c T_{\text{tables}} = 0.1004 \times 0.8414 = 0.08448$$

$$\beta_f = \beta_{\text{tables}}$$

Note that the damping factor is not altered by the scaling for frequency.

(e) Implement the filter using the program flowchart of Figure 10.7.

The step response for this filter is given in Figure 10.8.

10.1.4 Review of Design Assumptions

It is necessary to examine the assumptions which made the digital filter closely resemble the analog equivalent. This allows the selection of important design aspects such as register lengths, sampling rates, and any necessary preparation of the analog signal before conversion to the digital form.

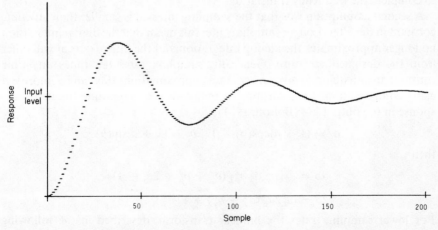

Figure 10.8 Step response of filter design for Example 10.2

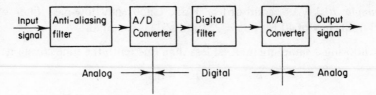

Figure 10.9   Anti-aliasing (low-pass) filter before A/D conversion

The first assumption was that the sampling rate was larger than any frequency component present at the input to the A/D converter. If they were present *aliasing* would occur. This means the high-frequency components appear as low-frequency components within the passband after sampling. Time and frequency domain representations of aliasing are given in Chapter 5.

The frequency domain interpretation is based on the fact that the product of two sinusoids gives both the sum and the difference frequency components. Sampling at intervals of T can be regarded as multiplication by all sinusoids of the form $\cos(2\pi n/t)$ where $n = 0, 1, \ldots, \infty$. In this way the frequency components above half the sampling frequency (the Nyquist frequency) are folded back as unwanted lower frequency components. These high-frequency components must be removed using anti-aliasing low-pass filters before the A/D conversion as shown in Figure 10.9. The frequency domain representation of aliasing gives an estimate of the amount of filtering required; the frequency components remaining above the Nyquist frequency will appear within the passband with the same magnitude.

It is of interest that use can be made of this scheme to provide selective band-pass filters. The A/D converter can act as a frequency translator, the anti-aliasing filter becoming a band-pass filter. Quadrature components are needed to complete the frequency translation.

A second assumption was that the sampling interval is smaller than any time constant in the filter. Longer sampling intervals mean that the digital integration no longer approximates the analog integration and the filter's output will differ from the designed response. Generally, sampling rates ten times any time constant are adequate to give an accurate approximation: this can be improved using compensation. Relationships to maintain the magnitude frequency response in the output are (Bolton, 1981):

$$a_y = (1 + a)\cos(b) - 1 \quad b_y = (1 + a)\sin(b)$$

then

$$\omega = \sqrt{[a_y^2 + b_y^2)/(a_y^2 + b_y^2 + 2a_y + 1)]}$$
$$2\beta = \omega[2a_y/(a_y^2 + b_y^2) + 1]$$

For lower sampling rates the bilinear transform, described in the following section, provides a more useful filter.

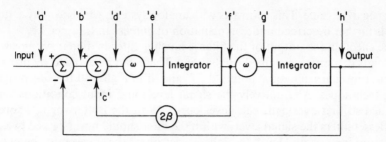

Figure 10.10   Signal points where underflow and overflow can occur

A third assumption was that the analog data were represented with an arbitrary degree of precision. Finite digital register lengths mean there must be some *magnitude quantization*. This aspect is also discussed in Chapter 12.

The input and output, A/D, conversion must have a sufficiently small quantization level to represent the signal with the desired accuracy. 8-bit conversions will give a level of 2^{-8} of 0.4% of the full analog range. However, the signal levels within the filter are also quantized; usually these effects are more severe than those input and output conversions. Various signals are labelled in Figure 10.10.

The dominant effect of quantization within the digital filter occurs at point 'e' of Figure 10.10. The difference signal, $r - e$ is multiplied by the coefficient, a, to give a component at 'e'. The coefficient is much less than unity, particularly for large sampling rates. Underflow at 'e' will occur when the original difference $r - e$ is $1/a$ times larger than the quantization level. This can cause a *deadband* in the output which is $1/a$ times the quantization level. An original quantization level of 0.4% can be expected to give a deadband of 4% even if the sampling rate is only ten times the time constant of the filter. This must be avoided.

One solution is to use longer registers within the filter, leaving the input and output data conversions unchanged. In practice use of 16-bit registers and time constants about ten times greater than the sampling interval provides workable operating levels. Larger sampling rates make the filter more ideal and reduce the aliasing filter requirements, however, *double precision* arithmetic may be needed to reduce the deadband effect. An alternative scheme is to cascade digital filters, the first having a faster sampling rate and smaller time constants. In effect this is a digital aliasing filter. Another approach is to use an instruction to detect whenever the signal at 'b' is greater than zero. If in these cases a least significant bit is added at 'g' the deadband will be removed and there will be only a minor distortion of the filter characteristic. This also removes all small-signal limit cycles (Bolton, 1981).

A fourth assumption was that *magnitude overload* does not occur. With digital systems magnitude overload causes a discontinuity in the output to the other extreme value. A slight overload in the positive direction will cause a large negative value. This contrasts with analog systems which usually saturate. This is further described under explanation of *glitches* in Chapter 12.

The effect on the output of the filter is much greater in digital systems where overloads can cause sustained oscillations. These can be avoided by providing a saturating characteristic at points 'd', 'f', and 'h' of Figure 10.10 using programming techniques. Alternatively the signal levels and input conditions can be confined so that overloads can never occur. With the Butterworth, Thomson, and Bessel filters the signal levels within the filter should never exceed twice the maximum input signal level. In higher-order filters the stages must be in order of ascending Q, with the signal passing through the more damped stages first. The 3 dB Chebyshev filters require an additional factor of 2.

Finally, it was assumed that the coefficients could be implemented precisely. With the type of digital filter described here the response never critically depends on the values of the coefficients (Bolton and Davis, 1981; Bolton, 1981). Having rounded the value of a coefficient for implementation it may be helpful to assess its affect on the filter's response by evaluating the pole locations for these new values. With some types of digital filters the response depends critically on proportionate changes in the coefficients to the point where the available transfer functions are severely limited. However, with the filters described so far the ability to realize the coefficients with precision provides stability advantages over corresponding analog types.

10.1.5   Extension of Low-pass Designs

Extension of the low-pass design to give a high-pass response uses the transform described in Chapter 9, that is, $s \rightarrow 1/s$, which means poles at $r/\pm\theta$ are relocated

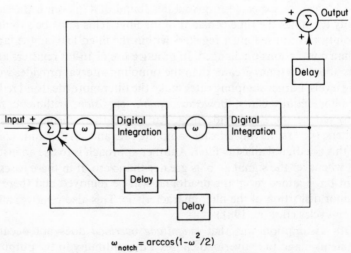

$$\omega_{notch} = \arccos(1 - \omega^2/2)$$

Figure 10.11   A notch filter structure

at $1/r\underline{/\pm\theta}$. The output from the digital filter is then taken from the point shown in Figure 10.6.

The transform to give a band-pass output is $s \to (s^2 + \omega^2_{centre})/s$. The transfer function at the bandpass output of Figure 10.6 approximates

$$T(s) = \frac{\omega s}{s^2 + 2\beta\omega s + \omega^2}$$

and this can be used to implement band-pass designs.

The notch output is quite deep at high sampling rates. At low sampling rates the special notch structure of Figure 10.11 is required to provide a deep notch (Bolton, 1981). Three delay terms are shown. Each is obtained by using the value from the previous loop. This means the notch addition is performed after the high-pass output is computed but before the low-pass output.

10.2 ANALYTIC TECHNIQUES FOR SAMPLED SYSTEMS

10.2.1 Introductory Remarks

At low sampling rates the comparison between analog and digital filters becomes inadequate for analysis or design. The transform which represents digital systems without approximation is the Z transform, in the same way as the Laplace operator s is used with continuous systems. The Z transform is now introduced and a filter design example given to serve as an introduction to specialist texts.

10.2.2 Sampling

When an A/D converter samples a value of a signal subsequent changes in the signal are not registered by the computer until the next sample is taken. It is as if the value of the signal is held constant during the time interval. However, holding the value of the sampled function constant is mathematically more complicated than representing the sampled output by a series of delta (δ) functions. The delta or impulse function, $\delta(t - T)$, is a function of time which is zero everywhere except at time $t = T$. At $t = T$ its value goes to infinity for an infinitesimally small duration. The area of a unit impulse is one unit. The function $2\delta(t - T)$ has an area, or integral, of two units, and so on. A sampling stage is described mathematically as

$$f^*(t) = \sum_{n=0}^{\infty} \delta(t - nT)f(t)$$

Sampling is represented in Figure 10.12.

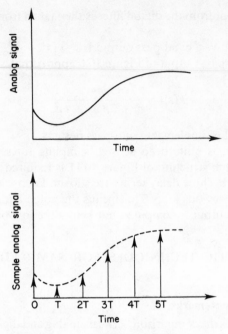

Figure 10.12 Sampling of an analog
signal at intervals of T

10.2.3 Convolution

Sampled systems can be used to show the significance of convolution when
determining the output of linear systems to known input functions. Consider a
system which has the response to a unit impulse given in Figure 10.13 (dotted
curve). The output of this system decays by 50% at each sampling interval,
which corresponds to a time constant $\tau = T/\ln 2$. The impulse response $g^*(t)$
is a series of functions:

$$g^*(t) = \delta(t) + 0.5\delta(t - T) + 0.25\delta(t - 2T) + \cdots$$

If the input function to this system is a sampled ramp the output at any time
is the sum of the components corresponding to each of the input samples.
This is because the system is linear for which the principle of superposition
applies.

When evaluating the output of $g^*(t)$ to $r^*(t)$ at, say, $t = 4T$ it is necessary to
sum the four components present. These are 4, 1.5, 0.5, and 0.125, which add to
6.125. This summation can be represented as;

$$c^*(4T) = \sum_{n=1}^{4} r^*(nT)g^*((4 - n)T)$$

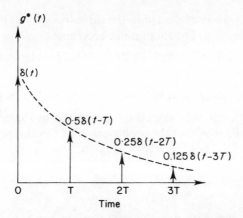

Figure 10.13   Impulse response of a system,
$g*(nT)$

where the function $g*$ has been turned about $t = 4T$. The output at any time $t = mT$ can be found using the more general convolution summation

$$c*(mT) = \sum_{n=0}^{m} r*(nT)g*((m - n)T)$$

The convolution summation lends itself to a numerical evaluation using a digital computer. This is particularly useful when it is not convenient to describe the functions in analytic form.

10.2.4   The Z-transform

All the functions used so far have been functions of time. The use of the Laplace and the Z-transforms involve functions of other variables which transform to functions of time. The transforms are devised so that the convolution operation in the time domain corresponds to multiplication in the Laplace or z domains.

The Z transform is defined so that a sampled function of time $f*(t)$ is transformed to a polynomial function of z^{-1}, $F(z)$. The power of z^{-1} is given by the time of the impulse and the coefficient of the term is given by its magnitude. In general if

$$f*(t) = \sum_{n=0}^{\infty} a_n \delta(t - nT)$$

then

$$F(z) = \sum_{n=0}^{\infty} a_n z^{-n}$$

The example previously evaluated in the time domain can be used to illustrate the use of the z domain. The ramp input becomes;

$$R(z) = 0z^0 + 1z^{-1} + 2z^{-2} + 3z^{-3} + 4z^{-4} + \cdots$$

and

$$G(z) = 1z^0 + 0.5z^{-1} + 0.25z^{-2} + 0.125z^{-3} + \cdots$$

The output function in the z domain, $C(z)$ is given by the product of these two functions. It can be seen that the coefficient of z^{-4} in this product is

$$\text{term in } z^{-4} = 1z^{-1} \times 0.125z^{-3} + 2z^{-2} \times 0.25z^{-2}$$
$$+ 3z^{-3} \times 0.5z^{-1} + 4z^{-4} \times z^0$$
$$= 6.125z^{-4}$$

The convolution summation in the time domain has been obtained using the product in the z domain. All of the output impulses are represented by the function $C(z)$; the output at nT being given by the coefficient of z^{-n}.

Most useful functions can be expressed as the ratio of finite polynomials in the z domain. This allows simple and convenient multiplication operations in the z domain. Remembering that

$$\frac{1}{1-x} = 1 + x + x^2 + x^3 + x^4 + \cdots$$

$$G(z) = \frac{1}{(2z)^0} + \frac{1}{(2z)^1} + \frac{1}{(2z)^2} + \frac{1}{(2z)^3} + \cdots$$

$$= \frac{1}{1 - 1/(2z)} = \frac{z}{z - \frac{1}{2}}$$

Transform tables are useful to find expressions such as those for $R(z)$:

$$R(z) = \frac{z}{(z-1)^2}$$

when the sampling interval $T = 1$ unit.

Functions in the z domain can be transformed into the time domain by obtaining the z domain function as a sum of terms in z^{-n}. Usually the function in the z domain contains a polynomial in the denominator so synthetic division is used to obtain the series of terms. An alternative technique for obtaining the inverse z-transform is the method of residues. This is very similar to the corresponding Laplace method. If a/α is the residue at a singular complex pole at z_p/β, the component from this pole at $t = nT$ is

$$2a|z_p|^{n-1} \cos[(n-1)\beta + \alpha]$$

It can be seen that if $|z_p| > 1$ and the pole lies outside the unit circle, the system is unstable. The unit circle gives the boundary of stability and corresponds to the $j\omega$ axis of the s-plane.

The z domain can be related to the Laplace domain using the fact that the operator z^{-1} in the z domain corresponds to the operator e^{-sT} in the Laplace domain. Often the equality $z = e^{sT}$ is inferred to relate the Laplace and the z domains. In fact the equality is not strictly correct since, for example, $e^{sT/2}$ gives a delay of $T/2$ whilst $z^{1/2}$ is not defined. The relationship between the Laplace and z domains can be inferred by comparing the responses in the time domain for given pole locations. When the impulse response from poles in the Laplace domain at $a \pm jb$ is sampled, the poles in the z domain for the corresponding response are at $e^a/\underline{\pm b}$.

In this sense $z = e^{sT}$. This is called *impulse invariance* and can be used as a design basis for filters. It should be remembered that the impulses do in fact differ because one is sampled, so the step, frequency, and phase responses will differ also. Figure 10.14 shows some examples of corresponding impulse invariant responses in the Laplace and Z domains.

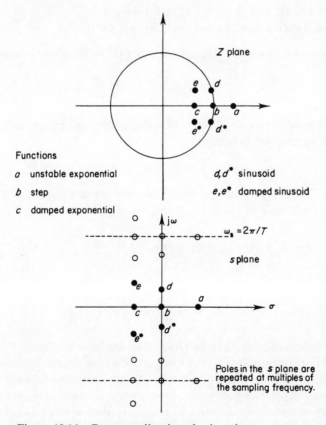

Figure 10.14   Corresponding impulse invariant responses
in the z- and s-planes

10.3 FILTER DESIGN USING THE BILINEAR TRANSFORM

If a transfer function $T(s)$ is given in the Laplace domain it can be implemented in the z domain using the *frequency warped bilinear* transform. This is

$$s = \omega \cot(\omega/2) \frac{z - 1}{z + 1}$$

where ω is normalized to the sampling interval. The implementation is approximate, however when the warping term is included the sinusoidal magnitude and phase response at ω is identical to the original Laplace domain function. In filter design ω is set equal to the cut-frequency so the salient properties of the filter are preserved.

Example 10.3. A second-order Butterworth digital filter is to be implemented having a cut-off frequency at 1 kHz (6.28 krad s^{-1}) using a sampling frequency of 10 kHz and the delay implementation.

To design such a filter we use the following steps:

(a) From filter tables the poles are at $1/\pm 135°$ which gives a transfer function

$$T(s) = \frac{\frac{1}{2}}{(s + 1/\sqrt{2})^2 + \frac{1}{2}}$$

This provides a gain of unity for arbitrarily small frequencies.

(b) The normalized frequency

$$\omega = \omega_c T = 6.28 \times 10^3 \times 10^{-4} = 0.628$$

(c) The frequency transformation required for the analog design can be combined with the frequency warping relationship using the substitution

$$s = \omega^2 \cot(\omega/2) \frac{z - 1}{z + 1} = 0.628 \times 1.934 \frac{z - 1}{z + 1} = 1.21 \frac{z - 1}{z + 1}$$

(d) Find $T(z)$ from

$$T(z) = \frac{0.137z^2 + 0.274z + \quad 1}{z^2 \quad + 0.333z + 0.262}$$

This implementation is given in block diagram form in Figure 10.15. It is the preferred implementation for very low sampling rate systems because it provides zeros in the stop band which give better attenuation in this region. However, for large sampling rates its coefficient values become critical and the signal magnitudes vary considerably throughout the filter. Also, at large sampling rates the provision of zeros at $z = -1$ is not necessary and they require additional computation.

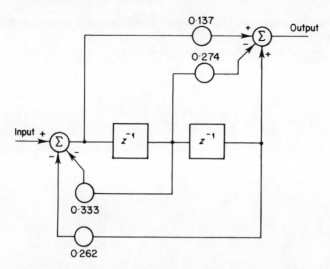

Figure 10.15 Block diagram of a delay implementation digital filter

Analog filters with carefully placed zeros in the stop band can be implemented digitally using either design technique and the corresponding advantages realized.

10.4 DISCUSSION

The introduction to digital filtering techniques presented here concentrates on the implementation of approximations to analog filters. Whilst this does provide a useful introduction to some of the properties of digital filters and to some useful filters, digital systems provide many unique features. These can be useful in special situations; and measurement systems are a collection of special situations. Although the techniques outlined here may find direct application it is more likely this treatment will be used as an introduction to more comprehensive treatments of the subject.

REFERENCES

Bogner, R. E., and Constantinides, A. G., (1975). *Introduction to Digital Filtering*, Wiley, Chichester.

Bolton, A. G., (1981). 'Design and implementation of digital filters for microprocessor implementation using fixed point arithmetic', *Proc. IEE*, **128**, Part G, No 5, 245–50.

Bolton, A. G. and Davis, B. R. (1981). 'Evaluation of coefficient sensitivities for second order digital resonators', *Proc. IEE*, **128**, Part G, No 3, 127–8.

Hamming, R. W. (1977). *Digital filters*, Prentice-Hall, Englewood Cliffs.

Hill, F. J. and Peterson, G. R. (1974). *Introduction to Switching Theory and Logical Design*, Wiley, New York.
Robinson, E. A. and Treital, S. (1964). 'Principles of digital filtering', *Geophysics*, **29**, 395–404.

Handbook of Measurement Science, Vol. 1
Edited by P. H. Sydenham
© 1982 John Wiley & Sons Ltd.

Chapter

11 D. M. MUNROE

Signal-to-Noise Ratio Improvement

Editorial introduction

A measurement system begins with sensing stages that couple to relevant measurands of the system under study. The power level of the information-bearing signals formed by the sensors is often very low and may be swamped by the unwanted noise signals that are present. Careful attention to sensor and circuit design and assembly, plus use of certain signal processing methods, makes it possible to greatly enhance the original signal-to-noise ratio to usable levels.

This chapter discusses the various strategies that are available providing a basis for their use (the best information is available in the literature of companies marketing such products). It is surprising to find that, after well over half a century of analog signal processing progress, there still exists no full length text that deals generally with signal recovery and enhancement in general instrumentation applications. This chapter is unique in this respect, apparently being the first to appear in a published text. It deals with this material at an extended theoretical depth.

It will be noticed that techniques, originally conceived for analog signals, are gradually being implemented in digital form, thus in many cases providing improved performance at comparable or less cost. This trend can be expected to increase with time as digitally oriented designers, seeking improved means of signal recovery, become more familiar with the principles already proved in the analog signal domain.

11.1 INTRODUCTION

Recovering or enhancing a signal or improving a signal-to-noise ratio simply means reducing the noise accompanying a signal. There are two basic ways of doing this:

(a) Bandwidth reduction, where the noise is reduced by reducing the system noise bandwidth (B_n). This approach works well if the frequency spectra of the noise and signal do not overlap significantly, so that reducing the noise bandwidth does not affect the signal. With random white noise the output noise is proportional to $\sqrt{B_n}$.

(b)   Averaging or integrating techniques, where successive samples of the
       signal are synchronized and added together. The signal will grow as the
       number (n) of added samples; with random white noise the noise will
       grow as $\sqrt{n}$.

In many applications there is significant overlap between the signal and noise
spectra and improving a signal-to-noise ratio must be done at the expense of
the response time or measurement time (T); with random white noise interference the output signal-to-noise ratio is proportional to $\sqrt{T}$. The bandwidth
reduction technique is best looked at from a frequency-domain point of view;
signal averaging and correlation techniques lend themselves to time-domain
analysis.

In this chapter, mathematics and theoretical considerations will be reduced
to a minimum; the reader is referred to Chapter 4 for additional theoretical
and background information. For further simplicity we will assume that all
noise processes are stationary and that both signal and noise are ergodic,
analog variables; we will not concern ourselves with digital signals or discrete-
time (sampled) signals except where such signals are involved in the enhancement techniques. In addition, only signal recovery techniques will be
considered. Further processing, such as least-squares polynomial smoothing
of a waveform or Fourier transformation to obtain a frequency spectrum, will
not be considered here.

We will start by reviewing some basic concepts, move on to discuss ways to
avoid *adding* noise (e.g. hum pick-up and preamplifier noise) and then discuss
instrumentational techniques to reduce the remaining noise content. Finally,
we will discuss some of the special considerations involved in recovering pulse
signals from photon (light), ion, or electron beams.

11.2   NOISE AND NOISE BANDWIDTH

Noise is an undesired signal. It usually becomes of interest when it obscures
a desired signal. Figure 11.1 shows the power spectral density (power/unit
bandwidth) of the most commonly encountered types of noise.

Deterministic noise can range from simple discrete-frequency components
such as power-line hum at harmonics of 50 or 60 Hz, to wide-band interference (RFI) caused by narrow, high-energy pulses from power-line switching
spikes, pulsed lasers, radar transmitters, and the like.

Stochastic (random) noise is found in most systems both as white noise,
where the power spectral density is independent of frequency, and also as $1/f$
or flicker noise, where the power spectral density decreases as frequency
increases. Power spectral density is usually measured in mean-squared-volts/Hz
or mean-squared-amperes/Hz; for noise, such specifications are usually
referred to as *spot noise* data and usually are a function of frequency. Notice

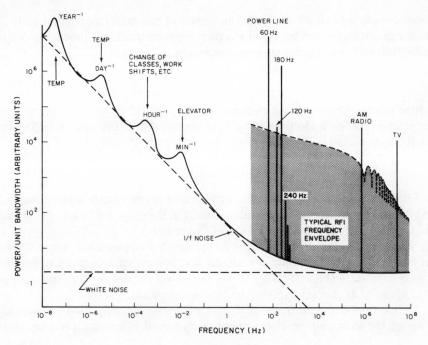

Figure 11.1   Environmental noise (reproduced by permission of EG&G Princeton Applied Research Corporation)

that for an r.m.s. voltage of e (volts) and a frequency range of Δf (Hz), the power spectral density, S, is given by

$$S = \frac{e^2}{\Delta f} = \left(\frac{e}{\sqrt{(\Delta f)}}\right)^2 \tag{11.1}$$

The quantity $e/\sqrt{(\Delta f)}$ is usually referred to as *voltage spectral density* and is measured in r.m.s.-volts/$\sqrt{\text{Hz}}$ (volts per root hertz). Similarly, we can refer to current spectral density specifications in units of r.m.s.-amperes/$\sqrt{\text{Hz}}$.

White noise is usually found in one of two forms: *Johnson* noise and *shot* noise. Johnson, or thermal, noise is caused by random motion of thermally agitated electrons in resistive materials, and the mean-square noise voltage is given by

$$e_n^2 = 4kTR\,\Delta f \tag{11.2}$$

where k is Boltzmann's constant (1.381×10^{-23} JK^{-1}), T is the absolute temperature (kelvin) and R is the resistance (ohm). Alternatively, from Ohm's law, the mean-square noise *current* is given by

$$i_n^2 = \left(\frac{e_n}{R}\right)^2 = \frac{4kT\Delta f}{R} \tag{11.3}$$

Shot noise is caused by the random arrival of electrons (see Section 11.10.2) at, for example, the electrodes of electron tubes or transistor junctions. A d.c. current, I, will have a noise-current component, i_n, given by

$$i_n^2 = 2AeI\Delta f \tag{11.4a}$$

where e is the charge of one electron ($\simeq 1.6 \times 10^{-19}$ C), A is the mean gain experienced by each electron and I is in amperes. In many cases (see Section 11.10.2), $A = 1$, so that

$$i_n^2 = 2eI\Delta f \tag{11.4b}$$

Flicker noise has many different origins and is not clearly understood but exhibits a $1/f^n$ power spectrum with n usually in the range 0.9 to 1.35. Note that *d.c. drift* is a very-low-frequency form of flicker noise.

What do we mean by *bandwidth*? In the simple low-pass filter circuit shown in Figure 11.2a, for example, we usually and somewhat arbitrarily define the *signal bandwidth* (Figure 11.2b) to be the *cut-off frequency*, f_c, where $e_o/e_i = 70.7\%$ (-3 dB) or $e_o^2/e_i^2 = 50\%$ (the half-power point).

Notice that frequencies above f_c will obviously pass (though attenuated) through the filter, and therefore are not really cut off. For noise, it is convenient

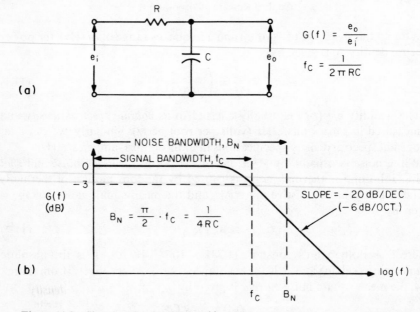

Figure 11.2   Signal and noise bandwidths of low-pass filter: (a) circuit; (b) Bode plot

to think in terms of an equivalent noise bandwidth, B_n, defined by the relationship

$$B_n = \frac{1}{G^2} \int_0^\infty |H(j\omega)|^2 \, df \tag{11.5}$$

where $H(j\omega)$ is the frequency response function of the system and G is a gain parameter suitably chosen to be a measure of the response of the system to some parameter of the signal: for low-pass systems (e.g. Figure 11.2b) G is usually taken to be the zero-frequency (d.c.) gain; for band-pass responses, G is usually made equal to the maximum gain.

Using the above definition, and taking G to be the zero-frequency gain (i.e. unity), we can readily calculate that for the simple RC filter shown in Figure 11.2a:

$$B_n = 1/4RC \text{ Hz} \tag{11.6}$$

Noise, of the stochastic form, is reviewed in relation to instrument systems in Fellgett and Usher (1980).

11.3 SIGNALS AND SIGNAL-TO-NOISE RATIO

Suppose we were to look at a complex waveform on an oscilloscope. What is the signal? Is it the complete waveform? The peak (or r.m.s. or average) amplitude? The depth of modulation? The implied frequency spectrum? The difference in time or amplitude between two features of the waveform? The answer, of course, is that the information-bearing signal could be any or none of the above. In this chapter, we will restrict ourselves to some commonly encountered types of signal where enhancement is often required. Together with the enhancement technique normally used, these are:

(a) base-band (d.c.) signals: low-pass filtering or autocorrelation;
(b) amplitude modulated signals: band-pass filtering or phase-sensitive detection;
(c) repetitive (not necessarily periodic) swept signals: signal averagers;
(d) photon, electron or ion beam signals: photon-counting systems.

The word *signal* is often used rather ambiguously to mean either the total signal being measured or a noise-free, information-bearing component of it. The following definitions should allow us to avoid such confusion. We will normally talk in terms of a *total signal* consisting of an r.m.s. signal component (S) accompanied by an r.m.s. noise component (N). Thus

$$\text{Signal-to-noise ratio, SNR} = S/N \tag{11.7}$$

Note that

$$\text{measurement uncertainty or inaccuracy} = \frac{1}{\text{SNR}} \qquad (11.8)$$

$$\text{Signal-to-Noise improvement ratio (SNIR)} = \frac{\text{SNR}_{\text{out}}}{\text{SNR}_{\text{in}}} = \frac{S_o/N_o}{S_i/N_i} \qquad (11.9)$$

For unity gain (i.e. $S_o = S_i$), band-limited white input noise of bandwidth B_{ni} and output noise bandwidth B_{no},

$$\text{SNIR} = N_o/N_i = \sqrt{(B_{ni}/B_{no})} \qquad (11.10)$$

11.4 NOISE MATCHING AND PREAMPLIFIER SELECTION

All preamplifiers add noise. Whether this additional noise is significant will depend, of course, upon the noise level from the signal source. Since uncorrelated noise adds vectorially (in an r.m.s. fashion), the preamplifier noise can be neglected if it is less than about one-third of the source noise:

$$\sqrt{[(1.0)^2 + (0.3)^2]} \simeq 1.0$$

We can think of a practical preamplifier as consisting of an ideal, noise-free amplifier with a (frequency-dependent) noise-voltage generator of voltage spectral density e_n (V/$\sqrt{\text{Hz}}$), and a noise-current generator of current spectral density i_n (A/$\sqrt{\text{Hz}}$), connected to its input as shown in Figure 11.3a. Figures 11.3b and 11.3c, respectively, show separately the gain seen by the amplifier internal noise voltage and current generators. Any input shunt capacitance (see Figure 11.3b) will decrease the input impedance and cause output noise that increases with frequency if Z_f is resistive.

The preamplifier noise may also be defined (Faulkner, 1966) in terms of an equivalent series noise resistance R_e, and an equivalent parallel noise resistance, R_i, where (from equation (11.2))

$$R_e = \frac{e_n^2}{4kT\Delta f} \text{ ohms}$$

and (from equation (11.3))

$$R_i = \frac{4kT\Delta f}{i_n^2} \text{ ohms}$$

We can also define the *noise figure* (NF) of the preamplifier to be (in dB)

$$\text{NF} = 10 \log_{10}\left(1 + \frac{R_e}{R_s} + \frac{R_s}{R_i}\right) \qquad (11.11)$$

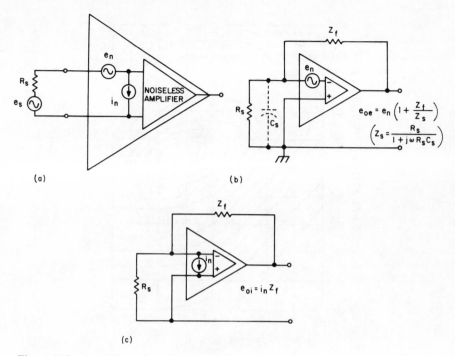

Figure 11.3 Amplifier noise: (a) equivalent circuit; (b) voltage noise; (c) current noise

A perfect or noiseless preamplifier would have a 0 dB noise figure. Figure 11.4 shows the noise figure contours that result when the noise figure of a practical preamplifier is plotted as a function of source resistance and frequency. Notice from equation (11.11), that with high source resistance, $R_e/R_s \to 0$, and

$$NF \simeq 10 \log_{10}\left(1 + \frac{R_s}{R_i}\right)$$

and the amplifier noise current, i_n, predominates. With low source resistances, the amplifier noise voltage, e_n, becomes the major noise source. Wherever possible, preamplifiers should be chosen so that their 3 dB noise-figure contour encloses the expected range of source resistance and frequency.

For a given preamplifier, the optimum source resistance, R_s, is given by

$$R_s(\text{opt}) = \frac{e_n}{i_n} = \sqrt{(R_e R_i)} \text{ ohms} \qquad (11.12)$$

Note that adding a series or parallel resistance between the signal source and the preamplifier always reduces signal and adds noise, and so cannot be used to obtain a *better match*.

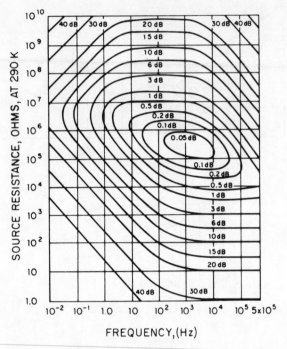

Figure 11.4 Typical noise figure contours for a
high input impedance preamplifier (reproduced by
permission of EG&G Princeton Applied Research
Corporation)

Preamplifiers can be classified in many ways; one basic division, for example,
is between *differential* input and *single-ended* input. All other things being
equal, a differential preamplifier generates 3 dB (41.4%) more noise than a
single-ended version. However, this disadvantage is significant only in situations
where preamplifier noise predominates and, in many cases, is outweighed by
the flexibility of a differential input and its ability to remove ground-loop
problems (see Section 11.5).

Transformers are often used to match very low source impedances (0.1 Ω–
1 kΩ). Figure 11.5 shows an amplifier with an optimum, $\sqrt{(R_e R_i)}$, source
resistance value of 1 MΩ being matched to a 100 Ω thermopile by means of a
100 : 1 voltage step-up transformer (10,000 : 1 impedance transformation).
Note that, in general, such noise matching does not result in the same circuit
values as would power matching; that is, the amplifier input resistance is not
normally equal to $\sqrt{(R_e R_i)}$. Transformers should be avoided if possible, since
they reduce frequency response, may pick up magnetically induced interference and may be microphonic.

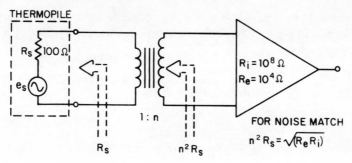

Figure 11.5   Transformer noise matching

For sources of approximately 100 Ω–10 kΩ, preamplifiers are available which use an input stage consisting of multiple bipolar transistors connected in parallel to provide a lower value of $\sqrt{(R_e R_i)}$. Such preamplifiers avoid the bandwidth constraints imposed by input transformers. For higher impedance sources (1 kΩ–100 MΩ), preamplifiers usually employ junction-FET's as input devices and are available as voltage preamplifiers, charge amplifiers (for use with capacitive transducers), or current-input (transresistance) amplifiers. (See Figure 11.6).

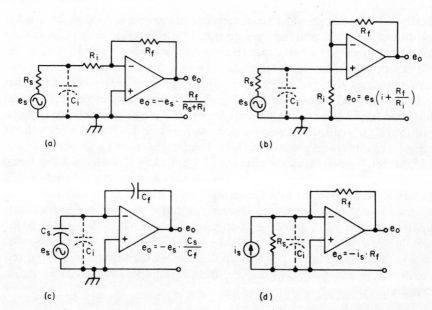

Figure 11.6   Amplifier configurations: (a) voltage, inverting; (b) voltage, non-inverting; (c) charge; (d) current input (trans-resistance)

In Figures 11.6a and 11.6b any cable capacitance or stray capacitance, C_i, will form a low-pass filter with the source resistance, R_s, having a -3 dB frequency given by

$$f_c = \frac{1}{2\pi R_i R_s C_i / (R_i + R_s)} \quad \text{(Figure 11.6a)}$$

or

$$f_c = \frac{1}{2\pi R_s C_i} \quad \text{(Figure 11.6b)}$$

In Figures 11.6c and 11.6d, such shunt capacitance appears at first sight to have no effect since it is effectively short-circuited by the virtual-ground input. However, as shown in Figure 11.3b, shunt capacitance will cause a deterioration in the output SNR and also (by introducing an additional pole into the loop gain) may cause *ringing* in the amplifier response, or even oscillation. By careful design (which usually includes adding a capacitor across the feedback resistor) these effects can be minimized and with high source impedance, commercial current and charge amplifiers usually provide significantly greater bandwidth than can corresponding voltage amplifiers.

11.5 INPUT CONNECTIONS; GROUNDING AND SHIELDING

Ideally, all grounds should have a zero-impedance connection to each other and to wet earth; in practice they do not. Due to voltage drops across their finite impedance to earth, capacitively or inductively coupled interference, and other reasons, each ground tends to be at a different potential from other nearby grounds. If two (or more) such adjacent grounds are connected together to form a ground loop (Figure 11.7a), then the potential difference between the grounds will cause a circulating current. The potential difference between grounds (e_{cm}), is called the *common-mode source* since it is common to both the signal (via loop 2) and ground (via loop 1) inputs of the preamplifier.

Figure 11.7b rearranges the circuit of Figure 11.7a and assumes the signal source, e_s, to be zero. Note that the low resistance of the coaxial cable shield (braid), R_{cg}, is in parallel with the series combination of the source resistance, R_s, the coaxial cable centre conductor resistance R_{cs} and the preamplifier input impedance, Z_{in}. Under normal circumstances $(R_s + R_{cs} + Z_{in}) \gg R_{cg}$ and $Z_{in} \gg (R_s + R_{cs})$, so that as shown in Figure 11.7c, the common-mode voltage dropped across R_{cg} is also applied across the preamplifier input terminals. More generally, with $e_s = 0$, the preamplifier input voltage, e_{in}, is given by

$$e_{in} = e_{cm} \frac{R_{cg}}{R_{cg} + R_{sg} + R_{pg}} \tag{11.13}$$

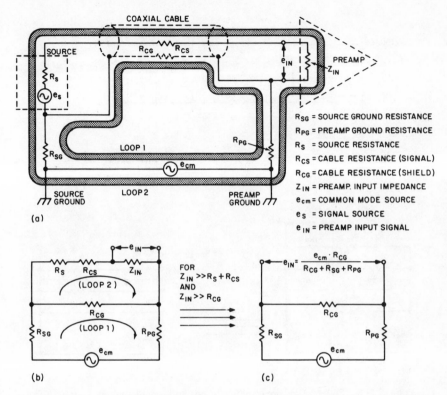

Figure 11.7 Ground loops: (a) physical occurrence; (b) schematic equivalent circuit; (c) reduced equivalent circuit

From equation (11.13), this common-mode input to the preamplifier can be removed (i.e. $e_{in} = 0$) by making:

(a) $e_{cm} = 0$. This can be attempted by grounding the source and preamplifier to the same ground point, and shielding to remove capacitively or inductively coupled interference, but the procedure is rarely completely successful.

(b) $R_{cg} = 0$. The usual approach here is to bolt both source and preamplifier chassis to a large metal plate. Unfortunately, it is fairly easy to develop large potential differences between points a centimetre or two apart on a large metal plate, such as a mounting rack.

(c) $R_{sg} = \infty$. *Floating* or disconnecting the source from ground is a good approach where practicable.

(d) $R_{pg} = \infty$. The preamplifier may also be floated—particularly if it is battery powered. Note that disconnecting the power-line ground from an instrument can be extremely dangerous. In many instruments R_{pg} consists of an internal 10 Ω–1 kΩ resistor that can be switched into the circuit to effectively float the amplifier input terminals.

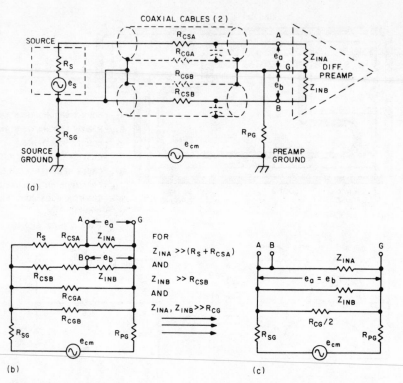

Figure 11.8 Differential preamplifier used with single-ended source: (a) physical occurrence; (b) schematic equivalent circuit; (c) reduced equivalent circuit

Figure 11.8 illustrates the use of a differential amplifier with an unbalanced (single-ended) source to eliminate or reduce ground-loop problems. As in Figure 11.7, the circuit simplification assumes that the input impedance of each side of the differential amplifier (Z_{inA} and Z_{inB}) is much larger than source or cable resistances. At low frequencies this differential connection results in equal common mode voltages at the amplifier's input terminals (A and B), and the amplifier's ability to discriminate against common-mode signals (i.e. its *common-mode rejection ratio*, CMRR or CMR), will determine the effectiveness of this configuration in suppressing ground-loop interference. At higher frequencies, the cable capacitances will act with the unequal resistances in the A and B input circuits to form unequal low-pass filters, so that e_A will no longer be equal to e_B and there will be a spurious differential (A–B) input to the preamplifier. Though cable resistances and capacitances are shown for convenience as lumped parameters, it should be remembered that in fact they are distributed. As shown in Figure 11.9, high-frequency unbalance

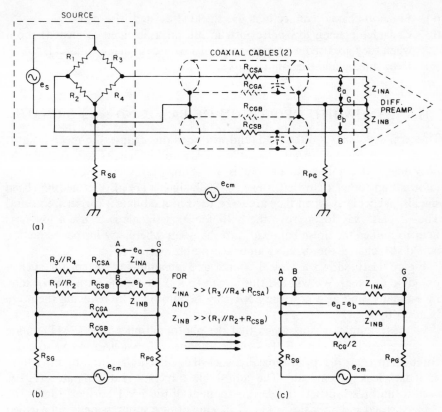

Figure 11.9 Differential preamplifier used with balanced source: (a) physical occurrence; (b) schematic equivalent circuit; (c) reduced equivalent circuit

problems can be avoided by using a balanced source. Specific comment on connecting stages together is to be found in (Morrison, 1977).

To end this section, here are a number of miscellaneous recommendations regarding good wiring and grounding practices.

(a) Keep cable lengths short; for differential connections, keep them equal and following the same route.

(b) Interference can be coupled into the ground (shield) or outer conductor of a coaxial cable. Consider coiling the cable to form an RF choke to suppress high-frequency interference of this kind, use a transformer, or use a balun (which allows d.c. continuity).

(c) Remember that a loop of wire acts as an antenna; reduce the area of such loops as much as possible.

(d) Separate low-level signals/cables from noisy ones. Where such cables must cross, cross them at right angles and with maximum separation.

(e) For non-coaxial connections use shielded twisted wire-pairs.
(f) Consider placing low-noise instruments in a shielded (screened) room when they are used with high-energy RF sources, such as pulsed lasers.
(g) Keep analog and digital grounds separate.

11.6 BANDWIDTH REDUCTION OF BASEBAND (d.c.) SIGNALS

The term *d.c. signal* is often used (and will be in this chapter) to mean a signal which has a frequency spectrum that includes zero frequency (d.c.). Technically, of course, a d.c. voltage or current is unvarying and, therefore, cannot carry information (other than that it exists). Such signals are also termed *baseband* signals, particularly when they are to be used to modulate a carrier frequency. The simplest way to improve the SNR for such signals is to use a low-pass filter to reduce the noise bandwidth to the point where any further reduction would also change the signal to an unacceptable extent.

Figure 11.10 shows a typical source and preamplifier system for such a pseudo-d.c. signal. We will use this circuit to show how the output SNR may by estimated and also how the SNR may be improved by reducing the noise bandwidth.

In this example, it is assumed that the photomultiplier tube (PMT) anode current consists of both a 5 Hz signal component ($i_s = 1$ nA r.m.s.) and a d.c. component ($I_{d.c.} = 5$ nA); typically, such d.c. currents are due to stray light and dark/leakage currents. The adjustable direct-current generator (I_{zs}) is used to null (zero offset) the d.c. component of the PMT current; that is, I_{zs} is made equal and opposite to $I_{d.c.}$. This kind of zero suppression is often called *background subtraction*. Notice that I_{zs} must be readjusted manually each time the background changes.

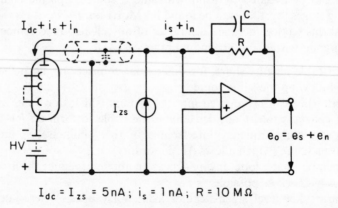

$$I_{dc} = I_{zs} = 5 \text{nA}; \quad i_s = 1 \text{nA}; \quad R = 10 \text{M}\Omega$$

Figure 11.10   d.c. measurement example

The coaxial cable connecting the PMT and preamplifier has capacitance. Notice that the virtual-ground input of this preamplifier configuration offers the following advantages in addition to those discussed in Section 11.4:

(a) With zero volts across it, the cable capacitance cannot be charged and the cable will, therefore, be less microphonic than otherwise.

(b) Since the PMT anode voltage is clamped at zero volts, the anode-to-last-dynode voltage is also held constant regardless of i_s (assuming that the dynode voltage remains fixed). Signal currents will, therefore, not change the PMT gain.

There are five, uncorrelated sources of noise in this circuit. These are:

(a) The d.c. component of the PMT current ($I_{d.c.}$) is produced by integrating anode pulses each of charge $Q = Ae$ where A is the mean PMT gain. The interval in time between successive pulses arriving at the anode is random and governed by *Poisson* statistics (see Section 11.10.2). Assuming no additional dynode noise in the PMT, then the r.m.s. value of the PMT *shot noise* current spectral density, i_{n1}, is given by

$$i_{n1} = \sqrt{(2AeI_{d.c.})} = \sqrt{A} \times \sqrt{(2eI_{d.c.})}$$

For $A = 10^6$ (say), $R = 10^7 \, \Omega$, $e \simeq 1.6 \times 10^{-19}$ C and $I_{d.c.} = 5$ nA, the resulting output noise voltage density, e_{n1}, is given by

$$e_{n1} = Ri_{n1} = 10^7 \times 10^3 \times \sqrt{2eI_{dc}} \simeq 10^{10} \times 4 \times 10^{-14}$$
$$= 4 \times 10^{-4} = 400 \, \mu V/\sqrt{Hz}$$

(b) For purposes of this example, we can assume that the zero-suppress current, I_{zs}, is obtained from a transistor current source circuit so that it has a shot noise current spectral density, i_{n2}, given by

$$i_{n2} = \sqrt{2eI_{zs}} = \sqrt{(2eI_{d.c.})}(= i_{n1}/10^3) = 4 \times 10^{-14} A/\sqrt{Hz}$$

Note that though $I_{d.c.} = I_{zs}$, the shot noise component from the PMT is much larger than that from the transistor current source. The resulting output noise voltage spectral density, e_{n2}, is given by

$$e_{n2} = Ri_{n2} \simeq 10^7 \times 4 \times 10^{-14} = 4 \times 10^{-7} = 400 \, nV/\sqrt{Hz}$$

(c) The feedback resistor, R, will generate (at $T = 290$ K) a Johnson noise output voltage density, e_{n3}, given by

$$e_{n3} = \sqrt{(4kTR)} \simeq 4 \times 10^{-7} = 400 \, nV/\sqrt{Hz}$$

(d) At 5 Hz, a typical value for the spot noise voltage density of the amplifier's internal noise voltage generator is 30 nV/$\sqrt{Hz}$. This amplifier voltage noise will experience unity gain (see Figure 11.3b) since Z_{in} (the PMT current source) is very high. The output noise voltage density, e_{n4}, due to this noise source is therefore given by

$$e_{n4} = 30 \, \text{nV}/\sqrt{\text{Hz}}$$

(e) At 5 Hz, a typical value for the spot noise current density, i_{n5}, of the amplifier internal noise current generator is 5 fA/$\sqrt{\text{Hz}}$. The resulting contribution, e_{n5}, to the amplifier output noise is given by

$$e_{n5} = Ri_{n5} = 10^7 \times 5 \times 10^{-15} = 5 \times 10^{-8} = 50 \, \text{nV}/\sqrt{\text{Hz}}$$

The total output noise voltage spectral density, e_n, is given by

$$e_n = \sqrt{(e_{n1}^2 + e_{n2}^2 + e_{n3}^2 + e_{n4}^2 + e_{n5}^2)}$$

Since $e_{n1}^2 \gg e_{n2}^2, e_{n3}^2, e_{n4}^2$, and e_{n5}^2, then

$$e_n \simeq e_{n1}$$

and the system is said to be *detector limited* or *shot noise limited*. An *electrometer*, an instrument characterized by extremely low leakage currents, is often used as a low-noise amplifier in d.c. measurements of this kind.

In Figure 11.10 the parallel resistor and capacitor in the feedback loop cause the circuit to act as low-pass filter of time constant RC seconds, so that the $-3 \, \text{dB}$ cutoff frequency is given by $1/2\pi RC$ and $B_n = 1/4RC$ (see Section 11.2).

If no discrete capacitor is connected across R, the typical stray capacitance will (say) be about $C = 2.5 \, \text{pF}$, so that $RC = 10^7 \times 2.5 \times 10^{-12} = 25 \, \mu\text{s}$ and $B_n = 10^4 \, \text{Hz}$. The output noise voltage (E_n) will, therefore, be

$$E_n = e_n \sqrt{B_n} = 4 \times 10^{-4} \times \sqrt{10^4} = 40 \, \text{mV}$$

The output signal

$$e_s = i_s R = 10^{-9} \times 10^7 \, \text{V} = 10 \, \text{mV}$$

Therefore,

$$\text{SNR} = \frac{S}{N} = \frac{10}{40} = \frac{1}{4}$$

The capacitance can be increased to 2.5 nF by adding discrete capacitors so that the noise bandwidth becomes $B_n = 10 \, \text{Hz}$. The $-3 \, \text{dB}$ corner frequency will now be at 6.4 Hz (i.e. $10 \times 2/\pi$) so that the signal (frequency is 5 Hz) is not significantly attenuated. The output noise voltage (E_n) is now reduced to

$$E_n = e_n \sqrt{B_n} = 4 \times 10^{-4} \times \sqrt{(10)} \simeq 1.26 \, \text{mV}$$

and

$$\text{SNR} = \frac{S}{N} = \frac{10}{1.26} \simeq \frac{8}{1}$$

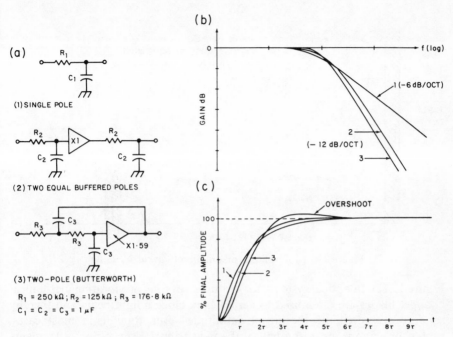

Figure 11.11 Low-pass filter characteristics: (a) filter types (all have same ENBW); (b) frequency responses; (c) time responses to voltage step input

so that (see equations (11.9) and (11.10))

$$\text{SNIR} = \frac{S_o/N_o}{S_i/N_i} = \sqrt{\frac{B_{ni}}{B_{no}}} = \sqrt{\frac{10000}{10}} \simeq \frac{32}{1}$$

The roll-off rate of a low-pass filter may be increased by adding more RC sections (see Figure 11.11a). Care should be taken in using some multipole filter configurations (Chebyshev or Butterworth, for example), since many such filters have undesirable overshoot characteristics (see Figure 11.11c). Notice that the term *time constant* (τ) is meaningful only in connection with a single RC filter section and, even then, does not adequately convey a sense of the response time of the filter. With a voltage-step input, for example, such a single RC section requires about five time-constant intervals for its output to rise to within 1 % of its final value.

11.7 AMPLITUDE-MODULATED SIGNALS; THE LOCK-IN AMPLIFIER

Most measurement systems are troubled by $1/f$ noise. By amplitude modulating the measurand (quantity to be measured) at some reference or carrier frequency, f_r, the output noise can often be reduced and d.c. drift problems avoided (see

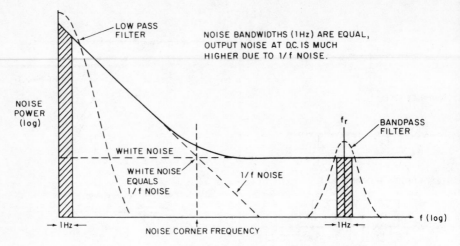

Figure 11.12   Amplitude modulation to avoid $1/f$ noise

Figure 11.12). In optical systems, for example, rotating or vibrating mechanical chopper blades are often used to periodically block a light beam and thereby square-wave modulate the signal amplitude—even though, in most cases, such chopping means losing half of the light (signal). Measuring instruments that respond only to the modulation provide *automatic background subtraction*; as with d.c. systems, however, the noise component of the background remains. Such modulation also allows the use of transformers to noise-match preamplifiers to low-resistance sources.

As with baseband signals and low-pass filtering, the SNR of a noisy amplitude-modulated signal can be improved by bandwidth reduction—in this case a *band-pass* filter is commonly used. In most applications, carrier frequencies are chosen from the 100 Hz–10 kHz range, where preamplifier and environmental noise is lowest; care should also be taken to avoid frequencies occupied by harmonics of the power-line frequency. A second-order band-pass filter (see Figure 11.13) is specified by its resonant or centre frequency, f_r, and its selectivity, Q (quality factor). For a given value of f_r, the higher the Q, the narrower the filter width.

The -3 dB frequencies are at $f_r \pm f_c$ and signal bandwidth (B_s) is defined by

$$B_s = 2f_c = f_r/Q \tag{11.14}$$

For a second-order band-pass, the signal bandwidth and the equivalent noise bandwidth (B_n) are related by

$$B_n = \tfrac{1}{2}\pi B_s$$

so that,

$$B_n = \pi f_r/2Q \tag{11.15}$$

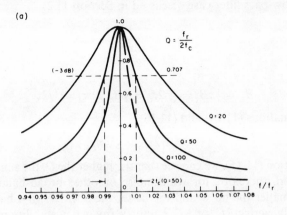

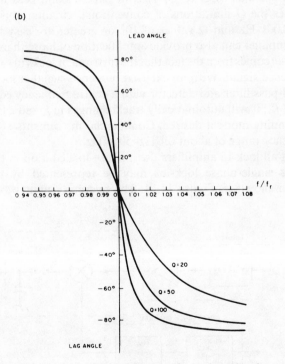

Figure 11.13 Characteristics of a second-order bandpass filter: (a) amplitude; (b) phase

The band-pass filter has associated with it an effective time constant, τ, where (as with the low-pass filter case discussed in Section 11.2)

$$f_c = \frac{1}{2\pi\tau} \tag{11.16}$$

so that,

$$B_n = \tfrac{1}{2}\pi B_s = \tfrac{1}{2}\pi 2 f_c = \tfrac{1}{2}\pi(2/2\pi\tau) = 1/2\tau \tag{11.17}$$

Also, from equations (11.14) and (11.16)

$$\tau = 1/2\pi f_c = Q/\pi f_r \tag{11.18}$$

From equation (11.15), we can see that the higher the Q, the smaller the noise bandwidth and, therefore, for white noise or other broad-band noise interference, the smaller the noise and the better the SNR. With a band-pass filter implemented by active RC (or LC) circuitry, frequency-stability problems limit the maximum practicable value of Q to about 100.

The *lock-in amplifier* (EG & G, 1) is in part, a band-pass filter–amplifier that overcomes the Q limitations of conventional circuits; noise bandwidths of less than 0.001 Hz and Q values of 10^8 or greater are easy to implement. The lock-in amplifier can also provide amplification of more than 10^9 (180 dB). The term *lock-in* comes from the fact that the instrument locks in to the frequency (f_r) of a reference signal. With an external reference signal, a lock-in acts as a tracking band-pass filter and detector with a centre frequency equal to that of the reference (f_r); it will automatically track changes in f_r and can be used in a frequency-scanning mode if desired. Commercial instruments are available to cover a frequency range of about 0.1 Hz–50 MHz.

Though not all lock-in amplifiers use a phase-locked loop in their reference channel, most single-phase lock-ins may be represented by the simplified block diagram shown in Figure 11.14. The reference input waveform to the

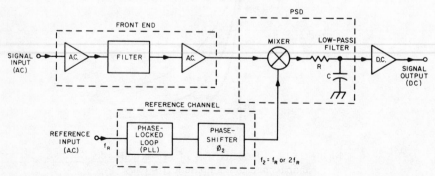

Figure 11.14   Basic lock-in amplifier (simplified) (reproduced by permission of EG&G Princeton Applied Research Corporation

lock-in may be of almost any waveshape and its zero crossings are used to define zero phase ($\phi_r = 0$). The output of the phase-locked loop circuit is a precise square-wave, locked in phase to the reference input, and at a frequency f_2. Normally, $f_2 = f_r$ (the reference frequency); most lock-ins also provide a second-harmonic mode where $f_2 = 2f_r$ and this mode is often used for derivative (signal rate of change) measurements.

All lock-ins use a *phase-sensitive detector* (PSD) circuit and all PSD circuits consist of nothing more than a *mixer* followed by a low-pass filter. The output of a mixer (e_3) is the product of its signal (e_1) and gating (e_2) inputs, that is, $e_3 = e_1 e_2$, and the phase difference between these two inputs can be precisely adjusted by the phase-shifter circuit in the reference channel. For use in lock-in amplifiers, mixer circuits must be capable of withstanding large amounts of noise (i.e. asynchronous signals, $f_1 \neq f_2$) without overloading. The term *dynamic reserve* is used to specify such noise overload performance (see Figure 11.15). Dynamic reserve is defined as the ratio of the overload level (peak value of an asynchronous signal that will just cause significant non-linearity), to the peak value of a full-scale synchronous signal. Dynamic reserve is often confused with *dynamic range*, which is the ratio of the overload level to the minimum detectable signal level.

The d.c. drift of both the mixer and d.c. amplifier may limit the minimum detectable signal and the gain of the d.c. amplifier should, therefore, be minimized to provide optimum output stability; a.c. gain should be used to provide most of the overall instrument gain required. Such a gain distribution is practicable and desirable for use with clean signals. With noisy signals, however, the a.c. gain must be reduced to provide increased dynamic reserve, and the d.c. gain increased proportionately. Most high-performance instruments

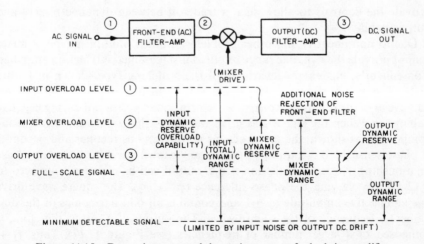

Figure 11.15   Dynamic range and dynamic reserve of a lock-in amplifier

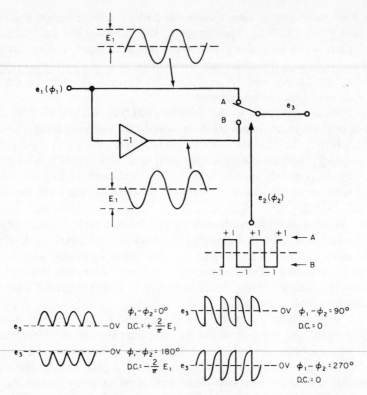

Figure 11.16   Switching mixer waveforms (reproduced by permission
of EG&G Princeton Applied Research Corporation)

provide the controls to allow such a trade-off between dynamic reserve and
output stability.

Due to non-linearity and other problems, a linear multiplier type of mixer
cannot provide the dynamic range required in a (commercial) lock-in amplifier.
Consequently, mixers are invariably of the switching type, shown in Figure
11.16. The switch shown in this figure will be in position A during positive
half-cycles of the square-wave drive waveform and in position B during negative
half-cycles. When the signal and drive waveforms have a common frequency
component, as shown, the mixer acts as a synchronous rectifier and produces
a phase-sensitive d.c. output. Outputs are shown for four different phase
relationships. Notice that the mixer d.c. output can be adjusted from zero to
$\pm(2/\pi)E$, by varying the phase-difference $(\phi_1 - \phi_2)$. The square-wave drive
has an effective amplitude of ± 1 and contains all odd harmonics of the fundamental frequency of the square-wave. The output of a mixer, therefore, is
composed of a large number of frequencies (see Figure 11.17). Thus, $f_1 +$
$f_2, f_1 + 3f_2, f_1 + 5f_2, \ldots$, are sum frequencies; $f_1 - f_2, f_1 - 3f_2, f_1 - 5f_2, \ldots$,

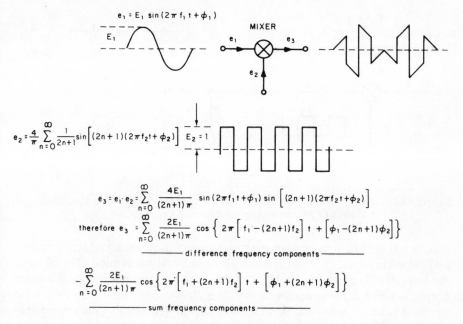

$$e_1 = E_1 \sin(2\pi f_1 t + \phi_1)$$

MIXER

$$e_2 = \frac{4}{\pi} \sum_{n=0}^{\infty} \frac{1}{2n+1}\sin\left[(2n+1)(2\pi f_2 t + \phi_2)\right]$$ $E_2 = 1$

$$e_3 = e_1 \cdot e_2 = \sum_{n=0}^{\infty} \frac{4E_1}{(2n+1)\pi} \sin(2\pi f_1 t + \phi_1) \sin\left[(2n+1)(2\pi f_2 t + \phi_2)\right]$$

$$\text{therefore } e_3 = \sum_{n=0}^{\infty} \frac{2E_1}{(2n+1)\pi} \cos\left\{2\pi\left[f_1 - (2n+1)f_2\right]t + \left[\phi_1 - (2n+1)\phi_2\right]\right\}$$

──────── difference frequency components ────────

$$- \sum_{n=0}^{\infty} \frac{2E_1}{(2n+1)\pi} \cos\left\{2\pi\left[f_1 + (2n+1)f_2\right]t + \left[\phi_1 + (2n+1)\phi_2\right]\right\}$$

──────── sum frequency components ────────

Figure 11.17   Switching mixer operation (reproduced by permission of EG&G Princeton Applied Research Corporation)

are difference frequencies. Note that when $f_1 = (2n + 1)f_2$, one of the difference-frequency components of the mixer output will be at zero frequency, or d.c.. A mixer will, therefore, produce a phase-sensitive d.c. output when $f_1 = (2n + 1)f_2$ and for $n > 0$, these outputs have magnitudes that are inversely proportional to their harmonic number (n) and are known as the *harmonic responses* of the mixer. Lock-ins respond to the *average full-wave rectified* value of the input signal but are usually calibrated in r.m.s.—a sinusoidal input or sinusoidal front-end response is assumed.

The PSD input (e_1) need not be sinusoidal. If e_1 were a synchronous square-wave signal, for example, such as that resulting from chopped-light experiments, then e_1 would contain a large number of synchronous components each of which would give rise to an output d.c. signal from the PSD.

In a perfect mixer, only synchronous inputs can cause a d.c. output. In practice, due to non-linearities of the mixer switching elements, a mixer can produce a d.c. output with high-level noise inputs; even with no (zero) input, capacitive feedthrough can cause a d.c. output. Such spurious d.c. outputs are normally negligible in amplitude. However, at higher frequencies (above 10 kHz typically), the magnitude of such a d.c. offset and its associated drift may become significant.

As we saw previously in Figure 11.17, the output of a mixer contains a large number of sinusoidal sum and difference frequency components. (The number

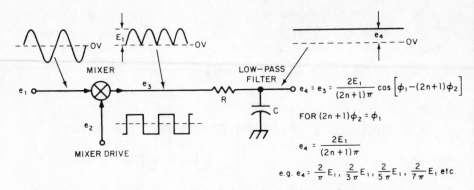

Figure 11.18   PSD operation with synchronous signal (reproduced by permission of EG&G Princeton Applied Research Corporation)

is large rather than infinite, since the squareness of the drive signal is not perfect, and the drive signal does not contain all higher odd harmonics). The effect of the low-pass filter (see Figure 11.18) is to remove all components of the mixer output which have frequencies beyond the filter cut-off. When the filter time constant is set normally (so that the filter cut-off frequency (f_c) is appreciably less than the fundamental frequency (f_2) of the mixer square-wave drive), the output of a PSD will contain only those difference frequency components having frequencies within (approximately) the equivalent noise bandwidth of the filter.

Suppose, as shown in Figure 11.19, that the PSD input (e_1) is asynchronous (noise) of frequency $f_1 = f_2 + \Delta f$. The resulting mixer sum and difference frequencies (ignoring harmonics for simplicity) will, therefore, be $2f_2 + \Delta f$ and Δf respectively. Only the Δf component may be low enough in frequency to pass through the low-pass filter and appear as output noise. Suppose we change the frequency of this input noise to $f_1 = f_2 - \Delta f$. The resulting sum and difference frequencies will respectively be $2f_2 - \Delta f$ and $-\Delta f (= \Delta f)$. Again, only the Δf component can appear as output noise and the low-pass filter 'cannot tell' whether its Δf input resulted from an $f_2 + \Delta f$ input to the mixer or an

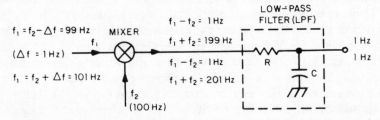

Figure 11.19   PSD operation with asynchronous (noisy) signal (reproduced by permission of EG&G Princeton Applied Research Corporation)

$f_2 - \Delta f$ input. In addition to its rectifying and phase-sensitive properties, therefore, the PSD filters noise as though it consisted of *band-pass* responses centered on all odd harmonics of f_2 (see Figure 11.20). Notice that each effective band-pass response consists of the output low-pass filter response and its mirror image; their centre-frequencies automatically track changes in the PSD drive frequency f_2.

Each band-pass response has an equivalent noise bandwidth determined by that of the low-pass filter. If the PSD input consists of white noise, the effect of the harmonic responses ($2n + 1 = 3, 5, 7, \ldots$) is to increase the PSD output noise by 11%. For square-wave signal inputs, the additional output noise (11%) caused by the harmonic responses is more than compensated for by the increase in signal (23%). If the PSD is used to measure a sinusoidal signal accompanied by white noise, a separate low-pass or band-pass filter, centred on f_2, may be used in front of the PSD to remove the harmonic responses and thus the additional 11% noise. The improvement in output SNR effected by the use of such front-end filtering on white noise is normally insignificant. With discrete frequency noise, however, front-end filters can be extremely helpful. By reducing the input noise before it reaches the PSD, the overload

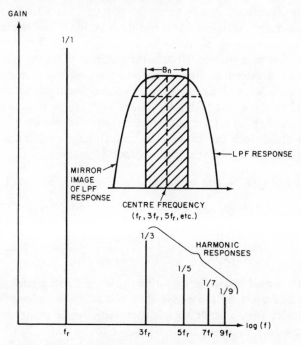

Figure 11.20 Frequency response of a PSD (reproduced by permission of EG&G Princeton Applied Research Corporation)

capability of the lock-in may be improved significantly beyond the dynamic reserve of the mixer and this additional dynamic range can also be used to provide increased output stability if desired. For this reason some lock-ins provide adjustable low-pass, high-pass, band-pass and band-reject (notch) filters, in addition to a flat frequency response (broad-band) mode. Applications and explanation of operation of PSD systems are also reviewed in Blair and Sydenham (1975).

Heterodyning front-ends are also available. With this approach, a fixed-frequency band-pass amplifier is used to protect the PSD and increase the overload capability of the lock-in. In order to use such a fixed-frequency filter, another mixer is used to heterodyne the input signal frequency up to the centre frequency of the filter; two such heterodyning schemes are shown in Figure 11.21. The advantage of this approach is that the instrument offers sinusoidal response (no harmonic responses) and overload performance approaching that of a tunable band-pass instrument, without the need for manual tuning. Because of the phase-shift characteristic (see Figure 11.13) of their front-end band-pass filters, however, manually tuned or heterodyning instruments cannot provide the phase stability of a broad-band or flat lock-in.

Figure 11.22 shows a simplified block diagram of a *two-phase* or *vector* lock-in amplifier. An additional quadrature (Q) output channel has been added, consisting of a mixer, low-pass filter, and d.c. amplifier. The reference channel provides quadrature gating inputs to the I (in-phase) and Q mixers. The *orthogonality* of the two mixer drives—that is, the accuracy of the 90° phase difference between them—is extremely important when measuring a small I signal in the presence of a large Q signal (or vice versa). Similarly, in a single-phase lock-in, the accuracy with which a 90° phase shift can be switched into or out of circuit (using the phase-quadrant switch), is equally important.

Most two-phase lock-ins provide a vector/phase circuit (usually as an option), which computes the vector magnitude M, where

$$M = \sqrt{(I^2 + Q^2)} = \sqrt{[(A \cos \phi)^2 + (A \sin \phi)^2]} = A$$

where A is the signal amplitude, and

$$\phi = \tan^{-1}\left(\frac{Q}{I}\right) = \tan^{-1}\left(\frac{A \sin(\phi_s + \phi_r)}{A \cos(\phi_s + \phi_r)}\right) = \phi_s + \phi_r$$

where ϕ_s is the signal phase shift relative to the reference signal phase, and ϕ_r is the phase offset set by the phase-shift controls. Two-phase lock-ins can, therefore, display their output signal in rectangular or polar form, with the phase controls (ϕ_r) allowing continuous vector rotation. Notice that asynchronous signals ($f_s \neq f_r$) with beat frequencies within the low-pass filter response will provide d.c. outputs and the instrument, therefore, acts as a *wave analyser*. Modern wave analysers are essentially vector lock-ins that are

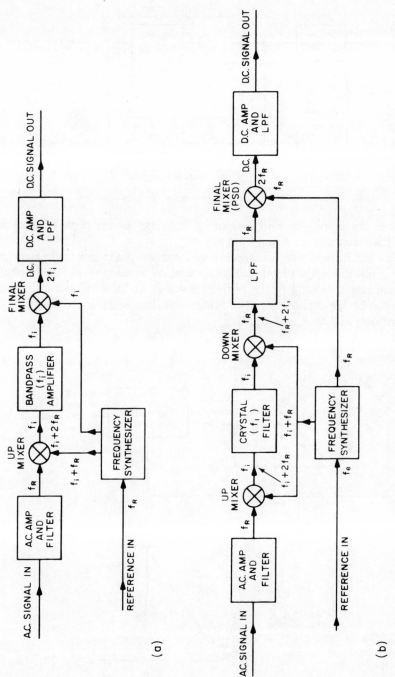

Figure 11.21   Heterodyning lock-in amplifiers: (a) single up-conversion; (b) double heterodyning

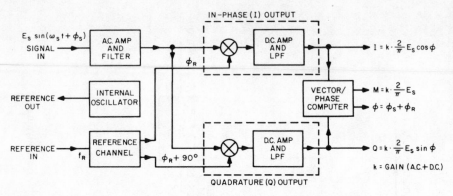

Figure 11.22   The two-phase (vector) lock-in amplifier

optimized for convenience in measuring frequency components of a signal rather than recovering a signal from noise.

Figure 11.23 shows a typical application for a two-phase lock-in. In such a.c. bridge applications, the phase shift (ϕ_r) can be set to zero, so that the in-phase (I) signal responds only to the bridge resistance and the quadrature output (Q) to the bridge capacitance. The bridge can then be balanced very simply by separately nulling R_s and C_s.

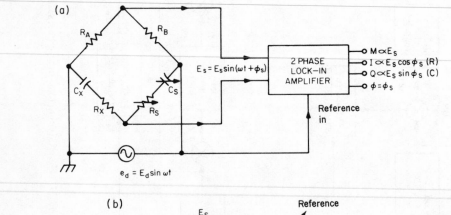

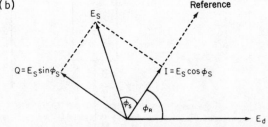

Figure 11.23   Alternating current bridge measurements with a two-phase lock-in amplifier: (a) system arrangement; (b) vector relationships

As in the case of d.c. measurements with preamplifiers or electrometers, the SNIR to be expected from a lock-in depends upon the input noise bandwidth (B_{ni}) to the lock-in, the noise bandwidth (B_{no}) of the lock-in, and the noise spectral characteristics. For random white noise and unity gain,

$$\text{SNIR} = \sqrt{(B_{ni}/B_{no})}$$

where for a -6 dB/octave rate of roll-off for the lock-in output low-pass filter, $B_{no} = 1/4RC$, or for a -12 dB/octave low-pass filter roll-off, $B_{no} = 1/8RC$.

11.8 SIGNAL AVERAGING

11.8.1 The Boxcar Averager

The *boxcar averager* (EG & G, 2) is a sampling instrument that is used to enhance repetitive signals. Also known as a *boxcar integrator* or *detector*, the boxcar takes only one sample during each signal occurrence or sweep, and requires a trigger signal at a fixed time interval prior to the beginning of each such sweep. The heart of any boxcar is the *gated integrator* circuit, shown in simplified form in Figure 11.24. This circuit is simply an RC low-pass filter gated by switch S_1 (the sampling gate). As shown, the gated integrator has unity d.c. signal gain.

If the gate is opened (i.e. S_1 *closed*) every T seconds for an aperture of t_g seconds, then the duty factor γ is given by $\gamma = t_g/T = t_g f$ where $f = 1/T$. When e_i is a voltage step, e_o will rise exponentially as shown in Figure 11.25, curve A. Notice that the *effective* time constant τ_{eff}, is much longer than the real (ungated) time constant, RC, and is given by

$$\tau_{eff} = \frac{RC}{\gamma} = \frac{RC}{t_g f} \tag{11.19}$$

As we saw previously for the PSD of a lock-in amplifier, the noise bandwidth (B_{no}) of the gated integrator is simply that of the low pass filter, that is $B_{no} = 1/4RC$.

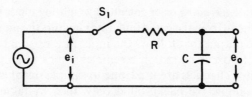

Figure 11.24   The gated integrator (simplified)

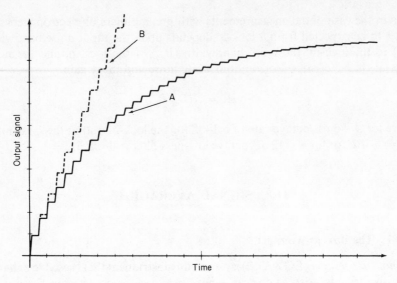

Figure 11.25   Gated integrator modes of operation: curve A, Exponential
averaging $\tau = RC/(t_s f)$; curve B, Linear summation

For input white noise limited to a bandwidth B_{ni} and unity gain, and where

$$B_{no} \ll B_{ni} \quad \text{and} \quad B_{no} \ll 1/t_g$$

then

$$\text{SNIR} = \sqrt{(B_{ni}/B_{no})} = \sqrt{(4RCB_{ni})} \tag{11.20}$$

The time resolution of a boxcar measurement will improve (shorten) as the gate duration (t_g) is decreased, until the point is reached where the input bandwidth (B_{ni}) limits the resolution. If we set

$$t_g = 1/(2B_{ni}) \quad \text{or} \quad B_{ni} = 1/(2t_g)$$

then we obtain the widely quoted formula

$$\text{SNIR} = \sqrt{(4RCB_{ni})} = \sqrt{(2RC/t_g)} \tag{11.21}$$

With pulsed signals, the ability of a boxcar to separate temporally the signal from (most of the) noise is usually of much greater significance than such theoretical white noise considerations.

In the *exponential averaging* or exponential weighting mode shown in Figure 11.24, the output signal from the gated integrator favours the most recent samples and provides a dc output that follows the input at a reasonable rate. In many ways this mode of boxcar operation resembles a lock-in amplifier; in fact, if two gated integrator channels are used, one to sample signal plus background, the other set to sample the background only, then by taking the difference between the two outputs, we have essentially built a lock-in amplifier with adjustable duty cycle.

Figure 11.26 shows a simplified schematic of a complete boxcar averager. When switch S_4 is moved to the A position, the circuit behaves as a true gated integrator rather than as a gated low-pass filter. In this mode, all samples have equal weight and, for a step input, the output will rise in a linear staircase fashion as shown in Figure 11.25, curve B. In this *linear summation* mode, the desired number of signal samples (m) is selected; after m triggers have occurred, switch S_3 is used to reset the integrator (discharge capacitor C). Since the signal samples will add linearly, and random noise samples will add vectorially, after m samples of a constant amplitude signal (S) plus white noise (N), and after maximizing the gate width to suit the signal waveshape, the output SNR is given by:

$$SNR_{out} = \frac{S_1 + S_2 + S_3 + \cdots + S_m}{\sqrt{(N_1^2 + N_2^2 + N_3^2 + \cdots + N_m^2)}} = \frac{mS}{\sqrt{(mN^2)}} = \frac{S}{N}\sqrt{m} = SNR_{in}\sqrt{m}$$

so that

$$SNIR = \frac{SNR_{out}}{SNR_{in}} \left(= \frac{SNR\,(m\,\text{samples})}{SNR\,(1\,\text{sample})} \right) = \sqrt{m} \qquad (11.22)$$

Notice that for this operating mode, it is easiest to think in terms of time averaging since the equivalent noise bandwidth of the gated integrator circuit is not constant but will decrease with increasing m.

The width (t_g) of the gating pulse is set by means of the aperture-duration controls and circuit, and the delay between receiving a trigger and sampling the following sweep is adjusted by means of the aperture-delay circuit. If the aperture delay is set manually to a constant value, then the boxcar is in a

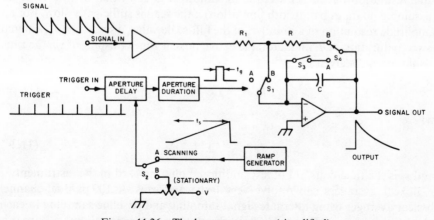

Figure 11.26   The boxcar averager (simplified)

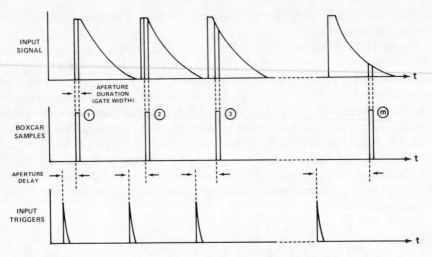

Figure 11.27   Boxar operation in scanning mode (reproduced by permission of
EG&G Princeton Applied Research Corporation)

stationary mode and will sample the same portion of each successive signal—thus
providing an output corresponding to the amplitude of the signal at that point.
Alternatively, in its *scanning mode*, the aperture delay can be slowly and continuously changed by a voltage ramp from the scan ramp generator, so that the
sampling aperture is slowly moved across the entire signal (see Figure 11.27).
In this mode, the boxcar output is a replica of the signal waveform and the boxcar
can be regarded as a time-translation device that can slow down and recover
fast waveforms.

In the scanning mode, the aperture duration (t_g) is not necessarily equal to the
time resolution but rather sets the maximum resolution that can be achieved
(assuming no input bandwidth limitation) if the scan is sufficiently slow. For an
amplitude resolution of within 1 % of the full-scale value, a scan time T_s, a signal
(sweep) duration of T, and a total effective boxcar time constant of τ_B, the time
resolution t_R, is given by

$$t_R = 5\tau_B T/T_S \tag{11.23}$$

where

$$\tau_B \simeq \sqrt{(\tau_{\text{eff}}^2 + \tau_f^2)} \tag{11.24}$$

and τ_f is the time-constant of any additional filtering used in the instrument.

Boxcar averagers can resolve very fast waveforms. A 100 ps dual-channel
boxcar averager using alternate signal sampling and baseline sampling in each
channel, is shown in Figure 11.28. Without its averaging capability, a boxcar is

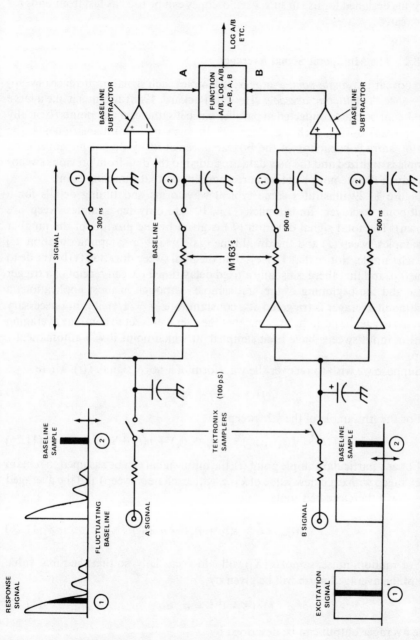

Figure 11.28 Dual channel, 100 ps boxcar system (reproduced by permission of EG&G Applied Research Corporation)

similar to a sampling oscilloscope and as shown in this figure, sample-and-hold plug-ins designed for use in such oscilloscopes can be used as fast front-ends for a boxcar.

11.8.2 The Multipoint Signal Averager

The boxcar is a *single-point* averager: it samples each signal occurrence (sweep) only once. A *multipoint averager* (Hewlett Packard, 1968) acts much like a large number of boxcars connected in parallel, since it samples many points (typically $2^{10} = 1024$) during each signal sweep. In such multipoint instruments, the analog storage capacitor of the boxcar is replaced by digital memory; each sample is digitized and the new data are added to the data from previous sweeps already in the memory location corresponding to that sampling point.

Figure 11.29 illustrates some typical waveforms and timing details for a multipoint averager; for simplicity $I = 10$ (i.e. only ten samples/sweep are shown). The total signal duration (T) is given by the product of the number of samples/sweep (I) and the dwell-time (gate width or sampling duration, t_g) of each sample. Notice that T is less than the total sweep duration (τ) by the dead time (t_d), and that there is usually a fixed delay time between receipt of a trigger pulse and the beginning of the first sample. Although in most applications a multipoint averager is triggered at a constant rate ($f = 1/\tau$), it is not necessary that the trigger be periodic. Assume that the averager is set to continue averaging until m input sweeps have been sampled, at which point it will automatically stop.

Suppose we wish to recover the waveform of a noisy signal, $f(t)$, where

$$f(t) = s(t) + n(t)$$

For the ith sample of the kth sweep,

$$f(t) = f(t_k + it_g) = s(t_k + it_g) + n(t_k + it_g) \tag{11.25}$$

For any particular sample point (i), the input signal can be assumed to remain unchanged with each new value of k (i.e. with each new sweep) and the averaged signal will therefore be simply

$$S(i)_{out} = \sum_{k=1}^{m} s(t_k + it_g) = ms(it_g) \tag{11.26}$$

For random noise, samples (X_i) will add vectorially, so that the r.m.s. value (σ) of the averaged noise will be given by

$$\left(\sum (X_i)^2 \right)^{1/2} = \sigma \sqrt{m} \tag{11.27}$$

The averager output can be described by

$$g(t_k + it_g) = ms(it_g) + \sigma \sqrt{m} \tag{11.28}$$

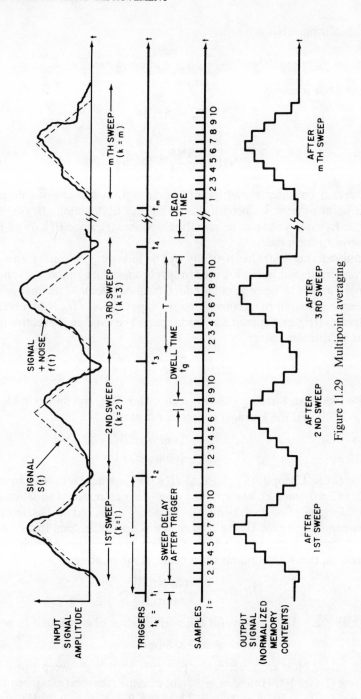

Figure 11.29 Multipoint averaging

so that the output SNR is

$$\text{SNR}_{\text{out}} = \frac{S_{\text{out}}}{N_{\text{out}}} = \frac{ms}{\sigma\sqrt{m}} it_g = \frac{s}{\sigma} it_g \sqrt{m} \tag{11.29}$$

The input SNR is simply

$$\text{SNR}_{\text{in}} = s(it_g)/\sigma \tag{11.30}$$

so that

$$\text{SNIR} = \frac{\text{SNR}_{\text{out}}}{\text{SNR}_{\text{in}}} = \sqrt{m} \tag{11.31}$$

In order to consider a multipoint averager from a frequency domain or filtering point of view, we need to know its transfer function: $H(j\omega)$. We can determine $H(j\omega)$ if we know the impulse response $h(t)$, since $H(j\omega)$ and $h(t)$ are a Fourier-transform pair.

We can determine $h(t)$ heuristically by the following reasoning. In a multi-point averager, trigger pulses are used to synchronize the signal sweeps and allow the signal samples to be coherently added (CO-ADDed). Mathematically, this action can be thought of as convolving the input signal, $f(t)$, with a train of m unit impulses (triggers) spaced τ seconds apart. The averager's effective impulse response is, therefore, given by

$$h(t) = \sum_{k=1}^{m} \delta(t - k\tau) \tag{11.32}$$

By Fourier transforming this expression for $h(t)$, we find (Childers and Durling, 1975) that the averager's transfer function is

$$|H(j\omega)| = \left| \frac{\sin(m\omega\tau/2)}{\sin(\omega\tau/2)} \right| \tag{11.33}$$

Notice (from L'Hôpital's rule) that $H(j\omega) = m$ whenever ωt is an integral multiple of 2π. Figure 11.30 shows the *comb filter* response of equation (11.33) for several values of m. Each band-pass response is centred at a harmonic (n/τ) of the sweep/trigger rate. (If the trigger rate is aperiodic, then this comb-filter concept becomes meaningless.) Since the peak transmission of each bandpass response is m, the -3 dB points must occur at $m/\sqrt{2}$, so that

$$|H(j\omega)| = \left| \frac{\sin(m\omega\tau/2)}{\sin(\omega\tau/2)} \right| = \frac{m}{\sqrt{2}} \tag{11.34}$$

from which the -3 dB bandwidth B for large values of m, is found to be

$$B = 0.886/(m\tau) \tag{11.35}$$

Large values of m are practicable, particularly at high sweep rates. With a trigger rate $(1/\tau)$ of 100 Hz and $m = 10^6$ for example, the total measurement time will be $m\tau = 10^6 \times 10^{-2} = 10^4$ s $\simeq 2.8$ h, and $B = 8.86 \times 10^{-5}$ Hz.

Thus far in this discussion of multipoint averagers, a *linear summation* mode of averaging has been assumed. That is, for the ith memory location, the average after m sweeps is given by

$$A_m = \sum_{k=1}^{m} f(t_k + it_g) = \sum_{k=1}^{m} I_k \tag{11.36}$$

where $I_k = f(t_k + it_g)$ is the value of the ith sample in the kth sweep.

This algorithm has the advantage of being simple to implement digitally. The output averaged signal, however, continually increases with each new sweep; manual scale changing is required to keep the displayed output at a useful size, yet within the bounds of the CRT screen. A seemingly more convenient algorithm would be to normalize the data in memory after each sweep, that is, implement

$$A_k = \frac{1}{k} \sum_{k=1}^{m} I_k = A_{k-1} + \frac{I_k - A_{k-1}}{k} \tag{11.37}$$

During each sweep, the data (A_{k-1}) in each memory location are compared with the new sample value I_k and the computed value of $(I_k - A_{k-1})/k$ is added to memory to form the new average value A_k. Because of practical difficulties in implementing a division by k during or after each sweep, the algorithm shown in equation (11.37) is often approximated by

$$A_k = A_{k-1} + \frac{I_k - A_{k-1}}{2^J} \tag{11.38}$$

where J is a positive integer selected automatically such that 2^J is the closest approximation to k. Notice for $k = 6$ for example, that the closest 2^J values are $2^2 = 4$ or $2^3 = 8$. Though this *normalized averaging* mode is very slightly slower than the summation mode in enhancing the signal, we can assume that SNIR $= \sqrt{m}$ for all practical purposes. Note that the discrepancy between k and 2^J increases as larger values of J are used to deal with very noisy signals. In compensation, this averaging mode provides a stable, constant-amplitude display from which the noise appears to shrink with time.

If we wish to recover and monitor slowly varying noisy signals, the algorithm of equation (11.38) can also be used for *exponential* averaging if J is made a manually selectable constant. When $J = 0$, then $2^J = 1$ and $A_k = I_k$; with this setting, the input signal may be monitored in real time, since it is digitized and stored without averaging. In general, selecting a value of J will establish an effective time constant, τ_J, where

$$\tau_J = \frac{t_g}{-\ln(1 - 2^{-J})} \tag{11.39}$$

or

$$2^{-J} = 1 - \exp(-t_g/\tau_J) \tag{11.40}$$

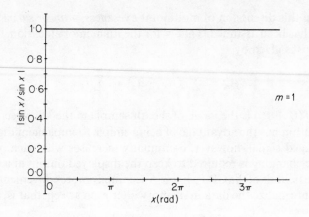

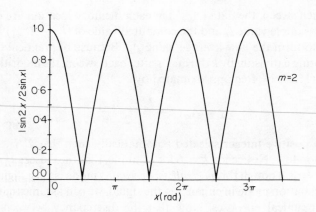

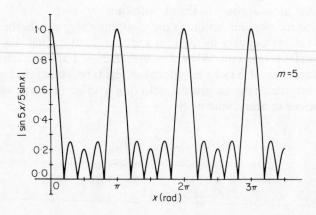

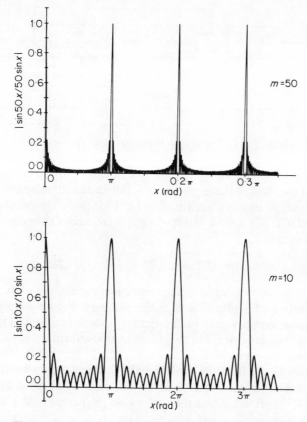

Figure 11.30 The comb filter action of a multipoint averager

The larger the value of J selected, the greater the signal enhancement, and the more slowly the averager responds to changes in the input signal. For a large number of sweeps (EG & G, 3) the SNIR is given by

$$\text{SNIR} \simeq \sqrt{2^{J+1}} \qquad (11.41)$$

Figure 11.31 shows the simplified block diagram of a multipoint averager. It is common to include a low-pass filter in the analog input channel with a $-3\,\text{dB}$ cut-off frequency (f_c) controlled by the dwell-time setting. A typical example might be

$$f_c \simeq 1/(2t_g)$$

that is, one-half of the sampling frequency. Such filters are used to improve the input SNR rather than as anti-aliasing filters.

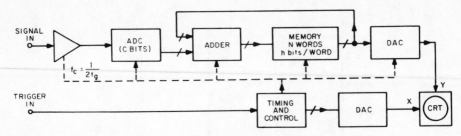

Figure 11.31   The multipoint signal averager (simplified)

The maximum number of sweeps (m_{max}) that can be digitized by an averager without data overflow, if the input signal is full-scale and noise-free, is given by 2^{h-c} where h is the memory size (bits) and c (bits) is the resolution of the A/D converter (ADC). For a 9-bit (8-bit + sign) ADC and a memory of N words, each of 28 bits, then

$$m_{max} = 2^{h-c} = 2^{28}/2^9 = 2^{19} = 524{,}288$$

In some instruments with *artifact-rejection* capability, each new signal sweep is digitized and placed in a buffer memory. Before adding the buffer contents to main memory, each buffer-memory location is checked for overflow; the buffer contents are discarded should an overflow (artifact) exist.

Suppose that the input SNR to an averager is 1 : 10; that is, the r.m.s. noise (σ) is ten times larger than the peak signal (S). The a.c. gain before the ADC, must be set such that the noise peaks do not exceed full scale. For Gaussian noise, it is 99.9% probable that the peak noise (N_p) amplitude is less than five times greater than the rms noise amplitude; that is, $N_p/\sigma \leqslant 5$, so that $N_p/S \leqslant 50$.

Assume that the input gain is set such that N_p is just equal (say) to the full-scale input level of a 9-bit ADC. Assume also that the resolution of the ADC is 2^9 ($= 512$), and the memory size $h = 2^{28}$, as before. Of these 9 bits, 6 bits ($= 2^6 = 64$) will be required as dynamic reserve (i.e. to handle the input noise), and only 3 bits ($= 2^3 = 8$) will be left to quantize the signal (S). In this example, then, the maximum number of sweeps before overflow would be

$$m_{max} = 2^{28}/2^3 = 2^{25} \simeq 3.4 \times 10^7$$

Under the conditions of this example, the output (vertical) resolution will not be limited to 3 bits. Random noise accompanying the signal will *dither* the ADC; that is, the noise will modulate the quantization levels of the ADC so as to provide a resolution that increases as m increases. Note, however, that without noise and with the same full-scale setting, the averager output would indeed have a 3-bit amplitude resolution. White noise can be added deliberately to signals that are clean, in order to improve resolution beyond that of the ADC (Horlick, 1975).

It is useful to compare boxcar and multipoint averagers. For dwell times of about 1 μs or longer, the multipoint averager typically needs less than one-thousandth of the measurement time needed by a boxcar to recover a waveform; on the other hand, the boxcar is the only choice for gate widths (dwell times) of 1 ns or less. For dwell times in the 1 ns–1 μs range, the choice is between a boxcar or a transient recorder interfaced to a multipoint averager. Such transient recorder–averager combinations are usually less time-efficient (i.e. $\tau \gg It_g$) than a multipoint averager alone, due to slow data transfer.

11.9 CORRELATION

For our purposes in this chapter, correlation analysis is a method of detecting any similarity between two time-varying signals (Honeywell). *Autocorrelation* consists of the point-by-point multiplication of a waveform by a delayed or time-shifted version of itself, followed by an integration or summation process. Mathematically, the autocorrelation function, $R_{xx}(\tau)$, of a time-varying function, $f(t)$, is given by

$$R_{xx}(\tau) = \lim_{T \to \infty} \frac{1}{2T} \int_{-T}^{T} f(t)f(t + \tau) \, dt \qquad (11.42)$$

where τ is the *lag value* or time shift between the two versions of $f(t)$.

Cross-correlation involves two waveforms and consists of the multiplication of one waveform, $f(t)$, by a time-shifted version of a second waveform, $g(t)$, followed by integration or summation. The cross-correlation function, $R_{xy}(\tau)$, is given by

$$R_{xy}(\tau) = \lim_{T \to \infty} \frac{1}{2T} \int_{-T}^{T} f(t)g(t + \tau) \, dt \qquad (11.43)$$

Notice that cross-correlation requires two input signals, as is also true in the case of signal averaging (where a synchronizing input is required in addition to the signal input). Also as with an averager, a cross-correlator preserves signal phase information. Unlike the averager, however, the cross-correlator output waveform, the *correlogram*, is affected by the waveform of the second input signal—an undesirable and unnecessary complication for signal recovery applications since a multipoint averager may be used. Cross-correlators are normally used in flow or velocity measurements; they are used but rarely for simple signal-recovery purposes. (Ignoring the lock-in amplifier and boxcar integrator, both of which can be regarded as a special type of cross-correlator.)

Phase information is lost in an autocorrelation function, as is also true for its Fourier transform, power spectral density. This lack of phase information means that in some cases, the input signal responsible for a given correlogram must be deduced by intelligent guesswork. For example, as shown in Figure 11.32, the autocorrelation function for band-limited white Gaussian noise is a

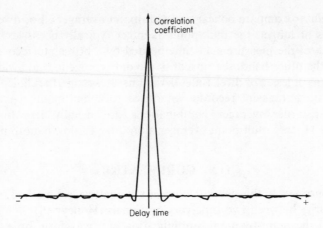

Figure 11.32   Correlation function of band-limited white
noise

spike-like peak at $\tau = 0$, with a width that would decrease as the noise band-width increases. A similar correlogram could have resulted from an input consisting of a single narrow pulse; the narrower the pulse width, the narrower the correlogram spike. Thus the pulse and the band-limited white noise have similar power-density spectra. (The difference between them is that frequency components of the noise have random phase relationships.)

A simplified block diagram of an autocorrelator is shown in Figure 11.33. The ADC will digitize the input signal once every lag interval or dwell time, t_g. Each such A/D conversion will require a conversion time, t_c, where $t_c \ll t_g$. The output digital *word* from the ADC, corresponding to the latest sample, provides one input to the digital multiplier and also is shifted, as word 0, into an N-word shift register. During this shift operation, the last word in the register, word $(N - 1)$, is shifted out and discarded and the former word $(N - 2)$ becomes the new word $(N - 1)$. The control and timing circuits

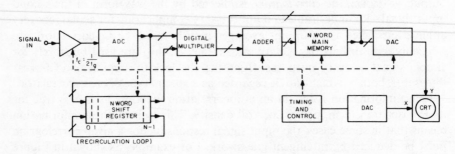

Figure 11.33   The autocorrelator (simplified)

then cause the shift register to step N times, recirculating its contents one full cycle, and requiring a time interval t_r. During t_r, each shift-register word (i.e. $(n - 1)$, $(n - 2)$, $(n - 3)$, ..., 3, 2, 1, 0) is applied sequentially to the other input of the multiplier, which multiplies each of these words by the word at its other input, word 0. Each multiplier output is added to the contents of the corresponding bin of the N-bin main memory; for instance, word 0 × word $(n - 16)$ → bin$(n - 16)$, word 0 × word 0 → bin 0. After many such cycles, the contents of bin i (for example) of the main memory, will be the sum of products formed by multiplying each new signal sample by an it_g delayed version of itself. Bin 0, for example, corresponds to the signal multiplied by itself with zero delay.

The minimum time between successive samples is $(t_c + t_r)$. When $t_g \geqslant (t_c + t_r)$, the correlator is said to be working in a *real-time* mode. When $t_g \leqslant (t_c + t_r)$, samples can no longer be taken every t_g seconds and the correlator is said to be in a *pseudo-real-time* mode. For $t_g \ll (t_c + t_r)$, the correlator is in a *batch* mode.

The process of autocorrelation involves the concept of sliding a waveform past a replica of itself. For random noise, the two waveforms will match at only one point as they slide by each other, that is at $\tau = 0$, when they are perfectly aligned. In contrast, a square-wave sliding by another square-wave will find a perfect match once in every period and will give rise to a triangular correlogram. More generally, signals that are periodic in time will produce a correlation function that is periodic in τ. Suppose, for example, that the input to an autocorrelator is $f(t) = A \cos(\omega t)$. Then

$$R_{xx}(\tau) = \lim_{T \to \infty} \frac{1}{2T} \int_{-T}^{T} A \cos(\omega t) A \cos(\omega t + \tau) \, dt = \tfrac{1}{2}A^2 \cos(\omega \tau) \quad (11.44)$$

Figure 11.34 shows the correlation function of a sine wave accompanied by band-limited white noise. Note that the peak value at $\tau = 0$ in this correlogram is the mean-squared value of the signal plus noise (i.e., $S + N$). The mean-squared value of the sinusoidal signal component is given by the peak value of

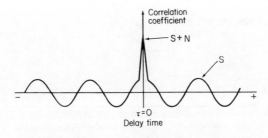

Figure 11.34   Correlogram of a noise sinewave

the sine wave at τ values where the white noise spike has damped to zero. The SNR of the correlator output can, therefore, be determined since

$$\frac{S}{(S + N) - S} = \frac{S}{N} = \text{SNR}$$

Most importantly, notice that the signal and noise components have been separated by virtue of their different positions on the τ-axis. It is this separating ability that makes correlation a powerful signal-recovery technique.

11.10 PHOTON (PULSE) COUNTING TECHNIQUES

11.10.1 Introduction

PMT's (photomultiplier tubes) are used to measure the intensity or flux of a beam of visible photons (see Figure 11.35). With its photo-cathode removed, the PMT becomes an EMT (electron multiplier tube) and is widely used to detect ions and electrons. One of the most important advantages of such detectors is that their high gain and low noise allow them to give one output pulse for each detected input particle. Since visible-light measurements are perhaps most commonplace, we will consider as an example a PMT and pulse-counting system of the type shown in Figure 11.36.

The probability of each incident photon causing an output pulse from the PMT is essentially equal to the *quantum efficiency*, ζ: typically ζ is between 5–25%. In addition to the photon-derived pulses (i.e. the signal pulses), there will be spurious noise pulses at the PMT output due to thermionic emission.

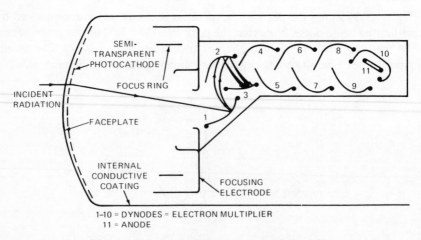

1–10 = DYNODES = ELECTRON MULTIPLIER
11 = ANODE

Figure 11.35   End-window photomultiplier tube

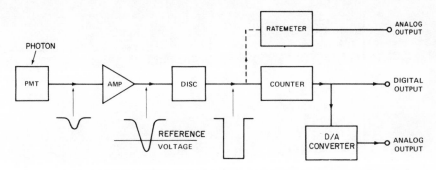

Figure 11.36   Typical photon counting system (reproduced by permission of EG&G Princeton Applied research Corporation)

Noise pulses caused by thermionic emission from the dynodes will experience less gain and will be smaller in amplitude than pulses due to cathode emission. The PMT output pulses are amplified and presented to a pulse-height discriminator circuit, where the peak amplitude of each pulse is compared to an adjustable threshold or reference voltage. Ideally, the discriminator will reject all dynode-derived noise pulses and accept all cathode-derived pulses; in practice, the PMT gain is statistical in nature and cathode and dynode-derived pulses have overlapping amplitude distributions. The discriminator will therefore accept *most* cathode-derived pulses and reject *most* dynode noise pulses. Each accepted input pulse will cause a standardized output pulse. Such pulse-height discrimination also reduces the effect of PMT gain variations with time and temperature.

Ratemeters are used to give a continuous analog output voltage which is proportional to the discriminator output pulse (count) rate. Alternatively, digital counter circuits can be used to accumulate output counts for a preselected measurement time. Such counters allow very long integration times and when a digital output is required, they can avoid the loss of resolution inherent in D/A conversion.

11.10.2   Poisson Statistics, Shot Noise, and Dark Counts

Suppose we use our PMT to detect photons emitted from a thermal light source such as a tungsten filament lamp. The time interval between successive photons impinging upon the PMT photocathode is random and governed by a *Poisson* distribution (see Figure 11.37a). The probability, P, of detecting n photons in a time t following the last photon is given by

$$P(n, t) = \frac{(\zeta R t)^n e^{-\zeta R t}}{n!} = \frac{N^n e^{-N}}{n!} \tag{11.45}$$

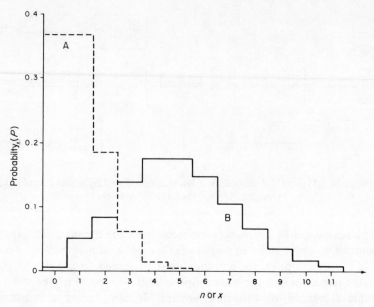

Figure 11.37 The Poisson distribution (reproduced by permission of EG&G Princeton Applied Research Corporation). Curve A, Probability of detecting n photons in time $t = (Rt)^n \exp(-Rt/n)$, $R = 10^8$ photons/s, $t = 10$ ns and so $Rt = 1$; $\sigma = \sqrt{(Rt)}$. Curve B, Probability of gain magnitude $P(x) = M^x e^{-M}/x!$, where x = dynode gain, M = mean dynode gain = 5 and $\sigma = \sqrt{M}$.

where R is the mean photon rate (photons/second) and $N = \zeta Rt$ is the signal (i.e. the mean number of photonelectrons emitted by the PMT photocathode during the time interval t). The noise, or uncertainty, in N is given by the standard deviation σ, where

$$\sigma = \sqrt{(\zeta Rt)} = \sqrt{N}$$

so that

$$\text{SNR} = \frac{N}{\sqrt{N}} = \sqrt{N} = \sqrt{(\zeta Rt)} \tag{11.46}$$

Notice that, as in all of the techniques examined in this chapter (with white noise), the SNR is again proportional to the square root of the measurement time (t). If we assume that there is no thermionic (dark) emission of electrons from the photocathode, then the photocathode (signal) current (in amperes) is given by

$$I_{pe} = \zeta Re \tag{11.47}$$

where e is the charge of an electron ($\simeq 1.6 \times 10^{-19}$ C). The signal-to-noise ratio (SNR_k) of the photocathode current (I_{pe}) is given by

$$SNR_k = \sqrt{(\zeta Rt)} = \sqrt{\left(\frac{\zeta Rte}{e}\right)} = \sqrt{\left(\frac{I_{pe}t}{e}\right)} \tag{11.48}$$

Now the measurement time t has associated with it a frequency range Δf, where

$$t = 1/2\Delta f,$$

so that

$$SNR_k = \sqrt{\left(\frac{I_{pe}}{2e\Delta f}\right)} = \sqrt{\left(\frac{I_{pe}^2}{2eI_{pe}\Delta f}\right)} = \frac{I_{pe}}{\sqrt{(2eI_{pe}\Delta f)}} \tag{11.49}$$

If we multiply both the numerator and denominator of equation (11.49) by the mean PMT gain A, then

$$SNR_k = \frac{AI_{pe}}{A\sqrt{(2eI_{pe}\Delta f)}} = \frac{I_a}{\sqrt{(2AeI_a\Delta f)}} = SNR_a \tag{11.50}$$

where I_a is the d.c. anode current and SNR_a is the signal-to-noise ratio of the anode current if thermionic emission and other dynode noise contributions are ignored. Note that the general expression for the shot noise of a d.c. current I is given by

$$\text{r.m.s. shot noise current} = \sqrt{(2AeI\Delta f)} \tag{11.51}$$

where A is the gain following the shot noise process. When $A = 1$, the expression simplifies to $\sqrt{(2eI\Delta f)}$. Notice also that shot noise was present in the light beam itself and that the PMT quantum efficiency (ζ) degrades the SNR by a factor of $\sqrt{\zeta}$.

In practice, with no input photons, the photocathode will emit electrons due to temperature effects. The dynodes will also emit thermionic electrons. The rate of such thermionic emission is reduced by cooling the PMT. Thermionic emission from the photocathode, that is, *dark counts*, can be further reduced by minimizing the cathode area and by selecting a photocathode material with no more red (long-wavelength) spectral response than is necessary.

If the photocathode emits electrons randomly at a dark count rate R_d, then the noise components of the cathode current will increase to $\sqrt{(\zeta Rt + R_d t)}$ and the signal-to-noise ratio of the cathode current will degrade to

$$SNR = \frac{\zeta Rt}{\sqrt{(\zeta Rt + R_d t)}} = \frac{\zeta R\sqrt{t}}{\sqrt{(\zeta R + R_d)}} \tag{11.50}$$

This will also be the PMT output SNR if dynode noise is assumed to be removed completely by pulse-height discrimination. For PMT's equipped with a high-gain first dynode with Poissonian statistics (see Figure 11.37b), this is a not-unreasonable approximation. Note that when a PMT is used in a non-counting or d.c. mode, as was discussed previously in Section 11.6, all of the output electrons resulting from spurious cathode emission (i.e. dark counts) and dynode emission are integrated by the anode or preamplifier time constant into a d.c. *dark current*, and the opportunity to remove dynode noise by pulse-height discrimination is lost.

11.10.3 Pulse-height Discrimination

Each electron emitted by a PMT photocathode will be amplified by the instantaneous value of the PMT gain. For a mean gain of 10^6, for example, each cathode electron will cause an average output charge q of $10^6 e$ coulomb. This charge q will accumulate at the anode during a time t given by the transit-time spread of the PMT. Typically, t will be about 10 ns, so that the resulting anode current pulse ($i_a = dq/dt$) will have a full width (t_w) between half-maximum amplitude points (FWHM) of about 10 ns also. The peak value, I_{pk}, of the anode-current pulse may be approximated by assuming the pulse to be rectangular, so that in our example,

$$I_{pk} \simeq \frac{q}{t_w} = \frac{10^6 e}{10 \times 10^{-9}} = \frac{10^6 \times 1.6 \times 10^{-19}}{10 \times 10^{-9}} = 16\,\mu A$$

In a photon-counting system, the anode-load resistor (R_a) of the PMT is kept small, usually 50–100 Ω. Therefore the time constant (τ_a) formed by the anode stray capacitance (C_a) will be small compared to t_w, and thus will not stretch the anode voltage pulse. Typically, $R_a = 50\ \Omega$ and $C_a = 20$ pF, so that $\tau_a = 1$ ns $\ll t_w$. The anode voltage pulse will then have the same shape as the anode current pulse, and a peak value of

$$E_{pk} = I_{pk} R_a = 16 \times 10^{-6} \times 50 = 0.8\,\text{mV}$$

It should be remembered that such pulse amplitudes depend upon the PMT gain which, in turn, depends upon the dynode gains—which are statistical. In the above example then, $E_{pk} = 0.8$ mV is the average pulse height to be expected. Actual pulse heights will be distributed above and below this value and the better the PMT, the narrower will be this distribution. A preamplifier is normally used to amplify the anode pulses to a suitable level for the pulse-height discriminator.

Notice that the PMT gain, the preamplifier gain, and the discriminator threshold controls may all be used to adjust the effective discrimination level. Figure 11.38 shows a typical count-rate variation with PMT high voltage (i.e. PMT gain) and a fixed discriminator threshold level. This is one of a family of curves that could be plotted for different threshold levels; similar curves

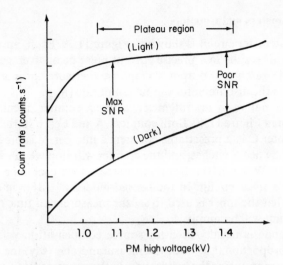

Figure 11.38 The counting plateau (reproduced by permission of EG&G Princeton Applied Research Corporation)

could be obtained by varying the preamplifier gain rather than that of the PMT, or by plotting count rate against discriminator threshold (Darland *et al.*, 1979). The upper curve in Figure 11.38 was plotted by allowing light to fall upon the PMT photocathode and slowly varying the PMT voltage (which is non-linearly related to the PMT gain). Notice that the steep slope at low PMT voltages begins to flatten and form a (not-quite-horizontal) *plateau* as proper focusing takes place in the PMT. The increasing slope at very high PMT bias voltages is due to increasing instability in the PMT.

The upper curve corresponds to $S + N$, since it is based on both signal and noise pulses. The lower curve was plotted with the PMT in darkness, and therefore represents noise (N) pulses only. Notice that typically the dark-count curve has no plateau; it has been suggested that the lack of a plateau is due to corona effects associated with microscopic protrusions from the dynode surfaces. Since the count rate is plotted on a logarithmic scale, the vertical distance between the two curves corresponds to

$$\log(S + N) - \log(N) = \log\left(\frac{S + N}{N}\right) = \log\left(1 + \frac{S}{N}\right) \simeq \log\left(\frac{S}{N}\right) \quad \text{for } \frac{S}{N} \gg 1$$

A commonly used method of setting the PMT high voltage for a given preamplifier gain and discriminator threshold level is, therefore, to select a point on the beginning of the counting plateau corresponding to maximum SNR.

11.10.4   Ratemeters and Counters

A simplified ratemeter circuit is shown in Figure 11.39. Each output pulse from the discriminator results in a precise current pulse being averaged by the low-pass filter. The ratemeter output voltage is, therefore, proportional to the average value of the discriminator output count rate (R_{sig}).

A digital alternative to the ratemeter, a timer–counter circuit, is shown in simplified form in Figure 11.40. Both counters, A and C, are started and stopped together. Counter C is a presettable counter: a number N is preset, usually by means of thumbwheel switches, and the counter will stop when its accumulated count equals N. When driven from an internal clock oscillator, this arrangement is called a *timer* circuit. In the *normal* mode, a 1 MHz internal clock is often used so that the timer is used to set the measurement time $t = N/R_{clk}$ μs. The output count will be simply $A = tR_{sig}$.

The ratio mode is usually used for source compensation where the signal count rate is proportional to both the measurand and (say) the intensity of a light source. By monitoring the light source with a separate PMT and amplifier–discriminator to produce a source-dependent count rate R_{sc}, R_{sig} can be normalized by R_{sc} to provide an output count that is independent of source fluctuations.

In the *reciprocal* mode, the system will measure the time (t, in μs) required for the cumulative signal counts to reach N; the smaller the signal count rate, the longer the elapsed time. If the dark count rate is negligible, then the measurement accuracy is $1/\text{SNR} = 1/\sqrt{(R_{sig}t)}$ and for a constant value of $R_{sig}t$ ($= N$), all measurements will have the same SNR and accuracy.

A *synchronous counting system* that acts like a *digital lock-in amplifier* to provide automatic background subtraction is shown in Figure 11.41. When the chopper blade blocks the input light, the output pulses from the amplifier–discriminator are, by definition, background noise; these pulses (N) are gated into counter B by the timing circuit—which is itself synchronized by the chopper reference signal. When the chopper blade allows light to reach the PMT, the discriminator output consists of signal-plus-background pulses

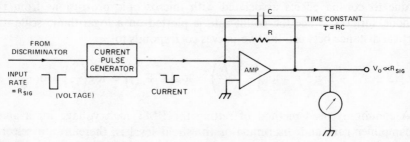

Figure 11.39   The ratemeter (simplified) (reproduced by permission of EG&G Princeton Applied Research Corporation)

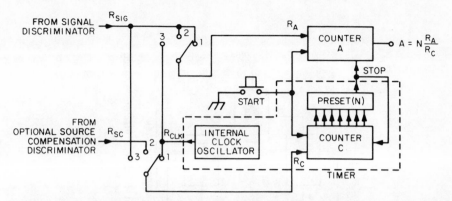

Figure 11.40 Counter–timer modes of operation. (1) Normal mode; $R_A = R_{sig}$, $R_c = R_{clk}$, $A = NR_{sig}/R_{clk} \propto R_{sig}$. (2) Ratio mode; $R_A = R_{sig}$, $R_c = R_{sc}$, $A = NR_{sig}/R_{sc} \propto R_{sig}/R_{sc}$. (3) Reciprocal mode; $R_A = R_{clk}$, $R_c = R_{sig}$, $A = NR_{clk}/R_{sig} \propto 1/R_{sig}$.

$(S + N)$ and these pulses are gated into counter A. After each measurement interval, an arithmetic circuit provides two outputs

$$A - B = (S + N) - N = S = \text{signal} \tag{11.52}$$

and

$$A + B = (S + N) + N = \text{total counts} \tag{11.53}$$

where A and B are the numbers of counts in counters A and B respectively. For Poissonian noise,

$$\text{SNR} = \frac{\text{signal}}{\sqrt{(\text{total counts})}} = \frac{A - B}{\sqrt{(A + B)}} \tag{11.54}$$

Suppose, for example, that $A = 10^6$ counts and $B = 9.99 \times 10^5$ counts then $S = A - B = 10^3$ counts and $\sqrt{(A + B)} = 1.41 \times 10^3$, so that

$$\text{SNR} = \frac{A - B}{\sqrt{(A + B)}} = \frac{10^3}{1.41 \times 10^3} = 0.71$$

and (in)accuracy

$$\frac{1}{\text{SNR}} = \frac{1}{0.71} = 141\%$$

or, expressed in words, the measurement is worthless! The $A + B$ output is important since it allows the measurement accuracy to be estimated in this way.

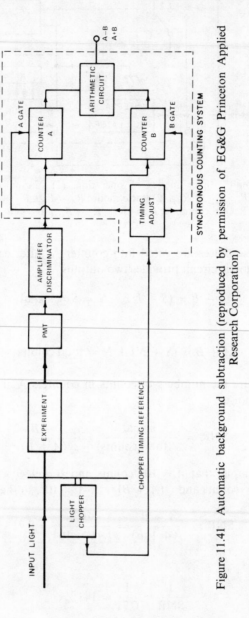

Figure 11.41 Automatic background subtraction (reproduced by permission of EG&G Princeton Applied Research Corporation)

11.10.5 Pulse Pile-up

The dynamic range of photon-counting measurements is limited at low light levels by PMT dark count and, at high light levels, by *pulse pile-up* in the PMT or electronics. As the mean rate (R) of photons arriving at a PMT photocathode increases, then so does the probability of two or more photons arriving with too short an interval between them to be resolved by the PMT.

The time-resolution of a PMT is effectively equal to its output pulse width t_w, and each output pulse from a PMT, therefore, occurs whenever an electron is emitted after a time greater than t_w following the previous electron. The probability of this happening is the same as that for zero photoelectron events in a time t_w (we can neglect dark counts at high light levels). As shown in Figure 11.37a and from equation (11.45) then

$$P(0, t_w) = \exp(-\zeta R t_w) \tag{11.55}$$

and the output count rate, R_o, is given by

$$R_o = P(0, t_w)R_i = \zeta R \exp(-\zeta R t_w) = R_i \exp(-R_i t_w) \tag{11.56}$$

The resulting PMT pulse pile-up error is given by

$$\varepsilon_{pmt} = \frac{R_i - R_o}{R_i} = \frac{R_i - R_i \exp(-R_i t_w)}{R_i} = 1 - \exp(-\zeta R t_w) \tag{11.57}$$

The PMT is a *paralysable* detector; that is, when the input count rate exceeds a certain value $(R_i = 1/t_w)$, the output count rate will begin to *decrease* for an increasing input count rate and will become zero when the PMT is completely paralysed (saturated) (See Figure 11.42).

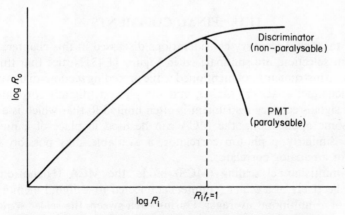

Figure 11.42 Counting Error due to pulse pile-up (reproduced by permission of EG&G Princeton Applied Research Corporation)

Discriminators and counters, on the other hand, are usually non-paralysable. Suppose a discriminator, for example, has a pulse-pair resolution or dead time t_d. That is, each time it accepts an input pulse, it cannot accept a new pulse until after a time t_d. Then for a measurement time t, an input pulse rate of R_i and an output rate R_o, the total number of output pulses, N_o, is given by

$$N_o = R_o t$$

and

$$\text{the total } dead \text{ time} = N_o t_d = R_o t t_d$$

so that

$$\text{total } live \text{ time} = t - R_o t t_d$$

The total number of input pulses accepted is therefore given by

$$N_o = R_o t = R_i(t - R_o t t_d)$$

so that

$$R_o = \frac{R_i}{1 + R_i t_d}$$

A modern, fast discriminator and counter have a dead time, t_d, of about 10 ns—similar to the time resolution, t_w, of a reasonably fast PMT. Notice, however, that any PMT pile-up will act as a *prefilter* to the discriminator; that is, such pile-up will decrease the input count rate to the discriminator. PMT pile-up usually provides the upper limit to the system dynamic range. Photon-counting systems cannot be used in pulsed light measurements where the peak photon rate (during the light pulse) will cause unacceptably high pulse pile-up errors.

11.11   FINAL COMMENTS

Many of the signal-recovery considerations discussed in this chapter, such as instrument selection, are summarized in Figure 11.43. Notice that this figure includes two instruments not mentioned in the preceding sections of this chapter: the multichannel analyser (MCA), and the photon (digital) correlator. The choice of signal-recovery instrument is often limited to that which is available and, in some applications, the MCA can be used in place of a multipoint averager. Similarly, a photon correlator, if available, may possibly be substituted for an analog correlator.

In its multichannel scaling (MCS) mode, the MCA (Nicholson, 1974) consists effectively of a scaler (counter) connected to a digital memory much like that of a multipoint averager. During each sweep, the scaler sequentially counts the number of input pulses during each dwell time and adds that number to the cumulative count in the corresponding memory address. By using a

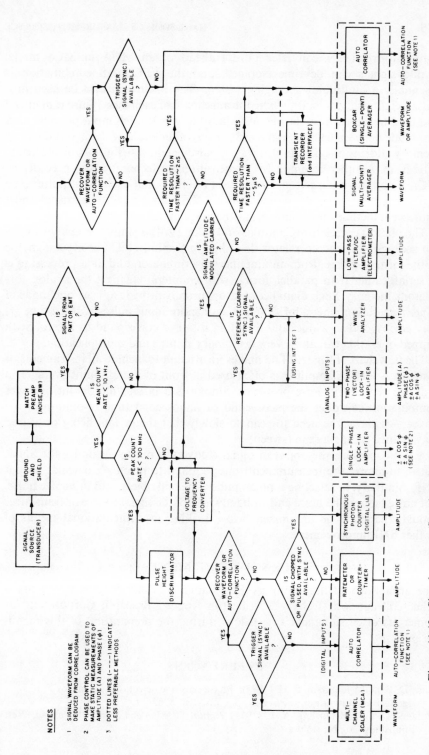

NOTES

1. SIGNAL WAVEFORM CAN BE DEDUCED FROM CORRELOGRAM

2. PHASE CONTROL CAN BE USED TO MAKE STATIC MEASUREMENTS OF AMPLITUDE (A) AND PHASE (φ)

3. DOTTED LINES (———) INDICATE LESS PREFERABLE METHODS

Figure 11.43   Signal recovery flowchart (reproduced by permission of EG&G Princeton Applied Research Corporation)

voltage-to-frequency converter (VFC) ahead of an MCA in MCS mode, analog signals can be time-averaged, and the VFC/MCS combination is essentially a multipoint averaging system. The MCA can also be used in a pulse-height analysis mode, where the amplitude of each input pulse is digitized and used to self-address the memory. In other words, each input pulse with an amplitude between 63.85% and 63.95% (say) of full-scale, will add one count to memory address No. 639. In this way, a pulse-height distribution, or spectrum, is built up. Another common MCA measurement technique is to precede the MCA by a time-to-amplitude converter (TAC), so that each input pulse to be digitized corresponds to a time interval. Low-level measurements of short fluorescent lifetimes, for example, may be made in this fashion.

The photon correlator (Cummins and Pike, 1974) is similar in many ways to the analog-input autocorrelator described in Section 11.9. The input signal is in the form of pulses, from an amplifier–discriminator, and data processing is in serial, rather than parallel, form—with counters replacing the analog correlator's memory. Such digital correlators usually provide at least one mode of *clipped* operation where, for example, n or more input pulses in a lag time (τ) may correspond to a one, and less than n pulses correspond to a zero. Such clipped operation can allow very fast binary shifting and multiplication.

The flowchart of Figure 11.43 makes no attempt to include all instruments or systems. A lock-in amplifier is often used in front of a multipoint averager in order to reduce $1/f$ noise problems. Similarly, a boxcar/multipoint averager combination can offer the picosecond or nanosecond time resolution of the boxcar—without the need to scan so slowly that the system being measured may change during a scan (sweep).

A last comment: the object in signal recovery is not to maximize the SNIR but to minimize the measurement time required to reach a particular output SNR. Similarly, in selecting a preamplifier, the real object is to minimize noise, not noise figure. The noise and/or bandwidth of the signal source or transducer should, therefore, be minimized before seeking instrumental means of further SNR improvement.

ACKNOWLEDGEMENTS

The author would like to thank Eric Faulkner, Hans Jorgensen and many other colleagues at EG & G Princeton Applied Research Corporation for numerous discussions and suggestions during the preparation of this manuscript.

REFERENCES

Blair, D. P. and Sydenham, P. H. (1975). 'Phase sensitive detection as a means to recover signals buried in noise', *J. Phys. E.: Sci. Instrum.*, **8**, 621–7.
Childers, D. G. and Durling, A. E. (1975). *Digital Filtering and Signal Processing*, West Publishing Co., New York.

Cummins, H. Z. and Pike, E. R. (Eds.) (1974). *Photon Correlation and Light Beating Spectroscopy*, Plenum Press, New York.

Darland, E. J., Leroi, G. E., and Enke, C. G. (1979). 'Pulse (photon) counting: determination of optimum measurement systems parameters', *Anal. Chem.*, **51**, 240–5.

EG&G 1. *Operating Manual for Model 124A Lock-in Amplifier*, EG&G Princeton Applied Research Corp., Princeton, US.

EG&G 2. *Operating Manual for Model 162 Boxcar Integrator*, EG&G Princeton Applied Research Corp., Princeton, US.

EG&G 3. *Operating Manual for Model 4203 Signal Averager*, EG&G Princeton Applied Research Corp., Princeton, US.

Faulkner, E. A. (1966). 'Optimum design of low-noise amplifiers', *Electron. Lett.*, **2**, 426–7.

Fellgett, P. B. and Usher, M. J. (1980). 'Fluctuation phenomena in instrument science', *J. Phys. E.: Sci. Instrum.*, **13**, 1041–6.

Hewlett-Packard Co. (1968). *Hewlett-Packard J.*, **19**, 1–16 (whole issue: articles on signal averaging).

Honeywell. 'Correlation and probability analysis', *Saicor Bulletin TB14*, Honeywell Test Instruments Div., US.

Horlick, G. (1975). 'Reduction of quantization effects by time averaging with added random noise', *Anal. Chem.*, **47**, 352–4.

Morrison, R. (1977). *Gounding and Shielding Techniques in Instrumentation*, Wiley, Chichester.

Nicholson, P. W. (1974). *Nuclear Electronics*, Wiley, New York. Chap. 4.

Handbook of Measurement Science, Vol. 1
Edited by P. H. Sydenham
© 1982 John Wiley & Sons Ltd.

Chapter

12   E. L. ZUCH

Signal Data Conversion

Editorial introduction

The majority of measurands are of analog form. Actuators and other output devices exist in both analog and digital form. Processing of signal information is often best performed using digital formats because of the increasing applicability of binary electronic circuitry. Interfacing analog and digital stages is, therefore, a most important part of measurement systems design. This chapter provides the basis of signal domain conversion.

12.1   DATA ACQUISITION SYSTEMS

12.1.1   Introduction

Data acquisition and conversion systems interface between the real world of physical parameters, which are analog, and the artificial world of digital computation and control. With current emphasis on digital systems, the interfacing function has become an important one; digital systems are used widely because complex circuits are low cost, accurate, and relatively simple to implement. In addition, there is rapid growth in use of minicomputers and microcomputers to perform difficult digital control and measurement functions.

Computerized feedback control systems are used in many different industries today in order to achieve greater productivity in our modern industrial societies. Industries which presently employ such automatic systems include steel making, food processing, paper production, oil refining, chemical manufacturing, textile production, and cement manufacturing.

The devices which perform the interfacing function between analog and digital worlds are analog-to-digital (A/D) and digital-to-analog (D/A) converters, which together are known as data converters. Some of the specific applications in which data converters are used include data telemetry systems, pulse code modulated communications, automatic test systems, computer display systems, video signal processing systems, data logging systems, and sampled-data control systems. In addition, every laboratory digital multimeter or digital panel meter contains an A/D converter.

12.1.2 Basic Data Acquisition System

Besides A/D and D/A converters, data acquisition and distribution systems may employ one or more of the following circuit functions:

(a) transducers;
(b) amplifiers;
(c) filters;
(d) non-linear analog functions;
(e) analog multiplexers;
(f) sample-holds.

The interconnection of these components is shown in the diagram of the data acquisition portion of a computerized feedback control system in Figure 12.1.

The input to the system is a *physical parameter* such as temperature, pressure, flow, acceleration, and position, which are analog quantities. The parameter is first converted into an electrical signal by means of a *transducer*; once in electrical form, all further processing is done by electronic circuits.

Next, an *amplifier* boosts the amplitude of the transducer output signal to a useful level for further processing. Transducer outputs may be microvolt or millivolt level signals which are then amplified to 1 to 10 volt levels. Furthermore, the transducer output may be a high-impedance signal, a differential signal with common-mode noise, a current output, a signal superimposed on a high voltage, or a combination of these. The amplifier, in order to convert such signals into a high-level voltage, may be one of several specialized types.

The amplifier is frequently followed by a low-pass *active filter* which reduces

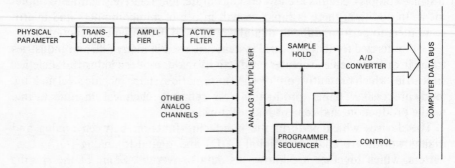

Figure 12.1   Data acquisition system

high-frequency signal components, unwanted electrical interference noise, or electronic noise from the signal. The amplifier is sometimes also followed by a special *non-linear analog function* circuit which performs a non-linear operation on the high-level signal. Such operations include squaring, multiplication, division, r.m.s. conversion, log conversion or linearization.

The processed analog signal next goes to an *analog multiplexer* which switches sequentially between a number of different analog input channels. Each input is in turn connected to the output of the multiplexer for a specified period of time by the multiplexer switch. During this connection time a *sample-hold* circuit acquires the signal voltage and then holds its value while an *A/D converter* converts the value into digital form. The resultant digital word goes to a computer data bus or to the input of a digital circuit.

Thus the analog multiplexer, together with the sample-hold, time shares the A/D converter with a number of analog input channels. The timing and control of the complete data acquisition system is done by a digital circuit called a *programmer-sequencer*, which in turn is under control of the computer. In some cases the computer itself may control the entire data acquisition system.

While this is perhaps the most commonly used data acquisition system configuration, there are alternative ones. Instead of multiplexing high-level signals, low-level multiplexing is sometimes used with the amplifier following the multiplexer. In such cases just one amplifier is required, but its gain may have to be changed from one channel to the next during multiplexing. Another method is to amplify and convert the signal into digital form at the transducer location and send the digital information in serial form to the computer. Here the digital data must be converted to parallel form and then multiplexed onto the computer data bus.

12.1.3 Basic Data Distribution System

The data distribution portion of a feedback control system, illustrated in Figure 12.2, is the reverse of the data acquisition system. The computer, based on the inputs of the data acquisition system, must close the loop on a process and control it by means of output control functions. These control outputs are in digital form and must, therefore, be converted into analog form in order to drive the process. The conversion is accomplished by a series of *D/A converters* as shown (also often called DAC's). Each D/A converter is coupled to the computer data bus by means of a register which stores the digital word until the next update. The registers are activated sequentially by a *decoder and control circuit* which is under computer control.

The D/A converter outputs then drive *actuators* which directly control the various process parameters such as temperature, pressure, and flow. Thus the loop is closed on the process and the result is a complete automatic process control system under computer control.

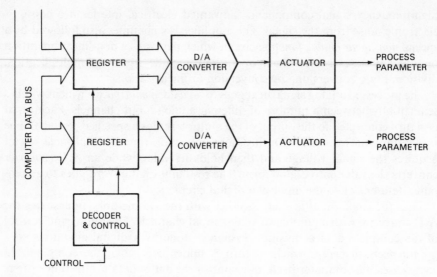

Figure 12.2   Data distribution system

12.2   QUANTIZING THEORY

12.2.1   Introduction

A/D conversion in its basic conceptual form is a two-step process: *quantizing* and *coding*. Quantizing is the process of transforming a continuous analog signal into a set of discrete output states. Coding is the process of assigning a digital code word to each of the output states. Some of the early A/D converters were appropriately called quantizing encoders.

12.2.2   Quantizer Transfer Function

The non-linear transfer function shown in Figure 12.3 is that of an ideal quantizer with eight output states; with output code words assigned, it is also that of a 3-bit A/D converter. The eight output states are assigned the sequence of binary numbers from 000 through to 111. The analog input range for this quantizer is 0 to +10 V.

There are several important points concerning the transfer function of Figure 12.3. First, the *resolution* of the quantizer is defined as the number of output states expressed in bits; in this case it is a 3-bit quantizer. The number of output states for a binary coded quantizer is 2^n, where n is the number of bits. Thus, an 8-bit quantizer has 256 output states and a 12-bit quantizer has 4096 output states.

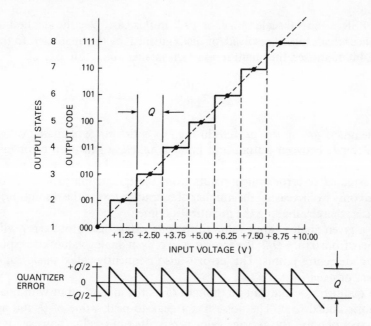

Figure 12.3   Transfer function and error of ideal 3-bit quantizer

As shown in the diagram, there are $2^n - 1$ analog decision points (or threshold levels) in the transfer function. These points are at voltages of $+0.625$, $+1.875$, $+3.125$, $+4.375$, $+5.625$, $+6.875$, and $+8.125$ V. The decision points must be precisely set in a quantizer in order to divide the analog voltage range into the correct quantizer values.

The voltages $+1.25$, $+2.50$, $+3.75$, $+5.00$, $+6.25$, $+7.50$, and $+8.75$ V are the centre points of each output code word. The analog decision point voltages are precisely halfway between the code word centre points. The quantizer staircase function is the best approximation which can be made to a straight line drawn through the origin and full scale point; notice that the line passes through all of the code word centre points.

12.2.3   Quantizer Resolution and Error

At any part of the input range of the quantizer, there is a small range of analog values within which the same output code word is produced. This small range is the voltage difference between any two adjacent decision points and is known as the analog quantization size, or *quantum*, Q. In Figure 12.3, the quantum is 1.25 V and is found in general by dividing the full scale analog range by the number of output states. Thus

$$Q = \text{FSR}/2^n \tag{12.1}$$

where FSR is the full scale range, or 10 V in this case. Q is the smallest analog difference which can be resolved, or distinguished, by the quantizer. In the case of a 12-bit quantizer, the quantum is much smaller and is found to be

$$Q = \frac{\text{FSR}}{2^n} = \frac{10 \text{ V}}{4096} = 2.44 \text{ mV} \tag{12.2}$$

If the quantizer input is moved through its entire range of analog values and the difference between output and input is taken, a sawtooth error function results, as shown in Figure 12.3. This function is called the quantizing error and is the irreducible error which results from the quantizing process. It can be reduced only by increasing the number of output states (or the resolution) of the quantizer, thereby making the quantization finer.

For a given analog input value to the quantizer, the output error will vary anywhere from 0 to $\pm Q/2$; the error is zero only at analog values corresponding to the code centre points. This error is also frequently called *quantization uncertainty* or *quantization noise*.

The quantizer output can be thought of as the analog input with quantization noise added to it. The noise has a peak-to-peak value of Q but, as with other types of noise, the average value is zero. Its r.m.s. value, however, is useful in analysis and can be computed from the triangular waveshape to be $Q/2\sqrt{3}$.

12.3   SAMPLING THEORY

12.3.1   Introduction

An A/D converter requires a small, but significant, amount of time to perform the quantizing and coding operations. The time required to make the conversion depends on several factors: the converter resolution, the conversion technique, and the speed of the components employed in the converter. The conversion speed required for a particular application depends on the time variation of the signal to be converted and on the accuracy desired.

12.3.2   Aperture Time

Conversion time is frequently referred to as *aperture time*. In general, aperture time refers to the time uncertainty (or time window) in making a measurement and results in an amplitude uncertainty (or error) in the measurement if the signal is changing during this time.

As shown in Figure 12.4, the input signal to the A/D converter changes by ΔV during the aperture time t_a in which the conversion is performed. The error

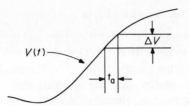

Figure 12.4 Aperture time and
amplitude uncertainty

can be considered an amplitude error or a time error; the two are related as
follows:

$$\Delta V = t_a \frac{dV(t)}{dt} \tag{12.3}$$

where $dV(t)/dt$ is the rate of change with time of the input signal.

It should be noted that ΔV represents the maximum error due to signal change,
since the actual error depends on how the conversion is done. At some point
in time within t_a, the signal amplitude corresponds exactly with the output code
word produced.

For the specific case of a sinusoidal input signal, the maximum rate of change
occurs at the zero crossing of the waveform, and the amplitude error is

$$\Delta V = t_a \frac{d}{dt} (A \sin \omega t)_{t=0} = t_a A\omega \tag{12.4}$$

The resultant error as a fraction of the peak-to-peak full scale value is

$$\varepsilon = \frac{\Delta V}{2A} = \pi f t_a \tag{12.5}$$

From this result the aperture time required to digitize a 1 kHz signal to 10
bits resolution can be found. The resolution required is one part in 2^{10} or
0.001.

$$t_a = \frac{\varepsilon}{\pi f} = \frac{0.001}{3.14 \times 10^3} = 320 \times 10^{-9} \tag{12.6}$$

The result is a required aperture time of just 320 ns! One should appreciate the
fact that 1 kHz is not a particularly fast signal, yet it is difficult to find a 10 bit
A/D converter to perform this conversion at any price! Fortunately, there is
a relatively simple and inexpensive way around this dilemma by using a sample-
hold circuit.

12.3.3 Sample-holds and Aperture Error

A sample-hold circuit samples the signal voltage and then stores it on a capacitor for the time-required to perform the A/D conversion. The aperture time of the A/D converter is, therefore, greatly reduced by the much shorter aperture time of the sample-hold circuit. In turn, the aperture time of the sample-hold is a function of its bandwidth and switching time.

Figure 12.5 is a useful graph of equation (12.5). It gives the aperture time required for converting sinusoidal signals to a maximum error less than one part in 2^n where n is the resolution of the converter in bits. The peak-to-peak value of the sinusoid is assumed to be the full scale range of the A/D converter. The graph is most useful in selecting a sample-hold by aperture time or an A/D converter by conversion time.

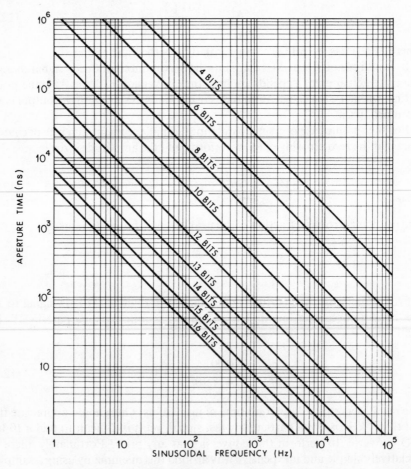

Figure 12.5   Graph for aperture error for sinusoidal signals

12.3.4 Sampled-data Systems and the Sampling Theorem

In data acquisition and distribution systems, and other sampled-data systems, analog signals are sampled on a periodic basis as illustrated in Figure 12.6. The train of sampling pulses in Figure 12.6b represents a fast-acting switch which connects to the analog signal for a very short time and then disconnects for the remainder of the sampling period.

The result of the fast-acting sampler is identical with multiplying the analog signal by a train of sampling pulses of unity amplitude, giving the modulated pulse train of Figure 12.6c. The amplitude of the original signal is preserved in the modulation envelope of the pulses. If the switch-type sampler is replaced by a switch and capacitor (a *sample-hold* circuit), then the amplitude of each sample is stored between samples and a reasonable reconstruction of the orginal analog signal results, as shown in Figure 12.6d.

A common use of sampling is in the efficient use of data processing equipment and data transmission facilities. A single data transmission link, for example, can be used to transmit many different analog channels on a sampled, time-multiplexed, basis, whereas it would be uneconomical to devote a complete transmission link to the continuous transmission of a single signal.

Likewise, a data acquisition and distribution system is used to measure and control the many parameters of a process control system by sampling the parameters and updating the control inputs periodically. In data conversion

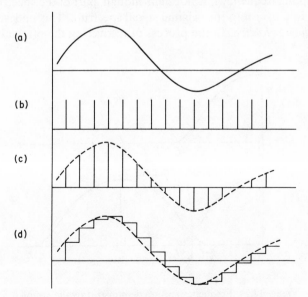

Figure 12.6 Signal sampling: (a) signal; (b) sampling pulses; (c) sampled signal; (d) sampled and held signal

systems it is common to use a single, expensive A/D converter of high speed and precision and then multiplex a number of analog inputs into it.

An important fundamental question to answer about sampled-data systems is this: 'How often must I sample an analog signal in order not to lose information from it?' It is obvious that all useful information can be extracted if a slowly varying signal is sampled at a rate such that little or no change takes place between samples. Equally obvious is the fact that information is being lost if there is a significant change in signal amplitude between samples.

The answer to the question is contained in the well known sampling theorem which may be stated as follows: *If a continuous, bandwidth-limited signal contains no frequency components higher than f_c, then the original signal can be recovered without distortion if it is sampled at a rate of at least $2f_c$ samples per second.*

12.3.5 Frequency Folding and Aliasing

The *sampling theorem* can be demonstrated by the frequency spectra illustrated in Figure 12.7. Figure 12.7a shows the frequency spectrum of a continuous bandwidth-limited analog signal with frequency components out to f_c. When this signal is sampled at a rate f_s, the modulation process shifts the original spectrum out to f_s, $2f_s$, $3f_s$,..., in addition to the one at the origin. A portion of this resultant spectrum is shown in Figure 12.7b.

If the sampling frequency f_s is not high enough, part of the spectrum centred about f_s will fold over into the original signal spectrum. This undesirable effect is called *frequency folding*. In the process of recovering the original signal, the

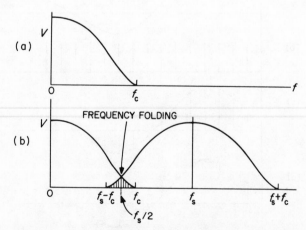

Figure 12.7   Frequency spectra demonstrating the sampling theorem: (a) continuous signal spectrum; (b) sampled signal spectrum

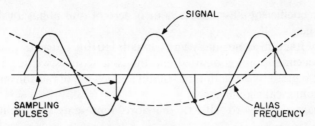

Figure 12.8   Alias frequency caused by inadequate sampling
rate

folded part of the spectrum causes distortion in the recovered signal which
cannot be eliminated by filtering the recovered signal.

From the figure, if the sampling rate is increased such that $f_s - f_c > f_c$,
then the two spectra are separated and the original signal can be recovered
without distortion. This demonstrates the results of the sampling theorem that
$f_s > 2f_c$. Frequency folding can be eliminated in two ways: first by using a high
enough sampling rate, and second by filtering the signal before sampling to
limit its bandwidth to $f_s/2$.

It must be appreciated that in practice there is always some frequency folding
present due to high-frequency signal components, noise, and non-ideal pre-
sample filtering. The effect must be reduced to negligible amounts for the
particular application by using a sufficiently high sampling rate. The required
rate, in fact, may be much higher than the minimum indicated by the sampling
theorem.

The effect of an inadequate sampling rate on a sinusoid is illustrated in
Figure 12.8; an *alias frequency* in the recovered signal results. In this case,
sampling at a rate slightly less than twice per cycle gives the low-frequency
sinusoid shown by the dotted line in the recovered signal. This alias frequency
can be significantly different from the original frequency. From the figure it is
easy to see that if the sinusoid is sampled at least twice per cycle, as required by
the sampling theorem, the original frequency is preserved.

12.4   CODING FOR DATA CONVERTERS

12.4.1   Natural Binary Code

A/D and D/A converters interface with digital systems by means of an appropriate digital code. While there are many possible codes to select, a few standard
ones are almost exclusively used with data converters. The most popular code
is *natural binary*, or *straight binary*, which is used in its fractional form to represent a number

$$N = a_1 2^{-1} + a_2 2^{-2} + a_3 2^{-3} + \cdots + a_n 2^{-n} \tag{12.7}$$

where each coefficient a assumes a value of zero or one. N has a value between zero and one.

A binary fraction is normally written as 0.110101, but with data converter codes the decimal point is omitted and the code word is written 110101. This code word represents a fraction of the full scale value of the converter and has no other numerical significance.

The binary code word 110101, therefore, represents the decimal fraction

$$(1 \times 0.5) + (1 \times 0.25) + (0 \times 0.125) + (1 \times 0.0625) + (0 \times 0.03125)$$

$$+ (1 \times 0.015625) = 0.828125 \quad \text{or} \quad 82.8125\%$$

of full scale for the converter. If full scale is $+10$ V, then the code word represents $+8.28125$ V. The natural binary code belongs to a class of codes known as *positive weighted* codes since each coefficient has a specific weight, none of which is negative.

The leftmost bit has the most weight, 0.5 of full scale, and is called the *most significant bit* (MSB); the rightmost bit has the least weight, 2^{-n} of full scale, and is, therefore, called the *least significant bit* (LSB). The bits in a code word are numbered from left to right from 1 to n.

The LSB has the same analog equivalent value as Q discussed previously, namely

$$\text{LSB (analog value)} = \text{FSR}/2^n \tag{12.8}$$

Table 12.1 is a useful summary of the resolution, number of states, LSB weights, and dynamic range for data converters from one to twenty bits resolution.

The *dynamic range* (DR) of a data converter in decibels (dB) is found as follows:

$$\begin{aligned} \text{DR} &= 20 \log 2^n = 20n \log 2 \\ &= 20n\,(0.301) = 6.02n \end{aligned} \tag{12.9}$$

where DR is dynamic range, n is the number of bits, and 2^n the number of states of the converter. Since 6.02 dB corresponds to a factor of two, it is simply necessary to multiply the resolution of a converter in bits by 6.02. A 12-bit converter, for example, has a dynamic range of 72.2 dB.

An important point to notice is that the maximum value of the digital code, namely all 1's, does not correspond with analog full scale, but rather with one LSB less than full scale, or $\text{FS}(1 - 2^{-n})$. Therefore a 12-bit converter with a 0 to $+10$ V analog range has a maximum code of 1111 1111 1111 and a maximum analog value of $+10\text{ V}(1 - 2^{-12}) = +9.99756$ V. In other words, the maximum analog value of the converter, corresponding to all ones in the code, never quite reaches the point defined as analog full scale.

Table 12.1   Resolution, number of states, LSB weight, and dynamic range for data
converters

Resolution bits n	Number of states 2^n	LSB weight 2^{-n}	Dynamic range (dB)
0	1	1	0.0
1	2	0.5	6.0
2	4	0.25	12.0
3	8	0.125	18.1
4	16	0.0625	24.1
5	32	0.03125	30.1
6	64	0.015625	36.1
7	128	0.0078125	42.1
8	256	0.00390625	48.2
9	512	0.001953125	54.2
10	1 024	0.0009765625	60.2
11	2 048	0.00048828125	66.2
12	4 096	0.000244140625	72.2
13	8 192	0.0001220703125	78.3
14	16 384	0.00006103515625	84.3
15	32 768	0.000030517578125	90.3
16	65 536	0.0000152587890625	96.3
17	131 072	0.00000762939453125	102.3
18	262 144	0.000003814697265625	108.4
19	524 288	0.0000019073486328125	114.4
20	1 048 576	0.00000095367431640625	120.4

12.4.2   Other Binary Codes

Several other binary codes are used with A/D and D/A converters in addition
to straight binary. These codes are *offset binary*, *two's complement*, *binary coded
decimal* (BCD), and their complemented versions. Each code has a specific
advantage in certain applications. BCD coding for example is used where
digital displays must be interfaced such as in digital panel meters and digital
multimeters. Two's complement coding is used for computer arithmetic logic
operations, and offset binary coding is used with bipolar analog measures.

Not only are the digital codes standardized with data converters, but so
also are the analog voltage ranges. Most converters use unipolar voltage ranges
of 0 to $+5$ V and 0 to $+10$ V although some devices use the negative ranges 0 to
-5 V and 0 to -10 V. The standard bipolar voltage ranges are ± 2.5 V, ± 5 V
and ± 10 V. Many converters today are pin-programmable between these
various ranges.

Table 12.2 shows straight binary and complementary binary codes for a
unipolar 8-bit converter with a 0 to $+10$ V analog FS range. The maximum
analog value of the converter is $+9.961$ V, or one LSB less than $+10$ V. Note

Table 12.2   Binary coding for 8-bit unipolar converters

Fraction of FS	+10 V FS	Straight binary	Complementary binary
$+FS - 1LSB$	$+9.961$	1111 1111	0000 0000
$+\frac{3}{4}FS$	$+7.500$	1100 0000	0011 1111
$+\frac{1}{2}FS$	$+5.000$	1000 0000	0111 1111
$+\frac{1}{4}FS$	$+2.500$	0100 0000	1011 1111
$+\frac{1}{8}FS$	$+1.250$	0010 0000	1101 1111
$+1$ LSB	$+0.039$	0000 0001	1111 1110
0	0.000	0000 0000	1111 1111

that the LSB size is 0.039 V as shown near the bottom of the table. The *complementary binary* coding used in some converters is simply the logic complement of straight binary.

When A/D and D/A converters are used in bipolar operation, the analog range is offset by half scale, or by the MSB value. The result is an analog shift of the converter transfer function as shown in Figure 12.9. Notice for this 3-bit A/D converter transfer function that the code 000 corresponds with -5 V, 100 with 0 V, and 111 with $+3.75$ V. Since the output coding is the same as before the analog shift, it is now appropriately called *offset binary coding*.

Table 12.3 shows the offset binary code together with *complementary offset binary*, *two's complement*, and *sign-magnitude binary* codes. These are the most popular codes employed in bipolar data converters.

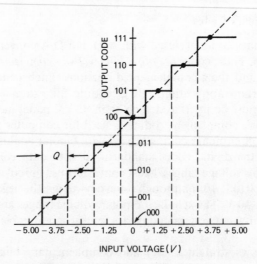

Figure 12.9   Transfer function for bipolar 3-bit
A/D converter

Table 12.3 Popular bipolar codes used with data converters

Fraction of FS	± 5 V FS	Offset binary	Comp. off. binary	Two's complement	Sign-mag. binary
$+FS - 1$ LSB	$+4.9976$	1111 1111	0000 0000	0111 1111	1111 1111
$+\frac{3}{4}FS$	$+3.7500$	1110 0000	0001 1111	0110 0000	1110 0000
$+\frac{1}{2}FS$	$+2.5000$	1100 0000	0011 1111	0100 0000	1100 0000
$+\frac{1}{4}FS$	$+1.2500$	1010 0000	0101 1111	0010 0000	1010 0000
0	0.0000	1000 0000	0111 1111	0000 0000	1000 0000*
$-\frac{1}{4}FS$	-1.2500	0110 0000	1001 1111	1110 0000	0010 0000
$-\frac{1}{2}FS$	-2.5000	0100 0000	1011 1111	1100 0000	0100 0000
$-\frac{3}{4}FS$	-3.7500	0010 0000	1101 1111	1010 0000	0110 0000
$-FS + 1$ LSB	-4.9976	0000 0001	1111 1110	1000 0001	0111 1111
$-FS$	-5.0000	0000 0000	1111 1111	1000 0000	—

* Sign magnitude binary has two code words for zero as shown here:

$$0+ \quad 1000\ 0000\ 0000$$
$$0- \quad 0000\ 0000\ 0000$$

The two's complement code has the characteristic that the sum of the positive and negative codes for the same analog magnitude always produces all zeros and a carry. This characteristic makes the two's complement code useful in arithmetic computations. Notice that the only difference between two's complement and offset binary is the complementing of the MSB. In bipolar coding, the MSB becomes the sign bit.

The *sign-magnitude binary code*, infrequently used, has identical code words for equal magnitude analog values except that the sign bit is different. As shown in Table 12.3 this code has two possible code words for zero: 1000 0000 or 0000 0000. The two are usually distinguished as $0+$ and $0-$, respectively. Because of this characteristic, the code has maximum analog values of $\pm (FS - 1LSB)$ and reaches neither analog $+FS$ nor $-FS$.

12.4.3 BCD Codes

Table 12.4 shows BCD and complementary BCD coding for a three-decimal digit data converter. These are the codes used with integrating type A/D converters employed in digital panel meters, digital multimeters, and other decimal display applications. Here four bits are used to represent each decimal digit. BCD is a positive weighted code but is relatively inefficient since in each group of four bits, only 10 out of a possible 16 states are utilized.

The LSB analog value (or quantum, Q) for BCD is

$$LSB(\text{analog value}) = Q = FSR/10^d \qquad (12.10)$$

Table 12.4   BCD and complementary BCD coding

Fraction of FS	+10 V FS	Binary coded decimal	Complementary BCD
$+FS - 1$ LSB	$+9.99$	1001 1001 1001	0110 0110 0110
$+\frac{3}{4}FS$	$+7.50$	0111 0101 0000	1000 1010 1111
$+\frac{1}{2}FS$	$+5.00$	0101 0000 0000	1010 1111 1111
$+\frac{1}{4}FS$	$+2.50$	0010 0101 0000	1101 1010 1111
$+\frac{1}{8}FS$	$+1.25$	0001 0010 0101	1110 1101 1010
$+1$ LSB	$+0.01$	0000 0000 0001	1111 1111 1110
0	0.00	0000 0000 0000	1111 1111 1111

where FSR is the full scale range and d is the number of decimal digits. For example if there are three digits and the full scale range is 10 V, the LSB value is

$$\text{LSB(analog value)} = 10 \text{ V}/10^3 = 0.01 \text{ V} = 10 \text{ mV} \tag{12.11}$$

BCD coding is frequently used with an additional overrange bit which has a weight equal to full scale and produces a 100% increase in range for the A/D converter. Thus for a converter with a decimal full scale of 999, an overrange bit provides a new full scale of 1999, twice that of the previous one. In this case, the maximum output code is 1 1001 1001 1001. The additional range is commonly referred to as $\frac{1}{2}$ *digit*, and the resolution of the A/D converter in this case is $3 - \frac{1}{2}$ digits.

Likewise, if this range is again expanded by 100%, a new full scale of 3999 results and is called $3 - \frac{3}{4}$ digits resolution. Here two overrange bits have been added and the full scale output code is 11 1001 1001 1001. When BCD coding is used for bipolar measurements another bit, a sign bit, is added to the code and the result is *sign-magnitude BCD* coding.

12.5   AMPLIFIERS AND FILTERS

12.5.1   Operational and Instrumentation Amplifiers

The front end of a data acquisition system extracts the desired analog signal from a physical parameter by means of a transducer and then amplifies and filters it. An amplifier and filter are critical components in this initial signal processing. The amplifier must perform one or more of the following functions: boost the signal amplitude, buffer the signal, convert a signal current into a voltage, or extract a differential signal from common mode noise.

To accomplish these functions requires a variety of different amplifier types. The most popular type of amplifier is an *operational amplifier* (op. amp.) which is a general purpose gain block with differential inputs. The op. amp. may be connected in many different closed-loop configurations, of which a few are

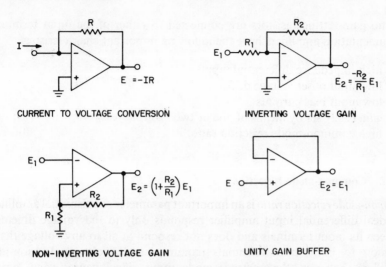

CURRENT TO VOLTAGE CONVERSION   INVERTING VOLTAGE GAIN

NON-INVERTING VOLTAGE GAIN   UNITY GAIN BUFFER

Figure 12.10   Operational amplifier configurations

shown in Figure 12.10. The gain and bandwidth of the circuits shown depend on the external resistors connected around the amplifier. An operational amplifier is a good choice in general where a single-ended signal is to be amplified, buffered, or converted from current to voltage.

In the case of differential signal processing, the *instrumentation amplifier* is a better choice since it maintains high impedance at both of its differential inputs and the gain is set by a resistor located elsewhere in the amplifier circuit. One type of instrumentation amplifier circuit is shown in Figure 12.11. Notice

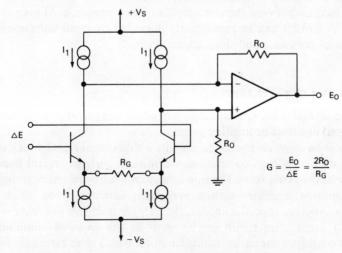

$$G = \frac{E_O}{\Delta E} = \frac{2R_O}{R_G}$$

Figure 12.11   Simplified instrumentation amplifier circuit

that no gain-setting resistors are connected to either of the input terminals. Instrumentation amplifiers have the following important characteristics:

(a)   high impedance differential inputs;
(b)   low input offset voltage drift;
(c)   low input bias currents;
(d)   gain easily set by means of one or two external resistors;
(e)   high common-mode rejection ratio.

12.5.2   Common-mode Rejection

Common-mode rejection ratio is an important parameter of differential amplifiers. An ideal differential input amplifier responds only to the voltage difference between its input terminals and does not respond at all to any voltage that is common to both input terminals (common-mode voltage). In non-ideal amplifiers, however, the common-mode input signal causes some output response even though small compared to the response to a differential input signal.

The ratio of differential and common-mode responses is defined as the common-mode rejection ratio (CMRR). *Common-mode rejection ratio of an amplifier is the ratio of differential voltage gain to common-mode voltage gain and is generally expressed in dB*:

$$CMRR = 20 \log_{10}(A_D/A_{CM}) \tag{12.12}$$

where A_D is differential voltage gain and A_{CM} is common-mode voltage gain. CMRR is a function of frequency and, therefore, also a function of the impedance balance between the two amplifier input terminals. At even moderate frequencies CMRR can be significantly degraded by small unbalances in the source series resistance and shunt capacitance.

12.5.3   Other Amplifier Types

There are several other special amplifiers which are useful in conditioning the input signal in a data acquisition system.

An *isolation amplifier* is used to amplify a differential signal which is superimposed on a very high common-mode voltage, perhaps several hundred or even several thousand volts. The isolation amplifier has the characteristics of an instrumentation amplifier with a very high common-mode input voltage capability. Another special amplifier, the *chopper stabilized amplifier*, is used to accurately amplify microvolt level signals to the required amplitude. This amplifier employs a special switching stabilizer which gives extremely low input offset voltage drift. Another useful device, the *electrometer amplifier*, has ultra-low input bias currents, generally less than one picoampere and is used to convert extremely small signal currents into a high-level voltage.

Morrison (1977) treats the subject of d.c. amplifiers.

12.5.4 Filters

A *low-pass filter* frequently follows the signal processing amplifier to reduce signal noise. Low-pass filters are used for the following reasons: to reduce man-made electrical interference noise, to reduce electronic noise, and to limit the bandwidth of the analog signal to less than half the sampling frequency in order to eliminate frequency folding. When used for the last reason, the filter is called a *pre-sampling filter* or *anti-aliasing* filter.

Man-made electrical noise is generally periodic, as for example in power line interference, and is sometimes reduced by means of a special filter such as a *notch filter*. Electronic noise, on the other hand, is random noise with noise power proportional to bandwidth and is present in transducer resistances, circuit resistances, and in amplifiers themselves. It is reduced by limiting the bandwidth of the system to the minimum required to pass desired signal components (refer to Chapter 11 for more detail).

No filter does a perfect job of eliminating noise or other undesirable frequency components, and therefore the choice of a filter is always a compromise. Ideal filters, frequently used as analysis examples, have flat passband response with infinite attenuation at the cut-off frequency, but are mathematical filters only and not physically realizable.

In practice, the systems engineer has a choice of cut-off frequency and attenuation rate. The attenuation rate and resultant phase response depend on the

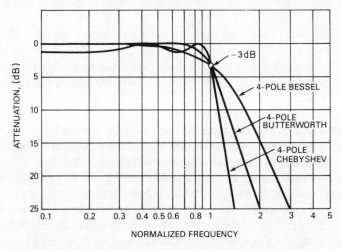

Figure 12.12   Some practical low-pass filter characteristics

particular filter characteristic and the number of poles in the filter function. Some of the more popular filter characteristics include Butterworth, Chebyshev, Bessel, and elliptic. In making this choice, the effect of overshoot and non-uniform phase delay must be carefully considered. Figure 12.12 illustrates some practical low-pass filter response characteristics. Their design is the subject of Chapters 9 and 10.

Passive *RLC* filters are seldom used in signal processing applications today due chiefly to the undesirable characteristics of inductors. Active filters are generally used now since they permit the filter characteristics to be accurately set by precision, stable, resistors and capacitors. Inductors, with their undesirable saturation and temperature drift characteristics, are thereby eliminated. Also, because active filters use operational amplifiers, the problems of insertion loss and output loading are also eliminated.

12.6 SETTLING TIME

12.6.1 Definition

A parameter that is specified frequently in data acquisition and distribution systems is *settling time*. The term settling time originates in control theory but is now commonly applied to amplifiers, multiplexers, and D/A converters.

Settling time is defined as *the time elapsed from the application of a full scale step input to a circuit to the time when the output has entered and remained within a specified error band around its final value.* The method of application of the input step may vary depending on the types of circuit, but the definition still holds. In the case of a D/A converter, for example, the step is applied by changing the digital input code whereas in the case of an amplifier the input signal itself is a step change.

The importance of settling time in a data acquisition system is that certain analog operations must be performed in sequence and one operation may have to be accurately settled before the next operation can be initiated. Thus a buffer amplifier preceding an A/D converter must have accurately settled before the conversion can be initiated.

Settling time for an amplifier is illustrated in Figure 12.13. After application of a full scale step input there is a small delay time following which the amplifier output slews, or changes at its maximum rate. *Slew rate* is determined by internal amplifier currents which must charge internal capacitances.

As the amplifier output approaches final value, it may first overshoot and then reverse and undershoot this value before finally entering and remaining within the specified error band. Note that settling time is measured to the point at which the amplifier output enters and remains within the error band. This error band in most devices is specified to either $\pm 0.1\%$ or $\pm 0.01\%$ of the full scale transition.

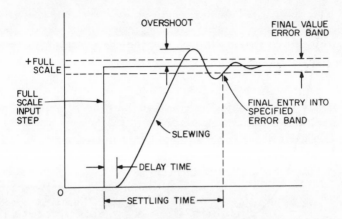

Figure 12.13 Amplifier settling time

12.6.2 Amplifier Characteristics

Settling time, unfortunately, is not readily predictable from other amplifier parameters such as bandwidth, slew rate, or overload recovery time, although it depends on all of these. It is also dependent on the shape of the amplifier open loop gain characteristic, its input and output capacitance and the dielectric absorption of any internal capacitances. An amplifier must be specifically designed for optimized settling time, and settling time is a parameter that must be determined by testing.

One of the important requirements of a fast-settling amplifier is that it have a single-pole open-loop gain characteristic, that is, one that has a smooth 6 dB per octave gain roll-off characteristic to beyond the unity gain crossover frequency—a first-order response. Such a desirable characteristic is shown in Figure 12.14.

It is important to note that an amplifier with a single-pole response can never settle faster than the time indicated by the number of closed-loop time constants

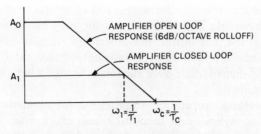

Figure 12.14 Amplifier single-pole open-loop gain characteristic

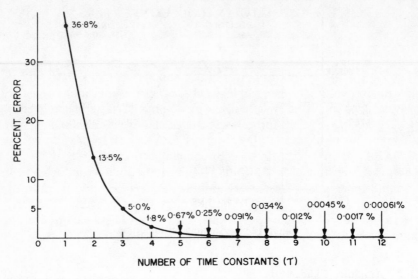

Figure 12.15   Output settling error as a function of number of time constants

to the given accuracy. Figure 12.15 shows output error as a function of the number of time constants τ where

$$\tau = 1/\omega = 1/2\pi f \tag{12.13}$$

and f is the closed loop 3 dB bandwidth of the amplifier.

Actual settling time for a good quality amplifier may be significantly longer than that indicated by the number of closed-loop time constants due to slew rate limitation and overload recovery time. For example, an amplifier with a closed-loop bandwidth of 1 MHz has a time constant of 160 ns which indicates a settling time of 1.44 μs (nine time constants) to 0.01 % of final value. If the slew rate of this amplifier is 1 V/μs, it will take more than 10 μs to settle to 0.01 % for a 10 V change.

If the amplifier has a non-uniform gain roll-off characteristic rather than a single-pole characteristic, then its settling time may have one of two undesirable qualities. First, the output may reach the vicinity of the error band quickly but then take a long time to actually enter it; second, it may overshoot the error band and then oscillate back and forth through it before finally entering and remaining inside it.

Modern fast-settling operational amplifiers come in many different types including modular, hybrid, and monolithic amplifiers. Such amplifiers have settling times to 0.1 % or 0.01 % of 2 μs down to 100 ns and are useful in many data acquisition and conversion applications.

12.7   DIGITAL-TO-ANALOG CONVERTERS

12.7.1   Introduction

D/A converters are the devices by which computers communicate with the outside analog world. They are employed in a variety of applications from CRT display systems and voice synthesizers to automatic test systems, digitally controlled attenuators, and process control actuators. In addition, they are key components inside most A/D converters. D/A converters are also referred to as DAC's and are termed *decoders* by communications engineers.

The transfer function of an ideal 3-bit D/A converter is shown in Figure 12.16. Each input code word produces a single, discrete analog output value, generally, but not always, a voltage. Over the output range of the converter 2^n different values are produced including zero; and the output has a one-to-one correspondence with input, which is not true for A/D converters.

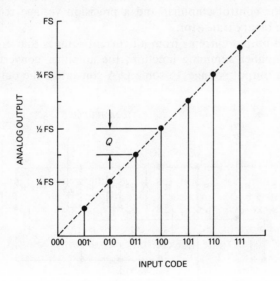

Figure 12.16   Transfer function of ideal 3-bit D/A
converter

There are many different circuit techniques used to implement D/A converters, but a few popular ones are widely used today. Virtually all D/A converters in use are of the *parallel type* where all bits change simultaneously upon application of an input code word; *serial type* D/A converters, on the other hand, produce an analog output only after receiving all digital input data in sequential form.

12.7.2 Weighted Current Source D/A Converter

The most popular D/A converter design in use today is the weighted current source circuit illustrated in Figure 12.17. An array of switched transistor current sources is used with binary weighted currents. The binary weighting is achieved by using emitter resistors with binary related values of $R, 2R, 4R, 8R, \ldots 2^n R$. The resulting collector currents are then added together at the current summing line.

The current sources are switched on or off from standard TTL semi-conductor device inputs by means of the control diodes connected to each emitter. When the TTL input is high the current source is on; when the input is low it is off, with the current flowing through the control diode. Fast switching speed is achieved because there is direct control of the transistor current, and the current sources never go into saturation.

To interface with standard TTL levels, the current sources are biased to a base voltage of $+1.2$ V. The emitter currents are regulated to constant values by means of the control amplifier and a precision voltage reference circuit together with a binary transistor.

The summed output currents from all current sources that are on go to an operational amplifier summing junction; the amplifier converts this output current into an output voltage. In some D/A converters the output current is

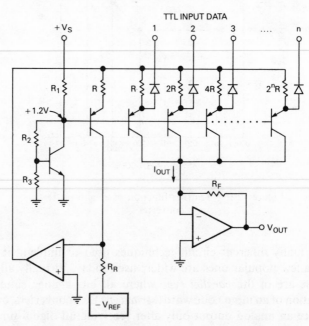

Figure 12.17   Weighted current source D/A converter

used to directly drive a resistor load for maximum speed, but the positive output voltage in this case is limited to about $+1$ V.

The weighted current source design has the advantages of simplicity and high speed. Both PNP and NPN transistor current sources can be used with this technique although the TTL interfacing is more difficult with NPN sources. This technique is used in most monolithic, hybrid, and modular D/A converters in use today.

A difficulty in implementing the higher resolution D/A converter designs to this concept is that a wide range of emitter resistors is required and very high value resistors cause problems with both temperature stability and switching speed. To overcome these problems, weighted current sources are used in identical groups, with the output of each group divided down by a resistor divider as shown in Figure 12.18.

The resistor network, R_1 through R_4, divides the output of group 3 down by a factor of 256 and the output of group 2 down by a factor of 16 with respect to the output of group 1. Each group is identical, with four current sources of the type shown in Figure 12.17, having binary current weights of 1, 2, 4, 8. Figure 12.18 also illustrates the method of achieving a bipolar output by deriving an offset current from the reference circuit which is then subtracted from the output current line through resistor R_o. This current is set to exactly one half the full scale output current.

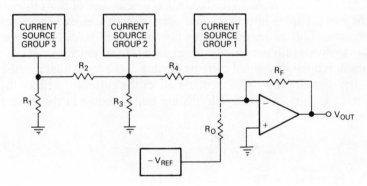

Figure 12.18   Current dividing the outputs of weighted current source groups

12.7.3 R–$2R$ D/A Converter

A second popular technique for D/A conversion is the R–$2R$ ladder method. As shown in Figure 12.19, the network consists of series resistors of value R and shunt resistors of values $2R$. The bottom of each shunt resistor has a single-pole double-throw electronic switch which connects the resistor to either ground or the output current summing line. As in the previous circuit, the output

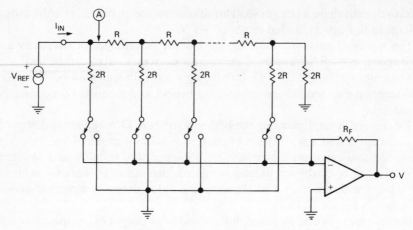

Figure 12.19 R–$2R$ ladder D/A converter

current summing line goes to an operational amplifier which converts current to voltage.

The operation of the R–$2R$ ladder network is based on the binary division of current as it flows down the ladder. Examination of the ladder configuration reveals that at point A looking to the right, one measures a resistance of $2R$; therefore the reference input to the ladder has a resistance of R. At the reference input the current splits into two equal parts since it sees equal resistances in either direction. Likewise, the current flowing down the ladder to the right continues to divide into two equal parts at each resistor junction.

The result is binary weighted currents flowing down each shunt resistor in the ladder. The digitally controlled switches direct the currents to either the summing line or ground. Assuming all bits are on, as shown in the diagram, the output current is

$$I_{\text{out}} = \frac{V_{\text{ref}}}{R}\left(\frac{1}{2} + \frac{1}{4} + \frac{1}{8} + \cdots + \frac{1}{2_n}\right) \tag{12.14}$$

which is a binary series. The sum of all currents is then

$$I_{\text{out}} = \frac{V_{\text{ref}}}{R}\left(1 - 2^{-n}\right) \tag{12.15}$$

where the 2^{-n} term physically represents the portion of the input current flowing through the $2R$ terminating resistor to ground at the far right.

The advantage of the R–$2R$ ladder technique is that only two values of resistors are required, with the resultant ease of matching or trimming and excellent temperature tracking. In addition, for high speed applications relatively low resistor values can be used. Excellent results can be obtained for high resolution D/A converters by using laser-trimmed thin film resistor networks.

12.7.4   Multiplying and Deglitched D/A Converters

The *R–2R* ladder method is specifically used for *multiplying type* D/A converters. With these converters, the reference voltage can be varied over the full range of $\pm V_{max}$ with the output being the product of the reference voltage and the digital input word. Multiplication can be performed in 1, 2, or 4 algebraic quadrants.

If the reference voltage is unipolar, the circuit is a one-quadrant multiplying DAC; if it is bipolar the circuit is a two-quadrant multiplying DAC. For four-quadrant operation the two current summing lines shown in Figure 12.19 must be subtracted from each other by operational amplifiers.

In multiplying D/A converters, the electronic switches are usually implemented with CMOS devices. Multiplying DAC's are commonly used in automatic gain controls, CRT character generation, complex function generators, digital attenuators, and divider circuits. Figure 12.20 shows two 14-bit multiplying CMOS D/A converters.

Another important D/A converter design takes advantage of the best features of both the weighted current source technique and the *R–2R* ladder technique. This circuit, shown in Figure 12.21, uses equal value switched current sources to drive the junctions of the *R–2R* ladder network. The advantage of the equal value current sources is obvious since all emitter resistors are identical and switching speeds are also identical. This technique is used in many ultra-high speed D/A converters.

One other specialized type D/A converter used primarily in CRT display systems is the *deglitched* D/A converter. All D/A converters produce output spikes, or *glitches*, which are most serious at the major output transitions of $\frac{1}{4}$FS, $\frac{1}{2}$FS and $\frac{3}{4}$FS as illustrated in Figure 12.22a.

Glitches are caused by small time differences between some current sources turning off and others turning on. Take, for example, the major code transition at half scale from $0111\cdots1111$ to $1000\cdots0000$. Here the MSB current source

Figure 12.20   CMOS 14-bit multiplying D/A converters

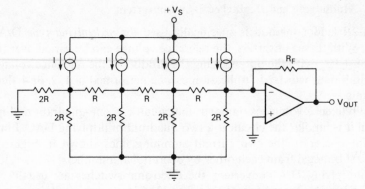

Figure 12.21   D/A converter employing R–$2R$ ladder with equal value
switched current sources

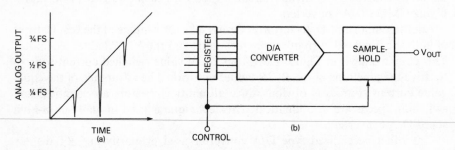

Figure 12.22   (a) Output glitches and (b) deglitched D/A converter

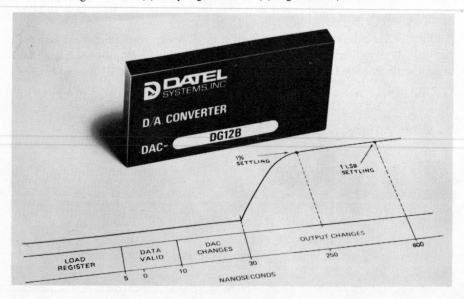

Figure 12.23   Modular deglitched D/A converter

turns on while all other current sources turn off. The small difference in switching times results in a narrow half-scale glitch. Such a glitch produces distorted characters on CRT displays.

Glitches can be virtually eliminated by the circuit shown in Figure 12.22b. The digital input to a D/A converter is controlled by an input register while the converter output goes to a specially designed sample-hold circuit. When the digital input is updated by the register, the sample-hold is switched into the hold mode. After the D/A converter has changed to its new output value and all glitches have settled out, the sample-hold is then switched back into the tracking mode. When this happens, the output changes smoothly from its previous value to the new value with no glitches present. Figure 12.23 shows a modular deglitched D/A converter which contains the circuitry just described.

12.8 VOLTAGE REFERENCE CIRCUITS

An important circuit required in both A/D and D/A converters is the voltage reference. The accuracy and stability of a data converter ultimately depends upon the reference; it must, therefore, produce a constant output voltage over both time and temperature.

The compensated Zener reference diode with a buffer-stabilizer circuit is commonly used in most data converters today. Although the compensated zener may be one of several types, the compensated *subsurface*, or *buried*, Zener is probably the best choice. These relatively new devices produce an avalanche breakdown which occurs beneath the surface of the silicon, resulting in better long-term stability and noise characteristics than with earlier surface breakdown Zeners. These reference devices have reverse breakdown voltages of about 6.4 volts and consist of a forward biased diode in series with the reversed biased Zener. Because the diodes have approximately equal and opposite voltage changes with temperature, the result is a temperature stable voltage. Available devices have temperature coefficients from 100 ppm/°C to less than 1 ppm/°C.

Some of the new IC voltage references incorporate active circuitry to buffer the device and reduce its dynamic impedance; in addition, some contain temperature regulation circuitry on the chip to achieve ultra-low temperature coefficient (tempco).

A popular buffered reference circuit is shown in Figure 12.24; this circuit produces an output voltage higher than the reference voltage. It also generates a constant, regulated current through the reference which is determined by the three resistors.

Some monolithic A/D and D/A converters use another type of reference device known as the *band-gap reference*. This circuit is based on the principle of using the known, predictable base-to-emitter voltage of a transistor to generate a constant voltage equal to the extrapolated band-gap voltage of silicon. This reference gives excellent results for the lower reference voltages of 1.2 or 2.5 V.

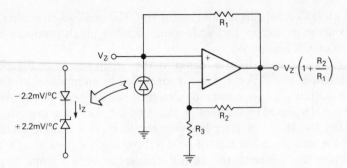

Figure 12.24   A precision, buffered voltage reference circuit

12.9   ANALOG-TO-DIGITAL CONVERTERS

12.9.1   Counter-type A/D Converter

A/D converters, also called ADC's or *encoders*, employ a variety of different circuit techniques to implement the conversion function. As with D/A converters, however, relatively few of the many originally devised circuits are widely used today. Of the various techniques available, the choice depends on the resolution and speed required.

One of the simplest A/D converters is the *counter*, or *servo*, type. This circuit employs a digital counter to control the input of a D/A converter. Clock pulses are applied to the counter and the output of the D/A is stepped up one LSB at a time. A comparator compares the D/A output with the analog input and stops the clock pulses when they are equal. The counter output is then the converted digital word.

While this converter is simple, it is also relatively slow. An improvement on this technique is shown in Figure 12.25 and is known as a *tracking* A/D converter, a device commonly used in control systems. Here an up-down counter controls the DAC, and the clock pulses are directed to the pertinent counter input depending on whether the D/A output must increase or decrease to reach the analog input voltage.

The obvious advantage of the tracking A/D converter is that it can continuously follow the input signal and give updated digital output data if the signal does not change too rapidly. Also, for small input changes, the conversions can be quite fast. The converter can be operated in either the *track* or *hold modes* by a digital input control.

12.9.2   Successive-approximation A/D Converters

By far, the most popular A/D conversion technique in general use for moderate to high-speed applications is the *successive-approximation* type A/D. This

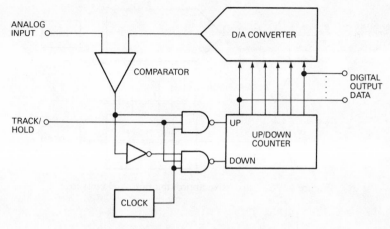

Figure 12.25   Tracking-type A/D converter

method falls into a class of techniques known as *feedback type* A/D converters, to which the counter type also belongs. In both cases a D/A converter is in the feedback loop of a digital control circuit which changes its output until it equals the analog input. In the case of the successive-approximation converter, the DAC is controlled in an optimum manner to complete a conversion in just n steps, where n is the resolution of the converter in bits.

The operation of this converter is analogous to weighing an unknown on a laboratory balance scale using standard weights in a binary sequence such as 1, $\frac{1}{2}$, $\frac{1}{4}$, $\frac{1}{8}$, ..., $1/n$ kilograms. The correct procedure is to begin with the largest standard weight and proceed in order down to the smallest one.

The largest weight is placed on the balance pan first; if it does not tip, the weight is left on and the next largest weight is added. If the balance does tip, the weight is removed and the next one added. The same procedure is used for the next largest weight and so on down to the smallest. After the nth standard weight has been tried and a decision made, the weighing is finished. The total of the standard weights remaining on the balance is the closest possible approximation to the unknown.

In the successive-approximation A/D converter illustrated in Figure 12.26, a *successive-approximation register* (SAR) controls the D/A converter by implementing the weighing logic just described. The SAR first turns on the MSB of the DAC and the comparator tests this output against the analog input. A decision is made by the comparator to leave the bit on or turn it off after which bit 2 is turned on and a second comparison made. After n comparisons the digital output of the SAR indicates all those bits which remain on and produces the desired digital code. The clock circuit controls the timing of the SAR. Figure 12.27 shows the D/A converter output during a typical conversion.

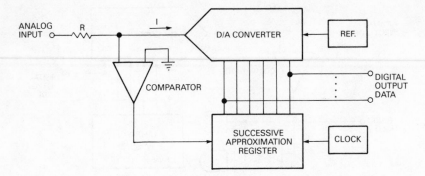

Figure 12.26   Successive approximation A/D converter

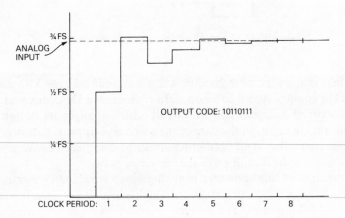

Figure 12.27   D/A output for 8-bit successive-approximation
conversion

The conversion efficiency of this technique means that high resolution conversions can be made in very short times. For example, it is possible to perform a 10-bit conversion in 1 μs or less. Of course the speed of the internal circuitry, in particular the D/A and comparator, are critical for high-speed performance.

12.9.3   The Parallel (Flash) A/D Converter

For ultra-fast conversions required in video signal processing and radar applications where up to 8 bits resolution is required, a different technique is employed; it is known as the *parallel* (also *flash*, or *simultaneous*) method and is illustrated in Figure 12.28. This circuitry employs $2^n - 1$ analog comparators to directly implement the quantizer transfer function of an A/D converter.

The comparator trip-points are spaced 1 LSB apart by the series resistor chain and voltage reference. For a given analog input voltage all comparators

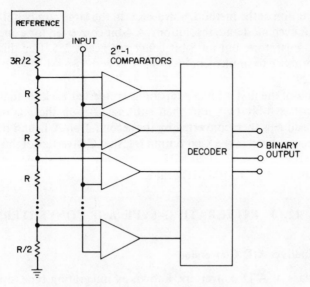

Figure 12.28   4-bit parallel A/D converter

biased below the voltage turn on and all those biased above it remain off. Since all comparators change state simultaneously, the quantization process is a one-step operation.

A second step is required, however, since the logic output of the comparators is not in binary form. Therefore an ultra-fast decoder circuit is employed to make the logic conversion to binary. The parallel technique reaches the ultimate in high speed because only two sequential operations are required to make the conversion.

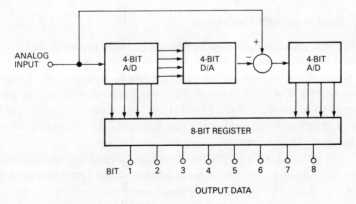

Figure 12.29   Two-stage parallel 8-bit A/D converter

The limitation of the method, however, is in the large number of comparators required for even moderate resolutions. A 4-bit converter, for example, requires only 15 comparators, but an 8-bit converter needs 255. For this reason it is common practice to implement an 8-bit A/D with two 4-bit stages as shown in Figure 12.29.

The result of the first 4-bit conversion is converted back to analog by means of an ultra-fast 4-bit D/A and then subtracted from the analog input. The resulting residue is then converted by the second 4-bit A/D and the two sets of data are accumulated in the 8-bit output register. Converters of this type achieve 8-bit conversions at rates of 20 MHz and higher, while single-stage 4-bit conversions can reach 50 to 100 MHz rates.

12.10   INTEGRATING-TYPE A/D CONVERTERS

12.10.1   Indirect A/D Conversion

Another class of A/D converters, known as integrating type, operates by an indirect conversion method. The unknown input voltage is converted into a time period which is then measured by a clock and counter. A number of variations exist on the basic principle such as *single-slope*, *dual-slope*, and *triple-slope* methods. In addition there is another technique—completely different—which is known as the *charge-balancing* or *quantized feedback* method.

The most popular of these methods are dual-slope and charge-balancing; although both are slow, they have excellent linearity characteristics with the capability of rejecting input noise. Because of these characteristics, integrating-type A/D converters are almost exclusively used in digital panel meters, digital multimeters, and other slow measurement applications.

12.10.2   Dual-slope A/D Conversion

The dual-slope technique, shown in Figure 12.30, is perhaps the best known. Conversion begins when the unknown input voltage is switched to the integrator input; at the same time the counter begins to count clock pulses and counts up to overflow. At this point the control circuit switches the integrator to the negative reference voltage which is integrated until the output is back to zero. Clock pulses are counted during this time until the comparator detects the zero crossing and turns them off.

The counter output is then the converted digital word. Figure 12.31 shows the integrator output waveform where T_1 is a fixed time and T_2 is a time proportional to the input voltage. The times are related as follows:

$$T_2 = T_1 E_{in}/V_{ref} \qquad (12.16)$$

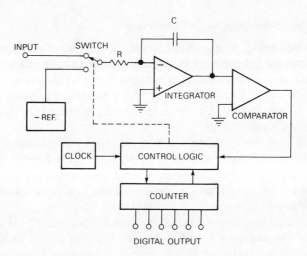

Figure 12.30   Dual-slope A/D converter

The digital output word, therefore, represents the ratio of the input voltage to the reference.

Dual-slope conversion has several important features. First, conversion accuracy is independent of the stability of the clock and integrating capacitor so long as they are constant during the conversion period. Accuracy depends only on the reference accuracy and the integrator circuit linearity. Second, the periodic noise rejection of the converter can be infinite if T_1 is set to equal the period of the noise. To reject 60 Hz power noise, therefore, requires that T_1 be 16.667 ms.

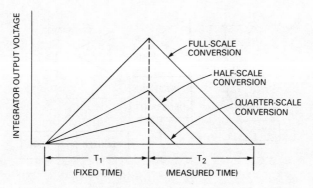

Figure 12.31   Integrator output waveform for dual-slope A/D converter

12.10.3   Charge-balancing A/D Conversion

The charge-balancing, or quantized feedback, method of conversion is based
on the principle of generating a pulse train with frequency proportional to the
input voltage and then counting the pulses for a fixed period of time. This
circuit is shown in Figure 12.32. Except for the counter and timer, the circuit
is a *voltage-to-frequency* (V/F) converter which generates an output pulse rate
proportional to input voltage.

The circuit operates as follows. A positive input voltage causes a current to
flow into the operational integrator through R_1. This current is integrated,
producing a negative going ramp at the output. Each time the ramp crosses zero
the comparator output triggers a precision pulse generator which puts out a
constant width pulse.

The pulse output controls switch S_1 which connects R_2 to the negative
reference for the duration of the pulse. During this time a pulse of current
flows out of the integrator summing junction, producing a fast, positive ramp
at the integrator output. This process is repeated, generating a train of current
pulses which exactly balances the input current—hence the name charge
balancing. This balance has the following relationship:

$$f = \frac{1}{\tau} \frac{V_{\text{in}}}{V_{\text{ref}}} \frac{R_2}{R_1} \qquad (12.17)$$

where τ is the pulse width and f the frequency.

A higher input voltage, therefore, causes the integrator to ramp up and down
faster, producing higher frequency output pulses. The timer circuit sets a fixed
time period for counting. Like the dual-slope converter, the circuit also inte-
grates input noise, and if the timer is synchronized with the noise frequency,

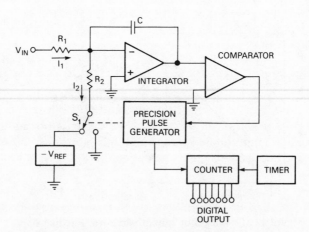

Figure 12.32   Charge-balancing A/D converter

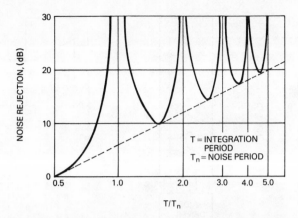

Figure 12.33 Noise rejection for integrating-type A/D
converters

infinite rejection results. Figure 12.33 shows the noise rejection characteristic of all integrating type A/D converters with rejection plotted against the ratio of integration period to noise period.

12.11 ANALOG MULTIPLEXERS

12.11.1 Analog Multiplexer Operation

Analog multiplexers are the circuits that time-share an A/D converter among a number of different analog channels. Since the A/D converter in many cases is the most expensive component in a data acquisition system, multiplexing analog inputs to the A/D is an economical approach. Usually the analog multiplexer operates into a sample-hold circuit which holds the required analog voltage long enough for A/D conversion.

As shown in Figure 12.34 an analog multiplexer consists of an array of parallel electronic switches connected to a common output line. Only one switch is turned on at a time. Popular switch configurations include 4, 8, and 16 channels which are connected in single (*single-ended*) or dual (*differential*) configurations.

The multiplexer also contains a decoder-driver circuit which decodes a binary input word and turns on the appropriate switch. This circuit interfaces with standard TTL inputs and drives the multiplexer switches with the proper control voltages. For the 8-channel analog multiplexer shown, a one-of-eight decoder circuit is used.

Most analog multiplexers today employ the CMOS switch circuits shown in Figure 12.35. A CMOS driver controls the gates of parallel-connected P-channel and N-channel MOSFET's. Both switches turn on together with the parallel connection giving relatively uniform on-resistance over the required

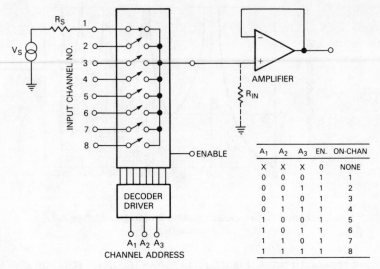

Figure 12.34   Analog multiplexer circuit

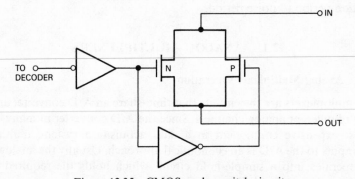

Figure 12.35   CMOS analog switch circuit

analog input voltage range. The resulting on-resistance may vary from about $50\ \Omega$ to $2\ k\Omega$ depending on the multiplexer; this resistance increases with temperature.

12.11.2   Analog Multiplexer Characteristics

Because of the series resistance, it is common practice to operate an analog multiplexer into a very high load resistance such as the input of a unity gain buffer amplifier shown in the diagram. The load impedance must be large compared with the switch on-resistance and any series source resistance in order to maintain high transfer accuracy. *Transfer error* is the input to output error of the multiplexer with the source and load connected; error is expressed as a per cent of input voltage.

Transfer errors of 0.1 % to 0.01 % or less are required in most data acquisition systems. This is readily achieved by using operational amplifier buffers with typical input impedances from 10^8 to 10^{12} Ω. Many sample-hold circuits also have very high input impedances.

Another important characteristic of analog multiplexers is *break-before-make* switching. There is a small time delay between disconnection from the previous channel and connection to the next channel which assures that two adjacent input channels are never instantaneously connected together.

Settling time is another important specification for analog multiplexers; it is the same definition previously given for amplifiers except that it is measured from the time the channel is switched on. *Throughput rate* is the highest rate at which a multiplexer can switch from channel to channel with the output settling to its specified accuracy. *Crosstalk* is the ratio of output voltage to input voltage with all channels connected in parallel and off; it is generally expressed as an input to output attenuation ratio in decibels.

As shown in the representative equivalent circuit of Figure 12.36, analog multiplexer switches have a number of leakage currents and capacitances associated with their operation. These parameters are specified on data sheets and must be considered in the operation of the devices. Leakage currents, generally in picoamperes at room temperature, become troublesome only at high temperatures. Capacitances affect crosstalk and settling time of the multiplexer.

12.11.3 Analog Multiplexer Applications

Analog multiplexers are employed in two basic types of operation; low-level and high-level. In *high-level multiplexing*, the most popular type, the analog signal is amplified to the 1 to 10 V range ahead of the multiplexer. This has the advantage of reducing the effects of noise on the signal during the remaining analog processing. In *low-level multiplexing* the signal is amplified after multiplexing; therefore, great care must be exercised in handling the low-level signal

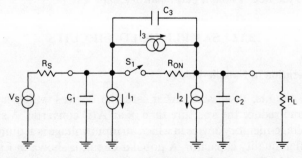

Figure 12.36   Equivalent circuit of analog multiplexer switch

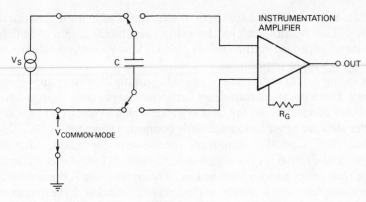

Figure 12.37   Flying-capacitor multiplexer switch

up to the multiplexer. Low-level multiplexers generally use two-wire differential switches in order to minimize noise pick-up. Reed relays, because of essentially zero series resistance and absence of switching spikes, are frequently employed in low-level multiplexing systems. They are also useful for high common-mode voltages.

A useful specialized analog multiplexer is the *flying-capacitor* type. This circuit, shown as a single channel in Figure 12.37 has differential inputs and is particularly useful with high common-mode voltages. The capacitor connects first to the differential analog input, charging up to the input voltage, and is then switched to the differential output which goes to a high input impedance instrumentation amplifier. The differential signal is, therefore, transferred to the amplifier input without the common mode voltage and is then further processed up to A/D conversion.

In order to realize large numbers of multiplexed channels, it is possible to connect analog multiplexers in parallel using the enable input to control each device. This is called *single-level multiplexing*. The output of several multiplexers can also be connected to the inputs of another to expand the number of channels; this method is *double-level multiplexing*.

12.12   SAMPLE-HOLD CIRCUITS

12.12.1   Operation of Sample-holds

Sample-hold circuits, discussed earlier, are the devices which store analog information and reduce the aperture time of an A/D converter. A sample-hold is simply a voltage-memory device in which an input voltage is acquired and then stored on a high quality capacitor. A popular circuit is shown in Figure 12.38.

A_1 is an input buffer amplifier with a high input impedance so that the source, which may be an analog multiplexer, is not loaded. The output of A_1 must be

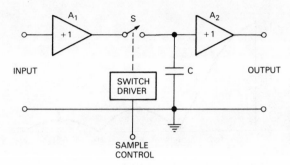

Figure 12.38   Popular sample-hold circuit

capable of driving the hold capacitor with stability and enough drive current to charge it rapidly. S_1 is an electronic switch, generally a FET, which is rapidly switched on or off by a driver circuit which interfaces with TTL inputs.

C is a capacitor with low leakage and low dielectric absorption characteristics; it is a polystyrene, polycarbonate, polypropylene, or Teflon type. In the case of hybrid sample-holds, the MOS type capacitor is frequently used.

A_2 is the output amplifier which buffers the voltage on the hold capacitor. It must, therefore, have extremely low input bias current, and for this reason a FET input amplifier is required.

There are two modes of operation for a sample-hold: *sample* or *tracking mode*, when the switch is closed; and *hold mode*, when the switch is open. Sample-holds are usually operated in one of two basic ways. The device can continuously track the input signal and be switched into the hold mode only at certain specified times, spending most of the time in tracking mode. This is the case for a sample-hold employed as a deglitcher at the output of a D/A converter, for example.

Alternatively, the device can stay in the hold mode most of the time and go to the sample mode just to acquire a new input signal level. This is the case for a sample-hold used in a data acquisition system following the multiplexer.

12.12.2   The Sample-hold as a Data Recovery Filter

A common application for sample-hold circuits is in *data recovery*, or *signal reconstruction*, filters. The problem is to reconstruct a train of analog samples into the original signal; when used as a recovery filter, the sample-hold is known as a *zero-order hold*. It is a useful filter because it fills in the space between samples, providing data smoothing.

As with other filter circuits, the gain and phase components of the transfer function are of interest. By an analysis based on the impulse response of a sample-hold and use of the Laplace transform, the transfer function is found to be

$$G_o(f) = \frac{1}{f_s} \frac{\sin(\pi f/f_s)}{\pi f/f_s} \exp(-j\pi f/f_s) \qquad (12.18)$$

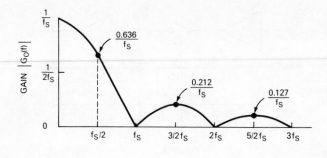

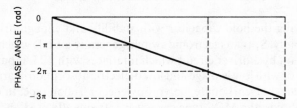

Figure 12.39   Gain and phase components of zero-order
hold transfer function

where f_s is the sampling frequency. This function contains the familiar $(\sin x)/x$ term plus a phase term, both of which are plotted in Figure 12.39.

The sample-hold is, therefore, a low-pass filter with a cut-off frequency slightly less than $f_s/2$ and a linear phase which results in a constant delay time of $T/2$, where T is the time between samples. Notice that the gain function also has significant response lobes beyond f_s. For this reason a sample-hold reconstruction filter is frequently followed by another conventional low-pass filter.

12.12.3   Other Sample-hold Circuits

In addition to the basic circuit of Figure 12.38, there are several other sample-hold circuit configurations which are frequently used. Figure 12.40 shows two such circuits which are closed-loop circuits as contrasted with the open-loop circuit of Figure 12.38. Figure 12.40a uses an operational integrator and another amplifier to make a fast accurate inverting sample-hold. A buffer amplifier is sometimes added in front of this circuit to give high input impedance. Figure 12.40b shows a high input impedance, non-inverting sample-hold circuit.

The circuit in Figure 12.38, although generally not as accurate as those in Figure 12.40, can be used with a diode-bridge switch to realize ultra-fast acquisition sample-holds.

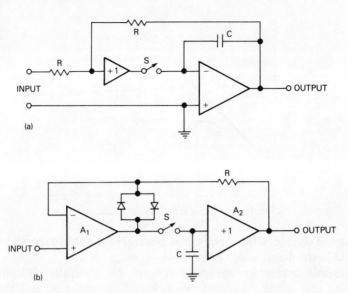

Figure 12.40   Accurate closed-loop sample-hold circuits: (a) inverting; (b) non-inverting

12.12.4   Sample-hold Characteristics

A number of parameters are important in characterizing sample-hold performance. Probably the most important of these is *acquisition time*. The definition is similar to that of settling time for an amplifier. It is the time required, after the sample-command is given, for the hold capacitor to charge to a full-scale voltage change and remain within a specified error band around final value.

Several hold-mode specifications are also important. *Hold-mode droop* is the output voltage change per unit time when the sample switch is open. This droop is caused by the leakage currents of the capacitor and switch, and the output amplifier bias current. *Hold-mode feedthrough* is the percentage of input signal transferred to the output when the sample switch is open. It is measured with a sinusoidal input signal and caused by capacitive coupling.

The most critical phase of sample-hold operation is the transition from the sample mode to the hold mode. Several important parameters characterize this transition. *Sample-to-hold offset* or *step error* is the change in output voltage from the sample mode to the hold mode, with a constant input voltage. It is caused by the switch transferring charge onto the hold capacitor as it turns off.

Aperture delay is the time elapsed from the hold command to when the switch actually opens; it is generally much less than a microsecond. *Aperture uncertainty* or *aperture jitter* is the time variation, from sample-to-sample, of the aperture delay. It is the limit on how precise is the point in time of opening the switch. Aperture uncertainty is the time used to determine the aperture error

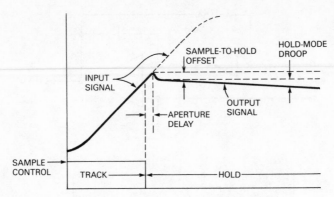

Figure 12.41   Some sample-hold characteristics

due to rate of change of the input signal. Several of the above specifications are illustrated in the diagram of Figure 12.41.

Sample-hold circuits are simple in concept, but generally difficult to fully understand and apply. Their operation is full of subtleties, and they must, therefore, be carefully selected and then tested in a given application.

12.13   SPECIFICATION OF DATA CONVERTERS

12.13.1   Ideal versus Real Data Converters

Real A/D and D/A converters do not have the ideal transfer functions discussed earlier. There are three basic departures from the ideal: *offset*, *gain*, and *linearity* errors. These errors are all present at the same time in a converter; in addition, they change with both time and temperature.

Figure 12.42 shows A/D converter transfer functions which illustrate the three error types. Figure 12.42a shows *offset error*, the analog error by which the transfer function fails to pass through zero. Next, in Figure 12.42b is *gain error*, also called *scale factor error*; it is the difference in slope between the actual transfer function and the ideal, expressed as a per cent of analog magnitude.

In Figure 12.42c *linearity error*, or non-linearity, is shown; this is here defined as the maximum deviation of the actual transfer function from the ideal straight line at any point along the function. It is expressed as a per cent of full scale or in LSB size, such as $\pm\frac{1}{2}$LSB, and assumes that offset and gain errors have been adjusted to zero.

Most A/D and D/A converters available today have provision for external trimming of offset and gain errors. By careful adjustment these two errors can be reduced to zero, at least at ambient temperature. Linearity error, on the other hand, is the remaining error that cannot be adjusted out and is an inherent characteristic of the converter.

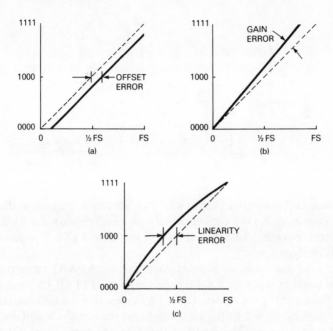

Figure 12.42   (a) Offset, (b) gain, and (c) linearity errors

12.13.2   Data Converter Error Characteristics

Basically there are only two ways to reduce linearity error in a given application. First, a better quality, higher cost converter with smaller linearity error can be procured. Second, a computer or microprocessor can be programmed to perform error correction on the converter. Both alternatives may be expensive in terms of hardware or software cost.

The linearity error discussed above is actually more precisely termed *integral linearity error*. Another important type of linearity error is known as *differential linearity error*. This is defined as the maximum amount of deviation of any quantum (or LSB change) in the entire transfer function from its ideal size of $FSR/2^n$. Figure 12.43 shows that the actual quantum size may be larger or smaller than the ideal; for example, a converter with a maximum differential linearity of $\pm\frac{1}{2}$ LSB can have a quantum size between $\frac{1}{2}$ LSB and $1\frac{1}{2}$ LSB anywhere in its transfer function. In other words, any given analog step size is $(1 \pm \frac{1}{2})$ LSB. Integral and differential linearities can be thought of as macro- and microlinearities, respectively.

Two other important data converter characteristics are closely related to the differential linearity specification. The first is *monotonicity*, which applies to D/A converters. Monotonicity is the characteristic whereby the output of a circuit is a continuously increasing function of the input. Figure 12.44a shows a

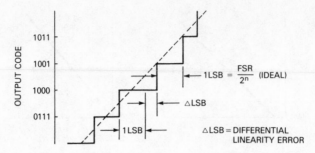

Figure 12.43   Defining differential linearity error

nonmonotonic D/A converter output where at one point, the output decreases as the input increases. A D/A converter may go nonmonotonic if its differential linearity error exceeds 1 LSB; if it is always less than 1 LSB, it assures that the device will be monotonic.

The term *missing code*, or *skipped code*, applies to A/D converters. If the differential linearity error of an A/D converter exceeds 1 LSB, its output can miss a code as shown in Figure 12.44b. On the other hand, if the differential linearity error is always less than 1 LSB, this assures that the converter will not miss any codes. Missing codes are the result of the A/D converter's internal D/A converter becoming nonmonotonic.

For A/D converters the character of the linearity error depends on the technique of conversion. Figure 12.45a, for example, shows the linearity characteristic of an integrating type A/D converter. The transfer function exhibits a smooth curvature between zero and full scale. The predominant type of error is integral linearity error, while differential linearity error is virtually nonexistent.

Figure 12.45b, on the other hand, shows the linearity characteristic of a successive approximation A/D converter; in this case differential linearity error

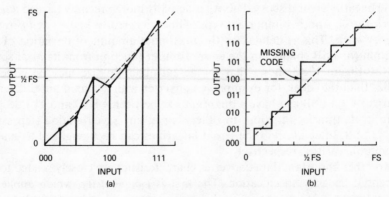

Figure 12.44   Important D/A converter characteristics: (a) Nonmonotonic
D/A converter; (b) A/D converter with missing code

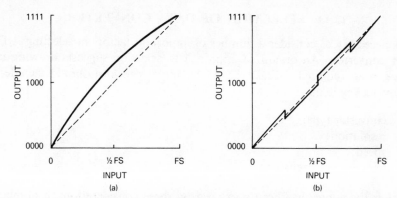

Figure 12.45   Linearity Characteristics of (a) integrating and (b) successive-approximation A/D converters

is the predominant type, and the largest errors occur at the specific transitions at $\frac{1}{2}$, $\frac{1}{4}$, and $\frac{3}{4}$ scale. This result is caused by the internal D/A converter non-linearity; the weight of the MSB and bit-2 current sources is critical in relation to all the other weighted current sources in order to achieve $\pm\frac{1}{2}$ LSB maximum differential linearity error.

12.13.3   Temperature Effects

Ambient temperature change influences the offset, gain, and linearity errors of a data converter. These changes over temperature are normally specified in parts per million (ppm) of full scale range per degree Celsius. When operating a converter over significant temperature change, the effect on accuracy must be carefully determined. Of key importance is whether the device remains monotonic, or has no missing codes, over the temperatures of concern. In many cases the total error change must be computed, that is, the sum of offset, gain, and linearity errors due to temperature.

The characteristic of monotonicity, or no missing codes, over a given temperature change can be readily computed from the *differential linearity tempco* (DLT) specified for a data converter. Assuming the converter initially has $\frac{1}{2}$LSB of differential linearity error, the change in temperature for an increase to 1 LSB is given by

$$\Delta T = \frac{2^{-n} \times 10^6}{2 \, \text{DLT}} \tag{12.19}$$

where n is the converter resolution in bits and DLT is the specified differential linearity tempco in ppm of FSR/°C. ΔT is the maximum change in ambient temperature which assures that the converter will remain monotonic, or have no missing codes.

12.14 SELECTION OF DATA CONVERTERS

It is necessary to consider a number of important factors in selecting A/D or D/A converters. An organized approach to selection suggests drawing up a checklist of required characteristics. An initial checklist should include the following key items:

(a) converter type;
(b) resolution;
(c) speed;
(d) temperature coefficient.

After the choice has been narrowed by these considerations, a number of other parameters must be considered. Among these are analog signal range, type of coding, input impedance, power supply requirements, digital interface required, linearity error, output current drive, type of start and status signals for an A/D, power supply rejection, size, and weight. These parameters should be listed in order of importance to efficiently organize the selection process.

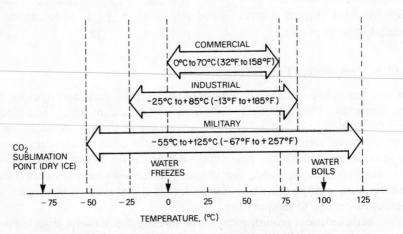

Figure 12.46   Standard operating temperature ranges for data converters

In addition, the required operating temperature range must be determined; data converters are normally specified for one of three basic ranges known in the industry as *commercial*, *industrial*, or *military*. These temperature ranges are illustrated in Figure 12.46. Further, the level of reliability must be determined in terms of a standard device, a specially screened device, or a military standard device.

Published works providing detail include Hoeschele (1968), Schmid (1970), Sheingold (1972), and Zuch (1979).

REFERENCES

Hoeschele, D. (1968). *Analog to Digital*; *Digital-to-Analog Conversion Techniques*, Wiley, New York.

Morrison, R. (1977). *DC Amplifiers in Instrumentation*, Wiley, New York.

Schmid, H. (1970). *Electronic Analog/Digital Conversions*, Van Nostrand Reinhold, New York.

Sheingold, D. H. (1972). *Analog-Digital Conversion Handbook*, Analog Devices, Norwood, USA.

Zuch, E. L. (Ed.) (1979). *Data Acquisition and Conversion Handbook*, Datel-Intersil, Inc., Mansfield, Mass.

Handbook of Measurement Science, Vol. 1
Edited by P. H. Sydenham
© 1982 John Wiley & Sons Ltd.

Chapter

13 R. W. GRIMES

Transmission of Data

Editorial introduction

Transferring information in electrical signal form over considerable distances began with telegraphic systems. It expanded over time into an extremely sophisticated methodology. Theoretical understanding was progressively improved providing a systematic and rigorous basis for implementing transmission systems capable of sending great quantities of data at selectably low error rates. The original digital signal methods of telegraphy were augmented by the analog signal methods of telephony and, later, radio communication.

Telemetry of measurement data followed on to meet scientific and specialized industrial demands. For the period until around 1960 analog systems of data transmission were more generally used. From then on, however, the impact of low cost, high reliability, digital electronic systems became evident as methods of data transfer used in the ever expanding size of digital data processing systems found their way into instrumentation systems in general.

This chapter comprises two parts. The first and major part is concerned with data transfer by digital means over extensive networks. It is compiled from the viewpoint of an extensive computer based communications network this being the most suitable viewpoint to present at the time of compilation when such methods are finding their way more and more into such areas as process plants, monitoring systems, and stand-alone instrument system modules.

The second provides an outline of more specific signal transmission practice. Analog methods, whilst still having a part to play, will undoubtedly decline in popularity as their inherent cost and relatively poor security of message transfer become greater penalties compared with digital alternatives.

13.1 INTRODUCTION

13.1.1 Data Communication

'The fundamental problem of communication is that of reproducing
at one point either exactly or approximately a message transmitted
from another point.'
                                                        C. E. Shannon

Data communication has only come into use on a large scale in the past ten years. Initially data terminals were used to input data traffic into large central *main frame* computers. The connections were made either on direct circuits or dial-up circuits through the use of the standard telephone exchange network. Advances in computer technology reduced the cost of computers; which were then provided in regional centres. There was an increase in the development of data base systems where customers had direct access to records in computers which could be updated and inspected using mainly visual display units (VDU). Methods initially developed for national *teletechnique* networks were gradually adopted for other applications such as process plant control and electrical supply network control.

13.1.2   International Standards for Data Communication

A number of international bodies are concerned with the setting of standards for all types of communications. The main body particularly concerned with operations in this field is the International Telegraph and Telephone Consultative Committee (CCITT) which cooperates with the International Standards Organization (ISO) and the International Electro-technical Commission (IEC) according to rules defined in Recommendation A20 of the CCITT. The CCITT has published many relevant works (CCITT, 1964, 1972, 1977a, b, c, d).

The CCITT has a responsibility for those aspects of data transmission which involve telecommunications networks or affect the performance of these networks. There are other topics such as standardization of the junction (interface) between the standardized *modems* and data terminal equipment which require agreement between CCITT and ISO. Standards also have to be set for operating speeds of telegraph services and signalling and line control procedures for international working of the Telex network. Many other bodies such as the Conference of European Postal and Telecommunications Administration (CEPT) and the International Federation of Information Processing (IFIP) contribute to the development of the international standards in the data field which are the primary responsibilities of CCITT and ISO.

Existing telecommunications networks have evolved until recent years without taking into account the requirement for data transmission which has developed rapidly. Existing data communications facilities, therefore, fall short of the ideal and specialized networks designed specifically for data communications are now being developed in several countries. Standards for modulation rates and interfaces for networks specialized for data are now being developed by CCITT and ISO.

13.2 TERMINAL EQUIPMENT AND CIRCUIT REQUIREMENTS

13.2.1 Summary

The transmission requirements of data communication circuits will depend on the needs of the terminal equipment. This section will give a brief summary of the signalling techniques and the types of terminal equipment provided.

13.2.2 Data Signalling Principles

Most telegraph machines, computers and other data processing devices operate with digital binary logic information which is transmitted in a coded sequence of events. Each element in the sequence has two possible states which are usually referred to as 1 or 0. In telegraph transmission the terms *mark* or *space* are used. The data terminal equipment presents these two binary states to a data transmission line as a direct current signal either as positive or negative potential, called *double current signalling*, or as the presence or lack of a potential, called *single current signalling*.

It is the function of the data communication circuit to convert these logic signals into a more suitable medium for use over communication circuits and to convert them back to logic signals with minimum distortion at the distant terminal.

Signalling rates

The element of binary coded data is known as a *bit* (binary digit) which has a binary value of 0 or 1. Most simple communication systems, for example a telegraph machine, transmit this binary data in a *serial binary form* (one bit at a time) and the data communication rate can be expressed in *bits per second* (*bit/s*).

Data codes

The sequence of binary coded data is sent in the forms of characters usually either five or seven bits long which normally conform to CCITT standards.

13.2.3 Asynchronous or Synchronous Transmission

Having connected the digital signal generating machine to the communications line it is necessary to coordinate the data received with the data sent.

With *asynchronous* transmission the characters are sent to line with a variable duration between characters (e.g. the random operation of the keyboard of a telegraph machine) in this case the receiving machine must know when a character is about to be received so the appropriate timing and sampling mechanisms can be initiated. With asynchronous transmission each character is preceded by a start element. This element will have the same duration as the normal signalling elements and would be opposite to the *at rest* or *steady state* line condition. The five-bit or eight-bit character is then followed by 1, 1.5 or 2 stop elements which are the same state or polarity as the at rest condition. The stop elements allow the receiving terminal time to process or print the character and set the receiver unit ready to receive the next character. Any discrepancies in speed between the transmitter and receiver are taken up during the stop element pause.

Asynchronous transmission is used for low speed data transmission up to about 1200 bit/s.

With *synchronous* transmission each character immediately follows the previous character without any start or stop elements and all the bits in the character are of equal duration. With this type of transmission the receiver must be kept in step with the transmitter and it is also necessary to know when transmission is about to commence so the separation between characters in the continuous bit stream can be recognized.

This type of transmission is used between computers and the more intelligent types of data terminals and operates at speeds greater than 600 bit/s. Blocks of data from 200 to 1000 characters in length are usually transmitted at a time, however, it is necessary to use special synchronizing characters at the start of each block so the receiving terminal can obtain *character* as well as *bit* synchronization.

13.2.4   Telegraph Distortion

Consider the signal in Figure 13.1. The characteristics of the transmission line will cause the detected signal at the receiving terminal to be distorted and the

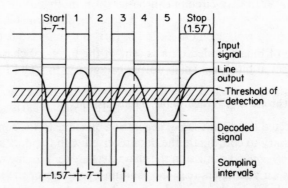

Figure 13.1   Distortion and subsequent detection of
an example digital signal sequence

timing of the start and the end of the pulses will vary from the ideal. The receiving mechanism will sample the pulses at an optimum point near the centre of the pulse, however, the sampling period will have a finite width of 1–3 ms.

If the total period of the pulse is 100% then the receiver should tolerate a variation of $\pm 50\%$ less than the ideal sampling period. A mechanical machine should tolerate $\pm 45\%$ distortion. Electronic equipment, with a shorter sampling time, could tolerate up to $\pm 49\%$ distortion.

13.2.5 Data Circuit Terminology

The following is a summary of the various types of data circuits and circuit configurations used:

One way circuit. Previously known as *half Duplex*, the transmission is in one direction only with no return path.

Either way circuit. Previously known as *Simplex*, transmission can occur in either direction but in only one direction at a time (equivalent to a telephone conversation).

Both way circuit. Previously known as *Duplex*, transmission can occur independently in either direction at the same time.

Backward channel control. Used on some data circuits when a lower speed channel, usually of 75 to 110 bit/s rate, is used on the return path to the sending station for supervisory and control purposes.

Multistation circuits. A number of stations can be connected to a single one way circuit so they all receive the traffic from the sending station.

Selective calling units (*SCU*). Devices to enable the outstation (on multistation circuits) to select and print a message with its own *header*.

Polled working. Both way circuits can be used on multistations, however the outstation cannot send messages until it receives its own identifying code. The control station *polls* each outstation in turn and so controls the incoming traffic.

Link procedures. On high speed circuits between computers or sophisticated terminals synchronous transmission is used with all messages occurring in blocks of fixed length. Each message block (or group of blocks) is acknowledged on the return path and so no traffic can be lost.

Concentrators. On switching networks, where a large number of low traffic terminals are in one area, the circuits are combined into one, or more, higher traffic circuit to the main exchange or message switching centre. Concentrators only connect circuits *through* or *forward* messages on to the main centre and do not perform any local switching.

13.2.6 Design of Data Networks

Many design considerations should be taken into account when selecting the optimum circuit configuration. When designing a network the designer should consider

(a) What are the acceptable delays in delivering data messages or setting up connections?

(b) How many outlets and inlets are required, are multistation facilities or polling facilities required?

(c) Are messages sent to more than one terminal and are message or circuit switching facilities required?

(d) If working in a computer interrogation mode what is the maximum permissible response time?

(e) What is the acceptable error rate?

(f) What message security procedures are required and what message handling procedures should be adopted?

(g) With line switching systems, what problems will occur if excessive calls are made to busy outlets?

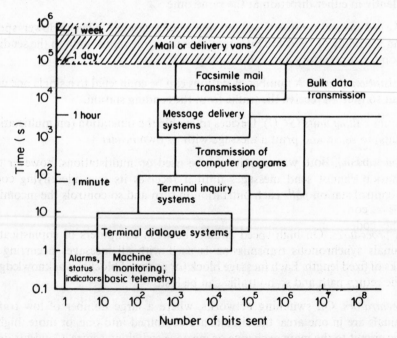

Figure 13.2   Delivery time and quantity of data of typical communications circuits (from Martin (1972a), reprinted by permission of Prentice-Hall, Inc., Englewood Cliffs, NJ)

(h) With message switching systems what is the maximum holding time of sent messages and how much message storage is required?

(i) What emergency back up facilities are available in the case of component or circuit failures?

(j) Are accounting and statistical facilities required?

(k) What are the network control station requirements such as traffic statistics, circuit control, alarms, message retrieval?

Figure 13.2 shows the delivery times and quantity of data for typical communication circuits.

13.3   GENERAL PRINCIPLES OF DATA TRANSMISSION

13.3.1   Introduction

Communication circuits were originally designed solely for voice communication and much of the design and development work in data transmission is based on the principle of designing data terminals and data translation and supervisory equipment to make the best use of the circuits and bandwidth that is available.

Voice frequency communication, because of the high level of redundancy, can tolerate frequency distortion, noise, and short interruptions on the circuit while still retaining a high level of credibility.

Telegraph transmission of plain spoken language has a certain amount of redundancy and a few individual characters can be incorrectly received without any appreciable change to the information in the message. Any errors in a data message, if not in plain language, could cause problems and faulty operation of the terminal equipment.

The basic data communication circuit is as shown in Figure 13.3. The transmitter is the input device which converts the internal information in the digital machine to a two-state, binary, sequential code.

The converter (for example a data modem) converts this code into some other coded state that is more suitable for transmission over the communication circuit. Examples are

(a) Voltage changes to convert the binary signals from the transmitter to a voltage level more suitable for d.c. transmission over a short physical circuit.

(b) Frequency changes where the transmitter's binary signals are converted either by amplitude, frequency or phase modulation to a frequency spectrum compatible with the normal derived voice frequency operating telephone circuit.

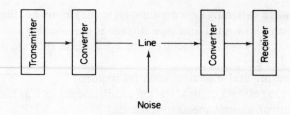

Figure 13.3   Schematic of data communication circuit

The converter at the distant terminal will convert the incoming signal to a series of binary bits acceptable to the receiver which then processes the binary code and performs some function in the receiving device. The CCITT conventions for data transmission are as shown in Table 13.1.

A low speed telegraph circuit at 75 bit/s can transmit information at almost the same information rate as a verbal communication. The telegraph channel will use much less of the frequency spectrum and it is possible to fit 24 telegraph circuits into one *voice frequency transmission* (VFT) circuit. However, the telegraph circuit is more prone to interference caused by noise and other line conditions.

Table 13.1   CCITT conventions for data transmission

	Digit 0	Digit 1
	'Start' signal in start–stop code Line available condition in telex switching 'Space' element of start–stop code Condition A	'Stop' signal in start–stop code Line idle condition in telex switching 'Mark' element of start–stop code Condition Z
Amplitude modulation	Tone-off	Tone-on
Frequency modulation	High frequency	Low frequency
Phase modulation with reference phase	Opposite phase to the reference phase	Reference phase
Differential two-phase modulation where the alternative phase changes are 0° or 180°	No-phase inversion	Inversion of the phase
Perforations	No perforation	Perforation

13.3.2 Intersymbol Interference

When a series of binary signals or bits is sent via a communication circuit the characteristics of the circuit must be such that the receiving terminal can detect whether the bit is a 0 or 1 for every bit combination occurring in the signal. The signals could be distorted sufficiently such that one binary bit could interfere with the following bit. This is known as intersymbol interference (ISI), as depicted in Figure 13.4, where a signal is sent through a circuit with a low frequency response.

Data circuits must be designed so they are adequately immune to the problems of intersymbol interference.

13.3.3 Direct Current Signalling

For short physical circuits direct current signalling can be used. Two basic types of digital signals are used for low speed telegraph machines:

(a) *Single current* (SC) signals where a continuous current on line is equivalent to a mark and no current is a space. The control terminal normally sends 40 mA for a mark and no current for a space. The signalling distance is restricted by the physical resistance of the cable pair. This type of signalling is suitable for one-way or either-way transmission.

(b) *Double current* (DC) signalling which sends a negative voltage (usually − 50 V) to line for a mark and a positive voltage (usually + 50 V) for a space. The signalling current is 20 mA. This is an *earth return* type signalling and each leg of the pair is used for one direction only so independent signals can be sent from each end at the same time. The circuit can be used for one-way, either-way, or both-way transmission.

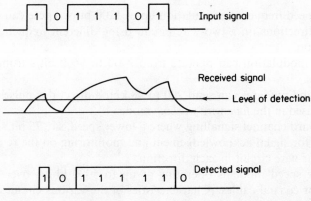

Figure 13.4 Distortion of a signal by intersignal interference

The signalling distance is restricted by the physical resistance of the cable pair; however, the distance can be extended with repeating relays.

13.3.4   VFT Systems

Low speed telegraphy operates at speeds of 50 and 75 bit/s and only requires a bandwidth of approximately 120 Hz to give an acceptable reproduction of the input pulse at the receiving terminal. It is possible to fit 24 telegraph circuits in the bandwidth of a normal voice frequency circuit. The standard system is the 24-channel frequency-modulated (FM) VFT system. Amplitude-modulated (AM) systems are no longer used as their performance is markedly inferior to the FM alternative.

13.3.5   Baseband Signalling Systems

A *baseband signalling* system can be described as a system wherein the signals sent to line are a series of binary d.c. pulses and there is no modulation of a sinusoidal signal as for AM and FM systems. These systems can operate up to at least 2 Mbit/s on cable pairs.

Baseband systems can change the voltage of the signals leaving the data terminal and in some cases baseband systems can convert the d.c. levels from the data terminal to a series of pulses. These signalling systems are used only in physical cables in city areas and not over trunk circuits.

A full description of baseband signalling techniques is give in Section 13.4.

13.3.6   Data Modems

Like VFT systems *data modems* convert the binary d.c. pulses into a signal form more suitable for transmission over VF or wider bandwidth circuits. Data modems can be provided in many versions:

(a)   Low speed frequency modulation up to 1200 bit/s. These can transmit in both directions on a two-wire circuit using different frequencies in each direction.

(b)   Phase modulation can operate from 2400 to 9600 bit/s using four-wire circuits.

(c)   Baseband modems can send 200 to 48 kbit/s direct d.c. pulses over local cable used in the *four-wire working* mode.

(d)   Backward channel signalling where a lower speed, say 75 bit/s, channel is added for signal acknowledgement and monitoring on the return path of the four-wire circuit in each direction.

(e)   Higher speed modems operating at up to 5000 kbit/s over microwave links or coaxial cables. (Characteristics of the various circuit carriers are discussed later in this chapter.)

13.3.7 Regenerative Repeaters

As a data signal is transmitted over a number of circuits the detected pulses can vary in timing (refer to Figure 13.1). Sometimes the signal is regenerated at some mid-point so that the repeated output signal can be given the correct timing between pulses. This is done by sampling the signal and transmitting a new signal at the time the sampling takes place—usually in the mid-point of the incoming pulse. The regenerated signal must always be at least half an element behind the incoming signal.

13.3.8 Time Division Multiplexing

Multiplexers are used to combine a number of low data speed circuits on to one high speed data circuit and so make more efficient use of the data circuits.

Consider the modem operating at 9600 bit/s in Figure 13.5. The inputs from circuits 1, 2, and 3 are combined on to one high speed circuit and the signals for each circuit are then separated at the other end. It should be noted that the input circuits to the time division multiplexer (TDM) must operate at some sub-multiple of the multiplexer's speed, also the inputs must be in synchronism with the multiplexer. For this reason the synchronizing clock in the multiplexer must control the incoming circuits and it would be difficult to operate this system with asynchronous signalling unless special circuitry is provided. Some elements in the multiplexer are required for control purposes.

13.3.9 Asynchronous Multiplexers

There are two methods of generating multiplexed signals from asynchronous signal inputs.

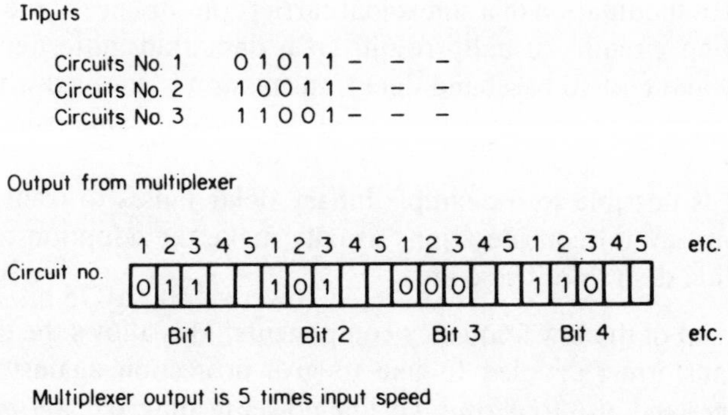

```
Inputs

       Circuits No. 1    0 1 0 1 1 –  –  –
       Circuits No. 2    1 0 0 1 1 –  –  –
       Circuits No. 3    1 1 0 0 1 –  –  –
```

Output from multiplexer

Multiplexer output is 5 times input speed

Figure 13.5   Time division multiplexing of three digital signal circuits

Code and speed independent multiplexers

These rely on sampling the input signal at a rate which is usually five or more times the maximum signalling rate. In this case the timing error on sampling at the input to the multiplexer is one half the sampling period. The modulating or signalling rate is five times the input rate and there would be a bandwidth restriction. It should be noted that some of the bandwidth is used for control purposes.

Code and speed dependent multiplexers

When the signalling code and the speed is known the input device on the multiplexer can sample and store the signal. As soon as the complete character is received the signal can be sent into a synchronous multiplexer at the same bit rate as the input signal. The input and output devices of the multiplexer must know the number of elements and the size of the stop and start elements. Although the system is restricted to fixed signalling speeds many more data channels can be accommodated in the same bandwidth. It should be noted that the signals are regenerated (see Section 13.3) when they pass through the multiplexer.

Character storage facilities must be allowed for at the input so that the multiplexer can cope with variation in the timing of the incoming characters.

13.4   BASEBAND SIGNALLING

13.4.1   Introduction

Baseband digital data transmission is a well established technique for data transmission over local circuits.

A baseband digital data signal can be defined as one that does not involve some form of modulation of a sinusoidal carrier; the absence of the modulating/demodulating circuitry usually results in a data transmitter/receiver of less complexity and cost so baseband data transmission is adopted where possible. Baseband data transmission can be used on physical cable pairs and coaxial cables.

While it is possible to use simple binary polar pulses to transmit the data information, several considerations usually make the adoption of some form of line coding desirable. These are:

(a)   Removal of the low frequency components. This allows the data signal to be transformer-coupled to line to give protection against longitudinal voltages and also to permit d.c. line power feeding, d.c. *wetting* in which a boost current is applied to break down insulating contact films, or auxiliary signalling.

(b) Sufficient transmitted information is needed to enable the extraction of a clock synchronizing signal at the receiver.

(c) The introduction of redundancy in the data to allow monitoring of the transmission performance and/or the transmission of line control information to be carried out.

(d) The available bandwidth. If shorter pulses are used in the coding a greater bandwidth is required.

(e) The shape of the signal spectrum to allow for minimum intersymbol interference.

(f) The average power transmitted to line and the line losses.

(g) The need for a receiver to be insensitive to polarity or line reversals.

13.4.2   Line Coding

Many different types of line codes are proposed for the transmission of data signals. Some of these types are shown in Figure 13.6. The advantages and disadvantages of each can be summarized as:

(1) *Non-return to zero (NRZ) binary*

This is the most common method for low speed telegraph transmission. Its advantages are that it uses simple technology and is suitable for asynchronous and synchronous operation. The disadvantages are the existence of a d.c. component, clock synchronizing problems, and that the signal is polarity sensitive.

(2) *Differentially coded NRZ binary*

A transition occurs for a 1 and no transition for a 0. Signalling characteristics are the same as for the NRZ binary system.

The advantage of this method is that the receiver is not polarity sensitive. Disadvantages are the existence of a d.c. component and clock synchronizing problems.

(3) *Alternate mark inversion (AMI)*

A pulse of half the bit width is sent for each 1 bit and each pulse is inverted in relationship to the previous pulse.

This has the advantage that a certain amount of error correction is available because the receiver can check to see if each pulse is of opposite polarity—less power is sent to line as the pulses are shorter. Its disadvantage is again the clock synchronizing problem—half-width pulses required greater bandwidth.

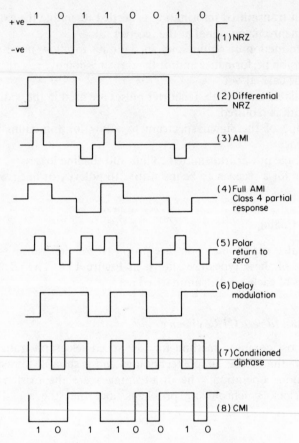

Figure 13.6   Examples of line coding methods

(4)   *Full AMI*

This is similar to AMI; however, a full width pulse is used.

A key advantage is that no d.c. component exists. Furthermore, error detection can be used. It uses a smaller bandwidth than half-AMI. Disadvantages are that it is sensitive to all zero combinations and that more power is sent to line than with the AMI method.

(5)   *Polar return to zero*

Here half-width pulses are used with one polarity for 0 and a different one for 1.

Advantages of this concept are that it is self-clocking, uses simple technology, and requires less power to line than NRZ binary. Disadvantages here are that the receiver must be polarity sensitive and a wider bandwidth is required.

(6) Delay modulation (Miller code)

A 1 gives a transition in the middle of a bit; a 0 does not give a transition unless followed by another 0 then a transition at the end of the bit period.

Advantages are that there are more transitions, even with a continuous zero, without increasing the bandwidth.

(7) Conditioned diphase (Manchester code)

Half-width pulses are sent with a polarity reversal for each bit. A 1 bit changes the polarity from the preceding bit. This is equivalent to phase modulation where a 1 bit changes the phase by 180° relative to the previous phase.

Advantages are that it is not polarity sensitive and the receiver can start sampling at the centre or the end of the bit and still get the same result. Additionally the receiver is self-clocking, some error detection is available, and as equal pulses of both polarities occur no d.c. component exists.

The disadvantage of this method is that a wider bandwidth is needed. This system is often used in storing and reading data from magnetic tape and disc systems. As there always occurs a transition for each bit it is easy to synchronize the clock on the incoming signals and to compensate for the slight speed variations which occur with electromechanical systems.

(8) CMI coded mark inversion

A 0 bit always has a transition from negative to positive in the centre of the pulse. A 1 bit is always full width and alternate 1's are inverted.

This has the advantages that the receiver is polarity independent for after a few bits are sent the receiver can be *trained* to know which polarity is meant to be negative. This is useful if special control pulses are required during a signal transmission. Error recognition is possible. Clocking is also possible but would not be as regular as for the conditioned diphase method. No d.c. component or d.c. drift problems arise.

13.4.3 Scrambling

The problems with d.c. components, clocking, and automatic gain control (AGC) when all zero combinations are sent can be alleviated by scrambling the message. That is, the transmitted bits are mixed with a continuous fixed pattern signal so that there is a continuous combination of 0's and 1's. The message is descrambled back to the original message at the receiver.

13.4.4 Frequency Spectrum of Baseband Signals

Referring to Chapter 4 a raised cosine-shaped pulse and the raised cosine shape on the filter cut-off give the best performance for pulse transmission. A number of tests have been carried out on some of the line signalling codes mentioned in Section 13.4.2. Measurements were made in each case with random character combinations and the power spectral curves are as shown in Figure 13.7.

It should be noted that the AMI, both normal and interleaved, gives a frequency response similar to the raised cosine curve, however, half-AMI (which also has a curve similar to a raised cosine) sends less power to line but requires twice the frequency spectrum. The type of signalling system to be used will depend on the attenuation of the line and the data rate required.

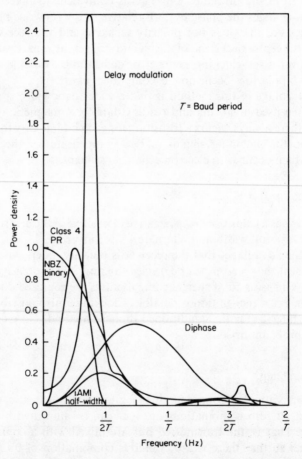

Figure 13.7 Power spectral densities of common line codes

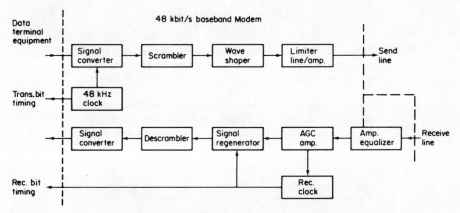

Figure 13.8   48 kbit/s baseband modem

13.4.5   Typical Baseband Modem

Figure 13.8 shows the layout of a typical baseband modem. The transmitted signals are first encoded and scrambled and then passed through a band-limiting filter to remove unwanted higher frequency components.

13.5   PHASE MODULATION

13.5.1   Introduction

With normal phase modulation the phase shift of the carrier is modulated by a variable analog signal. With digital data transmission we are concerned solely with phase shift keying where the carrier phase is not altered more than $\pm 180°$ at any time and the phase is shifted in definite fixed steps.

Phase modulation is used at data speeds greater than 2400 bit/s. In the simplest version a carrier frequency is shifted 180° from its previous value to indicate a change in the state of the bit (0 or 1).

It is more common to use four phase shifts and a 2400 bit/s signal can be generated if a 1200 Hz carrier is modulated with two bits at a time to give four combinations of phase shift at 0°, 90°, 180°, and 270°. This can be extended to eight combinations of phase shifts in multiples of 45° using three bits. By using combined amplitude and phase modulation four bits at a time can be selected and data will be transmitted at a bit speed of four times the carrier. For example, if a 2400 Hz carrier is used, the data rate will be 9600 bit/s which can be transmitted over a normal telephone circuit. Table 13.2 shows the phases used for various levels of modulation.

Table 13.2 Binary and phase values for various PM systems: (a) two-bit modulation; (b) three-bit modulation; (c) four-bit modulation, combined PM and AM modulation

(a)

Bit 1	Bit 2	Alternative A	Alternative B
0	0	$0°$	$45°$
1	0	$+90°$	$+135°$
1	1	$+180°$	$+225°$
0	1	$+270°$	$+315°$

(b)

Three-bit values			
Bit 1	Bit 2	Bit 3	Phase change[a]
0	0	1	$0°$
0	0	0	$45°$
0	1	0	$90°$
0	1	1	$135°$
1	1	1	$180°$
1	1	0	$225°$
1	0	0	$270°$
1	0	1	$315°$

(c)

	Bit 1	Bit 2	Bit 3	Bit 4	Phase change[a]
Bit 1 = 0	0	0	1	$0°$
	0	0	0	$45°$
Low level	0	1	0	$90°$
	0	1	1	$135°$
Bit 1 = 1	1	1	1	$180°$
	1	1	0	$225°$
High level	1	0	0	$270°$
	1	0	1	$315°$

[a] The phase change is the actual on-line phast shift in the transition region from the end of one signalling element to the beginning of the following signalling element.

The signal at the receiving terminal can be analysed by comparing the received signal with an internal fixed reference signal, however, there could be problems in maintaining synchronism. To avoid the problem of having to transmit a reference carrier phase against which the phase of the signal can be compared, the receiving clock is kept in synchronism by the timing of the phase shifts between symbols. This form of modulation is often referred to as differentially coherent phase shift keying (DCPSK).

Loss of synchronism can occur if a long string of binary 0's is transmitted because the absence of phase changes at the receiving end does not permit the recovery of timing information for automatic adjustment of the receiving clock.

It is worth noting that the probability of error in the detection process depends in some degree on the specific assignment of the binary code symbols for the various phase shifts. The most probable error is that of interpreting a particular phase change as one of the immediately adjacent possible phase changes. The coding of the symbols in Table 13.2 was assigned by CCITT so that symbols have a minimum of difference to adjacent phase change conditions. The code used is known as the *Gray code* which has only one bit changed for adjacent symbols as shown in Table 13.3.

A phase modulated signal with the instantaneous phase jumps as set out in Table 13.2 will have a wide frequency spectrum and it is normal to adopt measures to limit the spectrum of the signal. For example, after phase modulation, filtering may be used; alternatively, the carrier may be amplitude modulated and the phase jumps applied when the carrier is a minimum. A typical power spectrum for a 1200 baud four-phase modem (2400 bit/s) is shown in Figure 13.9. (Transmission rate of pulses is measured in *baud*, the number of pulses per second.)

Several techniques exist that can be used to demodulate the phase modulated signal, but when demodulating for example, a four-phase signal, the basic procedure is to divide the circle representing 360° into four quadrants as shown in Figure 13.10. The correct phase vectors are situated at the centres of the quadrants or *decision regions* and if the demodulated phase falls inside a given

Table 13.3   Gray code

Binary number	Gray coded number
000	000
001	001
010	011
011	010
100	110
101	111
110	101
111	100

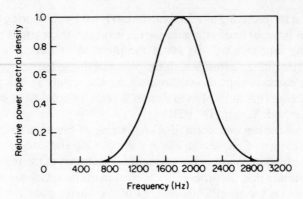

Figure 13.9 Relative power spectral density of a
particular four-phase data modem operating at 2400
bit/s

quadrant, the phase vector associated with that quadrant is assigned to be the
decision of the receiver. The added noise and the phase jitter to the signal is
represented by the vector $n(t)$.

13.5.2 Interference Effects on Phase Modulation

Normally if the noise causes an error in the detection of the phase shift, the
incorrectly detected phase shift will be one on either side of the correct one.
Consequently the bit error rate is minimized if the Gray code is used.

A further interesting point that emerges from consideration of operation of
the differentially coherent phase modulation system is that the comparison of
these two successive phases can be made before or after these phases are quantized in the receiver. If the phases are first quantized and the successive phases

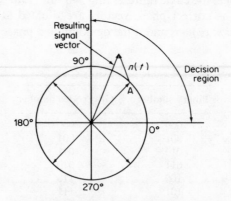

Figure 13.10 The decision regions of a
four-phase data modem

subtracted to give the phase change, an error in detection in a given phase will affect two successive subtractions and hence two *dibits* will be in error with probably only one bit in error per dibit due to coding of the phase changes (see Table 13.2). This means that errors will tend to occur in pairs. However if the successive phases of each dibit symbol are first subtracted and then quantitized to the changes shown in Table 13.2 then the likelihood of double errors is considerably reduced. The added noise can be represented by the vector, $\mathbf{n}(t)$, as shown in Figure 13.10. As long as the resulting phase vector falls inside the quadrant, the correct decision will be made. For Gaussian noise, the probability of an incorrect decision for four-phase modulation is plotted in Figure 13.11 as a function of the signal-to-noise ratio (SNR).

13.5.3   Synchronizing and Training Procedures for PM Systems

Before data can be transmitted over a communication circuit using PM modulation the following circuit conditions must be established:

(a)   the equalizers in the receiver must be adjusted to the line conditions;
(b)   the received signalling level must be adjusted;
(c)   the receiving clocks must be in synchronism with the transmitter;
(d)   the scrambler and descrambler must be in synchronism.

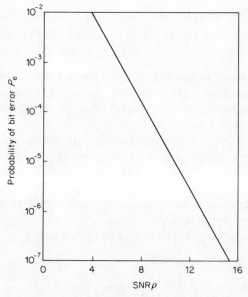

Figure 13.11   Probability of error (P_e) as a function of the signal-to-noise ratio (ρ) of a four-phase system: $P_e = \frac{1}{2}e^{-\rho}$

The data terminal indicates that it wishes to transmit by sending a *request to send* signal to the modem. The modem replies with a *ready to send* signal to the data terminal when all conditions have been met. This *line-up* procedure is done automatically in most modems over 4800 bit/s and is known as *synchronizing* or *training procedures*.

CCITT recommendations V27 and V29 (CCITT 1977a–d) describe standard training modes or turn-on sequences.

13.6 HIGH SPEED OR WIDEBAND DATA TRANSMISSION

13.6.1 Introduction

CCITT recommendations V35 and V36 specify standards for wideband data systems having bandwidth greater than 3.3 kHz. Data transmission at higher speeds requires a greater bandwidth. Two methods are used:

(a) direct baseband signalling on non-loaded cable pairs using local telephone circuits up to 48 kbit/s;
(b) Modulation of the data to operate in the basic 12-channel VF group (60 to 108 kHz). Speeds of up to 168 kbit/s can be obtained; however, the standard rate of 48 kbit/s is more common.

13.6.2 48 kbit/s Baseband System

The basic layout of a typical 48 kbit/s baseband local circuit is as shown in Figure 13.8. The transmitter has five basic elements:

(a) A signal converter to convert the binary series input from the data set to signal levels suitable for transmission through the system.
(b) A clock to control the signals from the data set to the modem so they are sent at the correct bit rate.
(c) An encoder/scrambler to generate the appropriate line code (Section 13.4.3) and to eliminate any d.c. component in the signal chain in *at rest* conditions. It also ensures that there are sufficient transitions to provide clock synchronism at the receiver terminal.
(d) A wave shaper to set the slope of the rectangular waves to reduce harmonics and intersignal interference.
(e) A limiter, output level adjuster, and line interface.

The receiver has six basic elements:

(a) The receiving line amplifier, filter, and line equalizer.
(b) An amplifier with built in AGC to compensate for line variations.
(c) A bit timing detector which uses the incoming transitions of the data signal to synchronize the receiving clock.

(d) A signal regenerator where the received signal is regenerated so that it is presented to the data set with the correct timing between elements and without variation in the timing caused by the line conditions.

(e) A decoder and descrambler to convert the signal back to the input code.

(f) A signal converter and data set interface to convert the signal to voltage levels acceptable to the data terminal equipment.

13.6.3 48 kbit/s Vestigal Sideband Transmission Modem

A 48 kbit/s modem working in the standard 60–108 kHz group bandwidth uses *vestigal sideband transmission* (VSB). In many cases a voice frequency channel is included.

A block schematic of a typical modem is as shown in Figure 13.12. The 48 kbit/s data signal is amplitude modulated by a 100 kHz carrier to produce a 36 kHz bandwidth in the range of 64–108 kHz which represents one sideband of the modulated carrier. This sideband and the 100 kHz carrier are sent to line. It is essential that the 100 kHz carrier be detected at the receiving terminal as this will be used to demodulate the sideband: absolute synchronous detection must be provided (*homochronous demodulation*).

The modem shown in Figure 13.12 allows for a voice frequency channel in the 104–8 kHz range which is combined with the data carrier at the transmitter hybrid. The signal is separated by band-pass filters and detected at the receiving

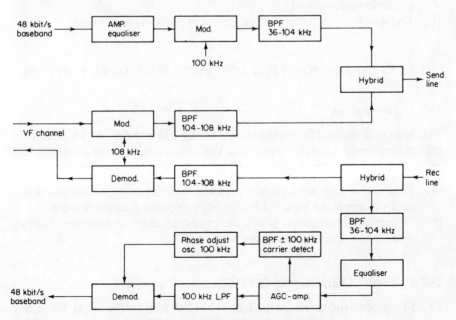

Figure 13.12 Schematic of 48 kbit/s VSB modem

terminal. The receiver uses AGC, line equalizers, and automatic phase compensation where the phase shift of the 100 kHz carrier is adjusted so that it is in correct phase relationship with the data sideband. The 100 kHz signal is then used to demodulate the carrier to produce the 48 kbit/s data.

13.6.4 Other Wideband Data Techniques

The vestigal sideband system of modulation described in the previous section is basically a direct modulation of the baseband signal and the system could also be used by wideband analog-type facsimile transmission with a bandwidth of about 35 kHz. The main problem with this system is that a reference carrier must be sent to provide homochronous detection and the reference carrier is at the upper or lower end of the wideband spectrum where it is most susceptible to the effects of distortion due to *group delay*.

Two other methods can be used for wideband data transmission. These methods are particularly suitable for synchronous bit-dependent data transmission. These techniques are as follows:

(a) Split the signal into two sections which are then modulated in quadrature with double sideband techniques and suppressed carrier. This is similar to the technique used in phase modulation. By using coherent signal detection with the carrier frequency at the centre of the band the effects of delay distortion are reduced.
(b) Use *duobinary* signalling with differential encoding of the data.

13.7 TRANSMISSION CHARACTERISTICS OF DATA CIRCUITS

13.7.1 Introduction

This section describes the characteristics of physical and derived VF circuits in the communication network and how the impairments of these circuits can effect data transmission. There are two basic classes of characteristics:

(a) Passive characteristics, such as the fixed make-up of the transmission line in, for example, its bandwidth frequency response, and attenuation.
(b) Dynamic characteristics which are caused by external influences such as noise, phase jitter or loss of synchronism.

13.7.2 Passive Characteristics of VF Circuits

CCITT recommendations M1020 and M1050 specify the ideal frequency response and group delay limits for a VF data circuit. These specifications, which were detailed by the sixth plenary assembly in 1976, supersede the earlier recommendation M102.

A power level of − 10 dB below normal speech level has been specified as the sending level for data transmission when a continuous tone is being transmitted to line. This level has been selected to prevent overloading of multichannel communication circuits which are used for both data and speech circuits. Although speech circuits can be set to transmit at 0dBm0 the average data level is between − 10 and − 15 dB.

Equalizers

Since speech is not sensitive to delay distortion, the telephone network has largely developed without this parameter being specified, but when data are transmitted over the network the delay distortion can very seriously degrade the signal especially for high data rates. Some form of compensation for the delay will then be needed. The effect of group delay distortion on data transmission is to cause interference between adjacent symbols of the received data. Normally this intersymbol interference is insufficient to cause errors by itself but rather it makes the data more susceptible to noise disturbances.

As linear distortion can cause intersymbol interference, it is better to adjust the equalizer to minimize this interference at the sampling time of the receiving data modem rather than attempt to fit the frequency response of the channel into some arbitrary criterion. Because the intersymbol interference is more easily represented in the time domain than the frequency domain, the form of the equalizer that best minimizes the intersymbol interference is a *transversal* filter which gives weighted versions of the original signal at a range of delays. In fact, it is this form of equalizer that lends itself to being made adaptive by observing the intersymbol interference in the data at the receiving terminal and adjusting the weighting of the various delays to minimize this interference.

13.7.3 Dynamic Characteristics

The noise, phase jitter frequency errors and other characteristics are also specified in CCITT recommendation M1020. The main dynamic characteristics are as follows:

Random noise

This is wideband noise generated by thermal effects and power induction and which usually remains at a fairly steady value with respect to time. The recommended limit is − 38 dB. Figure 13.13 shows typical random noise limits, and Figure 13.14 shows the relationship of SNR to error rate.

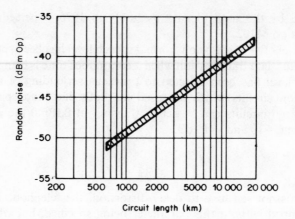

Figure 13.13   Random noise circuit performance

Impulse noise

In contrast to random noise this is quite variable with respect to time and can depend upon the time of day and the routing of the circuits. Sources of impulse noise are inductively or capacitively coupled dialling impulses and transients produced in exchanges by the operation of electromechanical switching equipment. Data modems can be designed to withstand the effects of this coupled impulse noise provided the network is maintained to specified limits. The recommendation is that the number of impulse noise peaks exceeding -21 dB should not be more than 18 in any 15-minute period.

Phase jitter

This is the regular or cyclic variation from correct phase of the received line signal. Low phase jitter depends upon good design and maintenance: in newer equipment tighter design specifications have been used. Phase jitter must be less than $15°$ peak-to-peak.

A phase hit

This is a sudden jump in the phase of the line signal received by the modem. These sudden jumps are generally caused by the switching of radio bearer circuits from normal to standby and vice versa. Phase hits are reduced by good design and good installation practices which ensure that, for example, there will be no difference in the propagation time between radio bearers.

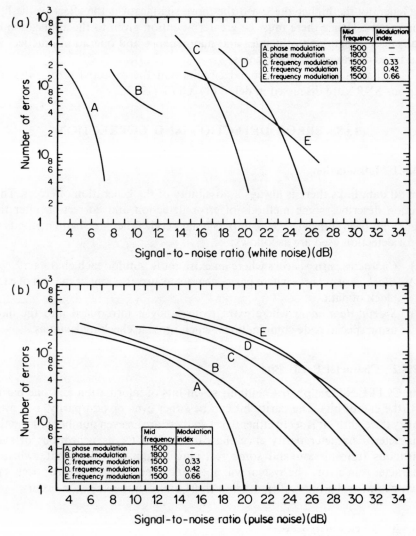

Figure 13.14   Relationship of signal-to-noise ratio to error rate: (a) white noise;
(b) pulse noise

Frequency asynchronism

This is due to differences in systems carrier supplied, for example a 1000 Hz tone
sent into a channel may be received at the distant end as 1005 Hz (the usual
limit). This distortion causes serious errors in low speed modems using frequency
shift keying. The solution to this is good maintenance practices with respect to
the synchronization of carrier supplies.

Generally the higher the speed the more significant is the effect of the line parameters. Hence there must be greater attention given to the application of transmission planning principles and maintenance and operation practices as the speed of transmission increases.

The characteristics of wideband circuits and the expected error rates for various SNR's are discussed in detail in CCITT (1972).

13.8 ERROR DETECTION AND CORRECTION

13.8.1 Introduction

On all data links there is always a possibility of the generation of errors. This section describes some methods of error detection and correction after the signal has been translated back into its binary code. The three main methods of error detection used are as follows:

(a) *Character parity checks* where an extra bit is sent for each character.
(b) *Block parity checks* where parity characters are also sent at the end of each block of data.
(c) *Special data codes* where extra redundancy is introduced into the data using special code combinations which provide checking facilities.

13.8.2 Character Parity Bits

The CCITT No. 5 alphabet contains seven bits of information and one parity bit (the eighth bit). The parity bit can be either even or odd parity. For even parity the eighth bit is set to either 0 or 1 so there are an even number of 1's in the character. Character parity checking can be used for asynchronous or synchronous transmission and some receiving terminals will indicate when a character is in error. The system will not work if more than one error occurs in a character or if a burst of errors or a short break in transmission occurs.

13.8.3 Block Parity Checks

With synchronous transmission a block of characters is sent at a time. Synchronizing characters are sent at the start of the block and a parity check character is sent at the end of the block. This system uses a two-coordinate parity check. A parity bit is allocated to check each character or column and a parity character at the end of the block acts as a longitudinal check and indicates if an even or odd number of 1's occur for each horizontal row in the block (see Figure 13.15).

If one error is indicated on the horizontal and vertical axes it is possible to detect which bit is in error and to correct this bit. Bit errors cannot be detected

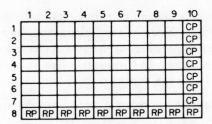

Figure 13.15 A fixed format of data:
nine seven-bit characters. CP, 'column'
parity bits; RP, row parity bits

if four errors occur on two horizontal and two vertical axes and they cancel out. The probability of quadratic errors of this type occurring is very low.

Error correction is not often used in block transmission on normal transmission circuits. It is not possible to correct more than one error in a block and errors often occur in short bursts.

The usual procedure is for the receiving terminal to request retransmission of the data block by sending a short message to the transmitter over the backward signalling channels.

Blocks of data transmitted over a high speed data circuit are usually numbered in sequence so the receiving terminal also checks block sequence numbers to detect long line breaks. The backward channel is used to acknowledge every block or every group of blocks.

13.8.4 Determination of Optimum Block Size

The terms given below are used in the formulations following:

m number of data bits in the forward message
n total number of bits in the forward message
S number of supervisory and synchronizing bits in forward message
K number of error detecting bits in the forward message
q number of bits in the backward path acknowledgement message
D_f propagation time of the forward data path
D_b propagation time of the backward data path
B_f data speed of forward channel
B_b data speed of backward channel
C computation time to validate the block

The *redundancy factor* can be expressed as the percentage ratio of control bits to data bits. The redundancy factor is given by

$$\frac{K + S}{m} \cdot 100$$

The redundancy can be made up of parity bits, start and stop elements for asynchronous characters, block control characters, and other non-message needs. When data are sent over high error rate circuits (e.g. space probes) a high redundancy factor is necessary to ensure error free transmission.

The optimum block size will depend on:

(a)   The bit error rate of the channel which will determine the number of retransmitted blocks.
(b)   The redundancy factor of the block.
(c)   The transmission speed which also determines how long a block is to be stored at the sending terminal before it is acknowledged and discarded.
(d)   The loop propagation time of the channel including the time of the acknowledgement on the return path.
(e)   The message length; there may be more than one block per message and some blocks will only be partly filled with valid data.
(f)   The storage capacity in the transmitter to hold blocks while waiting acknowledgement.
(g)   The message traffic rate for the channel which determines what spare time is available to repeat blocks.

The *storage time* T to hold a block after transmission while waiting acknowledgement is given by

$$T = D_f + D_b + q/B_b + C$$

In order to simplify the logic circuits, it is desirable that the decision on the block validity be made at the sending station during the transmission of the subsequent block which would also have to be stored; then T should be less than n/D_f. The total holding time including forward transmission is

$$T + n/B_f$$

The storage capacity necessary would then be for two blocks of minimum length n as determined above.

This is an important factor in the determination of block length for medium and high speed systems. However, no definite recommendation is possible until the speed of the forward and backward channels, the amount of redundancy (that is, the error-detection code arrangement), and the distance over which the system is to operate are each determined.

13.8.5   Probability of Block Errors

The probability of an error occurring in a block or a message is given in the approximate formula.

$$P_m = nP_b - \tfrac{1}{2}n(n - 1)P_b^2$$

where P_b is the probability of bit errors and P_m is the probability of a message error.

The effective speed of transmission for different data speeds and block sizes is shown in Figure 13.16. This takes into account the turnaround response and block retransmission due to errors and control characters.

An error rate better than 1 bit in 10^5 can be expected on a normal 9600 bit/s circuit. It is best to design for a higher than normal error, as is shown in Figure 13.16.

13.8.6 Other Types of Error Detection and Correction

The systems of character parity checks and block parity checks are satisfactory over normal communication channels with a bit error rate better than 1 bit in 10^4. High error rate circuits, such as low performance radio links, satellites, and space probes, require a higher level of redundancy. The technique is to select an error detection and correction code to give the most reliable operation with the least amount of redundancy. Bennett and Davey (1965), CCITT (1964, 1972),

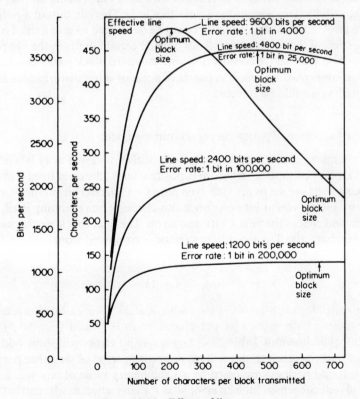

Figure 13.16   Effects of line errors

Duc (1973, 1974, 1975), and Lucky *et al.* (1968), describe various codes which can be used. Only the following few examples can be described here.

The m out of n code

For this code n bits of information are sent but only m 1's must occur in each character. An error is indicated if more or less than the prescribed m 1's are received. No error correction is possible. In general an m out of n code permits $n!/[m!(n-m)!]$ combinations out of 2^n possible ones.

A typical example of this code is the three out of five code used in multi-frequency code (MFC) signalling in telephone exchanges which gives ten permissible combinations out of a possible 32. The CCITT standard No. 3 alphabet uses a four out of seven code.

Null zone detection

The receiver makes a decision to recognize a 0 or a 1 depending on the position of the signal within an acceptable eye pattern area. This can output a positive or negative level to a decoder. A circuit could be developed to give a third position of null or zero if the received signal lies outside certain limits in the eye pattern. If this method was used in conjunction with standard block parity checks a high level of reliability would result as bursts of noise or other disturbances and low signalling levels could be rejected.

Interleaved sequence of parity checks (Hamming code)

Take a character 2^n bits long where n is a positive integer. Parity bits are allocated at all bit positions 2^m where m is positive with all values from 0 to n. The $n + 1$ parity bits are set to provide even parity on selected bits in the word such that parity bit x starts at bit position x and checks x bits counting itself, leaves out x bits and checks the next x bits, and so on. The final check bit checks all the bits in the character. The redundancy factor is computed from

$$\frac{n + 1}{2n - (n + 1)}$$

which gives 100% for 8-bit, 45% for 16-bit, and 23% for 32-bit characters.

An example with eight bits per character ($n = 3$) and four bits of data ($R = 100\%$) is shown in Table 13.4. Any single bit errors will show odd parity in bit 8 (2^n) at the receiver and will also appear in some of the other parity bits 1, 2 or 4. If odd parity is indicated by a 1 the binary value of bits 4, 2, 1 in that order will indicate which bit is in error so error correction can be carried out by reversing this bit. For example if bit 6 was in error bits 8, 4, and 2 will have odd

Table 13.4 Code for single-error correcting and double-error detecting

Decimal number	Position in sequence							
	1	2	3	4	5	6	7	8
0	0	0	0	0	0	0	0	0
1	1	1	0	1	0	0	1	0
2	0	1	0	1	0	1	0	1
3	1	0	0	0	0	1	1	1
4	1	0	0	1	1	0	0	1
5	0	1	0	0	1	0	1	1
6	1	1	0	0	1	1	0	0
7	0	0	0	1	1	1	1	0
8	1	1	1	0	0	0	0	1
9	0	0	1	1	0	0	1	1
10	1	0	1	1	0	1	0	0
11	0	1	1	0	0	1	1	0
12	0	1	1	1	1	0	0	0
13	1	0	1	0	1	0	1	0
14	0	0	1	0	1	1	0	1
15	1	1	1	1	1	1	1	1

parity and the binary value of 4, 2, 1 (110) indicates that bit 6 must be reversed. An error will be indicated if two bits are in error as one of the parity bits will always be odd but no error correction can be made. Three errors cannot always be detected.

This system, if incorporated with block transmission and longitudinal parity checks would be effective on high error rate circuits where one or more errors can occur on each block. Quadratic errors would also be detected. In most cases the error could be corrected thus eliminating the unacceptable time wasted caused by retransmissions on these high error circuits.

There are many varieties of interleaved error correction codes and many complex systems have been developed. The main aim is to achieve the greatest reliability with the minimum redundancy. Although these systems are seldom used on normal data transmission circuits where block error checking is sufficient there could exist special applications for use with low performance circuits. Some special cases, such as control data to stored-program controlled SPC exchanges where error free transmission is essential, could be considered.

Majority logic codes for error correction

By using parity indicators for different combinations of row and column checks in the same block of data more than one check can be carried out on each bit in the matrix. A decision to change a bit can then be made on the indication of the majority of the tests.

Research has been carried out on various combinations for majority logic codes to get the greatest number of checks with the minimum redundancy.

For further reading on all error detection and correction procedures see Bennett and Davey (1965), CCITT (1964, 1972), Duc (1973, 1974, 1975), Lucky *et al.* (1968), and Martin (1972a, b).

13.9   MESSAGE SWITCHING SYSTEMS

13.9.1   Introduction

Message switching networks, sometimes known as *store* and *forward networks*, are practical when one-way transmission of messages is all that is necessary. The main advantages over line switching are that speed and code conversions are practicable and no time is lost at the input terminal in attempting to set up a call to an outstation which may be busy. Disadvantages are the costs of providing message storage facilities and any special control procedures to ensure that messages are not lost.

Design considerations for message switching networks include:

(a)   Signalling speeds and limits of the switching system.
(b)   Maximum message storage capacity of the switching centre.
(c)   Maximum permissible delay in transferring messages (*cross office delay*).
(d)   Multistation and polled working requirements.
(e)   Special control procedures with the outstations, such as VDU procedures.
(f)   Address and message envelope structure.
(g)   Control station requirements for diverting and holding traffic, retrieving messages, message logs, and traffic reports.
(h)   Fall back procedures in the case of circuit and switching centre facilities.

Digital computers would mainly be used for control of all future message switching centres, along with some form of magnetic or logic storage of the messages. Systems could consist of one of more major message switching centres or a network of major and minor centres. If many outstations are required, particularly in the case of low traffic, the transmission and line terminations costs at the major centres lead to the need for small, low capacity, secondary switching centres or message concentrators. In these cases the remote centres could direct the traffic on higher speed, or more heavily loaded, circuits to the main central switching centres where most message analysis, editing and message storage would take place. Low level local switching could also be provided if necessary.

13.9.2 Message Security

There is a saying that a message is not lost if it is known that the message is lost. The originating terminal can always repeat the message if a lost or invalid message alarm is received. With point-to-point and circuit switching systems the sender can confirm delivery if some response or acknowledgement is received after transmission, for instance as in the answerback message of Telex machines. On links between computers and more sophisticated terminals, link procedures with block transmission, error detection, and block acknowledgement can be used. These link procedures are used for the transfer of messages between message switching centres but cannot be used on the circuits to low speed terminals. With *store* and *forward* switching there is no direct response in the return direction from the receiving terminal to the sending terminal so special message handling and monitoring procedures are required. Some methods are:

(a) *Sequential numbering of messages.* The switching centre will automatically check the sequence numbers of incoming messages and generate an alarm if there is a break in sequence. Sequence numbers will be generated on messages from the centre to the outstations.

(b) *Character parity checks.* These are applied on all characters in the message (if alphabet No. 5 used) with alarm generation when parity errors are detected.

(c) *Open line alarm.* A continuous space or no voltage on the incoming circuit will indicate a possible open circuit line causing an alarm to be generated.

(d) *No traffic alarms.* The input operator could be requested to send a test message at set intervals (say 30 minutes) if there is no traffic to be sent. An alarm can be generated if no messages are received indicating a circuit or machine failure.

(e) *Automatic test message.* Outgoing circuits from the switching centre test message can be sent at predetermined intervals if no traffic is waiting. The outstation operator would report a fault if no messages of any kind are received.

(f) *Polled station alarms.* With polled working, where more than one terminal is connected on a circuit, the switching centre will poll each station in turn and a message or a *no traffic* response will be received. An alarm will be generated is no response is received indicating a failure of the line or the terminal equipment.

(g) *Message storage alarms.* These can be generated when the storage area for undelivered messages exceeds specified limits.

(h) *Large queue alarms.* These can be given if a high rate of traffic on some outgoing circuits indicates a build-up of the message queue and possible delivery delays.

13.9.3   Control Station Functions

Control stations are required on message switching networks to handle the alarms and perform various traffic supervision functions. Some functions are:

(a)   receipt of alarms as described in Section 13.9.2;
(b)   open and close circuits to traffic;
(c)   divert traffic to other circuits;
(d)   obtain traffic reports, number of messages, queue sizes;
(e)   obtain circuit status reports;
(f)   get message logs of input and destination of messages;
(g)   retrieve messages from the message storage area;
(h)   perform accounting functions.

13.9.4   Message Concentrators or Low Traffic Switching Centres

It is often the practice to provide some form of buffering between low traffic outstations and message switching computers. The advantages are:

(a)   The occupancy of the main switching computer is reduced as time is not spent in handling and storing low speed messages on character by character basis.
(b)   The polling, circuit control, and local alarm procedures can be handled in the concentrator.
(c)   The costs of leasing long, low speed, low traffic circuits are reduced.
(d)   As the local terminals are connected over a short circuit, both way working could be economical and the concentrator could send status alarms and responses to each message directly to the outstation. Failure of these responses would indicate a fault condition, shortage of buffers or more, and the outstation would not send further traffic.
(e)   Local switching of messages which do not require processing in a central switching computer is possible; however, most concentrators would have limited storage.

13.9.5   Traffic Calculation

Methods used to dimension a message switching network are detailed in Chapter IV of (Martin, 1972a).

By using the traffic data pattern, calculations can be made on the response time for interrogation and response traffic, queuing calculations, number of terminals, grade of service on junctions, message distribution and buffer requirements, multidrop traffic calculations, and other applicable design parameters. Additional published material of relevance to the preceding part of this chapter includes Abramson and Kuo (1973), BSTJ (1975), Bell Telephone

Labs (1970), Martin (1967), McKinnon (1970), McKinnon *et al.* (1977), Miller (1978), Schwartz (1977), and Smith (1971).

13.10 SIGNAL BEARERS

13.10.1 Bearers as Subsystem Elements

Measurement systems consist of basic analog and digital subsystems interconnected to provide the required overall input–output relationships. It is important that the various subsystems be interfaced correctly if they are to perform as intended. But even with this condition satisfied, it should not be assumed that subsystems can be connected together without need to consider any other parameters in the interconnection process.

In practice the individual subsystem assemblies may be geographically separate—such as in the remote control of an offshore oil well by a shore-based computer, the recording of test data from a missile, the control of banking accounts by a central computer centre or the sensors of a refinery which send data to the central control room. Each of these require some form of data transmission system—in the instrumentation sense these are called *telemetry* systems.

When making connections it is also important, especially when noise sources are present that will interfere with the signal, to ensure that the signal is transferred from stage to stage without significant noise pick-up or signal degradation.

The bearer methods used should be considered as subsystem elements requiring as much consideration of their properties as is given to any part of the total measurement system.

13.10.2 Types of Transmission Bearers

Although there does exist the occasional application where data can be sent conveniently over mechanical or hydraulic forms of data channel, by far the most common method makes use of the electric form of signal. Either analog or digital information formats can be used on each of the kinds of electric signal bearer, the main criterion for selection being whether the chosen bearer method has adequate frequency bandwidth to convey the appropriate frequency spectrum required to be received.

The theoretical understanding and design methodology for electric signal bearers matured some years ago for analog signal working, the trend to greater use of digital signals being where the greater emphasis has been placed in more recent times.

The need to provide greater message security and wider bandwidths of operation has created interested in optical fibre communication methods. In the late 1970's experimental field evaluation systems were being brought into service on selected telephone routes and had been introduced in the control wiring of advanced military aircraft. There can be little doubt that optical fibre methods will be progressively used to a greater extent. Wolf's (1979) handbook provides a state-of-the-art review of this form of technology. Other relevant works are Elion and Elion (1978), Midwinter, (1979), and Miller and Chynoweth (1979).

Electric signals can be conveyed over bearers in which the signal is confined to a physical member—open wires, coaxial cables, and waveguides are considered here. Alternatively it might be practicable to make use of electromagnetic (EM) or acoustic wave propagation. In the case of EM radiation it is necessary to use a carrier frequency much higher than the highest frequency of the signal in order to obtain the desired propagation properties. Interconnections are discussed in Harper (1972).

Confined signal links

The simplest links are formed using an open-wire circuit (supported on insulators) or a multicore cable (such as is used in local telephone distribution).

Although apparently trivial, lines may, in fact, be an important part of the system. They are not as simple as they first appear because they have a limited frequency response that must be adequate for the signal bandwidth to be transmitted. Open-wire lines would not normally be used beyond 10 MHz. Above that coaxial cables are needed—these are useful to about 5000 MHz. When currents flow in a conducting line, magnetic and electric fields are set up around the wires. Figure 13.17 shows these plotted for the various kinds of cable. Open configurations radiate energy, the amount increasing with the frequency of the signal.

A line, is in reality, a distributed inductance and capacitance component which has losses due to the resistance of the wire and the resistance to ground. Figure 13.18 shows how lines can be considered as a lumped-element equivalent circuit which can be analysed more easily.

The equivalent circuit approach to modelling lines is discussed in texts on transmission lines, examples being Connor (1972), Johnson (1950), and ITT (1970). Models such as that of Figure 13.18b, provide an efficient approach to design for they provide means to evaluate the effect of the various parameters of construction on transmission performance.

Depending upon the factors that can be considered to be negligible for a particular case the equivalent can be reduced to simpler circuits. At very low frequencies (less than, say, 100 kHz) a medium length line may be represented by the series resistance of the cable shunted by the capacitance of the line (see

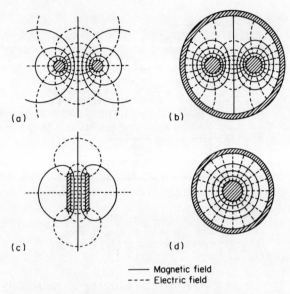

—— Magnetic field
---- Electric field

Figure 13.17 Magnetic and electric fields of common conducting bearers: (a) open, two-wire, line; (b) shielded pair; (c) parallel strip line; (d) coaxial cable

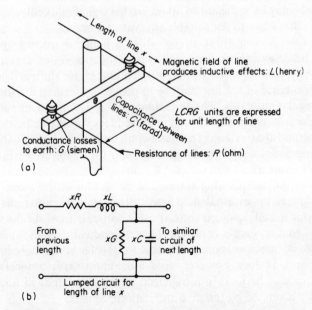

Figure 13.18 Open-wire line and a lumped equivalent model: (a) typical section of line; (b) lumped model of line length x

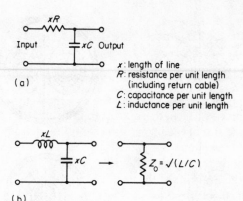

Figure 13.19 Reduced models of a line:
(a) long cable operated at low frequency;
(b) lossless cable operated at high frequency

Figure 13.19a). Typical cables may have a resistance of around 0.05 Ω/m and a capacitance of 100 pF/m. Hence a long length of shielded or open cable could provide a considerable shunting effect that attenuates and phase shifts the signal because of the effective first-order, low-pass, filter, so produced.

When connecting high output-impedance sensors to lines, as little as one metre of cable may be sufficient to attenuate the signal markedly. It is a matter of applying Ohm's law to the suitable equivalent circuit.

Because of the filtering effects of the cable the higher frequency signals transmitted will be degraded more than the low frequencies—for example, square waves become rounded as well as attenuated and phase shifted. The high frequency performance of the line may be improved by *loading* it with inductors placed at regular intervals. The inductance value is chosen to tune out the inherent capacitive reactance at the upper frequency where response begins to fall off, a method that extends the bandwidth some way beyond the inherent unloaded upper limit. This technique is used, for example, to broaden the bandwidth of submarine cables.

By virtue of the surrounding external shield acting as the second wire, the coaxial cable has no external field and, therefore, does not radiate energy. Because of this a well designed coaxial cable will pass from d.c. to microwave frequencies—that is, such a cable can have a bandwidth of about 5000 MHz. Coaxial cable is, therefore, potentially able to transfer much more information than open wires. It does, however, need a common earth connection (asymmetric) and cannot be used in a balanced mode. The bandwidth of practical coaxial cables is limited by resistive and dielectric losses.

When the losses of the line can be regarded as insignificant ($G = 0$, $R = 0$, in Figure 13.18b) the lumped-equivalent of the transmission lines reduces to L in series and C shunting, as shown in Figure 13.19b. The net result is, rather surprisingly, that the line exhibits only resistance of a fixed value when looking into the ends. This called the *characteristic impedance*, Z_0, for which

$$Z_0 = (\text{inductance per unit length/capacitance per unit length})^{1/2}$$

The line appears to be purely resistive and the Z_0 value is decided by the design of the line or cable, not by its length. As a general rule it is the ratio of conductor sizes not the absolute size of the line cross section that decides the value of Z_0. Examples are 600 Ω telephone lines and 75 Ω colour TV coaxial feeder cable. This means, in practice, that we can interconnect units on the basis of matching all connections to the Z_0 value of the cable without having to worry about the cable length. If this rule is observed, no high frequency energy will be *reflected* at the termination to change the information being transmitted.

However, if the line is very long matching must still be applied to obtain maximum transfer, but account must now be taken of losses. For example a typical 75 Ω coaxial cable will have losses of the order of 2 to 5 dB per 100 metres of length.

Details of parameters of the various forms of cable are obtained from manufacturers. In-house publications of the major national telecommunications administrations, such as British Telecom and Telecom Australia, are an excellent source of data where they are available. Practices related to cabling are covered in Harper (1972).

When greater bandwidth than a nominal 1000 MHz is required multiple coaxial cables may be used if the bandwidth can be split. In applications needing higher bandwidth for a single signal it becomes necessary to make use of *waveguides*.

Waveguides consist of precise pipework and convey travelling electromagnetic waves of very high frequency along the internal cavity. They cannot be used for low frequency transmission.

The cross-sectional area needed for a waveguide is inversely proportional to the design frequency. As a general rule of thumb the upper frequency limit of a waveguide is where the wavelength of the signal becomes one quarter of the guide aperture, millimetre wavelength signals (50 HGz or so) being the practical upper limit.

Beyond this, a still wider bandwidth is obtainable theoretically using optical fibre transmission elements which will pass radiation in the visible light region (10^{14} Hz to 10^{15} Hz). At the current state of technology, however, it is not possible to detect radiation cycles beyond far infra-red signals (around 10^{11} Hz). We cannot, as yet monitor individual cycles of light with electronic detectors.

Optical fibre transmission operates through modulation of light by high frequency signals. Waveguide theory and technology is explained in Collin (1966), Connor (1972), ITT (1970), Johnk (1975), Kennedy (1977), Ramo *et al.* (1965), and Saad (1971).

Radiation links

Electrical signals fed into open wires radiate energy out into the surrounding medium. As well as this radiated energy there also exists a *near field* that remains established, storing energy. This is the field associated with, say, an electro-magnet. As the frequency rises, the ratio of radiated energy to stored energy increases. For this reason it is possible to build efficient radio systems provided the frequency is kept above 100 kHz or so. Lower frequencies can be used as transmission systems but the power input needs rise enormously for the same distance radiated in free space. (The Omega navigation system uses extremely powerful VLF signals because of its reliability of reception and the ability to penetrate deep into the waters of the ocean.) Beyond the gigahertz frequency region, circuitry becomes impracticable with current technology. Even though the radiated energy link must operate at a very high frequency in order to operate efficiently it might not necessarily need to use the bandwidth available on the carrier.

Free space electromagnetic radiation transmission is a highly developed subject. Its theory and practice are presented in many publications, a selection of which is Hamsher (1967), ITT (1970), Kennedy (1977), and Picquenard (1974). Radio engineering as a total system is covered in Ross (1980).

Skin effect

The alternating magnetic field produced around a wire has the effect of causing the current flowing in the wire to flow at a greater density in the outer region of the wire: the higher the frequency the more pronounced is this so called *skin effect*. At very high frequencies so little current flows in the centre of the cable that the centre is often omitted completely, thus a tube is used as a conductor. The tube is also convenient for passing cooling fluid to remove heat losses. For example, at 1 MHz the majority of the current flows in a copper cable to a depth of only 60 μm whereas at 60 Hz the depth would be 8.6 mm. This also means that the effective resistance of a wire rises significantly with frequency—by factors of 100.

Optical and acoustic, free space, links

Visual-optical and infra-red beams are sometimes used as line-of-sight carriers carrying information as modulation of amplitude or of polarization angle. Commericial *opto-electronic* communication links are available, one example being designed to transmit television signals, plus operator voice channels, over relatively short distances. These find use where message security, from deliberate intervention or from environmental noise, is needed. It is sometimes more convenient to set up such systems for temporary use than a cable link.

Acoustic links have also been developed that use carriers ranging from 10 Hz to 10 MHz. Acoustic links are less directional, for a given level of hardware cost, than the optical equivalent, this being a function of wavelength.

Sources of general information on transmission systems usually contain chapters detailing the various kinds of bearers (see e.g. Bell Telephone Labs, 1970; Hamsher, 1967; Kennedy, 1977).

13.11 ANALOG COMPARED WITH DIGITAL SIGNAL TRANSMISSION

In theory an analog signal can assume any value between its saturation limits. In practice only a finite number of adequately definable states can be assigned due to such factors as the uncertainty of noise levels, the non-linearity characteristics of the system, and the drift performance with temperature change and time elapsed. It becomes very expensive to design analog systems with less than 1 in 1000 error levels. Thus analog systems typically are capable of conveying accurate information on from 200 to 1000 levels.

Digital signals, in the main, operate with the *binary* two-level system, the minimum number than can be used to convey information.

Consideration of the tolerance to error-producing physical effects shows that the two-state signal has a far greater probability of being received intact than does the multilevel analog alternative but that, because the 0 and 1 levels are not infinitely apart, there is always a finite probability of an error occurring. For this reason it is often still necessary to add redundancy into the system to gain an increased level of reliability.

The SNR of the analog signal will always degrade in transmission to some extent. It is far less possible to restore the signal, in a *repeater*, to its original state than is practicable for the binary case.

However, due to the larger numbers of information-bearing states, analog systems can convey more information in a given time. Thus an important decision to be made in choosing a system is the extent that data rate can be traded for error risk.

Analog signals are often more appropriate as much of the physical world is analog by nature. In such cases additional interfacing circuitry will be needed if digital transmission is to be used. The same applies to the receiver where analog actuators are often present.

Until the 1960's analog methods were preferred because, by default, digital hardware and methods had not then been as extensively developed. The sophistication of digital circuitry then proposed, brought with it high manufacturing cost. The introduction of integrated circuit microelectronic, mass production methods rapidly diminished cost as such a significant factor of design.

Digital and analog systems use the same bearers. A link created to transmit analog signals of a given bandwidth and voltage or current swing will be capable of transmitting a digital signal transmission provided the bandwidth and signal limits are appropriate.

Conversely, however, a system designed to transmit digital signals will usually not be suitable for conveying analog signals.

It is to be expected that the overwhelming availability of digital transmission experience and hardware will considerably influence designers to adopt an increasing number of digital methods as time progresses. There will, however, always be special cases where the analog alternative will be a better choice.

The concept of the *data highway* has its origins in the data *bus* system used in digital data processing machines. In the late 1960's this concept was adapted for computer control of process instrumentation systems (Collins, 1968) and today

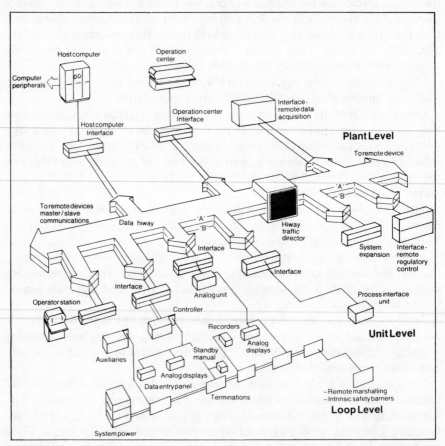

Figure 13.20   Architecture of data highway based, plant control system. (Reproduced by permission of Honeywell Inc.)

several process instrumentation manufacturers market (see e.g. Figure 13.20) digital data highways for use with extensive instrument systems such as are found in process plants.

Dominant reasons for adoption of the data highway approach were the needs to reduce installation cost, to increase flexibility to commissioning and later adjustments to plant operation, and to improve message reliability.

As the number of monitoring and control functions rose with development of plant sophistication so did the proportionate cost of cabling—it is not uncommon to need as many as a thousand control loops. Furthermore the need for operator redundancy in information transfer arose to reduce the risk of data transmission failures.

Further advantages of the data highway approach are that the control and monitoring functions can be arranged in a *distributed* manner and that diagnostic and control units can be accessed virtually anywhere in the system's *daisy-chain* highway. To increase security the highway bearer, basically a single coaxial cable, can be duplicated and run on different physical routes.

13.12 INTERFACING

It is usually necessary to consider how the measurement system block at the sending end should be interfaced to the transmission system and how the bearer should interface to the receiving end unit. Two parameters of general importance are matching and connection configurations.

Matching

Three basic matching criteria exist when connecting two stages together; Figure 13.21 summarizes these.

If the need is for maximum power transfer, as when driving an actuator from an output stage of an amplifier, the output impedance of the driving stage must equal the input impedance of the stage being driven, this being the case if the output impedance cannot be made small enough to be considered zero; which

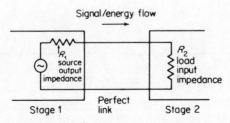

Figure 13.21 Summary of matching relationships

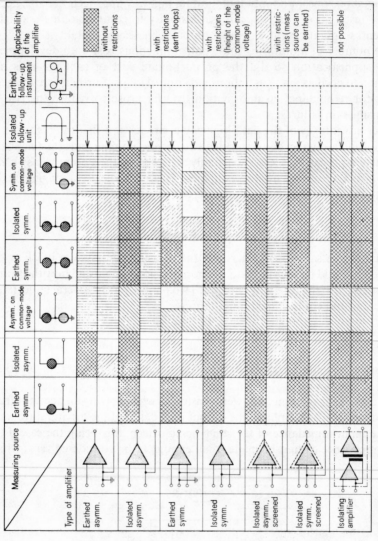

Figure 13.22   Chart showing commonly met source, amplifier, and output device connection arrangements. (Reproduced by permisson of Siemens Industries.)

is the most efficient method of driving a load from a source. When maximum voltage transfer is required, as occurs when a pick-up cartridge or other voltage generating transducer is used or when measuring a voltage in a circuit, the rule is to ensure that the connecting stage has a much higher input resistance than the output resistance of the stage producing the voltage signal. A factor of ten to one hundred times is usually sufficient.

The opposite situation, that is, loading a high output impedance stage with a low input impedance, arises when the maximum current transfer is required.

In many cases the appropriate buffer amplifier is required to provide the desired matching condition. In certain a.c. coupled systems—those which do not require a d.c. path between stages—a transformer can provide an adequate impedance match in an economic way. Transformers, however, can have limited frequency response and must be chosen carefully to suit the signal requirements.

In the above it is assumed that the transmission system is ideal. Consideration of the characteristics of various bearers shows that it is often necessary to interface bearers to transmitters and receivers with the appropriate impedance conversion.

Connection configurations

Output configuration of the various stages involved in instrumentation can take many forms depending on how the earth is connected and if the signal is symmetrically or asymmetrically connected. Six commonly encountered source output schemes are shown along the top of Figure 13.22. On the left-hand side are seven common kinds of amplifier connection (any other form of black box could be regarded similarly). On the right-hand side are leader lines that show a link between the output of the chosen amplifier and one of the two most commonly used instrument connections—fully isolated circuit with case only grounded, or one pole grounded to earth. Using the legend, the chart shows the applicability of connections between chosen combinations of source arrangement, amplifier, and output device. Not-possible situations usually arise because the earth connection shorts out one of the source arms.

Transmission bearers must be selected to suit the number of poles and earthing requirements for a given case.

13.13   PROCESS INDUSTRY TELEMETRY

Process plants such as oil refineries, paper mills, brick kilns, power stations, and aluminium refining plants are often monitored by using hundreds of sensors connected to the control-room area via analog instrumentation links. These are wired using shielded wire or coaxial cable. Because of the extremely high electrical noise levels of such plants and the low output signal levels of the sensors, these links could pick up significant noise thus degrading the sensor information. Over the years process instrument suppliers have, to some extent, standardized the design of control systems and their installation. The effect of noise pick-up by the cable has been reduced by several methods.

One strategy is to superimpose the information signal onto a standing bias current or voltage thus raising the wanted signal level above expected noise levels. Two systems commonly used transmit the signal range of the data through 4–20 mA d.c. or 10–50 mV d.c. systems. A 0–20 mA system is also common. Current transmission has the advantage that the circuit is of low impedance—a few ohms—which reduces the level of induced noise power.

An alternative method is to amplify the millivolt level signals produced by a sensor to span them in the range 0–10 V or higher, ready for transmission. The additional unit that interfaces the basic sensor, such as a thermocouple or a differential pressure cell, to a line with the appropriate sending signal format is known as a *transmitter*.

As a consequence of expanded use of digital transmission, mentioned in Section 13.11, it is to be expected that the analog methods mentioned above will gradually be displaced by transmitters that condition the sensor's analog signal into an appropriate digital form. The growing interest in silicon integrable sensors, those produced by the mass production methods of microelectronics, with on-board signal processing and format conversion will also bring about a marked change in the physical hardware of sensors and signal transmission in situations where a large number of sensors is involved.

13.14   SIGNAL TRANSMISSION IN EXPLOSIVELY HAZARDOUS ENVIRONMENTS

Often the sensor has to be placed at a location where an explosion could result from a spark or excessive overheating of a malfunctioning sensor circuit. The most obvious way of overcoming this is to place the whole unit in a *flame-proof* enclosure. This method, however, has disadvantages: the cost is high, and testing and maintenance difficult due to the need to shut off the power when the enclosure is opened.

An alternative, method is known as *intrinsic safety*. As inflammables require a specific level of energy to ignite them, explosion can be prevented by ensuring that the sensor stage cannot, under any conditions, provide enough ignition energy. No enclosures are needed and the circuit can be maintained whilst it is operating. Originally the concept was implemented by ensuring that the sensor circuitry could not draw, or produce via storage, more than a specified power level. This level was found by experiment in a test rig set up for the situation involved.

A more recent approach is to use *safety barriers*. At the entry point into the declared hazardous area the cables terminate into a shunting Zener diode and attenuator arrangement which ensures that the current and voltage entering the area are limited to safe values. Figure 13.23 a shows the circuit of a typical Zener barrier. Another safety device uses a solid-state closely-coupled electro-optic link which provides d.c. electrical isolation between its input and output, the information being transferred from a light-emitting diode mounted next to a silicon photodiode detector. These ensure that overvoltage or induced earth-loop currents cannot enter the isolated hazardous area.

A wide range of barrier devices is marketed. Figure 13.23b demonstrates, as an example, how a slide-wire displacement sensor can be interfaced from a safe to a hazardous area using the barrier of Figure 13.23a.

It must be made clear that there exists a bewildering range of intrinsic and other safety codes and that they differ widely from country to country. Barriers designed and manufactured by specialist companies (see e.g. MTL, 1976), are to be preferred to constructing one's own circuitry—in many cases it is manda-tory that the device be approved by the appropriate approvals body.

Other methods are also used, including use of a *pressurized* or *purged* en-vironment around the sensor so that flammable vapour cannot come into the sensor region. Another approach is to design the equipment so as to decrease the risk by reducing operating temperature levels using contactless devices and the like. A brief, but useful, introductory review of this subject is provided in Jones (1974) with a more extensive treatment being available in Maeison (1974). The IICA symposium proceedings (IICA, 1980), although centred around Australian standards (which are based on British sources), contains useful papers on intrinsic safety practice.

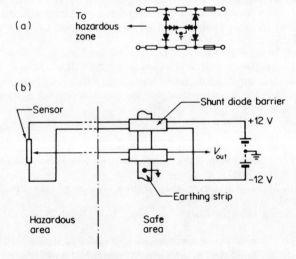

Figure 13.23   Shunt diode safety barrier: (a) one form of basic circuit; (b) use of units of (a) to couple slide-wire potentiometer into hazardous area

In areas where explosion or other catastrophic effects might conceivably occur it is the transmission bearers that need protection. It is not uncommon to make use of pressurized non-combustable gas in cable ducts or to place cables in sealed ducts. The case of multiple cables on different routes between the same nodes of a system has already been mentioned.

REFERENCES

Abramson, N., and Kuo, F. F. (1973). *Computer-Communication Networks*, Prentice-Hall, Englewood Cliffs.

Bell Telephone Labs (1970). *Transmission Systems for Communications*, Bell Telephone Laboratories, Merrimack Valley.

Bennett, W. R., and Davey, J. R. (1965). *Data Transmission*, McGraw-Hill, New York.

BSTJ (1975). 'Digital data system', *Bell. Syst. Tech J.*, **54**, 811–969 (whole May–June issue).

CCITT (1964). 'Data transmission', in *Blue Book*, Vol. VIII, International Telecommunications Union, Geneva.

CCITT (1972). *Green Book*, Vol. VIII, International Telecommunications Union, Geneva (whole book).

CCITT (1977a). 'Data transmission over the telephone network', in *Orange Book*, Vols. VIII.1, VIII.2, International Telecommunications Union, Geneva.

CCITT (1977b). 'Maintenance', in *Orange Book*, Vol. IV.1, International Telecommunications Union, Geneva.

CCITT (1977c). 'Telegraph technique', in *Orange Book*, Vol. VII, International Telecommunications Union, Geneva.

CCITT (1977d). 'Line transmission', in *Orange Book*, Volume III.1, International Telecommunications Union, Geneva.

Collin, R. E. (1966). *Foundations of Microwave Engineering*, McGraw-Hill, New York.

Collins, G. B. (1968). 'A survey of digital instrumentation and computer interface methods and developments', in *Industrial Measurement Techniques for On-line Computers*, *IEE Publ.* 43, June, pp. 1–8, 60–71.

Connor, F. R. (1972). *Wave Transmission*, Edward Arnold, London.

Duc, N. Q. (1973). 'Majority-logic decoding of block codes', *Aust. Telecom. Res.*, **7**, 22–31.

Duc, N. Q. (1974). 'Majority-logic decoding of product codes', *Aust. Telecom. Res.*, **8**, 31–8.

Duc, N. Q. (1975). 'Line coding techniques for baseband digital transmission', *Aust. Telecom. Res.*, **9**, 3–17.

Elion, G. R., and Elion, H. A. (1978). *Fiber Optics in Communications Systems*, Marcel Dekker, New York.

Hamsher, D. H. (1967). *Communication System Handbook*, McGraw-Hill, New York.

Harper, C. A. (1972). *Handbook of Wiring, Cabling and Interconnecting for Electronics*, McGraw-Hill, New York.

IICA (1980). *IICA Symp. on Intrinsic Safety, Melbourne, October 1980*. Institute of Instrumentation and Control, Melbourne.

ITT (1970). *Reference Data for Radio Engineers*, H. W. Sams, New York (many edns published).

Johnk, C. T. A. (1975). *Engineering Electromagnetic Fields and Waves*, Wiley, New York.

Johnson, W. C. (1950). *Transmission Lines and Networks*, McGraw-Hill, New York.

Jones, E. B. (1974). *Instrument Technology*, Vol. 1, Newnes-Butterworths, London.

Kennedy, G. (1977). *Electronic Communications Systems*, McGraw-Hill-Kogakusha, Tokyo.

Lucky, A. N., Salz, J., and Weldon, E. J. (1968). *Principles of Data Transmission*, McGraw-Hill, New York.

Maeison, F. C. (1974). *Electrical Instruments in Hazardous Locations*, Instrument Society of America, Pittsburgh.

Martin, J. (1967). *Design of Real-Time Computer Systems*, Prentice-Hall, Englewood Cliffs.

Martin, J. (1972a). *Systems Analysis for Data Transmission*, Prentice-Hall, Englewood Cliffs.

Martin, J. (1972b). *Introduction to Teleprocessing*, Prentice-Hall, Englewood Cliffs.

McKinnon, R. K. (1970). 'Development of data transmission services in Australia', *Telecom. J. Aust.*, **20**, 203–16.

McKinnon, R. K., Endersbee, B. A., and Boucher, J. M. (1977). 'Data communications', *Telecom. J. Aust.*, **27**, 14–23, 120–7.

Midwinter, J. E. (1979). *Optical Fibres for Transmission*, Wiley, Chichester.

Miller, G. M. (1978). *Modern Electronic Communications*, Prentice-Hall, Englewood Cliffs.

Miller, S. E. and Chynoweth, A. G. (1979). *Optical Fiber Telecommunications*, Academic Press, New York.

MTL (1976). *Shunt-Diode Safety Barriers*, PS300-6, Measurement Technology Ltd, Luton.

Picquenard, A. (1974). *Radio Wave Propagation: Philips Technical Library Series*, Wiley, New York.

Ramo, S., Whinnery, J. R., and Duzer, T. V. (1965). *Fields and Waves in Communications Electronics*, Wiley, New York.

Ross, J. F. (1980). *Handbook for Radio Engineering Managers*, Butterworths, London.

Saad, T. S. (1971). *Microwave Engineers' Handbook*, Artech House, Dedham (2 vols).

Schwartz, M. (1977). *Computer Communication Network Design and Analysis*, Prentice-Hall, Englewood Cliffs.

Smith, B. M. (1971). 'A brief introduction to digital transmission techniques', *Telecom. J. Aust.*, **21**, 111–19.

Wolf, H. F. (1979). *Handbook of Fibre Optics*, Garland STPM Press, New York.

Handbook of Measurement Science, Vol. 1
Edited by P. H. Sydenham
© 1982 John Wiley & Sons Ltd.

Chapter

14 P. ATKINSON

Closed-loop Systems

Editorial introduction

Measurements are clearly important to the formation of closed-loop control systems. This chapter, however, presents closed-loop system concepts from the opposite viewpoint because many instrument systems use closed-loop systems to improve performance in such ways as are seen in the feedback-connected operational amplifier and in the improvement of the time response of actuators. This review presents the elements of the more traditional linear control methods as these are very applicable (but too infrequently applied) to instrument systems design. It also includes discussion of the closed-loop digital systems now becoming popular due to the wholesale acceptance of the microprocessor as the prime data processor in even the simplest of instruments. An excellent complementary review, listing many instrument examples, is provided in Jones (1979).

14.1 INTRODUCTION

Modern closed-loop control systems can be traced back to James Watt who invented the first rotative steam engine (patented in 1781). Initially the speed of these engines was controlled manually using a throttle valve on the steam inlet. In 1788, Boulton, Watt's co-principal in a company manufacturing the new steam engines, had visited the Albion Mill and was able to describe to Watt a form of centrifugal governor being used to regulate the grinding speed of the mill-stones. Watt soon adapted the governor to measure and control the speed of his steam engines automatically (Atkinson, 1968). Since those days, closed-loop control and measurement have been inexorably linked; indeed to control a physical quantity accurately and rapidly in the presence of changing demand, changing internal parameters, and changing load, measurement of the control quantity coupled with feedback control is essential. Such measurement can, of course, only be accomplished by means of instruments. A further development down this path is that feedback itself is being used more and more in *measuring instruments themselves* to improve the accuracy of measurement. This notion is discussed further in Section 14.11.

In the nineteenth century, vast numbers of regulated steam engines were in operation in factories throughout Britain. These closed-loop systems were by no means perfect in that the precision of speed control was often rather poor. Engineers, with no real insight into system dynamics, attempted to improve performance by making smaller, lighter, and better lubricated governors. Much to their consternation this normally led to an unforeseen associated difficulty; the steam-engine speed then tended to oscillate or *hunt* continuously about the demanded value. This particular form of *instability*, as it is now called has plagued designers of feedback control systems of all kinds ever since. The tremendous practical importance of this difficulty stimulated no less a person than James Clerk Maxwell (1868) to investigate the problem in great depth. He brought mathematical insight to bear by relating the existence of instability to the presence of positive real parts in the complex roots of the characteristic equation for the system (see Section 14.4).

The 1914–18 war caused military engineers to realize that to win wars it is necessary to position heavy masses (e.g. ships and guns) precisely and quickly. Classic work was performed by N. Minorsky (1922) in the USA on automatic ship steering and H. L. Hazen (1934) defined a *servomechanism* for the *first time*. The concepts of automatic control, as they developed, are covered in Bennett (1979).

The problem of mechanical position control provides an ideal example to illustrate the need for feedback in control. Let us suppose we have the problem of controlling the angular position of a heavy rotatable mass; the resource of mechanical or electrical power assistance rather than total reliance on muscle power allows us the obvious advantages of rapid control. To simplify the problem, let us suppose that we have an ideal frictionless electric motor at our disposal and to achieve maximum acceleration of the rotatable load we couple the motor to the load through an ideal frictionless stepdown gearbox. It will be assumed that the motor produces a torque at the load which is directly proportional to its supply voltage. In order to control the supply voltage to the motor we connect it to an ideal power amplifier which receives at its input a control voltage v_i which is directly proportional to an angular positional signal θ_i; this signal is applied manually through a light handwheel connected to a position-to-voltage transducer (such as a rotary potentiometer). The notional arrangement is illustrated in Figure 14.1; this system will produce rapid acceleration of the rotatable mass in response to small and effortless motion applied manually to the handwheel. When the handwheel is at a nominal zero position the mass will cease accelerating; a change in the handwheel position in one direction will produce acceleration in one direction and a change in position in the opposite direction will produce acceleration in the other direction. A simple mathematical analysis is as follows.

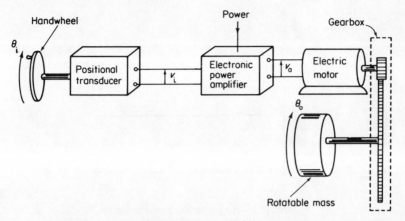

Figure 14.1 Position control system without feedback

Let the *effective* moment of inertia of the moving parts, referred to the position of the rotatable mass, be J. Also let

$$v_i = k_t \theta_i$$

$$v_a = k_a v_i$$

and the effective torque acting on the mass T be given by

$$T = k_m v_a$$

where k_t, k_a, and k_m are constants. Thus

$$T = k_m k_a k_t \theta_i = K\theta_i$$

where K is a composite system constant. Now in the calculus notation, according to Newton's second law of motion

$$J\frac{d^2\theta_o}{dt^2} = T = K\theta_i$$

The response to a small change in θ_i from zero to a fixed positive angle is shown in Figure 14.2.

In order to change the position of the mass from one fixed position to another in the shortest time we must first accelerate the mass and then retard it in such a way that at the instant the required position has been reached, both the velocity and acceleration are simultaneously reduced to zero. The manual control problem is extraordinarily difficult; it is analogous to steering a car with a trailer *backwards* along a desired path. Worse still, in many circumstances (e.g. control of the position of a gun turret) there will be load disturbance caused, for example, by wind gusts which will cause the mass to deviate from the desired position in a random manner.

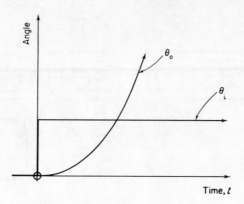

Figure 14.2 Response of position control
system without feedback

It would appear that the position control problem might be solved similarly
to the way in which Watt solved his speed control problem, that is, by means of
feedback of a measured value of the controlled variable. The controlled variable
θ_o can be measured by means of a transducer identical to that monitoring θ_i
so that a signal v_o (given by $v_o = k_t v_o$) is available for comparison with v_i.
The notion is now that if θ_i is made equal to the *required* position of the rotatable
mass, the amplifier can be fed by a difference signal $(v_i - v_o)$ which is given by

$$v_i - v_o = k_t \theta_i - k_t \theta_o = k_t \varepsilon$$

where ε is defined as the *positional error* between the required position θ_i
(termed the *command* or *input*) and the actual or output position θ_o. This error
signal is now amplified as before and applied to the motor. Thus a driving
torque will always be present as long as θ_o is different from θ_i. When they are
the same there will be no driving torque and the mass will hopefully stop
moving at the point where we want it to be. The notional practical arrangement
is shown in Figure 14.3. An analysis of this system is very revealing.

The effective torque T acting on the mass is no longer $K\theta_i$ but is now equal
to $K\varepsilon$. Thus again applying Newton's second law, we have

$$J\frac{d^2\theta_o}{dt^2} = K\varepsilon$$

But $\varepsilon = \theta_i - \theta_o$, hence

$$J\frac{d^2\theta_o}{dt^2} = K(\theta_i - \theta_o)$$

and so

$$J\frac{d^2\theta_o}{dt^2} + K\theta_o = K\theta_i$$

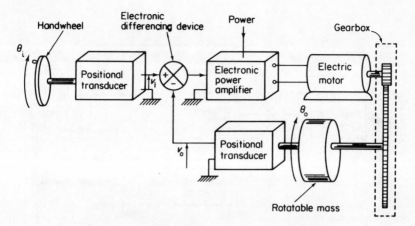

Figure 14.3   Position control system with feedback

This will immediately be recognized as the equation for simple harmonic motion. Hence a sudden displacement in the handwheel position will lead to continuous oscillations as illustrated in Figure 14.4. Mathematicians define this system as *critically stable* because its oscillations will neither increase in magnitude nor decay; control engineers tend to regard such a system as *unstable*. This system is exhibiting exactly the same type of behaviour as Watt's regulated steam engines and is entirely unsatisfactory for practical position control systems when compared with what can be achieved.

In non-ideal practical systems there is always some friction present which always acts against motion and this will cause the oscillation to decay eventually. There are, however, various forms of friction including *stiction* (the torque necessary to *just* cause motion), *Coulomb friction* (a constant torque independent of velocity) and *viscous friction* (a torque which is directly proportional to velocity). Stiction and Coulomb friction both cause undesirable side-effects (stiction producing *stick–slip* motion when the system is commanded

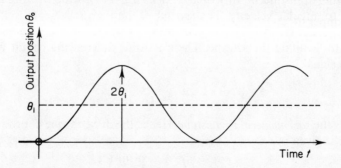

Figure 14.4   Response of position control system with feedback

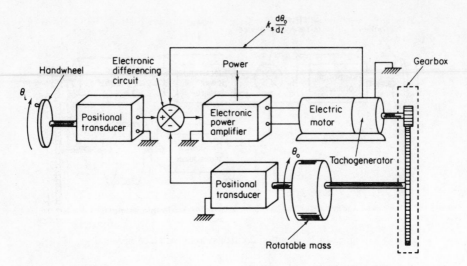

Figure 14.5   Position control system stabilized with velocity feedback

to follow a constant velocity input and Coulomb friction producing a constant offset in response to a constant input). It is thus essential to minimize stiction and Coulomb friction by the correct mechanical design and to ensure that the viscous component dominates or that a similar dominating effect is reproduced by other means. The effect of viscous friction on the differential equation of motion of the system is to add an extra term proportional to output angular velocity thus:

$$J\frac{d^2\theta_o}{dt^2} + F\frac{d\theta_o}{dt} + K\theta_o = K\theta_i$$

where F is the viscous frictional torque per unit angular velocity. Practically one may achieve the required viscous damping by either attaching a physical viscous damper to the rotating mass or by feeding back an extra signal (derived from another transducer—this time a *tachogenerator*) which is directly proportional to angular velocity. The second of these arrangements is shown in Figure 14.5; electric motor-tachogenerators in single units are commercially available to facilitate the scheme. The electronic differencing circuit will now produce a signal

$$k_t\varepsilon - k_s\frac{d\theta_o}{dt}$$

where k_s is the *tachogenerator constant*. Hence the drive torque T produced by the motor is given by

$$T = k_m k_a\left(k_t\varepsilon - k_s\frac{d\theta_o}{dt}\right)$$

Again assuming that all forms of friction are negligible and using Newton's second law:

$$J\frac{\mathrm{d}^2\theta_\mathrm{o}}{\mathrm{d}t^2} = k_\mathrm{m}k_\mathrm{a}\left(k_\mathrm{t}\varepsilon - k_\mathrm{s}\frac{\mathrm{d}\theta_\mathrm{o}}{\mathrm{d}t}\right)$$

$$= K\varepsilon - k_\mathrm{m}k_\mathrm{a}k_\mathrm{s}\frac{\mathrm{d}\theta_\mathrm{o}}{\mathrm{d}t}$$

Hence

$$J\frac{\mathrm{d}^2\theta_\mathrm{o}}{\mathrm{d}t^2} + k_\mathrm{m}k_\mathrm{a}k_\mathrm{s}\frac{\mathrm{d}\theta_\mathrm{o}}{\mathrm{d}t} + K\theta_\mathrm{o} = K\theta_\mathrm{i}$$

The term $k_\mathrm{m}k_\mathrm{a}k_\mathrm{s}$ can be regarded as the *equivalent viscous frictional constant F*.

The analysis of the response of this system to various inputs is important for two reasons; firstly the arrangement forms the basis of many recording instruments used in practice (e.g. X–Y recorders and X–t recorders) and secondly it represents the embodiment of the second-order system which is used as an important reference in the design of higher-order systems (see Section 14.2.5). The arrangement constitutes a basic position control *servomechanism*.

14.2 DETERMINATION OF THE DYNAMIC BEHAVIOUR OF A CLOSED-LOOP SYSTEM USING THE DIFFERENTIAL EQUATION

14.2.1 The Laplace Transform

It has been seen how the differential equation of a simple feedback control system may be derived by the application of physical laws (Newton's second law of motion in the example given). However, in order to determine the behaviour of the system in response to certain inputs we need to have available a method of analysis; the method of Laplace transforms (Gardner and Barnes, 1942; Goldman, 1966 and Chapter 4 of this volume) provides a suitable basis for this analysis.

The Laplace transform of a signal $\theta_\mathrm{i}(t)$ is formally defined as $\Theta_\mathrm{i}(s)$ in which s (p is also often used) is the complex variable $\sigma + \mathrm{j}\omega$ and

$$\Theta_\mathrm{i}(s) = \int_{0-}^{\infty} \theta_\mathrm{i}(t)\mathrm{e}^{-st}\,\mathrm{d}t$$

and

$$\theta_\mathrm{i}(t) = \frac{1}{2\pi\mathrm{j}} \int_{c-\mathrm{j}\omega}^{c+\mathrm{j}\omega} \Theta_\mathrm{i}(s)\mathrm{e}^{st}\,\mathrm{d}s$$

Here c is chosen to be larger than the real parts of all the singularities of $\Theta_\mathrm{i}(s)$.

Fortunately there is never any need to evaluate these integrals in practice because they have been tabulated in transform pairs to aid the rapid solution of differential equations.

In the absence of initial conditions we may transform derivatives by the rule

$$\mathscr{L} \frac{\mathrm{d}^n f(t)}{\mathrm{d}t^n} = s^n F(s)$$

and integrals by the rule

$$\mathscr{L} \int_0^t f(t)\,\mathrm{d}t = \frac{F(s)}{s}$$

in which $\mathscr{L}$ represents the operation of taking Laplace transforms and $F(s)$ is the Laplace transform of $f(t)$. In situations where the initial conditions are non-zero, then

$$\mathscr{L} \frac{\mathrm{d}^n}{\mathrm{d}t^n}[f(t)] = s^n F(s) - s^{n-1}f(0-) - s^{n-2}f^1(0-)\cdots - f^{n-1}(0-)$$

and

$$\mathscr{L} \int_0^t f(t)\,\mathrm{d}t = \frac{1}{s}F(s) + \frac{f^{-1}(0-)}{s}$$

where $f(0-), f^1(0-), \ldots, f^{n-1}(0-)$ are the values of the function and its $n-1$ derivatives and $f^{-1}(0-)$ is the value of the time integral of $f(t)$ *just prior* to the application of the signal at $t = 0$. It should be noted that the limit $t = 0-$ is used here, whereas in rigorous mathematical texts, in which the derivatives of discontinuous functions are not legitimate functions, the lower limit is quoted as $t = 0+$. However in practical analysis in which the unit impulse function $(\delta(t))$ is used, a more consistent methodology results by using a lower limit $t = 0-$.

A short table of transform pairs is included here for reference purposes in the examples that follow (Table 14.1). Other tables are given in Chapters 4 and 6. Notice that when control engineers use the *single-sided* Laplace transform, all the driving signals are considered to operate after $t = 0$; they are defined as zero prior to this instant and this may be conveniently represented as multiplying all the time functions by the unit step $u(t)$.

14.2.2 Analysis of the Simple Position Control System with Viscous Damping or Velocity Feedback

Returning to the simple position system described in Section 14.1, it is necessary to define the form of the input signal and the important initial conditions.

Table 14.1 Some functions and their Laplace transforms

$f(t)$	$F(s) = \mathcal{L}[f(t)]$
Unit impulse function $\delta(t)$	1
Unit step function $u(t)$	$\dfrac{1}{s}$
Ramp function $u(t)t$	$\dfrac{1}{s^2}$
Exponential delay $u(t)e^{-\alpha t}$	$\dfrac{1}{s + \alpha}$
Exponential rise $u(t)(1 - e^{-\alpha t})$	$\dfrac{\alpha}{s(s + \alpha)}$
$u(t)t\ e^{-\alpha t}$	$\dfrac{1}{(s + \alpha)^2}$
$u(t)\sin(\omega_{rt} t)$	$\dfrac{\omega_{rt}}{s^2 + \omega_{rt}^2}$
$u(t)e^{-\alpha t} \sin(\omega_{rt} t)$	$\dfrac{\omega_{rt}}{(s + \alpha)^2 + \omega_{rt}^2}$
$u(t)e^{-\alpha t} \cos(\omega_{rt} t)$	$\dfrac{s + \alpha}{(s + \alpha) + \omega_{rt}^2}$
$f(t - \tau)$	$e^{-s\tau}F(s)$

Assuming zero initial conditions we may transform both sides of the differential equation thus:

$$Js^2\Theta_o(s) + Fs\Theta_o(s) + K\Theta_o(s) = K\Theta_i(s)$$

Hence

$$\Theta_o(s) = \frac{K\Theta_i(s)}{Js^2 + Fs + K}$$

$$= \frac{(K/J)\Theta_i(s)}{s^2 + (F/J)s + K/J}$$

Consider the response to various input signals.

The unit impulse response

If $\theta_i(t) = \delta(t)$ then $\Theta_i(s) = 1$, and so

$$\Theta_o(s) = \frac{K/J}{s^2 + (F/J)s + K/J}$$

Let us take the case for which $F < 2\sqrt{(JK)}$, so that we must complete the square in the denominator; then

$$\Theta_o(s) = \frac{K/J}{(s + \alpha)^2 + \omega_{rt}^2}$$

where

$$\alpha = \frac{F}{2J} \quad \text{and} \quad \omega_{rt} = \left(\frac{K}{J} - \frac{F^2}{4J^2}\right)^{1/2}$$

This does not quite agree with any of the tabulated Laplace transforms, but with a slight modification it yields

$$\Theta_o(s) = \frac{K/J}{\omega_{rt}} \frac{\omega_{rt}}{(s + \alpha)^2 + \omega_{rt}^2}$$

which may be inverse transformed to yield

$$\theta_o(t) = u(t) \frac{K/J}{\omega_{rt}} e^{-\alpha t} \sin(\omega_{rt} t)$$

This is the *unit impulse response* of the system.

The unit step response

If $\theta_i(t) = u(t)$ then $\Theta_i(s) = 1/s$, thus

$$\Theta_o(s) = \frac{K/J}{s[s^2 + (F/J)s + K/J]}$$

This must be broken into partial fractions thus:

$$\Theta_o(s) = \frac{1}{s} - \frac{\alpha}{\omega_{rt}} \frac{\omega_{rt}}{(s + \alpha)^2 + \omega_{rt}^2} - \frac{s + \alpha}{(s + \alpha)^2 + \omega_{rt}^2}$$

Each term has a recognizable inverse Laplace transform

$$\theta_o(t) = u(t) - \frac{\alpha}{\omega_{rt}} u(t) e^{-\alpha t} \sin(\omega_{rt} t) - u(t) e^{-\alpha t} \cos(\omega_{rt} t)$$

$$= u(t)\{1 - e^{-\alpha t}\sqrt{[1 + (\alpha/\omega_{rt})^2]}\sin[\omega_{rt} t + \tan^{-1}(\omega_{rt}/\alpha)]\}$$

The unit ramp response

If $\theta_i(t) = u(t)$, then $\Theta_i(s) = 1/s^2$, and so

$$\Theta_o(s) = \frac{K/J}{s^2[s^2 + (F/J)s + K/J]}$$

This must be broken into partial fractions and yields a response

$$\theta_o(t) = u(t)\left(t - \frac{F}{K} + \frac{Ke^{-\alpha t}}{F\omega_{rt}}\sin(\omega_{rt}t + \phi)\right)$$

where

$$\phi = \tan^{-1}\left(\frac{2\alpha\omega_{rt}}{2\alpha^2 - K/J}\right)$$

The various responses are illustrated in Figure 14.6. The quantities M_{pt}, T_p, ε_{ss}, and ε_p form some of the basic means of performance specification.

14.2.3 The Concept of Damping Ratio and Undamped Natural Angular Frequency

The value of the damping term F relative to the terms J and K governs the dynamic behaviour of the system. There are four possibilities:

(a) $F = 0$ in which case the system will oscillate continuously with sinusoidal oscillations of angular frequency $\sqrt{(K/J)}$. This quantity is termed the natural *undamped* angular frequency. This response is termed *critically stable*.

(b) $F < 2\sqrt{(JK)}$ in which case the response contains an exponentially damped sinusoidal mode and will exhibit overshoot in response to a step input. This is called the *underdamped* response.

(c) $F = 2\sqrt{(JK)}$ for which the response is *critically damped*, that is, it does not quite overshoot in response to a step input.

(d) $F > 2\sqrt{(JK)}$ for which the response is a double exponential rise in response to a step input. This is called the *overdamped* response.

In Section 14.2.2 we concentrated on condition (b) because it is in practice the most important case because it allows the system to settle within a given tolerance band (Atkinson, 1968) around the desired value *faster* than any other.

It is convenient to non-dimensionalize the effect of the damping term F by relating it to the value F_c required to achieve critical damping where $F_c = 2\sqrt{(JK)}$. We define the damping ratio ζ as

$$\zeta = \frac{F}{F_c} = \frac{F}{2\sqrt{(JK)}}$$

so that for $\zeta < 1$ we have an underdamped system, for $\zeta = 1$ a critically damped system, and for $\zeta > 1$ an overdamped system.

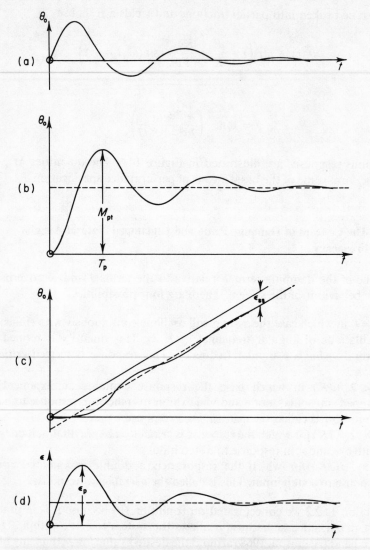

Figure 14.6   Time domain responses of a stabilized position control
system: (a) impulse, (b) step, (c) ramp output, (d) ramp error

The differential equation may be rewritten in terms of ω_n and ζ thus

$$\frac{d^2\theta_o}{dt^2} + 2\zeta\omega_n\frac{d\theta_o}{dt} + \omega_n^2\theta_o = \omega_n^2\theta_i$$

The response to a unit step input for various values of ζ is illustrated in Figure 14.7.

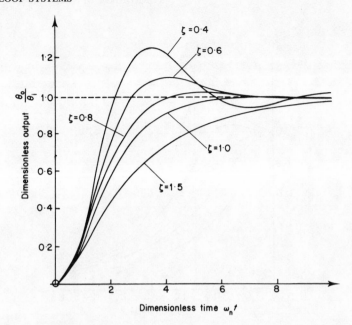

Figure 14.7   Dimensionless unit step response

14.2.4   Frequency Response

The *steady-state* behaviour of the system in response to a sinusoidal input is of considerable practical importance. When we refer to the *frequency response* of a system we mean the variation of the *phase* and *magnitude* of the *steady-state* output of the system as the frequency of the input sinusoid is varied over the range of interest.

The Laplace transform of the output of the system is related to the Laplace transform of the input by the equation

$$s^2\Theta_o(s) + 2\zeta\omega_n s\Theta_o(s) + \omega_n^2\Theta_o(s) = \omega_n^2\Theta_i(s)$$

For steady-state sinusoids we may substitute $j\omega = s$, where ω is the angular frequency of the input, and produce the following operational relationship between $\Theta_i(j\omega)$ and $\Theta_o(j\omega)$:

$$\frac{\Theta_o(j\omega)}{\Theta_i(j\omega)} = \frac{\omega_n^2}{(j\omega)^2 + 2\zeta\omega_n(j\omega) + \omega_n^2}$$

$$= \frac{\omega_n^2}{\omega_n^2 - \omega^2 + j2\zeta\omega_n\omega}$$

$$= \frac{1}{1 - (\omega/\omega_n)^2 + j2\zeta\omega/\omega_n}$$

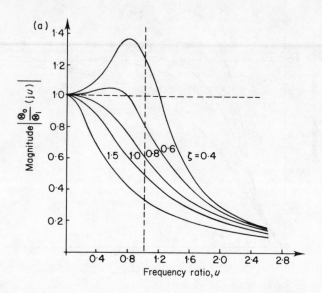

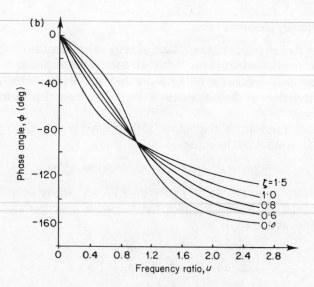

Figure 14.8   Frequency response characteristic: (a) magnitude; (b) phase

It is rather simpler to work in terms of a non-dimensional frequency ratio $u = \omega/\omega_n$ for which

$$\frac{\Theta_o(ju)}{\Theta_i(ju)} = \frac{1}{(1 - u^2) + j2\zeta u}$$

From this expression we can determine the modulus M, its peak value M_{pf} (if any), and the phase ϕ which are given by

$$M = \frac{1}{\sqrt{[(1 - u)^2 + (2\zeta u)^2]}}$$

$$M_{pf} = \frac{1}{2\zeta\sqrt{(1 - \zeta^2)}}$$

and

$$\phi = -\tan^{-1}\left(\frac{2\zeta u}{1 - u^2}\right)$$

The magnitude and phase characteristics are illustrated in Figure 14.8. The angular frequency at which the frequency response has its peak value is designated by the symbol ω_{rf}. It may be shown that

$$\omega_{rf} = \omega_n\sqrt{(1 - 2\zeta^2)}$$

14.2.5 Second-order Correlations

The time domain (step response) and frequency response of the second-order system are connected through *correlating equations*; these may be combined as graphs which are valuable in the approximate design of higher-order systems (Atkinson, 1968). These graphs are given in Figure 14.9.

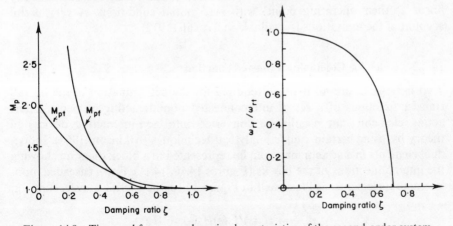

Figure 14.9 Time and frequency domain characteristics of the second-order system

14.3 TRANSFER FUNCTIONS AND THEIR USES

14.3.1 The Transfer Function Concept

In analysing the simple second-order position control system we used the Laplace transform to determine the response of the system to various commands. In Section 14.2.3, the relationship between the Laplace transform of the input to the servo $\Theta_i(s)$ and the Laplace transform of the output is expressed in the form

$$s^2\Theta_o(s) + 2\zeta\omega_n s\Theta_s(s) + \omega_n^2\Theta_s(s) = \omega_n^2\Theta_i(s)$$

We may recast this relationship in an alternative manner in which the transforms are expressed as a ratio called the *transfer function*:

$$\frac{\Theta_o(s)}{\Theta_i(s)} = \frac{\omega_n^2}{s^2 + 2\zeta\omega_n s + \omega_n^2}$$

This relates the output of the system to its input in a manner which is more convenient when handling complex systems. Rather than determine a complicated differential equation or set of differential equations, we may bypass this problem by combining the transfer functions of the basic elements by means of some very simple rules. The overall transfer function may then be used in various ways in order to analyse or design the system. A range of transfer functions for commonly-encountered devices has been derived in various textbooks (Thaler and Brown, 1960; Atkinson, 1968; Towill, 1970).

It should be understood, however, that although transfer function techniques are very useful in the analysis and design of moderately complicated systems, the methodology does have strict limitations; for example, it can only be applied to linear, time-invariant systems (or those which are substantially linear at their operating point) with zero initial conditions. A very useful account of the limitations is available in Truxall (1972).

14.3.2 Rules for Combining Transfer Functions

Two basic rules are all that are required for the determination of the overall transfer functions of a set of interconnected noninteracting elements. Interacting elements can usually be converted into non-interacting devices in theory by using certain simple analytical techniques (Atkinson, 1968). Individual elements in a system may then be represented in a block diagram showing the interconnections of the blocks (Figures 14.10, 14.11). For n cascaded, noninteracting elements, as illustrated in Figure 14.10, the overall transfer function

$$\frac{\Theta_o(s)}{\Theta_i(s)} = H_1(s)H_2(s)H_3(s)\cdots H_n(s)$$

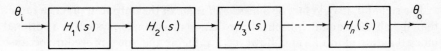

Figure 14.10   Elements in cascade

For the feedback combination illustrated in Figure 14.11

$$\frac{\Theta_o(s)}{\Theta_i(s)} = \frac{H_1(s)}{1 + H_1(s)H_2(s)}$$

Alternatively the inverse transfer function

$$\frac{\Theta_i(s)}{\Theta_o(s)} = H_1^{-1}(s) + H_2(s)$$

These transfer functions can be used in assessing the time domain response to any deterministic signal for zero initial conditions. The steady-state response to sinusoids may always be assessed by the substitution $s = j\omega$ where ω is the angular frequency.

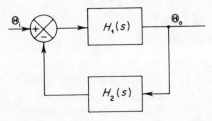

Figure 14.11   Elements in parallel

14.4   PERFORMANCE ASSESSMENT USING THE TRANSFER FUNCTION

When analysing linear continuous control systems, it is usually possible to represent any single input/single output system by a simplified block diagram as shown in Figure 14.12.

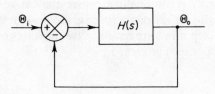

Figure 14.12   Notional single-input/ single-output system

The signal θ_o may in fact be a *measured value* of the output in practice rather than the physical quantity itself; the reduction of a complex system to a notional block diagram of this form is, however, a valuable analytical procedure because the concept of *unity feedback* often reduces unnecessary complication in the design process. In practice the measured value of the output is fairly tightly related to the actual value so that the procedure is normally fully justified. The *open-loop transfer function* $H(s)$ can normally be represented by the relationship

$$H(s) = \frac{K \exp(-sT_{L}) \prod_{i=1}^{p} (T_{eei} s^2 + T_{ei} s + 1)}{s^m \prod_{i=1}^{q} (T_{aai} s^2 + T_{ai} s + 1)}$$

where

K = the *error constant*

T_{L} = pure time delay

m = *type number* of the system

T_{eei} and T_{ei} are constants defining the *leads*

T_{aai} and T_{ai} are constants defining the *lags*

p is the number of leads

q is the number of lags

It should be noted that some or all of the leads may be *quadratic* as indicated or *simple* of the form $(1 + sT_{ei})$; equally well some or all of the lags may be *simple exponential* of the form $(1 + sT_{ai})$ or *quadratic*. Quadratic factors with real roots may be usefully factorized to simple factors.

It should also be noted that although pure time delay is only rarely encountered in servomechanisms it is often present in process control systems where the transport of material at a finite velocity is a necessary function of the system.

In the analysis of control systems we are concerned with many aspects of the behaviour: absolute stability, relative stability, response to commands, response to disturbances, and the sensitivity of these to parameter changes. It should be noted here that an *absolutely unstable system* is one whose output is unbounded for a bounded input; *a relatively stable system* is an absolutely stable system in which the transients in response to a command are not excessive and decay rapidly. Relative stability is nominally a rather qualitative concept but it can be quantified very readily in various ways (e.g. the peak overshoot and settling time in response to a step input are quantitative measures of relative stability).

The closed-loop response is governed by

$$\frac{\Theta_o(s)}{\Theta_i(s)} = \frac{H(s)}{1 + H(s)}$$

$$= \frac{K \exp(-sT_{L}) \prod_{i=1}^{p} (T_{eei} s^2 + T_{ei} s + 1)}{s^m \prod_{i=1}^{q} (T_{aai} s^2 + T_{ai} s + 1) + K \exp(-sT_{L}) \prod_{i=1}^{p} (T_{eei} s^2 + T_{ei} s + 1)}$$

When the denominator of the closed-loop transfer function is equated to zero we obtain the *characteristic equation* of the system. The roots of this equation in s govern the *modes* of the transient response of the system. Each negative real root gives rise to an exponential mode of the form $A \exp(-\alpha_1 t)$ and each complex conjugate pair of roots with a negative real part gives rise to an exponentially decaying sinusoidal mode of the form $B \exp(-\alpha_2 t)\sin(\omega_{rt} t + \phi)$. The existence of positive real roots or positive real parts of complex roots implies an unbounded output for a bounded input and consequent absolute instability. For systems containing no pure time delay, absolute instability may be predicted directly from the characteristic equation using the Routh–Hurwitz stability criterion (Atkinson, 1968; Towill, 1970). The variation of the roots with variation in the error constant K plotted in the s-plane is called the *root locus diagram*. The form of the root locus diagram for a system having an open-loop transfer function

$$H(s) = \frac{K}{s(1 + sT_1)(1 + sT_2)}$$

is shown in Figure 14.13.

The root locus concept and its application in the design process have been described by various authors including Truxall (1955), Thaler and Brown (1960), and Towill (1970). In root locus analysis the transfer function is expressed in terms of *poles* of the form $1/[s + (1/T_{ai})]$ and *zeros* of the form $[s + (1/T_{ei})]$ either of which may also occur in pairs with complex conjugate roots. Design of well conditioned closed-loop systems revolves around the insertion of extra open-loop zeros and/or poles coupled with adjustment of the error constant in such a way so as to force the closed-loop poles (governing the closed-loop modes) into positions where a particular one (or ones) *dominate* the response completely. Usually, this approach is only partially successful because subdominant modes tend to modify the behaviour to an extent which is not immediately predictable. However, the method is an excellent way of handling up

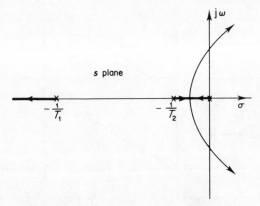

Figure 14.13   Typical root locus diagram

to sixth-order systems without pure time delay. Higher-order systems and those having pure time delay are probably best handled by frequency response techniques (as explained in Section 14.5).

The steady-state performance of a stable closed-loop system is determined by the type number m and the value of the error constant K. Figure 14.14 gives a summary showing how type 0, 1, and 2 systems respond to various typical command inputs.

In a particular application the lowest suitable type number should be used because the higher the type number, the more difficult it is to design the desired well conditioned system. Type 0 systems having a sufficiently large error constant will be quite adequate for many regulators where the object is to maintain the output at a constant level and tracking is unnecessary. Type 1 systems are perfectly satisfactory where dynamic accuracy is not vital (e.g. instrument servomechanisms, fin servomechanisms on guided weapons). However, in systems where dynamic accuracy under tracking conditions is vital (e.g. machine-tool control systems) it may be necessary to use type 2 systems (or at least type 1 systems with very large error constants).

The analytical determination of the response of a system (or even its root locus diagram) is normally impossibly complex for computation by *hand*. Analog or digital or hybrid computer simulation is invariably necessary in all but the simplest systems (Atkinson, 1972).

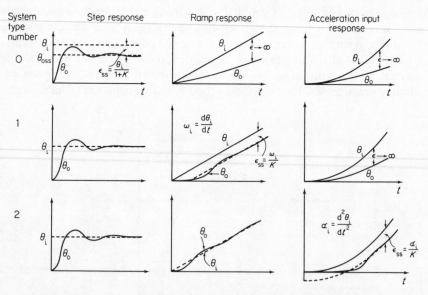

Figure 14.14   Responses of type 0, 1, and 2 Systems with various inputs

14.5 THE FREQUENCY RESPONSE

Whereas the calculation of the time domain response of control systems is extremely involved, even for quite simple systems, the frequency response is relatively easy to determine. Moreover the closed-loop stability can be determined from the open-loop frequency response by means of the Nyquist stability criterion (Atkinson, 1968; Towill, 1970; Thaler and Brown, 1960; Truxall, 1972). Also the closed-loop frequency response of a stable system may be rapidly assessed from the open-loop frequency response.

The open-loop frequency response must be determined from the expression

$$H(j\omega) = \frac{K \exp(-j\omega T_L) \prod_{i=0}^{p} [T_{eei}(j\omega)^2 + T_{ei}(j\omega) + 1]}{(j\omega)^m \prod_{i=0}^{q} [T_{aai}(j\omega)^2 + T_{ai}(j\omega) + 1]}$$

It is either necessary to calculate the magnitude and phase of this expression or to calculate the real and imaginary parts over the range of frequencies of interest. A simple example will serve to show that although this task is simple in theory, it is very arduous when performed by hand.

Consider the open-loop transfer function

$$H(j\omega) = \frac{4(1 + j\omega) e^{-j0.5\omega}}{j\omega(1 + j0.5\omega)(1 + j0.1\omega)}$$

for which

$$|H(j\omega)| = \frac{4\sqrt{(1 + \omega^2)}}{\omega\sqrt{[1 + (0.5\omega)^2]}\sqrt{[1 + (0.1\omega)^2]}}$$

and

$$\underline{/H(j\omega)} = \tan^{-1}\omega - \frac{\pi}{2} - 0.5\omega - \tan^{-1}(0.5\omega) - \tan^{-1}(0.1\omega)$$

Although the necessary computation required to evaluate these expressions is very time consuming when performed by hand, it can be performed very rapidly using a digital computer.

A *Nyquist diagram* is a Argand diagram on which $H(j\omega)$ is plotted over the required range of frequencies. An inverse Nyquist diagram is an Argand diagram on which $H^{-1}(j\omega)$ is plotted.

A *Bode diagram* shows a graph of $|H(j\omega)|$ expressed in decibels (dB) (i.e. $20 \log_{10}|H(j\omega)|$) versus angular frequency plotted on a logarithmic scale together with the phase $\underline{/H(j\omega)}$ plotted separately. Notionally, Bode diagrams may be plotted more simply than Nyquist diagrams when it is possible to make use of asymptotic approximations to the gain characteristic. Frequently the occurrence of quadratic lags and leads in transfer functions makes this very much less useful than it would appear at first sight. Nyquist, inverse Nyquist, and Bode diagrams representing the transfer function

$$H(j\omega) = \frac{K}{j\omega(1 + j\omega T_1)(1 + j\omega T_2)}$$

are shown in Figure 14.15 for various values of K.

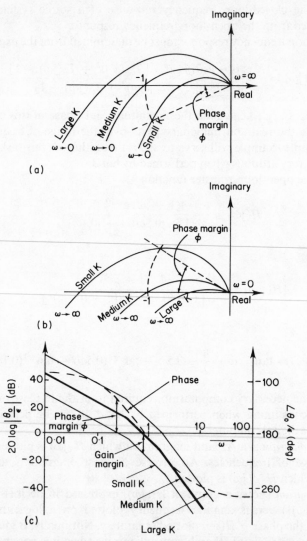

Figure 14.15 Frequency response representations for $\theta_o/\varepsilon = K/[j\omega(1 + j\omega T_1)(1 + j\omega T_2)]$: (a) Nyquist diagram; (b) inverse Nyquist diagram; (c) Bode diagram

The advantage of the frequency domain approach is the simple connection between the open-loop frequency response characteristic and the closed-loop frequency response. A locus of constant closed-loop modulus M on the Nyquist diagram or the inverse diagram is a *circle*. It is thus a simple matter to assess the value of the closed-loop resonance peak M_{pf} *or* to determine the value of K to give a specified value of M_{pf}. A selection of M-circles and inverse M-circles is shown in Figure 14.16.

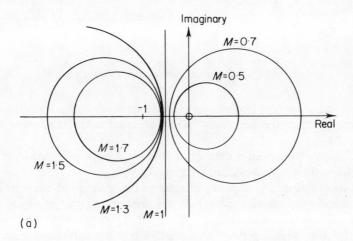

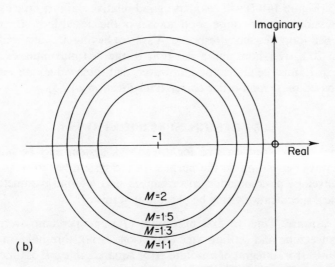

Figure 14.16 M-circles and inverse M-circles: (a) loci of constant closed-loop modules M on the Nyquist diagram; (b) loci of constant closed-loop modulus M on the inverse Nyquist diagram

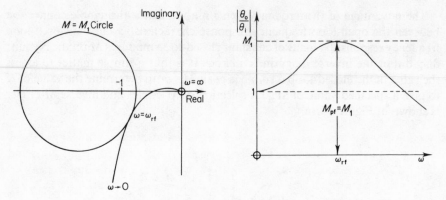

Figure 14.17   Defining the closed-loop resonant peak M_{pf}

The resonant angular frequency ω_{rf} in the frequency domain is the angular frequency at which the frequency locus just touches the maximum M-circle (Figure 14.17).

Translation of open-loop data on a Bode diagram to closed-loop data is best performed on a *Nichols chart* (Atkinson, 1968).

Open-loop frequency response characteristics give good measures of the relative stability of the closed-loop system. Systems having *gain margins* of at least 10 dB and *phase margins* of 40° to 50° (Figure 14.15) have adequate relative stability when operated on closed loop. A system having a peak value of $M = 1.3$ (Figure 14.17) will also have good relative stability on closed loop. It is also possible to make quite good guesses of the closed-loop step response from the open-loop frequency response. A system having M_{pf} of about 1.3 will have about 20% overshoot and a rise time to the first maximum of approximately $3/\omega_{rf}$. It must be understood, however, that these values are very rough estimates based on second-order correlations (Section 14.2.5).

14.6   DESIGN SPECIFICATIONS

The original design specifications for a feedback system may be phrased in numerous ways depending on the purpose of the system has to serve and the expected envelope of commands, disturbances, and internal parameter variations. Typical specifications may be phrased in terms of:

(a)   *Step response*. Time to first maximum (T_p) and per unit overshoot μ; tolerance zone and settling time (Atkinson, 1968); threshold, integral of absolute error, integral of absolute error squared, integral of time x absolute error (Towill, 1970). The system type number and error constant would also normally be specified in conjunction with step response criteria.

(b) *Ramp response.* Ramp peak error, settling time, steady-state error (or velocity error constant).
(c) *Frequency response.* Resonant angular frequency ω_{rf} (or bandwidth); resonance peak M_{pf}; phase margin and gain margin; type number and error constant.
(d) *Disturbance response.* Maximum permitted output to given disturbances over the expected range of disturbance frequencies.

Normally systems must meet a given set of specifications over the expected envelope of system parameter variations *and* in the presence of transducer and signal noise. Systems specifications may be translated from time domain to frequency domain and vice versa using second-order correlations which will give a very approximate guide to higher-order system behaviour.

14.7   MECHANISTIC MODELLING AND MODEL ORDER REDUCTION

When the specification for a system has been phrased in a suitable form the first task in the design procedure is to form a mechanistic model of the plant to be controlled. Mechanistic modelling may or may not be a difficult task, depending on the nature and complexity of the plant or process. Approximate linear models of frequently-encountered electromechanical, electronic, and electrohydraulic systems may be fairly easy to derive (Atkinson, 1968; Towill, 1970) although there will often be considerable doubt about the precise values of parameters. Linear models of some complex thermal and thermochemical processes may be more difficult to derive analytically (Maudsley, 1978). It is, however, a necessary part of the paper design stage of a control system to form a mechanistic model. Once a mechanistic model has been determined it is often necessary in all but the simplest cases to form a *reduced-order model* which is a sufficiently accurate representation of the original system to allow certain design procedures to progress. There have been various methods proposed for producing low-order models (Towill, 1973).

14.8   IDENTIFICATION

Although it is usually necessary to form a mechanistic model of the plant to be controlled so that a proper understanding of the plant dynamics can be obtained, this procedure is often essential because it is usually necessary to design the control system *before* the plant has actually been constructed. However, once the plant has been set up it is advisable to check that the proposed mechanistic model is actually an adequate description of the plant behaviour. This requires a practical method of *identification* and there are various possible methods (Davies, 1970; Graupe, 1972) including:

(a) the injection of a known impulse, step, or ramp signal and the analysis of the response;
(b) the injection of a series of sinusoids of various frequencies and the correlation between input and output;
(c) the injection of a pseudo-random binary input and the cross-correlation between input and output.

Method (b) has some very distinct advantages in that it is possible to identify certain non-linearities as well as the linear elements from one set of data. It is also a relatively simple procedure to fit a transfer function model to a set of frequency response data.

Identification is also necessary to ensure that the *bread-board* system and the development prototype are within the required specification and that all production models are *healthy* at the inspection/maintenance stage.

14.9 COMPENSATION

Having determined a mechanistic model of the plant (verified at the earliest possible stage by identification) in the form of a transfer function, the next problem in the design process is to decide on a control strategy and then determine the parameters of the elements required to implement that strategy. The decision regarding which strategy to use must be based on experience and intuition; indeed it may be necessary to investigate various strategies before a satisfactory one is obtained. The determination of the parameters of the elements required to implement the strategy can then proceed using one of the standard methods of design (preferably with the aid of a computer-aided design (CAD) suite—see Section 14.12) such as the Nyquist diagram, inverse Nyquist Diagram, Bode diagram (Atkinson, 1968), root locus diagram (Truxall, 1955, 1972; Thaler and Brown, 1960; Towill, 1970) or the coefficient plane (Towill, 1970, 1975; Ashworth, 1975).

The most naïve approach is to devise a proportional control scheme as illustrated in Figure 14.18. It is very rare that any but the simplest specification

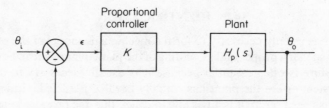

Figure 14.18 Simple proportional control

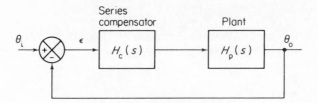

Figure 14.19   Series compensation

can be met by such an approach and the least one might expect is the need to use a series compensator having a transfer function $H_c(s)$ such that the overall closed-loop system is brought within the performance specification (Figure 14.19). The steady-state performance specification in response to both commands and disturbances will normally indicate the type number the system must possess (refer to Figure 14.14 for commands) and, therefore, the number of extra pure integrations which must be introduced into the loop.

Typical series compensators have the following transfer functions and effects:

(a)
$$H_c(s) = K\frac{1 + sT_1}{1 + sT_2}$$

where time constant $T_1 > T_2$ for *phase-lead* compensation which can be used to raise the error constant moderately for increased stiffness, and to raise the speed of response; and where $T_1 < T_2$ for *phase-lag* compensation which can be used to raise the error constant greatly for increased stiffness and to lower the speed of response marginally.

(b)
$$H_c(s) = \frac{K(1 + sT_i)}{s}$$

which is called *proportional plus integral* (P + I) control or two-term control, and can be used to increase the type number and stabilize the resultant system. The resultant system will inevitably have a slower response than a system with the same plant under proportional control.

(c)
$$H_c(s) = K_1 + K_2 s + \frac{K_3}{s}$$

which is called *three-term control* (P + I + D). This can be used to increase the type number, as with proportional plus integral action, but at the same time speed up the response. If the compensator is designed such that $K_1^2 > 4K_2 K_3$, then its transfer function may be rewritten as

$$\frac{K(1 + sT_1)(1 + sT_2)}{s}$$

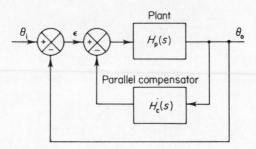

Figure 14.20 Parallel compensation

where T_1 and T_2 are real time constants.

(d)
$$H_c(s) = \frac{K(1 + sT_1)(1 + T_2)}{s(1 + sT_3)}$$

which is *proportional plus integral action with phase-lead*. This compensator can be realized physically using electronic components rather better than the three-term controller. Its action is very similar to that of the three-term controller.

(e)
$$H_c(s) = \frac{K(1 + sT_e + s^2 T_{ee})}{s(1 + sT_b)}$$

which produces control action similar to that of (c) and (d) but allows the numerator term in s to contain complex conjugate roots. This is sometimes of value in cancelling highly underdamped quadratic lags in the plant.

An alternative form of compensation involves the use of parallel elements in auxiliary feedback loops (Figure 14.20). Such an arrangement may have definite practical advantages over the series arrangement (D'Azzo and Houpis, 1966; Atkinson, 1968). Typical parallel compensators $H_c(s)$ have transfer functions and effects:

(a)
$$H_c(s) = k_s s$$

This is called *velocity* or *rate feedback* and can be used to raise the error constant of a system moderately and raise the speed of response.

(b)
$$H_c(s) = \frac{k_s T_s^2}{1 + sT}$$

This is called *transient velocity* or *rate feedback* and can be used to raise the error constant of a system greatly and to raise the speed of response.

(c)
$$H_c(s) = k_a s^2$$

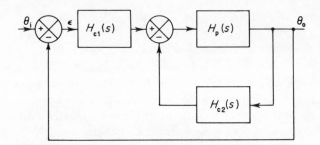

Figure 14.21   Combined series and parallel compensation

This is called *acceleration feedback*. Positive acceleration feedback can be used to raise the speed of response of a relatively stable system.

(d) $$H_c(s) = k_s s + k_a s^2$$

This is *combined velocity and acceleration feedback* and when properly designed can be used to raise the error constant and the speed of response.

It should be understood that while series compensators can usually be implemented very simply in electrically signalled systems by using cheap operational amplifiers in conjunction with resistance–capacitance networks, the implementation of parallel compensation usually involves the use of expensive transducers such as tachogenerators and rate gyroscopes.

Sometimes the use of series compensation alone, or parallel compensation alone, is insufficient when the performance specifications are stringent. It is then necessary to combine the advantages of both; parallel compensation can be used to linearize and speed up the plant and at the same time desensitize it to changes in parameters and to remove some of the effects of disturbances; series compensation is finally added to eliminate offsets in response to commands and disturbances and to bring the overall transient response within specification. The combination can be designed to be more powerful than either method used alone; the arrangement is shown in Figure 14.21.

14.10   SENSITIVITY ANALYSIS

One of the foremost problems in control engineering is that of the reduction of sensitivity of the system response to variations in the plant parameters caused perhaps by environmental changes, ageing, or by the variations attributable to the tolerances allowed to the hardware. Sensitivity is usually defined (Truxall, 1955) in terms of a sensitivity function S_p^T which denotes the sensitivity of the closed-loop transfer function $T(s)$ to variations in the plant transfer function $P(s)$ and which is expressed as

$$S_p^T = \frac{\partial T/T}{\partial P/P}$$

It has been shown by Horowitz (1959) that

$$S_p^T = \frac{1}{1 + L(s)}$$

where $L(s)$ is the open-loop transfer function. Thus if the open-loop frequency response function $L(j\omega)$ plotted on a Nyquist diagram never enters the unit circle, centred on the point $(-1, j0)$, then S_p^T is never greater than unity and the closed-loop system is always less sensitive than the open-loop system. Unfortunately design for this condition may well be impossible. Corresponding sensitivity functions have been derived by Horowitz (1963) for the transient response.

14.11   FEEDBACK INSTRUMENTS

It is interesting to consider the relationship between feedback control systems (Figure 14.22a), in which measuring instruments (in the form of transducers) are used to facilitate accurate control, with measuring instruments (Figure 14.22b) (again the form of transducers which are used in feedback loops to facilitate accurate measurement). The controller must be designed in exactly

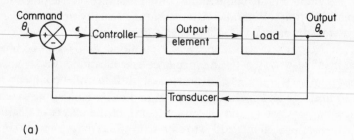

(a)

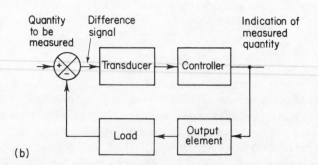

(b)

Figure 14.22   Feedback in control and instrumentation systems:
(a) the general feedback control system; (b) the general feedback
instrument

the same way in both systems and normally both must contain a series compensator designed to ensure adequate relative stability and small (preferably zero) steady-state error under static conditions. In the feedback instrument the object is to null the error signal between the quantity to be measured and its physical manifestation. The signal fed out from the controller is then an indication of the quantity to be measured. Jones (1977, 1979) discusses examples of such arrangements in some detail.

14.12 COMPUTER-AIDED DESIGN

Classical design of feedback systems, as explained in Section 14.9, used to involve the laborious hand computation and plotting of Nyquist diagrams, inverse Nyquist diagrams, Bode diagrams, and root locus diagrams. The necessary checks on the time domain performance are also very computationally involved and time consuming. Only the invention of the analog computer (Johnson, 1956; Atkinson, 1972) gave some relief to this latter problem.

When digital computers became readily available in the early 1960s, off-line computation of frequency response functions removed the need for hand computation. In the mid-1960's cathode ray displays were interfaced to digital computers and work began on the utilization of this new design aid. Iteractive design then became feasible and a new concept in design emerged. This concept places the control systems designer in a loop, the object of which is to design a control system to a given specification as indicated in Figure 14.23. The laboriousness of the design is removed and the designer is freed to practise his art, dealing with the parts of the problem which he does best, that is inventing the overall control strategy which he thinks is most likely to be the best in all engineering senses. Moreover, a rapid assessment of the time-domain performance can normally be made automatically and presented to the designer in the required graphical form.

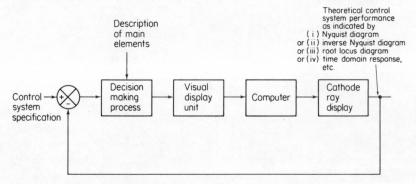

Figure 14.23   Control system design using CAD methods viewed as a feedback process

This automated form of design has not only freed designers from the labour of hand computation and graph plotting, it has encouraged them to try new schemes and to make their designs much more rugged in terms of sensitivity to plant uncertainty.

Computer-aided design is of great general importance in engineering as a whole, but is nowadays the key to rapid and successful feedback control systems design. There have been several conferences on the topic (such as the IEE conference on CAD at Southampton in 1969 and 1972 and the IEE conference on Computer-Aided Control System Design in Cambridge, 1973). A journal entitled *Computer Aided Design* is published quarterly in the UK and Rosenbrock (1974) covers computer-aided control system design. The availability of low-cost microprocessor-based computer systems is an additional attraction ensuring the more general application of this powerful tool.

14.13   CLOSED-LOOP SAMPLED DATA SYSTEMS

14.13.1   Introduction

The closed-loop systems considered in this chapter have all functioned on continuous data; however, many practical control systems function on sampled data (for example, process control systems controlled by digital computers). Usually the concept of an ideal impulse sampler is used in this work; the sampling process is defined as the generation of a sequence of impulses at the sampling instants, with the *area* or *strength* of each impulse equal to the original signal value at that time (Figure 14.24). The notation $x^*(t)$ represents the *ideal impulse sampled* version of the original signal $x(t)$ entering the sampler. It should, of course, be appreciated right from the start that the ideal sampler is a mathematical artifice and that in practical systems no impulses are actually present at any part of the real system. However, the artifice provides us with a valuable analytical tool.

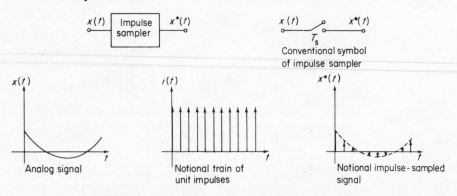

Figure 14.24   Action of a theoretical impulse sampler

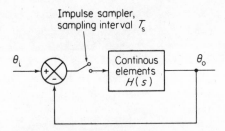

Figure 14.25 Closed-loop sampled data
system

Sometimes we are presented with closed systems which are effectively represented by the block diagram shown in Figure 14.25, in which the continuous elements have a transfer function $H(s)$. The precise analysis of this type of system may be very involved; however, in many practical situations the sampling interval T_s is short compared with the time for a transient oscillation of the whole system. Under these circumstances it is possible to replace the impulse sampler theoretically by a pure gain of value $1/T_s$ and analyse the circuit by normal continuous theory or simulation.

14.13.2 The Use of Hold Circuits

In the arrangement illustrated in Figure 14.25, the impulse sampler is placed in the error channel of the feedback system. This is a fairly normal situation in control systems; in a properly designed system the main elements then act as a low-pass filter which smooths the sampled error so that the output $\theta_o(t)$ follows the input $\theta_i(t)$ over the required profile of inputs.

Even in a system which has been designed correctly and which has a reasonably high sampling rate, the response will not be exactly the same as that of the equivalent linear system. It is often most economical to use low sampling rates, in which case the intersample ripple on the output may be very severe. In order to make the behaviour of the sampled-data system more like that of the continuous system and particularly to reduce intersampling ripple on the output, various forms of filter are used between the sampler and the continuous elements. The simplest type of filter is the *zero-order hold* or *clamp*. The actions of an impulse sampler with a zero-order hold are shown in Figure 14.26. The output is held at the previous sampled value until the next sampling instant. The transfer function of the zero-order hold takes the form:

$$\frac{1 - \exp(-sT_s)}{s}$$

It is shown in Stockdale (1962) that the total response of a sample and hold circuit approximates on average (ignoring harmonics) to a pure time delay of $T_s/2$. This approximation is valid down to quite low sampling rates and allows

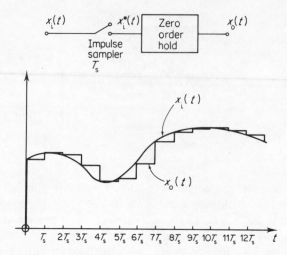

Figure 14.26   The action of the zero-order hold

design to be performed by classical methods in the frequency domain. The approximation can also be used for approximate transient analysis even at quite low sampling rates.

Further improvements in signal smoothing can be achieved by using *predictive hold* circuits which use the past two values to predict (or estimate) the slope of the curve to the next sampling instant.

14.13.3   Frequency Response Analysis of Sampled Data Systems

Although for many purposes frequency domain analysis may be employed for sampled-data systems using the approximate continuous equivalents described in Sections 14.13.1 and 14.13.2, at very low sampling rates, or in cases where more accurate analysis is required, it is necessary to use a more accurate technique. Linvill (1951) developed a formula for computing the frequency response of a sampled-data system by making a vector addition of all the harmonics produced by sampling.

Given a continuous signal of Laplace transform $E(s)$ then the Laplace transform $E^*(s)$ of the sampled signal is given by

$$E^*(s) = \frac{1}{T_s} \sum_{n=-\infty}^{+\infty} E(s + jn\omega_s)$$

where n is an integer and ω_s is the angular sampling frequency. The frequency response is then written in the usual way by substituting $s = j\omega$, that is,

$$E^*(j\omega) = \frac{1}{T_s} \sum_{n=-\infty}^{+\infty} E(j\omega + jn\omega_s)$$

Tou (1959) has shown how frequency response loci (i.e. Nyquist diagrams) can be constructed using this series and how absolute and relative stability can be determined. Further, Atkinson and Allen, (1973) have shown how Linvill's formula can be automated for use in interative computer-aided design.

14.13.4   Time Domain Analysis of Closed-loop Sampled Data Systems

The method of z-transforms (Chapter 10) provides a basis of time-domain analysis and stability analysis of a closed-loop sampled-data system. Its main limitation is that it provides information about signal amplitudes at the *sampling instants* only. It, therefore, provides no information regarding inter-sample ripple.

An impulse-sampled signal $\varepsilon^*(t)$ has a Laplace transform $\varepsilon^*(s)$ which contains s in the form $\exp(sT_s)$; the z-transform is obtained by making the substitution $z = \exp(sT_s)$. The z-transform can be represented as a series

$$\varepsilon(z) = \sum_{n=0}^{\infty} \varepsilon(nT_s)z^{-n}$$

where n is an integer.

We can interpret z^{-1} as a delay operator of T_s seconds and z^{-2} as a delay operator of $2T_s$ seconds and so on. The summation will take a general form

$$\varepsilon(z) = k_0 + k_1 z^{-1} + k_2 z^{-2} + k_3 z^{-3} + \cdots + k_n z^{-n}$$

Each term in the series contains the amplitude k_n at the sampling instant nT_s in the form $k_n z^{-n}$ (See Figure 14.27). The z-transforms for various functions of time are given in Table 14.2 for reference.

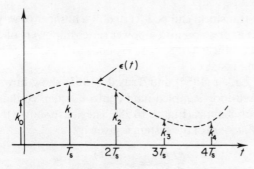

Figure 14.27 Illustrating z-transform series representation of a sampled signal

Table 14.2   Laplace and z-transforms of commonly met time functions

Time function $f(t)$	Laplace transform	z-transform
Unit step $u(t)$	$\dfrac{1}{s}$	$\dfrac{z}{z-1}$
Unit ramp $u(t)t$	$\dfrac{1}{s^2}$	$\dfrac{T_s z}{(z-1)^2}$
Acceleration function $u(t)t^2/2$	$\dfrac{1}{s^3}$	$\dfrac{T_s^2 z(z+1)}{2(z-1)^3}$
$u(t)\dfrac{t^n}{n!}$	$\dfrac{1}{s^{n+1}}$	$\displaystyle\lim_{\alpha\to 0}\dfrac{(-1)^n}{n!}\dfrac{\partial^n}{\partial\alpha^n}\left(\dfrac{z}{z-\exp(-\alpha T_s)}\right)$
$u(t)e^{-\alpha t}$	$\dfrac{1}{s+\alpha}$	$\dfrac{z}{z-\exp(-\alpha T_s)}$
$u(t)t\,e^{-\alpha t}$	$\dfrac{1}{(s+a)^2}$	$\dfrac{T_s z\exp(-\alpha T_s)}{[z-\exp(-\alpha T_s)]^2}$
$u(t)(1-e^{-\alpha t})$	$\dfrac{\alpha}{s(s+\alpha)}$	$\dfrac{[1-\exp(-\alpha T_s)]z}{(z-1)[z-\exp(-\alpha T_s)]}$
$u(t)\sin(\omega_{rt}t)$	$\dfrac{\omega_{rt}}{s^2+\omega_{rt}^2}$	$\dfrac{z\sin(\omega_{rt}T_s)}{z^2-2z\cos(\omega_{rt}T_s)+1}$
$u(t)e^{-\alpha t}\sin(\omega_{rt}t)$	$\dfrac{\omega_{rt}}{(s+\alpha)^2+\omega_{rt}^2}$	$\dfrac{z\exp(\alpha T_s)\sin(\omega_{rt}T_s)}{z^2\exp(2\alpha T_s)-2z\exp(\alpha T_s)\cos(\omega_{rt}T_s)+1}$
$u(t)\cos(\omega_{rt}t)$	$\dfrac{s}{s^2+\omega_{rt}^2}$	$\dfrac{z[z-\cos(\omega_{rt}T_s)]}{z^2-2z\cos(\omega_{rt}T_s)+1}$
$u(t)e^{-\alpha t}\cos(\omega_{rt}t)$	$\dfrac{s+\alpha}{(s+\alpha)^2+\omega_{rt}^2}$	$\dfrac{z^2-z\exp(-\alpha T_s)\cos(\omega_{rt}T_s)}{z^2-2z\exp(-\alpha T_s)\cos(\omega_{rt}T_s)+\exp(-2\alpha T_s)}$

Although the z-transform can be inverted in a number of ways, it is generally best to expand the expression into a power series in powers of z^{-1} by algebraic long division. The coefficient of z^{-n} corresponds to the value of the time function at the nth sampling instant.

Ragazzini and Zadeh (1952) and Truxall (1955) show how to calculate the response of a closed-loop sampled-data control system containing an impulse sampler in the error channel (Figure 14.28). They have shown that the z-transfer (or *pulse transfer function*) of the system is given by

$$\frac{\Theta_o(z)}{\Theta_i(z)} = \frac{G(z)}{1+HG(z)}$$

It should be noted that $HG(z) \neq H(z)G(z)$.

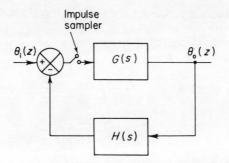

Figure 14.28 Sampled-data control system

The application of the closed-loop z-transform can be illustrated most readily by an example (Figure 14.29) in which the input is a unit step function and it is required to calculate the output.

$$H(s) = \frac{4}{s(s+1)} = \frac{4}{s} - \frac{4}{s+1}$$

From the table of z-transforms (Table 14.2)

$$H(z) = \frac{4z}{z-1} - \frac{4z}{z - \exp(-\alpha T_\mathrm{s})}$$

where $\alpha T_\mathrm{s} = 1$. Thus

$$H(z) = \frac{2.53z}{z^2 - 1.37z + 0.368}$$

and so

$$\frac{\Theta_\mathrm{o}(z)}{\Theta_\mathrm{i}(z)} = \frac{H(z)}{1 + H(z)} = \frac{2.53z}{z^2 + 1.16z + 0.368}$$

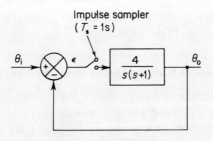

Figure 14.29 Sampled-data system for worked example

Now $\theta_i(t) = u(t)$. Hence again using the table of transforms:

$$\Theta_i(z) = \frac{z}{z - 1}$$

$$\Theta_o(z) = \frac{2.53z}{z^2 + 1.16z + 0.368} \frac{z}{z - 1}$$

$$= \frac{2.53z^2}{z^3 + 0.16z^2 - 0.793z - 0.368}$$

By algebraic long division this expression yields:

$$\Theta_o(z) = 2.53z^{-1} - 0.406z^{-2} + 2.03z^{-3} + 0.339z^{-4}$$
$$+ 1.37z^{-5} + 0.827z^{-6} + 1.06z^{-7} + 1.00z^{-8} + \cdots$$

As can be seen from this response, although the system is highly undamped, its output eventually converges towards the input.

14.13.5   Stability of Closed-loop Sampled Data Systems using the Z-transform

For a sampled-data control system the z-transfer function is given by

$$\frac{\Theta_o(z)}{\Theta_i(z)} = \frac{G(z)}{1 + HG(z)}$$

The stability of the system depends on the positions of the zeros of $[1 + HG(z)]$ in the s-plane. However, the transformation of $\exp(sT_s) = z$ has been made and the positions of the zeros can be mapped in the z-plane. The mapping, $z = \exp(sT_s)$, maps the imaginary axis of the s-plane into the unit circle about the origin of the z-plane and the left half of the s-plane into the interior of the unit circle (Tou, 1959). A sampled data control system will, thus, only be absolutely stable if all the zeros of $1 + HG(z)$ lie inside the unit circle. The direct application of this criterion may often be tedious and the use of a bilinear transform $z = (1 + w)/(1 - w)$ maps inside of the circle in the z-plane into the left-hand side of the w-plane. It is then possible to use the Routh–Hurwitz or the Nyquist stability criterion directly (Tou, 1959).

14.13.6   Compensation of Sampled Data Control Systems

Sampled-data systems offer a wider range of possibilities for compensation than continuous systems; series and parallel or series/parallel compensation of the kind described in Section 14.9, using continuous transfer functions, can be designed in a similar way using a Nyquist or inverse Nyquist diagram based on the Linvill approximation (Section 14.13.3). Alternative strategies involve

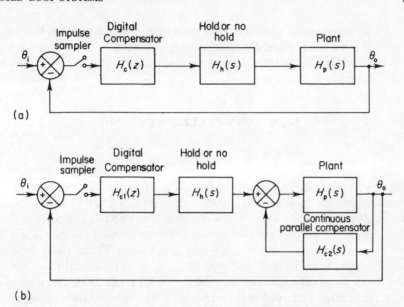

(a)

(b)

Figure 14.30 Forms of digital compensation: (a) series digital compensation; (b) series digital with parallel continuous compensation

series digital compensators alone (Figure 14.30a) or in combination with a parallel continuous compensator (Figure 14.30b).

The digital compensators can be arranged to produce phase-lead for which $H_c(z) = (z - a)/(z + b)$ with the zero lying to the *right* of the pole in the z-plane or phase lag for which $H_c(z) = (z + a)/(z - b)$ with the zero lying to the left of the pole. Many other forms of digital compensator could be envisaged which can be constructed from hard-wired logical circuitry or by means of software implemented on a digital computer (Frederick and Carlson, 1971); in these days of very low cost microprocessors, software implementation is clearly very advantageous. The arrangement illustrating control and digital compensation of a continuous process is shown in Figure 14.31; the error signal is generated

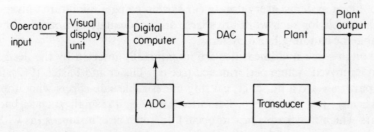

Figure 14.31 Control and compensation using a digital computer

by the computer and is then operated on by a suitable algorithm before it is outputed to the plant via a digital-to-analog converter (DAC). Frederick and Carlson (1971) have described how computer programs for typical algorithms can be written.

14.14 NON-LINEAR SYSTEMS

14.14.1 Introduction

Although certain non-linearities (Coulomb friction and stiction) were mentioned in connection with the positional servomechanism in Section 14.1, all the theory contained in this chapter so far has been concerned with the analysis and design of *linear* systems. In this connection we define linear systems as those which obey the principle of superposition (Gelb and Vander Velde, 1968). Although linear theory is an indispensable design aid, it must be realized that *all* practical systems do inevitably contain non-linearities. For example, every amplifier exhibits saturation so that when the error signal in a control system becomes large the control signal driving the plant will be limited in amplitude. The result of this will normally be that the response time of the system will be longer than predicted by means of linear theory based on the assumption that saturation is not present.

There are many types of non-linearity. They are distinguished in Gelb and Vander Velde (1968) as *explicit* or *implicit* non-linearities. Explicit implies that the output of the non-linearity is explicitly determined in terms of the required input variables, whereas implicit implies a more complicated relationship between input and output through, for example, an algebraic or differential equation. Then with explicit non-linearities we have to distinguish between *static* and *dynamic* forms, in which dynamic implies that the output is related not only through the input but also through the derivatives of the input. Among the explicit, static non-linearities we must again divide between *single-value* (memoryless) non-linearities such as saturation and dead-space, and *multiple-value* (memory) non-linearities such as hysteresis.

The analysis and design of non-linear systems is vastly more complex than that of linear systems, every non-linear system being a miniature universe; the tools available for such work are sparse and inadequate compared with those available for handling linear systems.

Although some non-linearities are intentionally inserted by the designer to obtain improved system performance, (see e.g. Thaler and Pastel, 1962), for the most part they are a nuisance, causing undesirable side-effects when the error signal is large (e.g. the effect of saturation) or when it is small (such as, backlash in gears which may cause *tick* to occur). In some circumstances (as with hysteresis) it may be possible to effectively eliminate the effects of the non-linearity

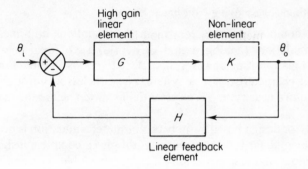

Figure 14.32 Non-linear element in high-gain feedback
loop

by means of a high-gain stable local feedback loop. This procedure is strongly recommended where it is possible. The procedure is based on the idea that the non-linear element has a variable static gain K. It is driven from a high-gain linear amplifier of gain G and feedback is applied through an element of gain H as shown in Figure 14.32. Analysis shows that the transfer function in such cases is given by

$$\frac{\theta_o}{\theta_i} = \frac{GK}{1 + HGK}$$

The method relies on making $HGK \gg 1$ for all possible values of K so that

$$\frac{\theta_o}{\theta_i} \simeq \frac{1}{H}$$

The effects of the non-linearity are thus made quite negligible *so long as none of the elements saturate.* The arrangement is only of use so long as the loop is relatively stable, so that proper design of the loop is essential. This is probably best performed in the frequency domain using the describing function method (see Section 14.14.4).

It was mentioned in Section 14.1 that stiction can cause stick–slip motion in servomechanisms following constant velocity (ramp) inputs. This problem can again be overcome by means of a high-gain local feedback loop. In this case velocity feedback is used around a motor–amplifier series combination; the gain of the amplifier is made very high so that the arrangement behaves like a perfect integrator of transfer function $1/k_s s$ where k_s is the velocity feedback constant. The arrangement is again identical to that shown in Figure 14.32; K represents the non-linear motor characteristic $H = k_s s$, and G is the high gain of the amplifier. There is usually very little problem in designing this loop to have adequate relative stability.

14.14.2   Methods of Studying Non-linear Systems

There are no completely satisfactory methods of studying non-linear systems. Computer simulations (analog, digital, or hybrid) are of great importance; to perform a simulation it is, however, necessary to choose a model to represent the important characteristics of a system. Nevertheless a thorough computer study of any non-linear system is an essential *final* step before the real system is built.

Trial and error design based entirely on computer simulation is not advisable because it may take far too long and may not give a complete insight into the system behaviour and the strategies which are available. The first step in designing a non-linear system is usually to attempt to linearize the essential non-linearities for small signals about a working point. For example, if we have a non-linear device with an input signal $x(t)$ and an output $y(t)$ such that $y(t) = kx^2(t)$ then $\partial y/\partial x = 2kx$; thus if we are working at some particular working point x_p, then for small departures from the working point, the effective gain of the device is $2kx_p$. For working points at values of x greater than x_p, the gain will be greater than $2kx_p$ and for working points at values of x less than x_p, the gain will be less. Linear systems design combined with a sensitivity analysis (see Section 14.10) and a computer simulation should again prove adequate.

Phase plane analysis (West, 1953; Atherton, 1975) has provided a very useful tool in the analysis of non-linear *second-order* systems subjected to step (and ramp) inputs. The technique involves the determination of the response in terms of its derivative of output (or error) plotted as a function of its output (or error). This plot is called a *phase plane diagram*; in formulating the relevant equations, time is removed explicitly and for many practical non-linearities where the non-linearities can be represented by linear segmented characteristics, the phase plane can be divided into various regions each of which corresponds to motion on a particular linear segment of the non-linearity (see Section 14.14.3).

For higher-order systems, on occasions where it is necessary only to determine whether or not the designed system will remain absolutely stable for the entire envelope of input signals, the methods of Lyapunov (La Salle and Lefschetz, 1961) or that of Popov (Aizerman and Gantmacher, 1964) will provide exact answers without solving the differential equations.

The method of describing functions (Gelb and Vander Velde, 1968; Atherton, 1975) based on the concept of quasi-linearization for a given class of input signals provides the main basis for the analysis and design of non-linear feedback control systems (see Section 14.14.4).

14.14.3   Phase Plane Analysis

Phase plane analysis considers any second-order differential equation of form:

$$\ddot{x} + A\dot{x} + Bx = F$$

where A, B, and F are not necessarily constant. A *phase portrait* consists of a number of *phase trajectories* in the $\dot{x}$- versus x-plane. If we define $y = \dot{x}$, then

$$\frac{dy}{dx} = \frac{dy/dt}{dx/dt} = \frac{F - Bx - Ay}{y}$$

This equation represents the *slope* of the phase trajectory in terms of functions of x and y. In general B, A, and F may be functions of x and y. To find the phase trajectory itself this equation must be integrated; sometimes this can be performed analytically but usually it is better done numerically using a digital computer or graphically using the method of isoclines (Thaler and Pastel, 1962).

By way of example, let us consider the simple viscously damped second-order sevomechanism described in Section 14.1 but with the additional Coulomb friction. Consider the response of the system with a *step input* and let C be the magnitude of the Coulomb frictional torque which always opposes motion. The instantaneous accelerating torque in $K\varepsilon$ and the retarding torque is $F\dot{\theta}_o + C \operatorname{sign} \dot{\theta}_o$, where $\operatorname{sign} \dot{\theta}_o$ is positive for $\dot{\theta}_o > 0$ and negative for $\dot{\theta}_o < 0$. Thus applying Newton's second law we have

$$J\ddot{\theta}_o = K\varepsilon - F\dot{\theta}_o - C \operatorname{sign} \dot{\theta}_o$$

Now $\varepsilon = \theta_i - \theta_o$, thus $\theta_o = \theta_i - \varepsilon$; also for a step input $d\theta_i/dt = 0$ and $d^2\theta_i/dt^2 = 0$ under steady-state conditions. It is possible to translate the above equation into the error form:

$$J\ddot{\varepsilon} + F\dot{\varepsilon} + K\varepsilon + C \operatorname{sign} \dot{\varepsilon} = 0$$

From this equation we can deduce

$$\dot{\varepsilon} = -\frac{K\varepsilon}{NJ + F} - \frac{C \operatorname{sign} \dot{\varepsilon}}{NJ + F}$$

in which N is the slope of the phase trajectory where it crosses the isocline. The first term defines the family of isoclines for the linear system whereas the second term introduces the effect of the non-linearity. The focal point is changed from $\varepsilon = +C/K$ to $\varepsilon = -C/K$ as $\dot{\varepsilon}$ changes from a negative value to a positive value. If a trajectory begins at $\varepsilon = A$ (where A is equivalent to the value of the input step) it transverses through the phase plane as shown in Figure 14.33. The *slope* of the phase trajectory is given by the value of N as each isocline is crossed. The determination of the passage of the trajectory through the isoclines to point B when $\dot{\varepsilon}$ becomes negative, thence to point D when it again becomes positive is a simple matter. Motion ceases at D because the generated torque is less than C.

The phase plane technique can be used to analyse the behaviour of the second-order system for a variety of commonly-encountered non-linearities (see e.g. Atherton, 1975); the main disadvantage of the method is that it cannot be extended to higher-order systems in a satisfactory manner.

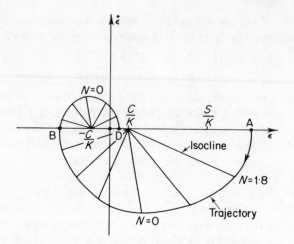

Figure 14.33   Phase trajectory for simple positional
servomechanism with Coulomb friction

14.14.4   The Method of Describing Functions

In its most simple form the describing function method is an extension of ordinary
transfer function analysis to take into account the effect of single non-linearities
in systems excited by sinusoidal input. It is particularly useful as a method for
predicting the amplitude and frequency of limit cycles but it can also be used to
assess relative stability by conventional frequency response methods (see
Section 14.5). In essence, the describing function of a non-linear element is the
gain of the element in terms of the ratio of its *fundamental component* output to a
sinusoidal input of given magnitude and frequency. Whereas the gain of a
linear element is only ever frequency dependent, the gain of a non-linear
element is always amplitude dependent and may also sometimes be frequency
dependent. Here we will confine our study to amplitude-dependent describing
functions. In order to illustrate this point a little better let us consider a very
common form of non-linearity, namely *saturation*. The output is linearly related
to the input for small positive or negative excursions of the input, but the output
reaches a limiting value for large excursions (see Figure 14.34). The response
of this non-linearity to a sinusoid will be sinusoid for small signals, a clipped
sinusoid for medium size signals, tending to a square wave of magnitude KE_s
for very large signals. The gain of the element, based on the ratio of fundamental
output to input thus is constant at a value K for small inputs, beginning to
decrease as the input goes beyond E_s and eventually trailing off towards zero
as the fundamental output tends towards its limiting value of $4KE_s/\pi$ as the
input tends towards infinity (see Figure 14.35).

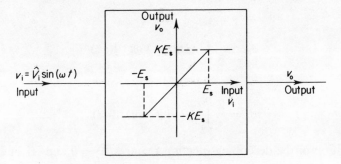

Figure 14.34   Illustrating a saturation non-linearity

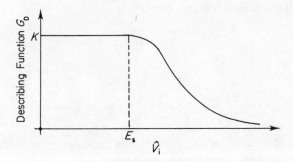

Figure 14.35   Describing function G_D for saturation
non-linearity

For memoryless non-linearities the describing function is purely real, while memory-type non-linearities introduce a phase shift, causing the describing function to be complex. In general if the fundamental output is expressed as a complex operator

$$V_o(j\omega) = \hat{V}_o \cos \theta + j\hat{V}_o \sin \theta$$

where $\hat{V}_o$ is the peak fundamental output and θ is the phase shift, then the describing function G_D is given by

$$G_D = \frac{\hat{V}_o \cos \theta}{\hat{V}_i} + j \frac{\hat{V}_o \sin \theta}{\hat{V}_i}$$

where $\hat{V}_i$ is the peak input.

To derive a describing function we must use the Fourier series representation for the output:

$$v_o(t) = \tfrac{1}{2}A_0 + A_1 \cos(\omega t) + A_2 \cos(2\omega t) + A_3 \cos(3\omega t) + \cdots$$
$$+ B_1 \sin(\omega t) + B_2 \sin(2\omega t) + B_3 \sin(3\omega t) + \cdots$$

where

$$A_N = \frac{2}{\pi} \int_0^\pi v_\mathrm{o}(t)\cos(N\omega t)\, \mathrm{d}(\omega t)$$

and

$$B_N = \frac{2}{\pi} \int_0^\pi v_\mathrm{o}(t)\sin(N\omega t)\, \mathrm{d}(\omega t)$$

The definition of the describing function requires $A_\mathrm{o} = 0$ and A_N and B_N for $N > 1$ to be negligible. The fact that most control systems contain low-pass filtering elements which filter out the harmonics to a substantial degree usually justifies these assumptions.

The describing function is then given by

$$\frac{B_1}{\hat{V}_\mathrm{i}} + \mathrm{j}\,\frac{A_1}{\hat{V}_\mathrm{i}} \quad \text{in coordinate form}$$

or

$$\frac{\sqrt{B_1^2 + A_1^2}}{\hat{V}_\mathrm{i}} \bigg/ \underline{\tan^{-1} \frac{A_1}{B_1}} \text{ in polar form}$$

The evaluation of the Fourier integrals may be quite simple, as in the case of the saturation non-linearity, or extremely difficult.

Fortunately the describing functions of some commonly encountered non-linearities have been tabulated by Thaler and Pastel (1962) and Gelb and Vander Velde (1968).

The describing function for the saturation element is given by

$$G_\mathrm{D} = \frac{2K}{\pi} \left[(\sin^{-1} R) + R\sqrt{(1 - R^2)} \right]\underline{/0^\circ}$$

The Nyquist stability criterion may be involved for the determination of the stability of systems containing a non-linear element as shown in Figure 14.36. The closed-loop transfer function of this system is given by

$$\frac{\Theta_\mathrm{o}}{\Theta_\mathrm{i}} = \frac{G_1(\mathrm{j}\omega)G_\mathrm{D}G_2(\mathrm{j}\omega)}{1 + G_1(\mathrm{j}\omega)G_\mathrm{D}G_2(\mathrm{j}\omega)}$$

The Nyquist criterion is based on the characteristic equation

$$1 + G_1(\mathrm{j}\omega)G_\mathrm{D}G_2(\mathrm{j}\omega) = 0$$

To avoid the need to plot a sheaf of Nyquist diagrams for every value of G_D we may reform the characteristic equation as

$$G_1(\mathrm{j}\omega)G_2(\mathrm{j}\omega) = -1/G_\mathrm{D}$$

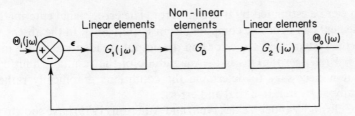

Figure 14.36   Non-linear feedback system

The encirclement of the locus $-1/G_D$ by the function $G_1(j\omega)G_2(j\omega)$ plotted as a polar diagram now indicates absolute instability. Relative stability can be assessed by treating the locus $-1/G_D$ as the *equivalent* of the critical point $(-1, j0)$ used in linear systems design. The amplitude and frequency of a limit cycle can readily be assessed by the point at which the frequency locus

$$G_1(j\omega)G_2(j\omega)$$

intersects the locus of $-1/G_D$. This is illustrated in Figure 14.37 which shows the Nyquist diagram of an *unstable* system that must *limit cycle* at angular frequency ω_c where

$$\underline{/G_1(j\omega_c)G_2(j\omega_c)} = -180°$$

from which ω_c may be calculated (generally by iterative trial and error). The amplitude of the limit cycle can now be determined from

$$|G_1(j\omega)||G_D||G_2(j\omega)| = 1$$

(which is the Nyquist amplitude condition for continuous oscillation). The value of G_D can now be determined and the value of $\hat{V}_i$, the signal input to the non-linear

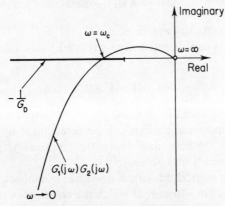

Figure 14.37   Nyquist diagram for a typical non-linear system showing $-1/G_D$ (unstable situation)

element, can be estimated by trial and error. The error signal entering the first linear element can hence be computed from $\hat{V}_i/|G_1(j\omega_c)|$. In the absence of any input to the system itself, the error and the output are identical in magnitude so that $\theta_o = \hat{V}_i/|G_1(j\omega_c)|$. This is the magnitude of the resultant limit cycle. In general it is necessary to determine the conditions graphically rather than analytically or by iterative trial and error.

Both Gelb and Vander Velde (1968) and Atherton (1975) have shown how the method of describing functions can be used to design compensating elements for non-linear systems and be extended to treat multiple non-linearities, to handle transient oscillations, dual sinusoidal inputs, and random inputs.

14.15 CONCLUDING REMARKS

Many lengthy textbooks have been written on closed-loop systems so that it is hardly surprising that the material contained in this chapter merely licks the surface of a vast and rapidly expanding subject. The material described in this chapter is all based on the transfer function approach because this certainly offers the designer the most comprehensive set of techniques. It should be understood however that many authors prefer to integrate the classical *transfer function* approach with the relatively modern *state variable* approach. In the state variable approach the system model is described in terms of n first-order differential equations, each equation being a separate description of the behaviour of a particular *state* and its connection with the other states and the driving inputs. This form of description allows the equations for the system states to be condensed into the form of a single vector $\mathbf{x}(t)$ and related to the driving input vector, $\mathbf{u}(t)$ by the equation

$$\dot{\mathbf{x}}(t) = \mathbf{A}\mathbf{x}(t) + \mathbf{B}\mathbf{u}(t)$$

where $\mathbf{A}$ and $\mathbf{B}$ are coefficient matrices.

The system outputs can then usually be related to the internal states by the vector equation

$$\mathbf{y}(t) = \mathbf{C}\mathbf{x}(t)$$

where $\mathbf{C}$ is another coefficient matrix and $\mathbf{y}(t)$ is the output vector. The solution to the first of these matrix equations can be determined by various powerful computerized matrix methods and hence the response $\mathbf{y}(t)$ for a given $\mathbf{u}(t)$ can be determined quickly and accurately.

Most of the theory associated with state-variable analysis emerged after the 1960 IFAC Conference in Moscow at which the concepts of controllability and observability were introduced (Kalman, 1960). Apart from their application in time-domain analysis, state-variable techniques form the basis of the Lyapunov stability analysis and of optimal control theory (Ogata, 1967). It can also be used in other design techniques (pole assignment, and pole assignment and de-coupling methods) using state vector feedback (Porter, 1969; Power and Simpson, 1978).

More recently (particularly in the UK) there has been a return to frequency response methods especially adapted to multi-input/multi-output systems; these newer methods are particularly suited to computer-aided design (Belletrutti and MacFarlane, 1971; Rosenbrock, 1974).

Returning to the more classical approach, statistical methods of handling linear and non-linear systems subjected to random inputs has been completely neglected in this chapter. The powerful concepts of probability description and spectral analysis have however been introduced in several texts (see e.g. Douce, 1963).

In order to obtain greater sensitivity and better accuracy of measurement it is likely that more and more instruments will incorporate feedback. The designers of instruments of all kinds will, therefore, have to become more familiar with the analysis and design of feedback systems.

REFERENCES

Aizerman, M. A. and Gantmacher, F. R. (1964). *Absolute Stability of Regulator Systems*, Holden-Day, San Francisco.

Ashworth, M. J. (1975). 'Computer aided design of tracking servomechanisms', *M.Phil. Thesis*, Council for National Academic Awards (UK).

Atherton, D. P. (1975). *Non-Linear Control Engineering*, Van Nostrand Reinhold, London.

Atkinson, P. (1968). *Feedback Control Theory for Engineers*, 1st edn., Heinemann, London.

Atkinson, P. (1972). *Feedback Control Theory for Engineers*, 2nd edn., Heinemann, London.

Atkinson, P. and Allen, A. J. (1973). 'Interactive design of sampled-data control systems', *IEE Conf. Publ. No.* 96, 141–8.

Belletrutti, J. and MacFarlane, A. G. J. (1971). 'Characteristic loci techniques in multi-variable control system design', *Proc. IEE*, **118**, 1291–7.

Bennett, S. (1979). *A History of Control Engineering: 1800–1930*, Peter Peregrinus, London.

D'Azzo, J. J. and Houpis, C. J. (1966). *Feedback Control System Analysis and Synthesis*, McGraw-Hill, New York. pp. 465–7.

Davies, W. D. T. (1970). *System Identification for Self-adaptive Control*, Wiley, London.

Douce, J. L. (1963). *An Introduction to the Mathematics of Servomechanisms*, EUP, London. pp. 132–67, 188–95.

Frederick, D. K. and Carlson, A. B. (1971). *Linear Systems in Communication and Control*, Wiley, New York.

Gardner, M. F. and Barnes, J. L. (1942). *Transients in Linear Systems*, Wiley, New York.

Gelb, A. and Vander Velde, W. E. (1968). *Multiple-input Describing Functions and Non-linear Systems*, McGraw-Hill, New York.

Goldman, S. (1966). *Laplace Transform Theory and Electrical Transients*, Dover, New York.

Graupe, D. (1972). *Identification of Systems*, Van Nostrand Reinhold, New York.

Hazen, H. L. (1934). 'Theory of servo-mechanisms', *J. Franklin Inst.*, **218**, 3, 279–331.

Horowitz, I. M. (1959). 'Fundamental theory of automatic linear feedback control systems', *Trans. IRE*, **AC-3**, 5–19.

Horowitz, I. M. (1963). *Synthesis of Feedback Systems*, Academic Press, New York, p. 241.

Johnson, C. L. (1956). *Analog Computer Techniques*, McGraw-Hill, New York (later edn. in 1963).

Jones, B. E. (1977). *Instrumentation, Measurement and Feedback*, McGraw-Hill, Maidenhead, England. pp. 190–215.

Jones, B. E. (1979). 'Feedback instruments', *J. Phys. E: Sci. Instrum.*, **12**, 145–58.

Kalman, R. E. (1960). 'On the general theory of control systems', *Proc. 1st Int. IFAC Congr. on Automatic and Remote Control*, 1960, Butterworths, London. pp. 481–92.

La Salle, J. P. and Lefschetz, S. (1961). *Stability of Liapunov's Direct Method*, Academic Press, New York.

Linvill, W. K. (1951). 'Sampled-data control systems studies through comparison of sampling with amplitude modulation', *Trans. AIEE*, **70**, Part II, 1779–88.

Maudsley, D. (1978). 'An Approach to the mathematical modelling of dynamic systems', *Measurement and Control*, **11**, 181–9.

Maxwell, J. Clerk. (1868). 'On governors', *Proc. R. Soc., Lond.*, **16**, 270–83.

Minorsky, N. (1922). 'Directional stability of automatically steered bodies', *J. Am. Soc. Naval Engng.*, **42**, 280–309.

Ogata, K. (1967). *State Space Analysis of Control Systems*, Prentice-Hall, Englewood Cliffs. pp. 437–587.

Porter, B. (1969). *Synthesis of Dynamical Systems*, Nelson & Sons, London.

Power, H. M. and Simpson, R. J. (1978). *Introduction to Dynamics and Control*, McGraw-Hill, London.

Raggazzini, J. R. and Zadeh, L. A. (1952). 'The analysis of sampled-data systems', *AIEE Trans.*, **71**, Part II, *Applications and Industry*, 225–232.

Rosenbrock, H. H. (1974). *Computer-aided Control System Design*, Academic Press, London.

Stockdale, L. A. (1962). *Servomechanisms*, Pitman, London. p. 256.

Thaler, G. J. and Brown, R. G. (1960). *Analysis and Design of Feedback Control Systems*, McGraw-Hill, New York.

Thaler, G. J. and Pastel, M. P. (1962). *Analysis and Design of Non-linear Control Systems*, McGraw-Hill, New York. pp. 292–359.

Tou, J. T. (1959). *Digital and Sampled-data Control Systems*, McGraw-Hill, New York.

Towill, D. R. (1970). *Transfer Function Techniques for Control Engineers*, Iliffe, London.

Towill, D. R. (1973). 'Low order modelling techniques: tools or toys?', *Computer Aided Control System Design Conf.*, IEE Conf. Publ. No. 96, 206–12.

Towill, D. R. (1975). 'Coefficient plane models for tracking system design', *Radio Electron. Engr.*, **45**, 465–71.

Truxall, J. G. (1955). *Automatic Feedback Control System Synthesis*, McGraw-Hill, New York.

Truxall, J. G. (1972). *Introductory System Engineering*, McGraw-Hill, Tokyo. pp. 82–5.

West, J. C. (1953). *Textbook of Servomechanisms*, EUP, London.

Index